Fuzzy Sets, Fuzzy Numbers, Fuzzy Modeling, and Their Applications in Management and Engineering

Contents

Fuzzy Sets, Fuzzy Numbers, Fuzzy Modeling, and Their Applications in Management and Engineering

Editor

Aleksandar Aleksić

MDPI

Basel • Beijing • Wuhan • Barcelona • Belgrade • Novi Sad • Cluj • Manchester

Editor
Aleksandar Aleksić
Faculty of Engineering
University of Kragujevac
Kragujevac, Central Serbia
Serbia

Editorial Office
MDPI
St. Alban-Anlage 66
4052 Basel, Switzerland

This is a reprint of articles from the Special Issue published online in the open access journal *Mathematics* (ISSN 2227-7390) (available at: https://www.mdpi.com/si/mathematics/Fuzzy_Set_Number_Appl).

For citation purposes, cite each article independently as indicated on the article page online and as indicated below:

Lastname, A.A.; Lastname, B.B. Article Title. *Journal Name* **Year**, *Volume Number*, Page Range.

ISBN 978-3-7258-0742-0 (Hbk)
ISBN 978-3-7258-0741-3 (PDF)
doi.org/10.3390/books978-3-7258-0741-3

Preface

In recent years, contemporary business operations have found themselves navigating a complex landscape influenced by market criteria that exhibit qualities of uncertainty, imprecision, and vagueness. The dynamic interplay of various factors, ranging from the disruptive impact of global pandemics to the ripple effects of geopolitical instability and the looming specter of climate change, has rendered traditional approaches to business management and engineering increasingly inadequate.

In this ever-shifting terrain, the reliability of existing quantitative and qualitative models has come into question. Unforeseen events continually challenge the efficacy of conventional methodologies, highlighting the inherent limitations of deterministic frameworks in accommodating the fluidity of real-world scenarios. The quest for exact values of variables, once considered a cornerstone of management and engineering models, has become an elusive pursuit, with past data offering scant reassurance for future predictions.

Amidst this uncertainty, the emergence of fuzzy sets theory has provided a beacon of hope, offering a robust framework for characterizing and navigating the complexities of ambiguity. By embracing the inherent fuzziness of real-world phenomena, fuzzy sets theory has transcended the constraints of traditional paradigms, offering a nuanced understanding of uncertainty that empowers practitioners to make informed decisions in the face of ambiguity.

It is within this context that this publication endeavors to shed light on the transformative potential of fuzzy sets theory in the realms of management and engineering. By delving into the latest advancements in fuzzy sets, fuzzy numbers, and fuzzy modeling, this publication aims to chart a course toward a more resilient and adaptive approach to problem-solving.

Moreover, this publication serves as a vital platform for researchers, both from academia and industry, to showcase their pioneering work and novel applications in the field. By fostering collaboration and knowledge exchange, it is our hope that this initiative will catalyze further innovation and inquiry, propelling the domains of management and engineering towards new frontiers of understanding and discovery.

In essence, this publication stands as a testament to the enduring spirit of inquiry and innovation that drives the fields of management and engineering forward. As we embark on this journey together, let us embrace the challenges of uncertainty with curiosity and courage, knowing that it is through our collective efforts that we will unlock the boundless potential of the future.

Aleksandar Aleksić
Editor

mathematics

Review

Industrial and Management Applications of Type-2 Multi-Attribute Decision-Making Techniques Extended with Type-2 Fuzzy Sets from 2013 to 2022

Aleksandar Aleksić * and Danijela Tadić

Faculty of Engineering, University of Kragujevac, 34000 Kragujevac, Serbia; galovic@kg.ac.rs
* Correspondence: aaleksic@kg.ac.rs

Abstract: The ongoing research in the field of decision-making can be analyzed from different perspectives. Research trends indicate that multi-attribute decision-making (MADM) methods have a significant impact on engineering and management scientific areas. Since many of the problems existing in the mentioned areas are associated with a certain level of uncertainty, type 2 fuzzy sets represent a common solution for the enhancement of conventional MADM methods. In this way, the decision-makers are encouraged to use linguistic expressions for the assessment of attributes' relative importance and their values. The purpose of this paper is to review a determination of attributes' relative importance, and their values, as well as the extension of ranking methods with type 2 fuzzy sets. The papers are systematically adjoined to groups consisting of hybrid models with the following characteristics: (1) indicating the procedure for modeling attribute relative importance and their values, (2) determining the extension of MADM methods with type 2 fuzzy sets to determine attributes' vector weights, and (3) the extension of MADM for attributes ranking with type 2 fuzzy sets. This study reviewed a total of 42 papers in the domain of engineering and management published from 2013 to 2023 in different journals indexed by the Springer, Science Direct, Emerald, Wiley, ProQuest, Taylor, and Francis research platforms.

Keywords: fuzzy multi-attributive decision making; the type 2 fuzzy sets; literature review

MSC: 03E72

Citation: Aleksić, A.; Tadić, D. Industrial and Management Applications of Type-2 Multi-Attribute Decision-Making Techniques Extended with Type-2 Fuzzy Sets from 2013 to 2022. *Mathematics* **2023**, *11*, 2249. https://doi.org/10.3390/math11102249

Academic Editor: Vassilis C. Gerogiannis

Received: 13 April 2023
Revised: 5 May 2023
Accepted: 7 May 2023
Published: 11 May 2023

1. Introduction

This paper provides insight into how certain MADM methods have been employed to bring solutions in the scientific areas of engineering and management with their enhancements and modifications. Decision-making represents one of the most important activities executed by the decision-makers (DMs) at the strategic, tactical, and operational levels in any company. The essential motivation of the DMs is to determine the best solution among the alternatives that lead to successful business results. Respecting their experience and the results of the best practice, DMs should consider many attributes that may conflict with each other. According to the stated, it may be considered that the management problems can be presented as multi-attribute decision-making problems (MADMs). Additionally, it may be assumed that MADM techniques strive to make the decision-making process more formalized [1], so the obtained solution seems to be less burdened by the bias of DMs. MADM is the discipline of operations research that has been widely studied by researchers and practitioners [1,2].

In recent decades, a large number of MADM techniques have been proposed and used for solving different area problems [1]. It should be underlined that the proposed MADM techniques are developed on different mathematical foundations, so they have different characteristics in finding the optimal solution.

In this paper, the classification of the analyzed MADM methods is performed according to [1,3–5] which is presented in Table 1.

Table 1. The classification of MADM techniques.

	Classification is Given by [5]	Classification is Given by [1]	Classification is Given by [3,4]
Weighted Aggregated Sum Product Assessment-WASPAS [6]	Utility-based		Other MADMs
Technique for Order Preference by Similarity to Ideal Solution-TOPSIS [7]	Distance-based	Normalizing models-additive types	Compromise
VIekriterijumsko KOmpromisno Rangiranje-VIKOR [8]	Distance-based	Normalizing models-additive types	Compromise
Complex Proportional Assessment-COPRAS [9]	Utility-based		Compromise
Multi-Objective Optimization on the basis of Ratio Analysis-MOORA [10]	Other		
Additive Ratio ASsessment-ARAS [11]	Other		Utility-based
Elimination et Choix Traduisant la Realité-ELECTRE [12]	Outranking	Normalizing models-additive types	Outranking
Analytical Network Process-ANP [13]	Pairwise comparison	Weighting models	Utility-based
Analytic Hierarchy Process-AHP [14]	Pairwise comparison	Weighting models	Utility-based
(An acronym in Portuguese for interactive and multi-criteria decision-making)-TODIM [15]	Outranking		Utility-based
Best Worst Method-BWM [16]	Pairwise comparison		Compromise
Multi-attributive border approximation area comparison method-MABAC [17]	Other		Compromise
Decision Making Trial and Evaluation Laboratory-DEMATEL [18]	Interaction based	Evaluating or choosing models	Other

As analysis covers different MADM methods, it may be noticed that there are more MADM methods than those considered by this research. The MADM methods enhanced with IT2FNs have different mathematical foundations. That is why scholars can adjoin them in different groups. A good example of this is BWM which can be interpreted as a pairwise comparison and/or compromise method. It can be concluded that there is no unique classification although scholars are trying to propose different frameworks.

The increase in the social and economic environment's complexity (the change of customer expectation, political change, business in a time of crisis, etc.), as well as the vagueness of the inherently subjective nature of human thinking, brings the inability to describe input data of the decision-making process with the crisp values. A lot of scholars believe that a more accurate assessment of uncertainties into the relative importance of attributes and their values may be obtained if the DMs use linguistic variables [19–21]. These variables are defined as words and/or sentences in a natural or artificial language [22]. It may be suggested that the concept of linguistic variables is useful in dealing with complex situations. The shortcoming of this assessment approach is that the words might not have a clear and well-defined meaning since the DMs may have different subjective perceptions or personalities. In this research, the authors' attention is focused on the application of type 2 fuzzy sets (T2FSs) which were introduced by Zadeh [23] for the modeling of uncertainties. These T2FSs represent the extension of the concept of type 1 fuzzy sets (TFSs) [24] that are characterized by a primary and a secondary membership function with an additional dimension of membership function. T2FSs can deal with the fuzziness and uncertainty characteristics of decision-making problems more accurately and effectively compared to T1FSs. It should be emphasized that in real-world applications, interval type 2 fuzzy numbers (IT2FNs) [25] are widely applied. The IT2FNs represent the special version of generalized T2FSs. It may be assumed that the handling of uncertainties by using T2FNs means making fewer assumptions and making fewer assumptions provides more realistic solutions to real-life decision-making problems. IT2FNs are the most frequently used T2FSs [25] because of their easiness and reduced computational effort in comparison with T2FSs. Just a few authors [26] have used other fuzzy numbers, such as Gaussian interval type 2 fuzzy numbers (GIT2FN). Therefore, many real-life situations can be described by employing IT2FNs, so the calculation effort is decreased but the preciseness of the obtained data is satisfied.

The literature contains a variety of research MADM methods that have been extended with IT2FNs (interval type 2 fuzzy multi-attribute decision making-IT2FMADMs). This research attempts to document the exponentially grown interest in IT2FMADM techniques and provide a state-of-the-art review of relevant literature where the treated problems have been solved within the last ten years. In literature, solving many complex management and engineering problems is based on using IT2FMADMs (see Table 2). To present the data in Table 2 in a concise manner, the titles of MADM methods and techniques are extended with the prefix IT2F denoting that the mentioned methods and techniques are enhanced with IT2F numbers. As the research domain of industrial and management applications is considered, the literature sources containing IT2FMADM explanations and applications are comprehensively reviewed, employing academic databases of Springer, Science Direct, Emerald, Wiley, ProQuest, Taylor, and Francis. It is worth mentioning that some papers containing adequate MADM techniques are not considered due to different application domains. The other criteria for filtering research were the enhancement of MADM techniques with IT2FNs. Table 2 denotes the papers with the research focus on industrial and management applications with IT2FMADMs.

Table 2. A brief explanation of IT2FMADM ranking techniques.

Authors	Year	Research Focus	Rank of Alternatives
Celik et al. [27]	2013	The satisfaction of customers with public transportation	IT2FTOPSIS
Baležentis and Zeng [28]	2013	Selection of manager for research and development	IT2FMULTIMOORA
Ghorabaee et al. [29]	2014	Supplier selection	IT2FCOPRAS
Chen and Hong [30]	2014	The selection of a system analysis engineer	IT2FTOPSIS
Qin et al. [31]	2015	Metro station dynamic risk assessment	IT2FTOPSIS
Kilic and Kaya [32]	2015	Evaluation and selection of investment projects	IT2FTOPSIS
Abdullah and Zulkifli [33]	2015	Human resource management problem	IT2FDEMATEL
Cebi and Otay [34]	2015	Cement factory selection	IT2FTOPSIS
Qin et al. [35]	2015	Evaluation of the high-tech risk investment project	IT2FVIKOR

Table 2. *Cont.*

Authors	Year	Research Focus	Rank of Alternatives
Ghorabaee et al. [36]	2015	Selecting a suitable hydroelectric power station project	IT2FVIKOR
Özkan et al. [37]	2015	Determining the best electrical energy storage technology	IT2FTOPSIS
Liao [38]	2015	Evaluation of materials	IT2FTOPSIS
Sang and Liu [39]	2015	Green supplier selection in the automotive industry	IT2FTODIM
Ghorabaee [19]	2016	Selecting the suitable robot for its production process	IT2FVIKOR
Celik et al. [40]	2016	Green Logistic Service Providers Evaluation	IT2FELECTRE
Ghorabaee et al. [41]	2016	Green supplier selection	IT2FWASPAS
Buyoukozkan et al. [42]	2016	Evaluation of Knowledge Management Tools	IT2FTOPSIS
Qin et al. [43]	2017	Green supplier selection	IT2FTODIM
Gorener et al. [44]	2017	Supplier selection in a high-stake aviation company	IT2FTOPSIS
Deveci et al. [45]	2017	Airline new route selection	IT2FTOPSIS
Mousakhani et al. [46]	2017	Green supplier evaluation	IT2FTOPSIS
Soner et al. [47]	2017	Selecting the right hatch cover design in maritime transportation industry	IT2FVIKOR
Zhong and Yao [48]	2017	Supplier selection	IT2FELECTRE
Deveci et al. [49]	2018	Selection for car-sharing station	IT2FWASPAS
Celik and Akyuz [20]	2018	Selecting the appropriate ship loader type	IT2FTOPSIS
Debnath and Biswas [50]	2018	The supplier selection problem	IT2FAHP
Meng et al. [51]	2019	Risk assessment of supply chain in social commerce	IT2FTODIM
Đurić et al. [52]	2019	The software failure analysis	IT2FCOPRAS
Dinçer et al. [53]	2019	Evaluate the financial service performance in E7 economies	IT2FMOORA
Xu et al. [54]	2019	Green supplier selection	IT2FAHP Sort II
Wu et al. [55]	2019	Green supplier selection	IT2FVIKOR
Aleksic et al. [56]	2019	Ranking failures in a recycling center	IT2FTOPSIS
Yucesan et al. [57]	2019	Green supplier selection	IT2FTOPSIS
Dorfeshan and Mousavi [58]	2020	Aircraft maintenance planning	IT2FMABAC
Bera et al. [59]	2020	Supplier selection	IT2FTOPSIS
Mohamadghasemi et al. [26]	2020	Selection of conveyors	IT2FELECTRE
Ayyildiz et al. [60]	2020	Credit application	IT2FELECTRE
Yang et al. [61]	2020	Choosing the best investment option	IT2FTOPSIS
Kiraci and Akan [21]	2020	Aircraft selection	IT2FTOPSIS
Pourmand et al. [62]	2020	Water Resources Management	IT2FTOPSIS
Özdemir and Üsküdar [63]	2020	Strategy selection	IT2FTOPSIS
Deveci et al. [64]	2020	Offshore wind farm development	IT2FTOPSIS
Mirnezami et al. [65]	2021	Project cash flow evaluation	IT2FTODIM
Sharaf [66]	2021	Solar power systems	IT2FTOPSIS
Komatina et al. [67]	2021	Evaluation of different risk factors	IT2FTOPSIS
Karagöz et al. [68]	2021	Facility location	IT2FARAS
Celik et al. [69]	2021	Green supplier selection	IT2FTODIM

Table 2. *Cont.*

Authors	Year	Research Focus	Rank of Alternatives
Zhang et al. [70]	2022	The subway station's risks	IT2FTOPSIS
Komatina et al. [71]	2022	Supplier selection	IT2FMABAC
Aleksić et al. [72]	2022	Evaluation and ranking of failures in the automotive industry	IT2FVIKOR
Ecer [73]	2022	Green supplier selection in-home appliance manufacturer	IT2FAHP

The scientific objective of the research is to provide insight into the used MADM techniques enhanced with interval type 2 fuzzy numbers (IT2FNs) and applied in solving management and engineering problems. At the same time, the utilitarian objective of the research is to provide answers on conducting MADM techniques steps regarding different approaches and their execution considering the process of decision-making and mathematical operations. In this way, scholars can think about different approaches to the MADM steps execution in their future work. It is worth mentioning that this research is scoped to papers containing hybrid MADM for modeling attributes' weights and values, determining weights and values, and their ranking. As denoted methods are used for ranking, it should be noticed that other MADM mainly used for determining the attributes' relative importance are analyzed in Section 2.3.

The motivation for this research comes from the fact the literature does not suggest the answers to the following questions:

(1) Which IT2FMADM techniques are being used frequently in (i) industrial engineering and (ii) computer science?
(2) Which characteristics of IT2FNs are mostly employed?
(3) Which method is mostly used for the aggregation of DMs' assessment into unique opinion?
(4) Which type of study is executed on these IT2FMADM techniques (distance between two IT2FNs, method of defuzzification, a method for the comparison of IT2FNs, etc.)? This paper provides a systematic survey that provides answers to the identified gap in the literature.

2. Materials and Methods

It is known that fuzziness and vagueness in the relative importance of attributes and their values exist in many MADM problems. Dealing with uncertainties by employing T2FNs means making fewer assumptions during the decision-making process, so it should lead to more realistic solutions to real-life decision-making problems. This can be seen as a main advantage of T2FNs over T1FSs. On the other hand, employment of the T2FSs results in the need to solve very complex mathematical calculations which is their main shortcoming.

The majority of scholars employ interval type 2 triangular fuzzy numbers (IT2FNs) and interval type 2 trapezoidal fuzzy numbers (IT2FTrFNs) [36,45]. Handling uncertainties by using these IT2FNs demands less complex computational calculations compared to IT2FNs of higher order. Generally, it may be said that there are no official guidelines on how to choose the appropriate shape of membership functions and this problem may be analyzed as a task itself. The same approach is valid for the rest of the two IT2FNs' characteristics–granularity and domain. The number of linguistic expressions that are used for describing the uncertainty depends on the scale and the complexity of the problem. In real decision-making problems, it is necessary to set the fine gradation; in other words, it is necessary to use a larger number of linguistic expressions that are used for defining the relative importance and values of alternatives. Many scholars propose IT2FMADM techniques based on 3 or 5 linguistic expressions for describing the attributes' relative importance and more than 5 linguistic expressions for describing the alternatives' value [21,56]. Almost all the research found in the referent literature supports the definition of the IT2FNs' domain on the set of real numbers with different domains, as it is further explained.

Many scholars believe that when it comes to real decision-making problems [1], a group of DMs should assess the attributes' relative importance according to which the assessment is brought, as well as their values. In this case, the assessment of the attributes' relative importance and their value is stated as a fuzzy group decision-making problem. In situations where more DMs participate in the decision-making process, it is necessary to aggregate their opinion in the unique assessment. The aggregation of DMs' assessments into the unique assessment can be given by using various aggregation operators. The selection of the aggregation operator is based on an assumption of the DMs' importance.

In the course of an easier understanding of the analyzed papers, firstly, the basic considerations on type 2 fuzzy sets and arithmetic operations on IT2FNs are presented in the next section.

2.1. Basic Consideration of Type 2 Fuzzy Sets

In this section, a brief review of some definitions of type-2 fuzzy sets and IT2FSs [74,75] is presented.

Definition 1. *A type 2 fuzzy set,* $\tilde{\tilde{A}}$ *in the universe of discourse X can be represented by a type-2 membership function* $\mu_{\tilde{\tilde{A}}}$ *as follows:*

$$\tilde{\tilde{A}} = \left\{ (x,u), \mu_{\tilde{\tilde{A}}}(x,u) \middle| \forall x \in X, \forall u \in J_x \subseteq (0,1), 0 \leq \mu_{\tilde{\tilde{A}}}(x,u) \leq 1 \right\}, \tag{1}$$

Definition 2. *If X is a set of real numbers, then a type-2 fuzzy set and an interval type-2 fuzzy set in X are called a type-2 fuzzy number and an interval type-2 fuzzy number, respectively.*

Definition 3. *As trapezoidal fuzzy numbers are well known to the wider audience, it is worth mentioning that triangular fuzzy numbers are their special case. It is the case when there is just one modal value. While the upper membership function and lower membership function of* $\tilde{\tilde{A}}$ *are two triangular type-1 fuzzy numbers,* $\tilde{\tilde{A}}$ *is referred to as a triangular interval type-2 fuzzy number,* $\tilde{\tilde{A}} = \left(\tilde{A}^U, \tilde{A}^L\right)$ *so that:*

$$\tilde{\tilde{A}} = \left(\tilde{A}^U, \tilde{A}^L\right) = \left(\left(a_1^U, a_2^U, a_3^U, \alpha\right), \left(a_1^L, a_2^L, a_3^L, \beta\right)\right), \tag{2}$$

where the lower and upper bounds in the domain are denoted as a_1^U, a_3^U*, respectively, and* a_1^L, a_3^L*, respectively. The modal values are* a_2^U*, respectively, and* a_2^L*, respectively. The values of the membership function are defined as* $(\alpha, \beta) \in [0,1]$

Definition 4. *Let us consider two IT2TFNs,* $\tilde{\tilde{A}}$ *and* $\tilde{\tilde{B}}$

$$\tilde{\tilde{A}} = \left(\left(a_1^U, a_2^U, a_3^U, \alpha_1\right), \left(a_1^L, a_2^L, a_3^L, \beta_1\right)\right), \ \tilde{\tilde{B}} = \left(\left(b_1^U, b_2^U, b_3^U, \alpha_2\right), \left(b_1^L, b_2^L, b_3^L, \beta_2\right)\right) \tag{3}$$

The arithmetic operations are introduced by Mendel [75]. The addition operation, which is denoted as $\tilde{\tilde{A}} + \tilde{\tilde{B}}$, *can be defined as:*

$$\tilde{\tilde{A}} + \tilde{\tilde{B}} = \begin{pmatrix} \left(a_1^U + b_1^U, a_2^U + b_2^U, a_3^U + b_3^U; \min(\alpha_1, \alpha_2), \min(\beta_1, \beta_2)\right) \\ \left(a_1^L + b_1^L, a_2^L + b_2^L, a_3^L + b_3^L; \min(\alpha_1, \alpha_2), \min(\beta_1, \beta_2)\right) \end{pmatrix}, \tag{4}$$

The subtraction operation, which is denoted as $\tilde{\tilde{A}} - \tilde{\tilde{B}}$, *can be defined as:*

$$\tilde{\tilde{A}} - \tilde{\tilde{B}} = \begin{pmatrix} \left(a_1^U - b_3^U, a_2^U - b_2^U, a_3^U - b_1^U; \min(\alpha_1, \alpha_2), \min(\beta_1, \beta_2)\right) \\ \left(a_1^L - b_3^L, a_2^L - b_{32}^L, a_3^L - b_1^L; \min(\alpha_1, \alpha_2), \min(\beta_1, \beta_2)\right) \end{pmatrix}, \tag{5}$$

The multiplication operation, which is denoted as $\tilde{\tilde{A}} \cdot \tilde{\tilde{B}}$, *can be defined as:*

$$\tilde{\tilde{A}} \cdot \tilde{\tilde{B}} = \begin{pmatrix} \left(a_1^U \cdot b_1^U, a_2^U \cdot b_2^U, a_3^U \cdot b_3^U; \min(\alpha_1, \alpha_2), \min(\beta_1, \beta_2)\right) \\ \left(a_1^L \cdot b_1^L, a_2^L \cdot b_2^L, a_3^L \cdot b_3^L; \min(\alpha_1, \alpha_2), \min(\beta_1, \beta_2)\right) \end{pmatrix}, \tag{6}$$

The division operation, which is denoted as $\tilde{\tilde{A}} : \tilde{\tilde{B}}$, *can be defined as:*

$$\tilde{\tilde{A}} : \tilde{\tilde{B}} = \begin{pmatrix} \left(a_1^U : b_3^U, a_2^U : b_2^U, a_3^U : b_1^U; \min(\alpha_1, \alpha_2), \min(\beta_1, \beta_2)\right) \\ \left(a_1^L : b_3^L, a_2^L : b_2^L, a_3^L : b_1^L; \min(\alpha_1, \alpha_2), \min(\beta_1, \beta_2)\right) \end{pmatrix}, \tag{7}$$

Definition 5. *Let us discuss triangular interval type-2 fuzzy numbers,* $\tilde{\tilde{A}}$ *and crisp value k:*

$$k \cdot \tilde{\tilde{A}} = \begin{pmatrix} \left(k \cdot a_1^U, k \cdot a_2^U, k \cdot a_3^U; \alpha_1\right), \\ \left(k \cdot a_1^L, k \cdot a_2^L, k \cdot a_3^L; \beta_1\right) \end{pmatrix}, \tag{8}$$

$$\left(\tilde{\tilde{A}}\right)^{-1} = \begin{pmatrix} \left(\frac{1}{a_3^U}, \frac{1}{a_2^U}, \frac{1}{a_1^U}; \alpha_1\right) \\ \left(\frac{1}{a_3^L}, \frac{1}{a_2^L}, \frac{1}{a_1^L}; \beta_1\right) \end{pmatrix}, \tag{9}$$

2.2. Determining the Relative Importance of Attributes and Their Values

This section is used for the clarification of elements that are needed for determining the relative importance of attributes and their values. This issue is scoped to the linguistic expressions and basic features used for modeling type 2 fuzzy sets. Additionally, the different approaches to defining the weights vector are discussed. As a part of activities needed for determining the relative importance of attributes and their values, the fuzzy group decision-making problem may be employed, so it is also discussed (see Table A1 in Appendix A). The grouping of data presented in Table A1 is based on the following. Firstly, the features of IT2FNs are presented, then the aggregation procedures are denoted. In the end, the proposed IT2FMADM techniques or proposed approaches for the determination of attributes' weights vectors are presented.

It is worth mentioning that the fuzzy group decision-making problem is used for determining criteria values too (see Table A2 in Appendix B). The grouping of data presented in Table A2 is taking into account features of IT2FNs and the aggregation procedures.

The analysis of both tables from Appendices A and B are presented in Section 3, results, and discussion of the research.

2.3. Determining of Attributes' Weight

The weights vector of attributes can be determined by using the different approaches. The activities needed for this are summarized in Figure 1.

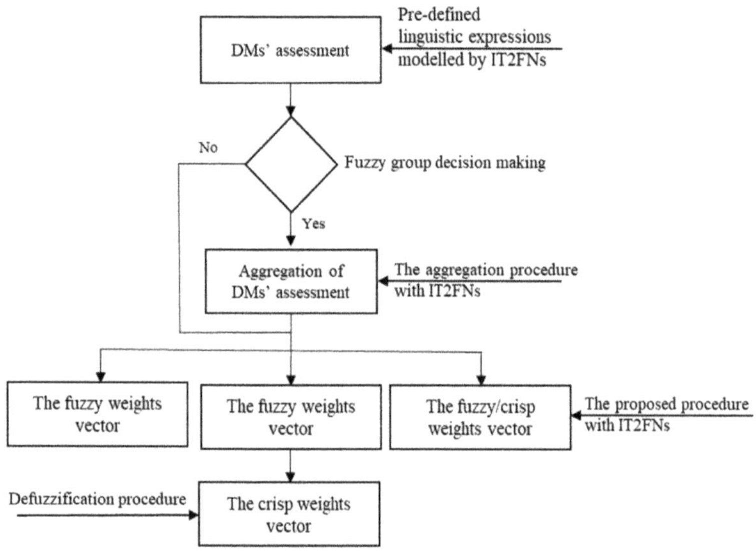

Figure 1. A flowchart of activities needed for determining attributes' weights.

In much conducted research, the weights of attributes are based on using aggregation operators, Delphi techniques, IT2FAHP, and IT2FBWM, as is presented in the next sections. The flowchart starts with the DMs' assessment whether it is a single decision-making or group decision-making approach. If the research is based on group decision-making, then three different paths are possible to be executed. First, scholars may decide to continue with the fuzzy weights vector obtained by applying the different aggregation procedures (e.g., fuzzy arithmetic mean, fuzzy geometric mean). The other option is to perform the defuzzification procedure and continue with the crisp weights vector. The third option is to perform the proposed procedure with IT2FNs (e.g., IT2FAHP, IT2FBWM) and to continue with the fuzzy/crisp weights vector.

2.3.1. The Assessment in a Direct Way

A significant number of scholars suggest that is it appropriate to determine the weights vector in a direct manner [19,26,27,34–36,39,42,45,46,49,59,68,70].

In conventional MADM, the weights vector is given as normalized. Therefore, some authors have performed the normalization of assessed attributes' relative importance [21,38,43,51,65] by using a linear normalization procedure. In this way, the weights of attributes are described by IT2FNs. The normalized weights vector can be given [56] by using the procedure for the comparison of IT2FNs [74], so in this way, the weights of attributes are described by crisp values.

Some scholars believe that obtaining the weights of attributes can be delivered through the several rounds where DMs are making their assessment [60,67]. Hence, the weights vector can be given by using the Delphi technique that is extended with IT2TFNs [60,67].

2.3.2. Interval Type 2 Fuzzy Analytical Hierarchy Process-IT2FAHP

A significant number of scholars think that DMs can make their assessment in a more precise way if they analyze each pair of attributes by analogy to AHP [14]. In the literature, there are many papers with the fuzzy pair-wise matrix of the attributes' relative importance described with IT2FNs [76]. It can be said that the fuzzy pair-wise comparison matrix is consistent only if the appropriate pair-wise comparison matrix is consistent. That is why many scholars have transformed the fuzzy pair-wise comparison matrix into a corresponding pair-wise comparison matrix by using different defuzzification procedures, such as (i) the center of area method [77], (ii) the proposed procedure by Kahraman et al. [76], and (iii) the proposed procedure by Debnath and Biswas [50]. In all of the papers, the consistency check is determined by applying the eigenvector method by analogy to the conventional AHP method [14]. The weights vector is given by the application of (i) synthetic analysis [78] extended with IT2TFNs and (ii) procedure based on the employment of fuzzy geometric mean [76].

2.3.3. Interval Type 2 Fuzzy Best Worst Method-IT2FBWM

In the treated scientific area, there are many papers where the determining of attributes' weights is based on IT2FBWM [55,57,69,72]. In IT2FBWM, all attributes are compared regarding the best and worst items by using pre-defined linguistic expressions. In this way, two fuzzy pair-wise comparison matrices, whose elements are IT2FNs, are constructed. A fuzzy nonlinear optimization model to obtain the weights vector of attributes is proposed, by analogy to the existing procedure [55]. The consistency level of the comparisons can be calculated as defined [16]. It can be said that IT2FBWM is somewhat similar to the IT2FAHP, although many scholars think that IT2FBWM has certain advantages compared to IT2FAHP [72]. This advantage [79] is manifold: (i) there are fewer needed data compared to a full pairwise comparison matrix, and (ii) the obtained results of the BWM application seem to be more consistent than those of the AHP.

The next section provides the analysis of the proposed IT2FMADMs which are denoted in Table 2 into defined categories [4].

2.4. Analysis of Ranking Multi-Atrubutive Decision-Making Methods

The analyzed methods are joined together based on the criteria provided by [4]. Those groups are (1) the utility-based IT2FMADM, (2) the outranking IT2FMADM, (3) the compromise IT2FMADM, and (4) the other IT2FMADM.

The utility-based IT2FMADM

2.4.1. Interval Type 2 Fuzzy Additive Ratio Assessment-IT2FARAS

While applying IT2FARAS in the scope of research [68], the weighted normalized fuzzy decision matrix can be constructed by using a linear normalization procedure [74] and fuzzy algebra rules [75]. The optimality function of benefit/cost attributes as well as the utility degree of benefit/cost attributes can be calculated by using the proposed formula in conventional ARAS which is enhanced with fuzzy operations. By using the defuzzification procedure proposed by [76], the crisp values of the utility degree of benefit/cost attributes can be given. In the mentioned research, the rank of considered attributes is given by using the normalized appraisal score.

2.4.2. Interval Type 2 Fuzzy Multi-Objective Optimization on the Basis of Ratio Analysis-IT2FMOORA (IT2FMULTIMOORA)

The method IT2FMOORA is employed to evaluate financial service performance [53]. In the mentioned research, the fuzzy decision matrix is constructed, and by applying the defuzzification procedure [74], the fuzzy decision matrix is transformed into a decision matrix. The normalized decision matrix is given by using a vector normalization procedure [80]. The rank of alternatives is obtained by using the procedure proposed in conventional MOORA. The method IT2FMULTIMOORA is employed for the selection of a manager for the research and development department in a telecommunication company [28]. In the presented research, the elements of the decision fuzzy matrix are set through a weighted geometric average operator. The normalized fuzzy decision matrix is obtained by applying the linear normalization procedure. The fuzzy positive and fuzzy negative ideal solutions are determined according to the veto concept. The rank of alternatives is based on conventional MULTIMOORA combined with fuzzy algebra rules.

2.4.3. Interval Type 2 Fuzzy "An Acronym in Portuguese for Interactive and Multi-Criteria Decision-Making"-IT2FTODIM

IT2FTODIM is employed for the different problems in the treated scientific area [39,43,51,65,69]. While applying the IT2FTODIM, the normalized decision matrix can be obtained by using: (i) the procedure proposed by Chen and Lee [74] in [39,69] and (ii) the linear normalization procedure enhanced with IT2FNs [65]. The weighted normalized fuzzy decision matrix can be given by using fuzzy algebra rules [69]. In the research presented by Meng et al. [51], the weighted fuzzy decision matrix is constructed using fuzzy algebra rules. It is assumed that in the course of decreasing the calculation complexity, it is necessary to transform the fuzzy decision matrix into a decision matrix [69]. The dominance degree can be determined according to the procedure proposed in conventional TODIM. Additionally, the dominance degree of each alternative can be determined by applying the proposed distance measure between two IT2FNs [39]. In the research presented by Qin et al. [43], the dominance degree of each alternative is based on a new distance measure proposed in this function. The dominance degree of each alternative can be based on distance [43]. The calculation of the Euclidean distance between two IT2FNs [70] is applied by Mirnezami et al. [65].

In all analyzed papers, the overall dominance degree of each alternative is obtained according to the procedure proposed in conventional TODIM.

2.4.4. Interval Type 2 Fuzzy Analytical Hierarchy Process-IT2FAHP

IT2FAHP is employed for solving different problems in engineering and management [50,73]. In the mentioned research, the rank of alternatives is determined by the procedure proposed by Kahraman et al. [76]. Other research based on the AHP framework [54] employs IT2FAHPSort II for the ranking of green suppliers.

The outranking IT2FMADM

2.4.5. Interval Type 2 Fuzzy Elimination et Choix Traduisant la Realité-IT2FELECTRE

IT2FELECTRE is used in several papers [26,40,48,60]. In the mentioned research, the fuzzy weighted decision matrix can be constructed by respecting fuzzy algebra [26,48]. The weighted normalized fuzzy decision matrix can be obtained by using a linear normalization procedure [74] and fuzzy algebra rules [40,60]. By applying the defuzzification procedure [74], the fuzzy decision matrix can be transformed into a decision matrix. Determining the concordance and dis-concordance sets is based on the procedure proposed in conventional ELECTRE: (i) the α-based distance method [81] in [48], (ii) the distance between two (IT2FNs) in [23], and (iii) the procedure proposed in conventional ELECTREE [40,60]. In all of the analyzed papers, the concordance dominance matrix is constructed similarly just as in conventional ELECTRE. Additionally, the rank of alternatives is based on the dis-concordance dominance matrix.

The compromise IT2FMADM

2.4.6. Interval Type 2 Fuzzy Complex Proportional Assessment-IT2FCOPRAS

Several papers have proposed the IT2FCOPRAS method for obtaining the solution to the treated problem [29,52]. In all of the mentioned papers, the weighted fuzzy decision matrix is constructed by using fuzzy algebra rules [75]. In the research presented by Ghorabaee et al. [29], the fuzzy optimality function of benefit/cost attributes is determined. Their crisp values are given by using the defuzzification procedure proposed by Kahraman et al. [76]. In the research presented by Đurić et al. [52], the fuzzy decision matrix is transformed into a decision matrix by using the defuzzification procedure [76]. In all of the presented papers, the rank of alternative is given according to the procedure in conventional COPRAS.

2.4.7. Interval Type 2 Fuzzy Multi-Attributive Border Approximation Area Comparison Method-IT2FMABAC

This method is employed in several papers [58,71]. The aggregated fuzzy decision matrix is considered by Komatina et al. [71] where the aggregation of attribute values is performed by using the order averaging operator extended with IT2TFNs [82]. The fuzzy decision matrix is stated in the other analyzed paper [58]. In both papers, the weighted no-aggregated/aggregated normalized decision matrix is given by using the procedure proposed by Chen and Lee [74] and the fuzzy algebra rules [75]. Additionally, the border approximation area matrix (BAA) is given by applying a fuzzy geometric mean [58,71].

Dorfeshan and Mousavi [58] have transformed the fuzzy decision matrix into the decision matrix by using the defuzzification procedure given by Kahraman et al. [76].

Komatina et al. [71] have proposed the determination of belonging to BAA areas based on their procedure. In this case, criteria function values for each supplier are determined by using Euclidean distance and fuzzy algebra rules [71]. The rank of suppliers is given by analogy to the procedure of conventional MABAC.

2.4.8. Interval Type 2 Fuzzy Technique for Order Preference by Similarity to Ideal Solution-IT2FTOPSIS

As the TOPSIS method is widely used, IT2FTOPSIS has also been employed many times for finding an appropriate solution to research problems. While applying IT2FTOPSIS, the normalized fuzzy decision matrix can be given by using (i) the linear normalization procedure extended with IT2FNs [38], (ii) the linear normalization procedure [74] in [27,35,45,57,59,61,64,66,67,70], and (iii) the center area method in [32].

Fuzzy positive ideal solution (FPIS) and fuzzy negative ideal solution (FNIS) can be determined (i) by the applied procedure [83] in [61,67,70] and (ii) by the veto concept [56,66]. For a similar purpose, Yang et al. [61] employed the distance proposed in Chen [84] and Liu and Jin [85]. Euclidean distance between two IT2TFNs is applied in Komatina et al. [67].

According to mentioned authors' suggestions, the weighted normalized fuzzy decision matrix [38] and the weighted fuzzy decision matrix [20,21,27,34,37,42,44,45,57,59,66,70] can be given by using fuzzy algebra rules.

To decrease the scope of calculations, some authors believe that it is necessary to transform the fuzzy decision matrix into a decision matrix. This transformation can be applied by employing the defuzzification procedures: (i) proposed by Kahraman et al. [76] in [20,38,64], (ii) using the center area method [32], and (iii) proposed by Lee and Chen [86] in [21,30,34,37,42,44–46,59,62,63].

The closeness coefficient can be determined by using Hamming distance [46]. Determining the closeness coefficient can be: (i) based on the α-level [87] in [70], (ii) based on the α-level [81] in [35], (iii) a procedure proposed in conventional TOPSIS extended with IT2TrFNs and defuzzification procedure [76], and (iv) based on similarity measures [66].

In the rest of the analyzed research, the authors suggest the employment of Euclidean distance. The rank of the alternative is determined according to the values of the closeness coefficient.

2.4.9. Interval Type 2 Fuzzy VIekriterijumsko KOmpromisno Rangiranje-IT2FVIKOR

The method IT2FVIKOR has been used several times in the treated scientific area. While performing the calculations based on IT2FVIKOR, the normalized fuzzy decision matrix can be given by using the procedure proposed by Chen and Lee [74] in [72]. In this case, the weighted normalized fuzzy decision matrix is constructed by applying the fuzzy algebra rules.

Determining FPIS and FNIS can be based on the veto concept [72]. Respecting the procedure proposed by Kuo and Liang [83], FPIS and FNIS are determined in delivered research [19,47,55]. Another procedure for determining FPIS and FNIS is introduced by Ghorabaee et al. [36], so it is based on the previously defined procedure [83].

The group utility value can be calculated by using the proposed procedure extended with IT2FNs [72]. The distance developed by Chen and Lee [74] is used for determining the minimum individual regret value.

According to Ghorabaee et al. [19] and Ghorabaee et al. [36], the group utility value and minimum individual regret can be determined by applying the procedure proposed in conventional VIKOR extended with IT2TrFNs. Additionally, there is an assumption introduced by Qin et al. [35] that the group utility value and minimum individual regret can be determined by applying the defined procedure [88]. Soner et al. [47] have determined the group utility value and minimum individual regret by the proposed distance measures.

The rank of alternative is given according to the fuzzy VIKOR index by combining MADM and IT2FNs [74] by Ghorabaee et al. [19]. The rank of alternative is given according to the crisp VIKOR index in the rest of the analyzed papers.

The compromise solution is given by using the procedure proposed in conventional VIKOR combining two conditions [35,55,72].

The other IT2FMADM

2.4.10. Interval Type 2 Fuzzy Weighted Aggregated Sum Product Assessment-IT2FWASPAS

Two papers have proposed the application of the IT2FWASPAS method [41,49]. In the mentioned research [49], the normalized fuzzy decision matrix is constructed according to the procedure proposed by Chen and Lee [74]. On the other hand, Ghorabaee et al. [41] have used the linear normalization procedure. WSM measures are used for determining the attributes' rank [41,49], and corresponding scalar values of WSM measures are obtained by using the procedure proposed in [41].

2.4.11. Interval Type 2 Fuzzy Decision-Making Trial and Evaluation Laboratory-IT2FDEMATEL

One research paper has proposed the application of the IT2FDEMATEL method [33]. The rank of alternatives is obtained through the conventional DEMATEL procedure enhanced with the IT2TrFNs.

3. Results and Discussion of the Research

The first part of the discussion is appointed to the comparative analysis of the treated IT2FMADM techniques. Having in mind the classification given by [5], the pairwise comparison IT2FMADM and all other IT2FMADM can be compared.

The pairwise comparison IT2FMADM (e.g., IT2FAHP, IT2FBWM) employs the relative importance of the attributes which represent the element of the decision matrix. All other MADM techniques support obtaining the values of the attributes by the DMs' assessment or through the evidence data, resulting in fuzzy values or crisp values.

If utility-based IT2FMADM techniques (e.g., IT2FCOPRAS) are compared to all other IT2FMADM techniques, it may be considered that their main advantage is the decreased complexity of calculations needed for the normalization of data. Their main disadvantage is that in the process of alternatives' ranking, the type of the attribute must be considered carefully since there are going to be distinguished as cost and benefit types.

If outranking-based IT2FMADM techniques (e.g., IT2FTODIM, IT2FELECTRE) are compared to distance-based IT2FMADM techniques (e.g., TOPSIS, VIKOR), it may be considered that their main advantage is the decreased complexity of calculations needed for obtaining the rank of alternatives. On the other hand, the employment of distance-based IT2FMADM techniques has an advantage over the employment of outranking-based IT2FMADM techniques in terms of obtaining a compromise solution.

It should be noticed that the mentioned IT2FMADM techniques are developed on different mathematical foundations, so it makes their comparative analysis very complex. Their application can be determined by the domain of interest and the preferences of scholars. The analysis and discussion of the proposed research are scoped to the following criteria: (1) the analysis of the number of DMs participating in the delivered research (Figures 2 and 3); (2) the features of IT2FNs used for modeling the relative importance of attributes as well as their values (Figures 4–7); (3) the frequency of IT2FMADM for determining the attributes weights' vector (Figure 8); and (4) the frequency of IT2FMADM for determining the rank of alternatives (Figure 9). The analyzed characteristics are granulation and the domain of IT2FNs. The shape of the membership function is not discussed since many authors have employed the trapezoidal membership function.

The domain of the analyzed research is scoped to the areas of engineering and management. Figure 2 provides insight into how the decision-making process within the research for describing the relative importance of the attributes has been conducted. Similarly, Figure 3 explains the decision-making process within the research for describing the attributes' values.

Figure 2 indicates that the problem of determining the relative importance of the attributes is set as a single decision-making problem. From the analytical perspective, it should be noticed that this approach includes reaching a consensus while there are more DMs. This is suitable when there are some rules on how to make an assessment in compliance with described guidelines. As a significant number of engineering and management problems do not have clearly described assessment guidelines, it is appropriate to use a group decision-making approach to obtain a more precise assessment. Everyday business operations are exposed to increasing complexity and uncertainty which applies to different industries, so companies strive to develop managers and decision-makers to overcome difficulties. However, due to resource scarcity and organizational culture, it is not always possible to have senior managers that are oriented to group thinking and sharing responsibility. Additionally, the individual decision-making process is less complex from a mathematical point of view and can be executed more efficiently compared to group decision making which, in practice, demands more time for collecting input data.

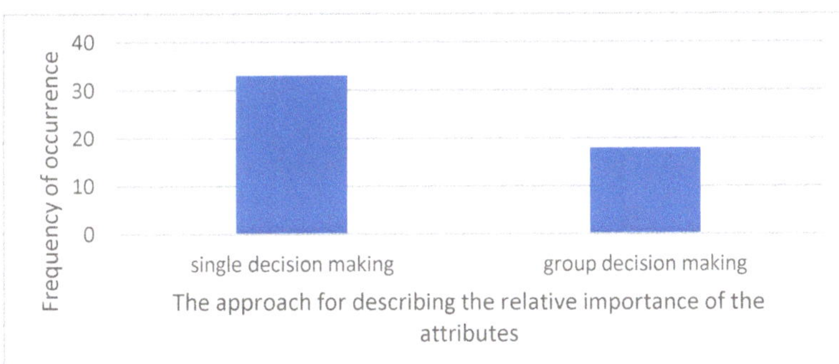

Figure 2. The decision-making process within the research for describing the relative importance of the attributes.

The problems in engineering and management often exist in the presence of uncertainty due to changes in the business, organizational structure, and market conditions. As is shown in Figure 3, while assessing attributes' values, it is expected that more research will be conducted through the single decision-making approach.

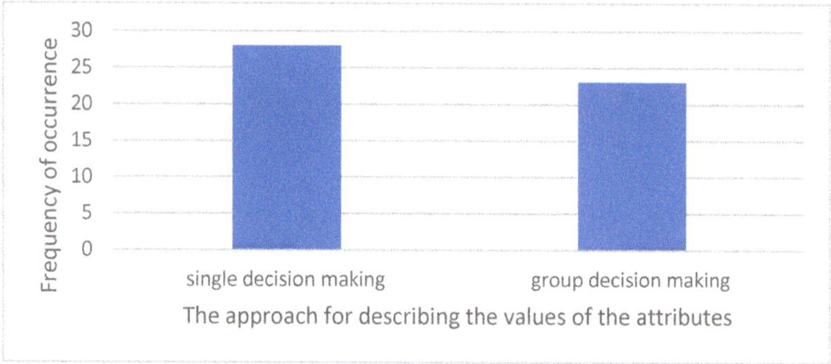

Figure 3. The decision-making process within the research for describing the attributes' values.

Further analysis is oriented to the granulation of linguistic expressions used for describing the relative importance of the attributes (Figure 4) and the attributes' value (Figure 5).

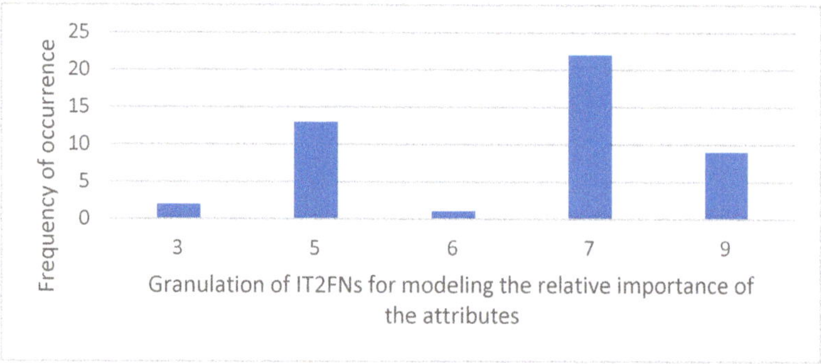

Figure 4. Granulation of linguistic expressions for describing the relative importance of the attributes.

The granulation is associated with the size of the treated problem. While problems that include a lower number of attributes can be described by using at least three linguistic expressions, large-scale problems may employ more linguistic expressions. The engineering and management research analysis shows that problems with a larger number of attributes indicate the employment of nine expressions. The majority of researchers, however, have employed five or seven expressions believing that these numbers would be suitable.

Figure 5 explains the frequency of usage of five, seven, or nine linguistic expressions for determining the attributes' value. The selection of the appropriate number of expressions can be described in a similar way that is analyzed in determining the relative importance of the attributes. It is easy to see that many authors propose seven linguistic expressions for determining the attributes' value. It is worth mentioning that some authors have employed more linguistic expressions for describing uncertainties in attributes' values [52,56]. It may be used as a reference for further research.

Figures 6 and 7 present the applied domains within the research for describing the relative importance of the attributes and attributes' value, respectively.

Determining the domain applied within the research for describing the relative importance of the attributes, and the attributes' value can be set as a task itself.

The analysis of Figures 6 and 7 clearly shows that many authors propose the domain on an interval between 0 and 1. It is expected since the employment of this domain decreases the calculation complexity in determining the weights' vector of treated attributes and there is no need to conduct the normalization procedure of the fuzzy decision matrix. On the other hand, some methods, such as IT2FAHP and IT2FBWM, do not support the employment of the domain between 0 and 1. This represents the main constraint of the analyzed domain.

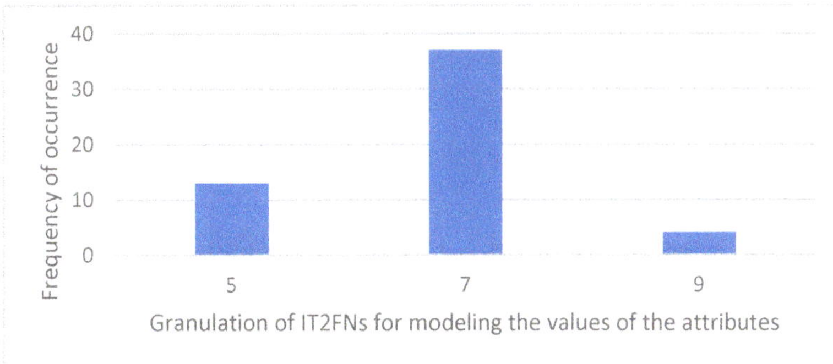

Figure 5. Granulation of linguistic expressions for describing the attributes' value.

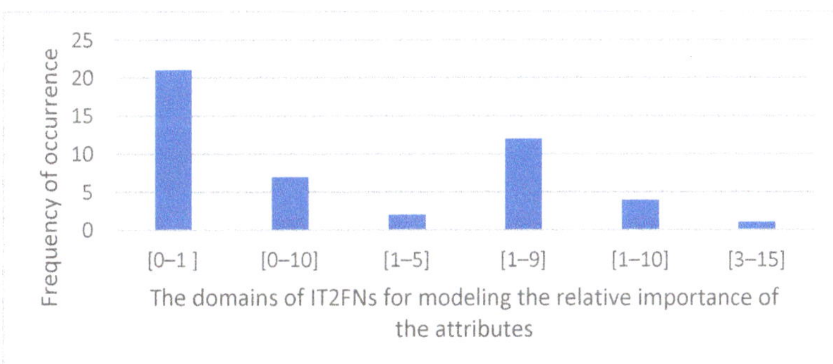

Figure 6. The domains applied within the research for describing the relative importance of the attributes.

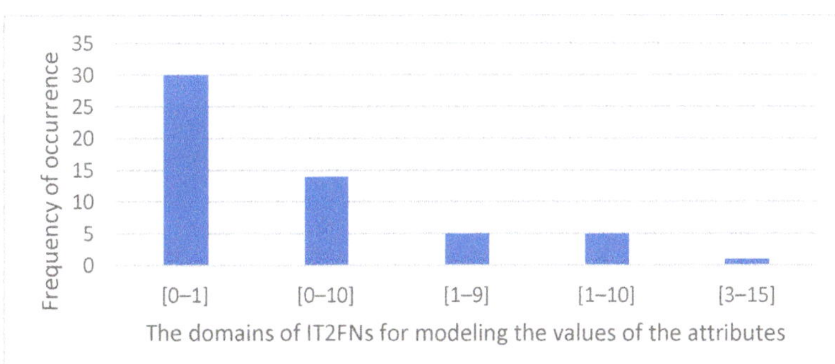

Figure 7. The domains applied within the research for describing the attributes' value.

The methods used for determining the attributes' weights vector are presented in Figure 8. Due to MADM's suitability for the named purpose, many authors employ IT2FAHP and IT2FBWM to determine the attributes' weights vector. IT2FAHP is a well-known method, and it can be smoothly applied while the attributes have a hierarchical structure. The main lack of this method is the need for obtaining well-defined input data in a matrix shape if the treated problem is large scale. That practically means that the decision-maker could be fully loaded, and his/her focus could be questioned. This is related to the consistency of the assessment. Obtaining input data for the IT2FBWM is less complex compared to the IT2FAHP. On the other hand, IT2FBWM implies the need for more complex calculations compared to IT2FAHP.

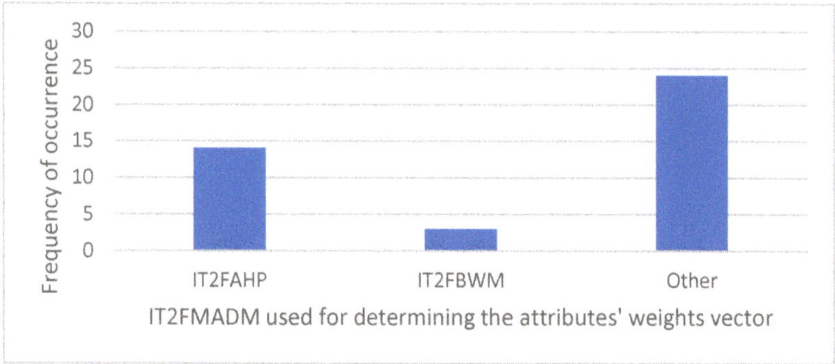

Figure 8. The methods used for determining the attributes' weights vector.

The majority of authors use different fuzzy operators for determining the attributes' weights vector (Figure 8). The advantage of this approach can be explained since the complexity of calculations is significantly decreased compared to IT2FBWM and IT2FAHP. According to the authors' opinion, the main lack of applying fuzzy operators could be a slightly decreased preciseness of the assessment compared to the named MADM methods.

Figure 9 denotes the frequency of the methods' appearances in the executed research presented in Table 2.

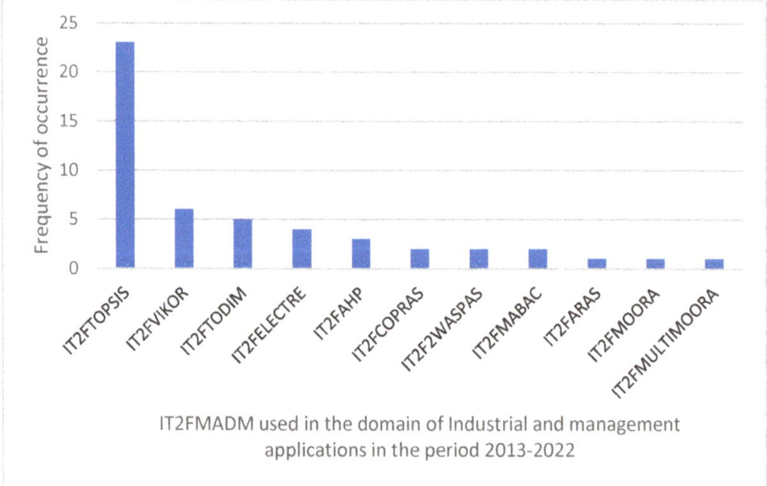

Figure 9. A brief explanation of the usage of IT2FMADM techniques.

Figure 9 denotes the frequency of IT2FMADM for determining the rank of alternatives in the domain of engineering and management from 2013 to 2023 according to the Springer, Science Direct, Emerald, Wiley, ProQuest, Taylor, and Francis platforms. It is easy to see that IT2FTOPSIS has been used most frequently.

Each of the analyzed MDM is employed on a different mathematical foundation. Therefore, it is inappropriate to compare the results delivered by their calculations. The proposed analysis cannot tell if IT2FTOPSIS is the most suitable for the application in the presented domain. Future research could confirm or dispute this.

4. Conclusions

The starting point of this research is an intention to provide insight into research in the field of MADM encompassing the application of IT2 fuzzy sets in the domain of engineering and management within the period between 2013 and 2023. If compared to papers of authors that have performed reviews of MADM applications in the field of management, the following may be stated. While Mardani et al. [1] have conducted the review on two decades from 1994 to 2014, this research covers the decade between 2013 and 2022. Mardani et al. [1] have covered a similar field of MADM application in terms of management and business emphasizing the employment of conventional MADM techniques and MADM techniques enhanced with type 1 fuzzy sets. The focus of the review [1] was to present the frequency of occurrence of each MADM technique and the trend of technique application. Celik et al. [22] conducted a review of papers employing IT2FMADM techniques between 2007 and 2015, embracing different application domains. This research [22] proposed the frequency of occurrence of each MCDM and the trend of their application. Compared to the mentioned review papers [1,22], our research sets a focus on the analysis of IT2FNs features used for modeling the relative importance and values of the attributes, as well as the application frequency IT2FMADM techniques.

The main contribution of the research to the literature may be summarized as follows: (1) it determines the two-stage MADM techniques that have been integrated with IT2FNs; (2) it represents two application areas, engineering and management; (3) the trend in research of IT2FMADM will remain stable in the future; (4) within the presented research, the sample of 41 papers in the treated areas is analyzed according to the following features: (i) the membership function shapes, (ii) the granulation, (iii) the domains of IT2FNs, (5) the frequency of IT2FADM employed for ranking the attributes' weights, as well as the frequency of IT2FADM employed for determining the alternative rank are analyzed.

The theoretical implications of the research are oriented to the exploitation of results within future research in this application. Different authors will have kinds of recommendations on how to determine the relative importance and values of attributes in different engineering and management problems.

The main constraint of the research is the size of the sample since papers are derived from the search covering Springer, Science Direct, Emerald, Wiley, ProQuest, Taylor, and Francis research platforms. Future research should expand the search and cover more different scientific databases. Additionally, future research should cover the research of different domains of IT2FADM and compare them with the obtained data. Bearing in mind the number of papers in the previous decade per year, it may be concluded that there is an ongoing trend that the number of research papers employing IT2FMADM is increasing. The research hotspots in the domain are oriented to sustainability and risk management, while industrial applications are oriented to industrial engineering applications.

Author Contributions: Conceptualization, A.A. and D.T.; methodology, A.A. and D.T.; formal analysis, A.A. and D.T.; investigation, A.A. and D.T.; resources, A.A.; data curation, D.T.; writing—original draft preparation, A.A. and D.T.; visualization, A.A. and D.T. All authors have read and agreed to the published version of the manuscript.

Funding: This research received no external funding.

Data Availability Statement: Not applicable.

Conflicts of Interest: The authors declare no conflict of interest.

Appendix A

Table A1. The linguistic expressions and corresponding of IT2FNs and their features which are used to describe the relative importance of attributes.

Authors	Type of IT2FNs	Granulation/The Domain	The Aggregation Operators	The Determination of Attribute Weights
Celik et al. [27]	IT2TrFN	5/[0–10]	Fuzzy arithmetic mean	Fuzzy weights vector
Baležentis and Zeng [28]	IT2TrFN	9/[0–1]	-	Crisp weights vector
Chen and Hong [30]	IT2TrFN	7/[0–1]	Method for comparison of IT2FNS combined with arithmetic mean	Weight attributed to the largest variable/Crisp weights vector
Ghorabaee et al. [29]	IT2TrFN	7/[0–1]	Fuzzy arithmetic mean	Fuzzy weights vector
Abdullah and Zulkifli [33]	TrFN	9/[1–9]	-	FAHP and fuzzy geometric mean/defuzzification are performed by using the centroid defuzzification method [89]/crisp weights vector
Ghorabaee [19]	IT2TrFN	7/[0–1]	Fuzzy arithmetic mean	Fuzzy weights vector
Kilic and Kaya [32]	IT2TrFN	5/[1–9]		IT2FAHP and fuzzy geometric mean/defuzzification are performed by using the center of area method [77]/crisp weights vector
Cebi and Otay [34]	IT2TrFN	7/[0–1]	Fuzzy arithmetic mean	Fuzzy weights vector
Qin et al. [31]	IT2TFN	5/[0–10]	-	Fuzzy weights vector
Qin et al. [43]	IT2TrFN	7/[0–10]	Type 2 fuzzy weighted aggregation method	KM algorithm [90]
Özkan et al. [37]	IT2TrFN	5/[1–9]		IT2FAHP and fuzzy geometric mean/aggregation performed by using fuzzy arithmetic mean/fuzzy weights vector
Liao [38]	IT2TrFN	5/[0–1]	-	Fuzzy weights vector
Sang and Liu [39]	crisp			Crisp weights vector
Ghorabaee et al. [36]	IT2TrFN	7/[0–1]	Fuzzy arithmetic mean	Fuzzy weights vector
Qin et al. [35]	IT2TrFN	7/[0–10]	-	Fuzzy weights vector
Ghorabaee et al. [41]	IT2TrFN	7/[0–1]		Entropy method/fuzzy weights vector
Celik et al. [40]	IT2TrFN	7/[0–10]	Fuzzy arithmetic mean	-
Buyoukozkan [42]	IT2TrFN	7/[0–1]	Fuzzy arithmetic mean	Fuzzy weights vector
Gorener et al. [44]	IT2TrFN	5/[1–9]	-	IT2FAHP and fuzzy geometric mean/fuzzy weights vector
Deveci et al. [45]	IT2TrFN	7/[0–1]	Fuzzy arithmetic mean	Fuzzy weights vector
Mousakhani [46]	IT2TrFN	7/[0–1]	Fuzzy geometric mean	Fuzzy weights vector
Soner et al. [47]	IT2TrFN	9/[1–10]		IT2FAHP and fuzzy geometric mean/fuzzy weights vector
Zhong and Yao [48]	IT2TrFN	7/[0–1]	-	The information entropy/crisp weights vector
Deveci et al. [49]	IT2TrFN	5/[0–10]	Fuzzy arithmetic mean	Fuzzy weights vector
Celik and Akyuz [20]	IT2TrFN	9/[1–10]	-	IT2FAHP and fuzzy geometric mean/defuzzification procedure [76]/crisp weights vector

Table A1. *Cont.*

Authors	Type of IT2FNs	Granulation/The Domain	The Aggregation Operators	The Determination of Attribute Weights
Debnath and Biswas [50]	IT2TrFN	5/[1–9]	-	IT2FAHP and fuzzy geometric mean/the proposed defuzzification procedure/fuzzy weights vector
Meng et al. [51]	IT2TrFN	7/[0–1]		The linear normalization procedure
Xu et al. [54]	crisp	/	/	AHP/crisp vector weights
Dinçer et al. [53]	IT2TrFN	7/[0–1]	-	IT2DEMATEL combined with IT2FANP and defuzzification procedure [76]/crisp weights vector
Wu et al. [55]	IT2TrFN	9/[1–9]	-	IT2FBWM and fuzzy geometric mean and defuzzification by using the centroid area method [91]
Aleksic et al. [56]	IT2TrFN	3/[1–5]	Fuzzy averaging mean	Ranking of IT2FNs [74]/crisp weights vector
Yucesan et al. [57]				BWM/crisp weights vector
Đurić et al. [52]	IT2TrFN	3/[1–5]		IT2FAHP and fuzzy geometric mean/fuzzy weights vector
Dorfeshan and Mousavi [58]	IT2TrFNs	7/[0–1]		IT2FWASPAS/crisp weights vector
Bera et al. [59]	IT2TrFN	7/[0–1]	-	Fuzzy weights vector
Mohamadghasemi et al. [26]	(GIT2FN)	7/[3–15]	-	Crisp weights vector
Ayyildiz et al. [60]	IT2TrFN	9/[1–10]	-	IT2FAHP and fuzzy geometric mean/defuzzification procedure/linear normalization procedure/crisp weights vector
Kiraci and Akan [21]	IT2TrFN	5/[1–9]		IT2FAHP and fuzzy geometric mean/defuzzification are performed by using the center of area method [77]/arithmetic mean/crisp weights vector
Pourmand et al. [62]	IT2TrFN	7/[0–1]	-	IT2FTOPSIS combined with the ranking of IT2FNs [74] and linear normalization procedure/crisp weights vector
Özdemir and Üsküdar [63]	IT2TrFN	5/[1–9]	-	IT2FAHP and fuzzy geometric mean/fuzzy weights vector
Deveci et al. [64]	IT2TFN	7/[0–1]	Fuzzy arithmetic mean	Fuzzy weights vector
Mirnezami et al. [65]	-			-
Komatina et al. [67]	IT2TFN	9/[0–1]	-	IT2FDelphi technique
Karagöz et al. [68]	IT2TFN	7/[0–1]	Fuzzy arithmetic mean	Fuzzy weights vector
Kaya and Aycin [92]	IT2TrFN	5/[1–9]		IT2FAHP and fuzzy geometric mean/fuzzy weights vector
Celik et al. [69]	IT2TrFN	9/[1–10]		IT2FBWM based on [55]/fuzzy weights vector
Zhang et al. [70]	IT2TrFN	5/[0–10]	Fuzzy arithmetic mean	Fuzzy weights vector
Sharaf [66]	IT2TrFN	7/[0–1]	Fuzzy arithmetic mean	Fuzzy weights vector

Table A1. *Cont.*

Authors	Type of IT2FNs	Granulation/The Domain	The Aggregation Operators	The Determination of Attribute Weights
Komatina et al. [71]	IT2TFN	5/[1–9]		IT2FAHP and fuzzy geometric mean/fuzzy weights vector
Aleksić et al. [72]	IT2TFN	6/[1–9]	Geometric mean	IT2FBWM [55]
Ecer [73]	IT2TFN	5/[1–9]	-	IT2FAHP and fuzzy geometric mean/fuzzy weights vector

Appendix B

Table A2. The determining attributes' values.

Authors	Type of IT2FNs	Granulation/The Domain	The Aggregation Operators
Celik et al. (2013) [27]	IT2TrFN	5/[0–10]	Fuzzy arithmetic mean
Baležentis and Zeng (2013) [28]	IT2TrFN	9/[0–1]	The weighted geometric average operator
Chen and Hong (2014) [30]	IT2TrFN	7/[0–10]	Fuzzy arithmetic mean
Ghorabaee et al. (2014) [29]	IT2TrFN	7/[0–10]	Fuzzy arithmetic mean
Abdullah and Zulkifli (2015) [33]	IT2TrFN	5/[0–1]	-
Cebi and Otay (2015) [34]	IT2TrFN	7/[0–10]	Fuzzy arithmetic mean
Qin et al. (2015) [31]	IT2TrFN	7/[0–1]	-
Ghorabaee et al. (2016) [19]	IT2TrFN	7/[0–1]	Fuzzy arithmetic mean
Kilic and Kaya (2015) [32]	IT2TrFN	7/[0–1]	Fuzzy arithmetic mean
Özkan et al. (2015) [37]	IT2TrFN	5/[1–9]	Fuzzy arithmetic mean
Liao (2015) [38]	IT2TrFN	5/[0–1]	-
Sang and Liu (2015) [39]	IT2TrFN	7/[0–1]	Fuzzy arithmetic mean
Qin et al. (2015) [35]	IT2TrFN	5/[0–10]	
Ghorabaee et al. (2015) [36]	IT2TrFN	7/[0–1]	Fuzzy arithmetic mean
Ghorabaee et al. (2016) [41]	IT2TrFN	7/[0–1]	Fuzzy arithmetic mean
Buyoukozkan (2016) [42]	IT2TrFN	7/[0–1]	fuzzy arithmetic mean
Celik et al. (2016) [40]	IT2TrFN	7/[1–10]	Fuzzy arithmetic mean
Qin et al. (2017) [43]	TrFN	7/[0–1]	-
Soner et al. (2017) [47]	IT2TrFN	7/[0–10]	Fuzzy arithmetic mean
Deveci et al. (2017) [45]	IT2TrFN	7/[0–10]	Fuzzy arithmetic mean
Gorener et al. (2017) [44]	IT2TrFN	7/[0–1]	Fuzzy arithmetic mean
Zhong and Yao (2017) [48]	IT2TrFN	7/[0–1]	-
Mousakhani (2017) [46]	IT2TrFN	7/[0–10]	Fuzzy geometric mean
Debnath and Biswas (2018) [50]	IT2TrFN	5/[1–9]	-
Celik and Akyuz (2018) [20]	IT2TrFN	7/[0–1]	-
Deveci et al. (2018) [49]	IT2TrFN	9/[0–10]	Fuzzy arithmetic mean
Meng et al. (2019) [51]	IT2TrFN	7/[0–1]	-
Dinçer et al. (2019) [53]	IT2TrFN	7/[0–1]	Fuzzy arithmetic mean
Xu et al. (2019) [54]	IT2TrFN	5/[0–1]	-

Table A2. *Cont.*

Authors	Type of IT2FNs	Granulation/The Domain	The Aggregation Operators
Yucesan et al. (2019) [57]	IT2TrFN	-/[0–1]	Fuzzy arithmetic mean
Aleksic et al. (2019) [56]	IT2TrFN	7/[0–1] and 5/[0–1]	-
Đurić et al. (2019) [52]	IT2TrFN	5/[0–1] and 7/[0–1]	-
Wu et al. (2019) [55]	IT2TrFN	7/[0–10]	The interval type 2 fuzzy weighted average operator
Dorfeshan and Mousavi (2020) [58]	IT2TrFN	7/[0–1]	-
Bera et al. (2020) [59]	IT2TrFN	7/[0–1]	Fuzzy arithmetic mean
Mohamadghasemi et al. (2020) [26]	GIT2FN	7/[3–15]	-
Ayyildiz et al. (2020) [60]	IT2TrFN	9/[1–10]	Fuzzy arithmetic mean
Kiraci and Akan (2020) [21]	IT2TrFN	7/[0–1]	Fuzzy arithmetic mean
Pourmand et al. (2020) [62]	IT2TrFN	7/[0–1]	-
Özdemir and Üsküdar (2020) [63]	IT2TrFN	7/[0–1]	Fuzzy arithmetic mean
Deveci et al. (2020) [64]	IT2TFN	7/[0–10]	-
Mirnezami et al. (2021) [65]	IT2TrFN	7/[0–1]	-
Sharaf (2021) [66]	IT2TrFN	7/[0–1]	-
Zhang et al. (2022) [70]	IT2TrFN	5/[0–10]	
Komatina et al. (2021) [67]	IT2TFN	7/[1–9]	-
Karagöz et al. (2021) [68]	IT2TFN	7/[1–10]	Fuzzy arithmetic mean
Kaya and Aycin (2021) [92]		7/[0–10]	-
Celik et al. (2021) [69]	IT2TrFN	9/[1–10]	-
Komatina et al. (2022) [71]	IT2TFN	7/[1–9]	-
Aleksić et al. (2022) [72]	IT2TFN	5/[1–10]	
Ecer, F. (2022) [73]	IT2TFN	5/[1–9]	-

References

1. Mardani, A.; Jusoh, A.; Zavadskas, E.K. Fuzzy Multiple Criteria Decision-Making Techniques and Applications—Two Decades Review from 1994 to 2014. *Expert Syst. Appl.* **2015**, *42*, 4126–4148. [CrossRef]
2. Behzadian, M.; Khanmohammadi Otaghsara, S.; Yazdani, M.; Ignatius, J. A State-of the-Art Survey of TOPSIS Applications. *Expert Syst. Appl.* **2012**, *39*, 13051–13069. [CrossRef]
3. Danesh, D.; Ryan, M.J.; Abbasi, A. Multi-Criteria Decision-Making Methods for Project Portfolio Management: A Literature Review. *Int. J. Manag. Decis. Mak.* **2018**, *17*, 75–94. [CrossRef]
4. Sousa, M.; Almeida, M.F.; Calili, R. Multiple Criteria Decision Making for the Achievement of the UN Sustainable Development Goals: A Systematic Literature Review and a Research Agenda. *Sustainability* **2021**, *13*, 4129. [CrossRef]
5. Zayat, W.; Kilic, H.S.; Yalcin, A.S.; Zaim, S.; Delen, D. Application of MADM Methods in Industry 4.0: A Literature Review. *Comput. Ind. Eng.* **2023**, *177*, 109075. [CrossRef]
6. Zavadskas, E.K.; Turskis, Z.; Antucheviciene, J.; Zakarevicius, A. Optimization of Weighted Aggregated Sum Product Assessment. *Elektron. Elektrotechnika* **2012**, *122*, 3–6. [CrossRef]
7. Hwang, C.-L.; Yoon, K. Methods for Multiple Attribute Decision Making. In *Multiple Attribute Decision Making: Methods and Applications A State-of-the-Art Survey*; Hwang, C.-L., Yoon, K., Eds.; Lecture Notes in Economics and Mathematical Systems; Springer: Berlin/Heidelberg, Germany, 1981; pp. 58–191, ISBN 978-3-642-48318-9.
8. Opricovic, S. Multicriteria Optimization of Civil Engineering Systems. PhD. Thesis, Faculty of Civil Engineering, University of Belgrade, Belgrade, Serbia, 1998.
9. Zavadskas, E.K.; Kaklauskas, A.; Šarka, V. The New Method of Multicriteria Complex Proportional Assessment of Projects. *Technol. Econ. Dev. Econ.* **1994**, *1*, 131–139.
10. Brauers, W.K.M.; Zavadskas, E.K. The MOORA Method and Its Application to Privatization in a Transition Economy. *Control Cybern.* **2006**, *35*, 445–469.

11. Zavadskas, E.K.; Turskis, Z. A New Additive Ratio Assessment (ARAS) Method in Multicriteria Decision-making. *Ukio Technol. Ekon. Vystym.* **2010**, *16*, 159–172. [CrossRef]
12. Roy, B. Classement et choix en présence de points de vue multiples. *Rev. Franjaise D'informatique Rech. Opérationnelle* **1968**, *2*, 57–75. [CrossRef]
13. Saaty, T.L. Decisions with the Analytic Network Process (ANP). In Proceedings of the International Symposium on the Analytic Hierarchy Process, Vancouver, BC, Canada, 12–15 July 1996.
14. Saaty, T.L. The Modern Science of Multicriteria Decision Making and Its Practical Applications: The AHP/ANP Approach. *Oper. Res.* **2013**, *61*, 1101–1118. [CrossRef]
15. Gomes, L.; Lima, M. TODIM: Basics and Application to Multicriteria Ranking of Projects with Environmental Impacts. *Found. Comput. Decis. Sci.* **1992**, *16*, 113–127.
16. Rezaei, J. Best-Worst Multi-Criteria Decision-Making Method. *Omega* **2015**, *53*, 49–57. [CrossRef]
17. Pamučar, D.; Ćirović, G. The Selection of Transport and Handling Resources in Logistics Centers Using Multi-Attributive Border Approximation Area Comparison (MABAC). *Expert Syst. Appl.* **2015**, *42*, 3016–3028. [CrossRef]
18. Fontela, E.; Gabus, A. *DEMATEL, Innovative Methods*; Battelle Geneva Research Institute: Geneva, Switzerland, 1974.
19. Ghorabaee, M.K. Developing an MCDM Method for Robot Selection with Interval Type-2 Fuzzy Sets. *Robot. Comput. Integr. Manuf.* **2016**, *37*, 221. [CrossRef]
20. Celik, E.; Akyuz, E. An Interval Type-2 Fuzzy AHP and TOPSIS Methods for Decision-Making Problems in Maritime Transportation Engineering: The Case of Ship Loader. *Ocean. Eng.* **2018**, *155*, 371–381. [CrossRef]
21. Kiracı, K.; Akan, E. Aircraft Selection by Applying AHP and TOPSIS in Interval Type-2 Fuzzy Sets. *J. Air Transp. Manag.* **2020**, *89*, 101924. [CrossRef]
22. Celik, E.; Gul, M.; Aydin, N.; Gumus, A.T.; Guneri, A.F. A Comprehensive Review of Multi Criteria Decision Making Approaches Based on Interval Type-2 Fuzzy Sets. *Knowl. Based Syst.* **2015**, *85*, 329–341. [CrossRef]
23. Zadeh, L.A. The Concept of a Linguistic Variable and Its Application to Approximate Reasoning-III. *Inf. Sci.* **1975**, *9*, 43–80. [CrossRef]
24. Zimmermann, H.-J. *Fuzzy Set Theory—And Its Applications*; Springer Science & Business Media: Berlin/Heidelberg, Germany, 2011; ISBN 978-94-010-0646-0.
25. Derelï, T.; Baykasoglu, A.; Altun, K.; Durmuşoğlu, A.; Türkşen, İ. Industrial Applications of Type-2 Fuzzy Sets and Systems: A Concise Review. *Comput. Ind.* **2011**, *62*, 125–137. [CrossRef]
26. Mohamadghasemi, A.; Hadi-Vencheh, A.; Lotfi, F.H.; Khalilzadeh, M. An Integrated Group FWA-ELECTRE III Approach Based on Interval Type-2 Fuzzy Sets for Solving the MCDM Problems Using Limit Distance Mean. *Complex Intell. Syst.* **2020**, *6*, 355–389. [CrossRef]
27. Celik, E.; Bilisik, O.N.; Erdogan, M.; Gumus, A.T.; Baracli, H. An Integrated Novel Interval Type-2 Fuzzy MCDM Method to Improve Customer Satisfaction in Public Transportation for Istanbul. *Transp. Res. Part E* **2013**, *58*, 28–51. [CrossRef]
28. Baležentis, T.; Zeng, S. Group Multi-Criteria Decision Making Based upon Interval-Valued Fuzzy Numbers: An Extension of the MULTIMOORA Method. *Expert Syst. Appl.* **2013**, *40*, 543–550. [CrossRef]
29. Ghorabaee, M.K.; Amiri, M.; Sadaghiani, J.S.; Goodarzi, G.H. Multiple Criteria Group Decision-Making for Supplier Selection Based on COPRAS Method with Interval Type-2 Fuzzy Sets. *Int. J. Adv. Manuf. Technol.* **2014**, *75*, 1115. [CrossRef]
30. Chen, S.-M.; Hong, J.-A. Fuzzy Multiple Attributes Group Decision-Making Based on Ranking Interval Type-2 Fuzzy Sets and the TOPSIS Method. *IEEE Trans. Syst. Man Cybern. Syst.* **2014**, *44*, 1665–1673. [CrossRef]
31. Qin, Y.; Zhang, Z.; Liu, X.; Li, M.; Kou, L. Dynamic Risk Assessment of Metro Station with Interval Type-2 Fuzzy Set and TOPSIS Method. *J. Intell. Fuzzy Syst.* **2015**, *29*, 93–106. [CrossRef]
32. Kiliç, M.; Kaya, İ. Investment Project Evaluation by a Decision Making Methodology Based on Type-2 Fuzzy Sets. *Appl. Soft Comput.* **2015**, *27*, 399–410. [CrossRef]
33. Abdullah, L.; Zulkifli, N. Integration of Fuzzy AHP and Interval Type-2 Fuzzy DEMATEL: An Application to Human Resource Management. *Expert Syst. Appl.* **2015**, *42*, 4397–4409. [CrossRef]
34. Cebi, F.; Otay, İ. Multi-Criteria and Multi-Stage Facility Location Selection under Interval Type-2 Fuzzy Environment: A Case Study for a Cement Factory. *Int. J. Comput. Intell. Syst.* **2015**, *8*, 330–344. [CrossRef]
35. Qin, J.; Liu, X.; Pedrycz, W. An Extended VIKOR Method Based on Prospect Theory for Multiple Attribute Decision Making under Interval Type-2 Fuzzy Environment. *Knowl.-Based Syst.* **2015**, *86*, 116–130. [CrossRef]
36. Ghorabaee, M.K.; Amiri, M.; Sadaghiani, J.S.; Zavadskas, E.K. Multi-Criteria Project Selection Using an Extended VIKOR Method with Interval Type-2 Fuzzy Sets. *Int. J. Inf. Tech. Dec. Mak.* **2015**, *14*, 993–1016. [CrossRef]
37. Özkan, B.; Kaya, İ.; Cebeci, U.; Başlıgil, H. A Hybrid Multicriteria Decision Making Methodology Based on Type-2 Fuzzy Sets for Selection among Energy Storage Alternatives. *Int. J. Comput. Intell. Syst.* **2015**, *8*, 914–927. [CrossRef]
38. Liao, T.W. Two Interval Type 2 Fuzzy TOPSIS Material Selection Methods. *Mater. Des.* **2015**, *88*, 1088–1099. [CrossRef]
39. Sang, X.; Liu, X. An Interval Type-2 Fuzzy Sets-Based TODIM Method and Its Application to Green Supplier Selection. *J. Oper. Res. Soc.* **2016**, *67*, 722–734. [CrossRef]
40. Celik, E.; Gumus, A.; Erdogan, M. A New Extension of the ELECTRE Method Based upon Interval Type-2 Fuzzy Sets for Green Logistic Service Providers Evaluation. *J. Test. Eval.* **2016**, *44*, 20140046. [CrossRef]

41. Ghorabaee, K.M.; Zavadskas, E.K.; Amiri, M.; Esmaeili, A. Multi-Criteria Evaluation of Green Suppliers Using an Extended WASPAS Method with Interval Type-2 Fuzzy Sets. *J. Clean. Prod.* **2016**, *137*, 213–229. [CrossRef]
42. Büyüközkan, G.; Parlak, I.B.; Tolga, A.C. Evaluation of Knowledge Management Tools by Using An Interval Type-2 Fuzzy TOPSIS Method. *Int. J. Comput. Intell. Syst.* **2016**, *9*, 812–826. [CrossRef]
43. Qin, J.; Liu, X.; Pedrycz, W. An Extended TODIM Multi-Criteria Group Decision Making Method for Green Supplier Selection in Interval Type-2 Fuzzy Environment. *Eur. J. Oper. Res.* **2017**, *258*, 626–638. [CrossRef]
44. Görener, A.; Ayvaz, B.; Kuşakcı, A.O.; Altınok, E. A Hybrid Type-2 Fuzzy Based Supplier Performance Evaluation Methodology: The Turkish Airlines Technic Case. *Appl. Soft Comput.* **2017**, *56*, 436–445. [CrossRef]
45. Deveci, M.; Demirel, N.Ç.; Ahmetoğlu, E. Airline New Route Selection Based on Interval Type-2 Fuzzy MCDM: A Case Study of New Route between Turkey- North American Region Destinations. *J. Air Transp. Manag.* **2017**, *59*, 83–99. [CrossRef]
46. Mousakhani, S.; Nazari-Shirkouhi, S.; Bozorgi-Amiri, A. A Novel Interval Type-2 Fuzzy Evaluation Model Based Group Decision Analysis for Green Supplier Selection Problems: A Case Study of Battery Industry. *J. Clean. Prod.* **2017**, *168*, 205–218. [CrossRef]
47. Soner, O.; Celik, E.; Akyuz, E. Application of AHP and VIKOR Methods under Interval Type 2 Fuzzy Environment in Maritime Transportation. *Ocean. Eng.* **2017**, *129*, 107–116. [CrossRef]
48. Zhong, L.; Yao, L. An ELECTRE I-Based Multi-Criteria Group Decision Making Method with Interval Type-2 Fuzzy Numbers and Its Application to Supplier Selection. *Appl. Soft Comput.* **2017**, *57*, 556–576. [CrossRef]
49. Deveci, M.; Canıtez, F.; Gökaşar, I. WASPAS and TOPSIS Based Interval Type-2 Fuzzy MCDM Method for a Selection of a Car Sharing Station. *Sustain. Cities Soc.* **2018**, *41*, 777–791. [CrossRef]
50. Debnath, J.; Biswas, A. Analytic Hierarchy Process Based on Interval Type-2 Intuitionistic Fuzzy Sets with Their Application to Multicriteria Decision Making. *Intell. Decis. Technol.* **2018**, *12*, 359–370. [CrossRef]
51. Meng, Q.; Liu, X.; Song, Y.; Wang, W. An Extended Generalized TODIM Method for Risk Assessment of Supply Chain in Social Commerce under Interval Type-2 Fuzzy Environment. *J. Intell. Fuzzy Syst.* **2019**, *37*, 8551–8565. [CrossRef]
52. Đurić, G.; Mitrović, Č.; Komatina, N.; Tadić, D.; Vorotović, G. The Hybrid MCDM Model with the Interval Type-2 Fuzzy Sets for the Software Failure Analysis. *J. Intell. Fuzzy Syst.* **2019**, *37*, 7747–7759. [CrossRef]
53. Dinçer, H.; Yüksel, S.; Martínez, L. Interval Type 2-Based Hybrid Fuzzy Evaluation of Financial Services in E7 Economies with DEMATEL-ANP and MOORA Methods. *Appl. Soft Comput.* **2019**, *79*, 186–202. [CrossRef]
54. Xu, Z.; Qin, J.; Liu, J.; Martínez, L. Sustainable Supplier Selection Based on AHPSort II in Interval Type-2 Fuzzy Environment. *Inf. Sci.* **2019**, *483*, 273–293. [CrossRef]
55. Wu, Q.; Zhou, L.; Chen, Y.; Chen, H. An Integrated Approach to Green Supplier Selection Based on the Interval Type-2 Fuzzy Best-Worst and Extended VIKOR Methods. *Inf. Sci.* **2019**, *502*, 394–417. [CrossRef]
56. Aleksic, A.; Runic Ristic, M.; Komatina, N.; Tadic, D. Advanced Risk Assessment in Reverse Supply Chain Processes: A Case Study in Republic of Serbia. *Adv. Prod. Eng. Manag.* **2019**, *14*, 421–434. [CrossRef]
57. Yucesan, M.; Mete, S.; Serin, F.; Celik, E.; Gul, M. An Integrated Best-Worst and Interval Type-2 Fuzzy TOPSIS Methodology for Green Supplier Selection. *Mathematics* **2019**, *7*, 182. [CrossRef]
58. Dorfeshan, Y.; Mousavi, S.M. A Novel Interval Type-2 Fuzzy Decision Model Based on Two New Versions of Relative Preference Relation-Based MABAC and WASPAS Methods (with an Application in Aircraft Maintenance Planning). *Neural Comput. Appl.* **2020**, *32*, 3367–3385. [CrossRef]
59. Bera, A.K.; Jana, D.K.; Banerjee, D.; Nandy, T. Supplier Selection Using Extended IT2 Fuzzy TOPSIS and IT2 Fuzzy MOORA Considering Subjective and Objective Factors. *Soft Comput.* **2020**, *24*, 8899–8915. [CrossRef]
60. Ayyildiz, E.; Taskin Gumus, A.; Erkan, M. Individual Credit Ranking by an Integrated Interval Type-2 Trapezoidal Fuzzy Electre Methodology. *Soft Comput.* **2020**, *24*, 16149–16163. [CrossRef]
61. Yang, Y.-Y.; Liu, X.-W.; Liu, F. Trapezoidal Interval Type-2 Fuzzy TOPSIS Using Alpha-Cuts. *Int. J. Fuzzy Syst.* **2020**, *22*, 293–309. [CrossRef]
62. Pourmand, E.; Mahjouri, N.; Hosseini, M.; Nik-Hemmat, F. A Multi-Criteria Group Decision Making Methodology Using Interval Type-2 Fuzzy Sets: Application to Water Resources Management. *Water Res. Manag.* **2020**, *34*, 4067–4092. [CrossRef]
63. Özdemir, Y.S.; Üsküdar, A. Strategy Selection by Using Interval Type-2 Fuzzy Mcdm and an Application. *J. Eng. Res.* **2020**, *8*, 172–189. [CrossRef]
64. Deveci, M.; Cali, U.; Kucuksari, S.; Erdogan, N. Interval Type-2 Fuzzy Sets Based Multi-Criteria Decision-Making Model for Offshore Wind Farm Development in Ireland. *Energy* **2020**, *198*, 117317. [CrossRef]
65. Mirnezami, S.A.; Mousavi, S.M.; Mohagheghi, V. An Innovative Interval Type-2 Fuzzy Approach for Multi-Scenario Multi-Project Cash Flow Evaluation Considering TODIM and Critical Chain with an Application to Energy Sector. *Neural Comput. Appl.* **2021**, *33*, 2263–2284. [CrossRef]
66. Sharaf, I.M. An Interval Type-2 Fuzzy TOPSIS for MAGDM Applied to Solar Power Systems. *Pak. J. Stat. Oper. Res.* **2021**, *17*, 559–575. [CrossRef]
67. Komatina, N.; Djapan, M.; Ristić, I.; Aleksić, A. Fulfilling External Stakeholders' Demands—Enhancement Workplace Safety Using Fuzzy MCDM. *Sustainability* **2021**, *13*, 2892. [CrossRef]
68. Karagöz, S.; Deveci, M.; Simic, V.; Aydin, N. Interval Type-2 Fuzzy ARAS Method for Recycling Facility Location Problems. *Appl. Soft Comput.* **2021**, *102*, 107107. [CrossRef]

69. Celik, E.; Yucesan, M.; Gul, M. Green Supplier Selection for Textile Industry: A Case Study Using BWM-TODIM Integration under Interval Type-2 Fuzzy Sets. *Environ. Sci. Pollut. Res.* **2021**, *28*, 64793–64817. [CrossRef]
70. Zhang, Z.; Zhao, X.; Qin, Y.; Si, H.; Zhou, L. Interval Type-2 Fuzzy TOPSIS Approach with Utility Theory for Subway Station Operational Risk Evaluation. *J. Ambient. Intell. Hum. Comput.* **2022**, *13*, 4849–4863. [CrossRef]
71. Komatina, N.; Tadić, D.; Aleksić, A.; Jovanović, A.D. The Assessment and Selection of Suppliers Using AHP and MABAC with Type-2 Fuzzy Numbers in Automotive Industry. *Proc. Inst. Mech. Eng. Part O* **2022**, 1748006X221095359. [CrossRef]
72. Aleksić, A.; Milanović, D.D.; Komatina, N.; Tadić, D. Evaluation and Ranking of Failures in Manufacturing Process by Combining Best-Worst Method and VIKOR under Type-2 Fuzzy Environment. *Expert Syst.* **2022**, *40*, e13148. [CrossRef]
73. Ecer, F. Multi-Criteria Decision Making for Green Supplier Selection Using Interval Type-2 Fuzzy AHP: A Case Study of a Home Appliance Manufacturer. *Oper. Res. Int. J.* **2022**, *22*, 199–233. [CrossRef]
74. Chen, S.-M.; Lee, L.-W. Fuzzy Multiple Attributes Group Decision-Making Based on the Interval Type-2 TOPSIS Method. *Expert Syst. Appl.* **2010**, *37*, 2790–2798. [CrossRef]
75. Mendel, J.M. Type-1 Fuzzy Sets and Fuzzy Logic. In *Uncertain Rule-Based Fuzzy Systems: Introduction and New Directions*, 2nd ed.; Mendel, J.M., Ed.; Springer International Publishing: Cham, Switzerland, 2017; pp. 25–99, ISBN 978-3-319-51370-6.
76. Kahraman, C.; Öztayşi, B.; Uçal Sarı, İ.; Turanoğlu, E. Fuzzy Analytic Hierarchy Process with Interval Type-2 Fuzzy Sets. *Knowl. Based Syst.* **2014**, *59*, 48–57. [CrossRef]
77. Bellman, R.E.; Zadeh, L.A. Decision-Making in a Fuzzy Environment. *Manag. Sci.* **1970**, *17*, B141–B164. [CrossRef]
78. Chang, D.-Y. Applications of the Extent Analysis Method on Fuzzy AHP. *Eur. J. Oper. Res.* **1996**, *95*, 649–655. [CrossRef]
79. Badri Ahmadi, H.; Kusi-Sarpong, S.; Rezaei, J. Assessing the Social Sustainability of Supply Chains Using Best Worst Method. *Resour. Conserv. Recycl.* **2017**, *126*, 99–106. [CrossRef]
80. Pomerol, J.-C.; Barba-Romero, S. *Multicriterion Decision in Management*; International Series in Operations Research & Management Science; Springer: Boston, MA, USA, 2000; Volume 25, ISBN 978-1-4613-7008-6.
81. Figueroa-García, J.C.; Chalco-Cano, Y.; Román-Flores, H. Distance Measures for Interval Type-2 Fuzzy Numbers. *Discret. Appl. Math.* **2015**, *197*, 93–102. [CrossRef]
82. Zhou, S.; John, R.I.; Chiclana, F.; Garibaldi, J.M. On Aggregating Uncertain Information by Type-2 OWA Operators for Soft Decision Making. *Int. J. Intell. Syst.* **2010**, *25*, 540–558. [CrossRef]
83. Kuo, M.-S.; Liang, G.-S. A Soft Computing Method of Performance Evaluation with MCDM Based on Interval-Valued Fuzzy Numbers. *Appl. Soft Comput.* **2012**, *12*, 476–485. [CrossRef]
84. Chen, T.-Y. A Signed-Distance-Based Approach to Importance Assessment and Multi-Criteria Group Decision Analysis Based on Interval Type-2 Fuzzy Set. *Knowl. Inf. Syst.* **2013**, *35*, 193–231. [CrossRef]
85. Liu, P.; Jin, F. A Multi-Attribute Group Decision-Making Method Based on Weighted Geometric Aggregation Operators of Interval-Valued Trapezoidal Fuzzy Numbers. *Appl. Math. Model.* **2012**, *36*, 2498–2509. [CrossRef]
86. Lee, L.-W.; Chen, S.-M. Fuzzy Multiple Attributes Group Decision-Making Based on the Extension of TOPSIS Method and Interval Type-2 Fuzzy Sets. In Proceedings of the 2008 International Conference on Machine Learning and Cybernetics, Kunming, China, 12–15 July 2008; pp. 3260–3265. [CrossRef]
87. Wang, Y.-M.; Elhag, T.M.S. Fuzzy TOPSIS Method Based on Alpha Level Sets with an Application to Bridge Risk Assessment. *Expert Syst. Appl.* **2006**, *31*, 309–319. [CrossRef]
88. Qin, J.; Liu, X. Multi-Attribute Group Decision Making Using Combined Ranking Value under Interval Type-2 Fuzzy Environment. *Inf. Sci.* **2015**, *297*, 293–315. [CrossRef]
89. Wang, Y.-M. Centroid Defuzzification and the Maximizing Set and Minimizing Set Ranking Based on Alpha Level Sets. *Comput. Ind. Eng.* **2009**, *57*, 228–236. [CrossRef]
90. Karnik, N.N.; Mendel, J.M. Centroid of a Type-2 Fuzzy Set. *Inf. Sci.* **2001**, *132*, 195–220. [CrossRef]
91. Mendel, J.M.; Wu, H. New Results about the Centroid of an Interval Type-2 Fuzzy Set, Including the Centroid of a Fuzzy Granule. *Inf. Sci.* **2007**, *177*, 360–377. [CrossRef]
92. Kayapinar Kaya, S.; Aycin, E. An Integrated Interval Type 2 Fuzzy AHP and COPRAS-G Methodologies for Supplier Selection in the Era of Industry 4.0. *Neural Comput. Appl.* **2021**, *33*, 10515–10535. [CrossRef]

mathematics

MDPI

Article

Sustainable Supplier Selection and Order Allocation Using an Integrated ROG-Based Type-2 Fuzzy Decision-Making Approach

Mehdi Keshavarz-Ghorabaee

Department of Management, Faculty of Humanities (Azadshahr Branch), Gonbad Kavous University, Gonbad Kavous 49717-99151, Iran; m.keshavarz@gonbad.ac.ir

Abstract: The sustainable Supplier Evaluation and Selection and Order Allocation (SSOA) problem has received significant attention in supply chain management due to its potential to enhance a company's performance, improve customer satisfaction, and reduce costs. In this study, an integrated methodology is proposed to address the SSOA problem. The methodology combines multiple techniques to handle the uncertainties associated with supplier evaluation, including a new ranking method based on the concept of Radius of Gyration (ROG) for interval type-2 fuzzy sets. The methodology also incorporates both subjective weights obtained using the Simple Multi-Attribute Rating Technique (SMART) and expert preferences, and objective weights calculated using the Method based on the Removal Effects of Criteria (MEREC) method to determine the weights of evaluation criteria. Some criteria for sustainable development are used to evaluate supplier performance, resulting in type-2 fuzzy sets, which are evaluated using the Weighted Aggregated Sum Product Assessment (WASPAS) method. The ROG-based ranking method is employed to calculate the relative scores of suppliers. Finally, a multi-objective decision-making (MODM) mathematical model is presented to identify suitable suppliers and allocate their order quantities. The methodology is demonstrated in a sustainable SSOA problem and is shown to be efficient and effective, as the ROG-based ranking method allows for more accurate supplier performance evaluation, and the use of the criteria highlights the importance of sustainability in supplier selection and order allocation. The methodology's practicality is further supported by the analysis conducted in this study, which demonstrates the methodology's ability to handle the uncertainties associated with supplier evaluation and selection. The proposed methodology offers a comprehensive approach to the SSOA problem that can effectively handle the uncertainties in supplier evaluation and selection and promote sustainable practices in supply chain management.

Keywords: sustainability; supplier selection; order allocation; SSOA; MCDM; type-2 fuzzy sets

MSC: 90B50; 90C70; 91B06; 03B52

Citation: Keshavarz-Ghorabaee, M. Sustainable Supplier Selection and Order Allocation Using an Integrated ROG-Based Type-2 Fuzzy Decision-Making Approach. *Mathematics* **2023**, *11*, 2014. https://doi.org/10.3390/math11092014

Academic Editor: Aleksandar Aleksic

Received: 29 March 2023
Revised: 20 April 2023
Accepted: 21 April 2023
Published: 24 April 2023

1. Introduction

Sustainable supplier selection and order allocation (SSOA) has become increasingly important for companies to achieve sustainable development and maintain their competitiveness in the global market. The supply chain is a crucial aspect of a company's operations, and suppliers play a critical role in the sustainability of the supply chain. Therefore, evaluating and selecting sustainable suppliers has become a critical strategic decision for companies [1,2]. Sustainable suppliers are those who are committed to environmentally friendly, socially responsible, and economically viable practices in their operations. By evaluating and selecting such suppliers, organizations can ensure that their suppliers' performance aligns with their own sustainability goals in these three dimensions. Consumers and stakeholders are becoming increasingly conscious of the environmental, social, and economic impacts of the products and services they use. Partnering with sustainable suppliers can enhance an organization's reputation as a socially responsible and environmentally conscious business while contributing to the economic development of the community in which they operate. By engaging with suppliers who promote ethical and fair labor practices, organizations can support social sustainability and help ensure the well-being of workers throughout the supply chain. For example, if a supplier uses unsustainable practices or materials that are subject to regulatory scrutiny, the organization may face legal or reputational risks, so collaboration with sustainable suppliers can help mitigate these risks and ensure continuity in the supply chain [3,4].

Sustainability issues in the supply chain have gained increasing attention in recent years due to the increasing awareness of the negative impact of business operations on the environment and society. Sustainable SSOA aims to identify and assess suppliers based on their sustainability performance, including environmental, social, and economic aspects. Evaluating suppliers based on sustainability criteria enables companies to reduce risks associated with supply chain disruptions and ensure a reliable supply of goods and services. It also helps companies meet their sustainability goals and enhance their reputation and brand image [5,6]. Order allocation is another critical aspect of the SSOA problem, which involves determining the most efficient and effective way of allocating orders among chosen suppliers. Companies need to consider both sustainability and efficiency factors when allocating orders to suppliers. Allocating orders to sustainable suppliers enables companies to reduce their environmental impact and promote social responsibility while ensuring the quality and reliability of goods and services. The SSOA problem is complex, and companies face numerous challenges when evaluating and selecting sustainable suppliers and allocating orders. The problem of sustainability in supply chains is further compounded by various types of complexity and uncertainty. Static complexity could emerge from the consideration of multi-echelon and multi-tier supply chains, where there are numerous nodes and interdependent processes that need to be taken into account. Dynamic complexity may also come into play when considering a supply chain in a multi-period context, where demand, supply, and other factors constantly change over time. In addition to these, there may be technological complexity related to the use of advanced manufacturing processes and digital technologies, as well as social complexity related to dealing with diverse stakeholders and communities [7–9]. All of these types of complexity can increase the uncertainty associated with sustainability issues, making it challenging for organizations to manage their supply chains effectively and achieve their sustainability goals. Therefore, companies need to develop effective models and methods to address the SSOA problem. The development of advanced mathematical models and decision-making tools has facilitated the evaluation and selection of sustainable suppliers and order allocation, considering multiple criteria and uncertainty [10,11].

Multi-criteria decision-making (MCDM) approaches can help organizations to define criteria, weight criteria, evaluate suppliers, generate alternatives, and make decisions about which sustainable suppliers to select for supply chain management (SCM) [12,13]. By using these approaches, organizations can make informed and objective decisions that promote environmental sustainability, social responsibility, and overall business success MCDM methods are flexible and adaptable to different contexts and situations. They can be used to evaluate suppliers in different industries, regions, and supply chain contexts, and can be customized to suit the specific needs and requirements of an organization [14,15].

The uncertainty of information is a common challenge in the SSOA problems. There are several methods that can be used to handle uncertainty and improve the accuracy and reliability of the evaluation and selection process [16]. In the context of sustainable supply chain management, the reliability of decisions refers to the degree to which the decisions made by a company in evaluating and selecting sustainable suppliers can be trusted to be accurate, consistent, and unbiased over time. It is important for companies to make reliable decisions in sustainable supply chain management because these decisions have significant impacts on the environment, society, and the economy. The reliability of decisions is closely linked to the quality of the data and information used to make those decisions. If the data used to evaluate and select suppliers are uncertain or imprecise, the resulting decisions will also be unreliable [17–19]. Uncertain information refers to information that is incomplete or unpredictable, where there is a lack of clarity about the outcome or the likelihood of different scenarios. Imprecise information, on the other hand, refers to information that is not precise or exact, where there is a degree of ambiguity or vagueness in the data. Fuzzy logic can be used to deal with uncertain and imprecise information [20,21]. It allows for a more flexible approach to decision-making, where criteria and weights are assigned based on linguistic variables [22]. Fuzzy MCDM is important for the evaluation and selection of sustainable suppliers for SCM because it can help to handle uncertainty and imprecision in the evaluation process [23]. Decision-makers may have different opinions and preferences regarding the importance of different criteria, and the criteria themselves may be vague and imprecise. Fuzzy MCDM allows decision-makers to represent criteria and weights in linguistic variables, which can be more intuitive and meaningful than numerical values. It also allows decision-makers to incorporate qualitative information, such as sustainability practices and social responsibility, into the evaluation process. This is important because sustainability performance may not be easily quantifiable, and qualitative information may be critical in evaluating suppliers' sustainability practices. Evaluating and selecting sustainable suppliers typically involves multiple criteria. These criteria may have different levels of importance and may be interdependent [24,25]. Fuzzy MCDM can handle this by allowing decision-makers to evaluate and rank suppliers based on multiple criteria simultaneously. Fuzzy MCDM can also be used to identify the most critical criteria and their relative weights, which can help decision-makers to prioritize the criteria and suppliers based on their importance. This process is very important in the SSOA problem [26,27].

In fuzzy logic, a type-2 fuzzy set is an extension of the traditional type-1 fuzzy set that allows for more uncertainty and ambiguity in the definition of the set [28]. Type-2 fuzzy sets can be more useful in situations where there is a lot of uncertainty or imprecision in the definition of a concept or in the data being used to represent that concept [29]. However, they can also be more computationally intensive to work with and require more complex algorithms and techniques for inference and decision-making. Type-2 fuzzy sets are well-suited for problems involving uncertainty due to imprecise or incomplete data, or when there are multiple sources of uncertainty that need to be modeled. Type-1 fuzzy sets, on the other hand, are often used when the data are well-defined and there is little uncertainty [30]. By developing type-2 fuzzy sets for linguistic variables, we can capture the uncertainty and imprecision in the evaluation process. This can lead to more accurate evaluations of supplier performance, which can help in making better-informed supplier selection decisions in the supply chain [31–33]. Type-2 fuzzy sets have been applied to several real-world problems in different fields [34–36].

This study proposes a methodology to address the sustainable SSOA problem by integrating multiple techniques. First, a new ranking method based on the concept of Radius of Gyration (ROG) is introduced for interval type-2 fuzzy sets to handle the uncertainty in supplier evaluation. To determine the weights of evaluation criteria, both subjective weights obtained using the Simple Multi-Attribute Rating Technique (SMART) and expert preferences, and objective weights calculated using the Method based on the Removal Effects of Criteria (MEREC) method are combined [37,38]. Then, using sustainability criteria, a type-2 fuzzy decision-matrix, combined weights, and the Weighted Aggregated Sum Product Assessment (WASPAS) method [39], supplier performance is evaluated as type-2 fuzzy sets. The ROG-based ranking method is employed to calculate the relative scores of suppliers. Finally, a multi-objective decision-making (MODM) mathematical model is presented to identify suitable suppliers and allocate their order quantities. The application of the proposed methodology is demonstrated in a sustainable SSOA problem, highlighting the methodology's effectiveness and applicability. The analysis conducted in this study demonstrates the practicality and efficiency of the proposed approach. By integrating multiple methodologies, this methodology can effectively handle the uncertainty in supplier evaluation and selection. Additionally, the use of the ROG-based ranking method allows for more accurate supplier performance evaluation, resulting in better supplier selection decisions. The proposed approach also takes into account sustainability criteria, emphasizing the importance of sustainability in supplier selection and order allocation.

The remainder of the paper is structured as follows. Section 2 provides an extensive literature review, discussing some of the recent studies pertaining to the SSOA problem. Section 3 outlines the proposed methodology, which encompasses the ROG-based ranking method, a step-by-step procedure for evaluating suppliers, and an approach to solving the MODM model of the sustainable SSOA problem. The results and discussion concerning the proposed methodology are presented in Section 4, where an example of the sustainable SSOA problem is illustrated, along with a sensitivity analysis. Finally, Section 5 presents the concluding remarks, summarizing the key findings of the study and highlighting future research directions.

2. Literature Review

The SSOA is an essential problem for many organizations as they directly impact the quality of the end product and the overall efficiency of the supply chain. Selecting the right suppliers and allocating orders optimally can help organizations reduce costs, increase profits, and maintain a competitive edge in the market. Over the years, several studies have been conducted on the SSOA problem, with a focus on different aspects such as sustainability, risk, and uncertainty. In this section, some of the recent studies on this topic, highlighting their contributions, are reviewed.

Esmaeili-Najafabadi et al. [40] enhanced the process of SSOA within a centralized supply chain, by devising a mixed integer nonlinear programming (MINLP) mathematical model that incorporates two precautionary measures aimed at mitigating disruption risks. The investigation revealed that as the likelihood of disruptions increases, the variables that influence decisions regarding SSOA undergo alterations.

Moheb-Alizadeh and Handfield [41] proposed a sustainable supplier management tool by simultaneously tackling the challenges of sustainable SSOA. These issues have received limited attention in the literature. They developed an MODM model that is comprehensive, considering multiple periods, products, and transportation modes, as well as discount and shortage conditions. They select the preferred solution based on the data envelopment analysis (DEA) super efficiency score of all purchasing firms. The proposed approach was applied to a real-world case study in the automotive industry.

Hosseini et al. [42] developed an efficient solution for managing supply chain disruptions by developing a resilient SSOA approach. The researchers proposed a graphical model to obtain the likelihood of disruption scenarios for the supplier selection problem and a stochastic MODM model to help with decision-making on when and how to use both reactive and proactive strategies in SSOA.

Kellner and Utz [43] devised a decision support approach that helps purchasing managers build mid-term supplier portfolios while weighing purchasing costs, supplier sustainability and overall supply risk trade-offs. To achieve this, the researchers developed an MODM model that prioritized supplier sustainability, selected the suppliers with the lowest costs, and reduced supply risk. They used the ε-constraint method to deal with the MODM model.

Duan et al. [44] presented an integrated model for green SSOA that can aid companies in cutting costs, enhancing their green performance, and gaining a competitive edge by combining the alternative queuing method (AQM), linguistic Z-numbers, and an MODM model. The study employed the step-weight assessment ratio analysis technique to determine the weights of criteria, and an extended AQM to rank the given suppliers while establishing an MODM model to find the optimal order quantity for the selected suppliers based on their scores.

Mohammed et al. [45] developed a hybrid approach based on MCDM and MODM techniques for sustainable SSOA. The authors put forward an integrated approach based on the fuzzy analytic hierarchy process (AHP) and fuzzy Technique for Order Preference by Similarity to Ideal Solution (TOPSIS) to evaluate and rank suppliers based on three sets of criteria and created an MODM model for selecting suppliers and determining optimal order quantities.

Safaeian et al. [46] proposed an MODM model for SSOA that takes into account incremental discounts in a fuzzy environment. The researchers utilized the Zimmermann fuzzy approach to transform the MODM model into a single objective format, which was then solved using Genetic Algorithm and Non-dominated Sorting GA (NSGA). Finally, the methodology's effectiveness and performance were evaluated and discussed.

Alegoz and Yapicioglu [47] developed a hybrid approach for SSOA that takes both qualitative and quantitative criteria into account. The goal was to identify appropriate suppliers and make optimal order allocations. To achieve this, the researchers used trapezoidal type-2 fuzzy AHP, fuzzy TOPSIS and goal programming. The study also compared the use of MCDM methods regarding their effectiveness.

The purpose of the study made by Mari et al. [48] was to establish resilient criteria for SSOA in an uncertain environment, aiming to mitigate low probability disruption risks that can have a high impact and enhance supply chain resilience. To accomplish this aim, the study proposed a possibilistic fuzzy MODM model and an interactive fuzzy optimization methodology to help organizations balance resilience and cost in their supply chains.

Laosirihongthong et al. [49] introduced a comprehensive approach for assessing suppliers based on sustainability indicators, as well as allocating purchase orders among the top-ranked suppliers. To achieve this goal, the researchers used a mixed-methods approach and utilized the fuzzy AHP to rank suppliers. Furthermore, the study devised a cost-minimization method for allocating purchase orders. The findings of the study demonstrated that both economic and environmental factors are essential considerations.

Feng and Gong [50] proposed an integrated approach for green SSOA in the automobile manufacturing industry using MODM and the linguistic entropy weight method (LEWM). The LEWM was used to analyze the performance and select qualified green suppliers on each evaluation criterion. The order allocation model aimed to minimize carbon emission and total cost and maximize supply value. The study found that the proposed framework could effectively deal with green SSOA for automobile manufacturers.

Khoshfetrat et al. [51] established an MODM model for a sustainable SSOA problem in the automotive industry that considers various criteria in a fuzzy environment. To achieve this goal, the study combined the evaluation process of suppliers, which used the AHP method, with the process of order allocation to determine the ideal quantity needs to be purchased from each supplier in each period. Furthermore, the study provided a sensitivity analysis to analyze the best suppliers and their allocated orders.

Jia et al. [52] The study addressed the issue of uncertain factors, such as emissions, supply capacity, per-unit cost, demand, and minimum order quality, whose probability distributions were imprecise, by estimating their distributions. The study proposed a robust MODM model for sustainable SSOA, which optimized four conflicting objectives while considering the sustainability dimensions. The proposed model effectively balanced multiple objectives and solved the sustainable SSOA problem by structuring ambiguous distribution sets.

Wong [53] studied the complicated issue of selecting eco-friendly suppliers. To address this problem, the study created a fuzzy goal programming model that considered various factors such as suppliers' dynamic risk, importance functions, and green market segmentation. The effect of different ratios of environmentally conscious consumers was studied and a solution was proposed to incorporate market incentives and result in mutually beneficial outcomes for the environment and the economy.

The aim of the research carried out by You et al. [54] was to create a unique framework for SSOA that could benefit organizations in accomplishing sustainable development goals. To deal with the uncertainty involved in evaluating the sustainable performance of suppliers, the researchers employed Double Hierarchy Hesitant Linguistic Term Sets (DHHLTSs). They proposed an extended approach to select efficient and sustainable suppliers and established a linear MODM model to apportion rational order quantities among the selected suppliers, taking quantity discounts into account.

Rezaei et al. [55] proposed an integrated approach for SSOA in lean manufacturing companies by utilizing both MODM and MCDM techniques. The study was conducted in four phases. Firstly, relevant leanness criteria were identified from previous research. Secondly, the AHP method was employed to evaluate these criteria for supplier selection. Next, a fuzzy AHP method was used to choose suppliers based on the lean supplier selection criteria. Finally, an MODM mathematical model was developed to determine the optimal allocation of orders.

Kaviani et al. [56] developed a new approach that combined fuzzy multi-objective optimization and intuitionistic fuzzy AHP to tackle the SSOA problem. They started by using intuitionistic fuzzy AHP to establish the key criteria weights for evaluating suppliers and then utilized a fuzzy MODM mathematical model to determine the optimal order quantity for each supplier. The authors concluded that their innovative decision-making tool could handle decision-makers' uncertainty and had demonstrated practical usefulness.

Rezaei et al. [57] addressed the issue of risk and uncertainty in SSOA for closed-loop supply chain (CLSC) networks and reverse logistics. They proposed a two-stage model based on stochastic programming that uses a conditional value-at-risk (CVaR) risk measurement tool to assess both risk-averse and risk-neutral scenarios. The goal of the study was to explore how changes in key problem parameters affect a company's sourcing strategies. The researchers recommended that firms consider purchasing from spot markets and backup suppliers to mitigate uncertainties.

Wang et al. [58] devised a model based on the analytic network process (ANP) and integer programming that leverages MCDM techniques to optimize the SSOA problem. The researchers aimed to evaluate how different emission trading schemes (ETS) scenarios could affect a company's overall cost structure and the creation of a low-carbon supply chain, taking into account the carbon competitiveness of suppliers by factoring in the carbon embedded in raw materials and carbon emission trading schemes.

Çalık [59] developed a framework for managing the sustainable SSOA problem in the agricultural machinery industry in Turkey. To achieve this, an MODM mathematical model was developed, which included sustainability dimensions. The weight of the criteria was determined using an approach based on the AHP method and interval type-2 fuzzy sets. The proposed approach offered an integrated model that considered the integration of quantitative and qualitative evaluation criteria, taking into account varying preferences.

Khalili Nasr et al. [60] proposed a two-stage model to deal with the SSOA problem in CLSCs that could minimize waste. In the first stage, a fuzzy Best-Worst Method (BWM) was used to evaluate suppliers based on various criteria. In the second stage, a linear MODM model was used to design a CLSC network incorporating vehicle scheduling, inventory-location-routing, and quantity discounts. To solve the MODM model a fuzzy goal programming approach was proposed.

Kaur and Prakash Singh [61] presented a multi-stage hybrid model for the SSOA problem that would account for risks and disruptions arising from positive and negative events, such as natural/man-made disasters and Industry 4.0, and optimize the distribution of orders to suppliers over multiple periods in a manner that would minimize costs as well as the disruption risk. The proposed model involved supplier segmentation and evaluation using the DEA, fuzzy AHP, and TOPSIS. Moreover, the risk related to each supplier was assessed using the model.

Islam et al. [62] developed a new two-stage approach to handle SSOA problems with uncertain demand. The study introduced a Relational Regressor Chain method for demand forecasting in the first stage. In the second stage, suitable suppliers and order quantities from each supplier were determined based on the forecasted demand and an MODM model. To obtain efficient solutions ε-constraint and weighted-sum methods were employed. The outcomes indicated the efficiency of the proposed method over the other methods in terms of forecasting accuracy.

Rezaei et al. [63] devised an effective framework for SSOA in a centralized supply chain while considering collaboration between the supplier and buyer and the strategies for risk reduction. The study employed MINLP models and risk reduction strategies such as protected suppliers, emergency stock, reserving additional capacity, backup suppliers, and geographical separation. It also employed the risk priority number constraint and Failure Mode and Effects Analysis (FMEA) technique to account for suppliers' reliability.

Firouzi and Jadidi [64] proposed a fuzzy MODM model for the SSOA problem that could manage the uncertainties brought about by disasters in Japan. The researchers acknowledged that such catastrophes could have unfavorable effects on businesses and markets, resulting in increased demand for certain goods or a reduction in the suppliers' ability to provide them in the appropriate quantity, quality, and time. To effectively consider decision makers' preferences, the study used a weighted additive function to solve the MODM model with parameters defined by fuzzy sets.

Li et al. [65] presented an inclusive mathematical model to assist in SSOA while considering both qualitative and quantitative factors in the risk management of supply chains. The study noted the emerging trend of environmental considerations in this field and highlighted the importance of dynamic SSOA. The presented model included the preliminary selection of suppliers based on the risk value assessed through quantitative and qualitative methods. This was followed by developing an MODM model for dynamic SSOA.

Yousefi et al. [66] developed a two-stage hybrid approach that could be utilized to select efficient suppliers, allocate orders, and determine prices in a coordinated supply chain. The first stage of the proposed model employed the DEA technique and an MODM mathematical model to evaluate suppliers and minimize costs simultaneously. The second stage of the proposed approach utilized the order quantity specified in the first stage, the bargaining game, the Nash equilibrium concept, and a quadratic programming model to determine the price.

Beiki et al. [67] introduced a new approach to tackle the SSOA problem by combining an MODM model with the language entropy weight method. The authors emphasized the need to improve the collaboration between potential suppliers and supply chain practitioners to achieve sustainable development goals. An MODM model based on three objectives was developed, aiming to maximize procurement value while minimizing total cost and carbon emissions. The language entropy weight method was utilized in the study to evaluate suppliers based on their sustainability performance.

Esmaeili-Najafabadi et al. [68] proposed a multi-objective model for integrated SSOA in a centralized supply chain based on a risk-averse decision-maker and the risks of disruption. Two types of risks including local disruption risks and regional disruption risks were considered in the study. Risk-averse and risk-neutral and models were developed, and the decision maker's behavior was analyzed using two risk assessment tools, value-at-risk (VaR) and CVaR. The model was solved using the particle swarm optimization (PSO) algorithm.

Mohammed et al. [69] aimed to put forward a new technique for SSOA that takes into account green and resilience aspects by devising an integrated framework. The proposed framework was based on calculating importance weights using the AHP method, assessing suppliers using the TOPSIS method, and applying an MODM mathematical model with the ε-constraint method to solve the problem. The purpose of the study was to support companies in augmenting their supply chain resilience while fulfilling their environmental responsibilities.

Hosseini et al. [70] developed a solution methodology for the SSOA challenges under uncertainty. An integration of the evidential reasoning and BWM was used to propose an approach for the evaluation of suppliers based on sustainability dimensions. Stochastic programming and dynamic programming were utilized to solve the MODM model, and its results were compared with some other techniques. The effect of uncertainties in suppliers' availability, quantity discounts, and demand was examined through a sensitivity analysis.

Ali et al. [71] devised a comprehensive method for SSOA in a sustainable supply chain under uncertainty. The study utilized a fuzzy AHP approach to compute the criteria weights and a fuzzy TOPSIS technique to assess the performance of suppliers and ascertain their final ranks. Then an MODM model based on goal programming was used for allocating the optimum order quantity to the selected suppliers. The results of the study and analyses indicated that the suggested model was able to deal with uncertainties associated with the SSOA problem.

Goodarzi et al. [72] aimed to develop a model that integrated a decision-making approach to evaluate green suppliers and allocate optimal orders while accounting for uncertainty. The fuzzy Delphi method was employed to refine supplier evaluation criteria and use green and resilient indexes were for the prioritization of suppliers. The gray Correlation method and TOPSIS were utilized to analyze the results.

Liaqait et al. [73] proposed a decision-support framework based on the integration of fuzzy MCDM techniques, demand forecasting, and MODM mathematical models. The focus of the research was on a multi-modal transportation network to demonstrate the effect of transportation on travel time, the supply chain's total cost, and environmental impact. The findings of the proposed model showed that the multi-modal transportation network had a substantial impact on the supply chain's travel time, total cost, and environmental impact.

The study of Gai et al. [74] aimed to present an integrated two-stage MCDM approach that incorporated both quantitative and qualitative analyses for dealing with the challenges of SSOA in green supply chain management. In the first stage, the evaluation of green suppliers was made using linguistic Z-Numbers and the MULTIMOORA (Multi-Objective Optimization on the basis of a Ratio Analysis plus the full Multiplicative form) method. In the second stage, an MODM mathematical model was employed to determine the number of orders allocated to the preferred suppliers.

Aouadni and Euchi [75] developed a hybrid solution methodology for SSOA based on the best-worst method and TOPSIS technique in the first phase to find a robust ranking of suppliers and to use the Linear Programming approach in the second phase to determine the weight of the objective function. The study applied the methodology to a real case of the Tunisian Electric Society, and the experimental results showed that the proposed model provided effective gains concerning solution quality.

The purpose of the study made by Galankashi et al. [76] was to tackle the problem of merging agile manufacturing with purchasing and supplier selection. The authors reviewed past research thoroughly and utilized the AHP method to finalize the criteria for choosing agile suppliers. They utilized the criteria to evaluate suppliers using a fuzzy AHP and established an MODM model based on multiple periods for allocating orders. A sensitivity analysis was conducted to provide more practical and comprehensible results.

Liu et al. [77] put forward a linear MODM model to help manage supply chains through the efficient selection of suppliers and allocation of orders. The study introduced a modified BWM method to assess and prioritize suppliers. The authors used fuzzy variables to find the amount of raw material order quantities. The goal programming method was employed to solve the MODM model that included four objective functions. The study illustrated that the proposed model yielded lower costs and better criteria in comparison to other models.

The purpose of the study carried out by Bai et al. [78] was to address the neglect of net-zero emissions and carbon neutrality in the SSOA problems of supply chain management. They introduced an MODM mathematical model that can assess various procurement policies and provide practical and theoretical insights. A case of an energy trading platform was used for the implementation and assessment of the model. The results indicated the importance of purchasing fossil fuels or attaining net zero through carbon emissions sequestration and carbon offsets.

Ahmad et al. [79] developed an approach to deal with the SSOA problem in a two-echelon make-to-order supply chain. The focus of the study was on determining the acceptable tolerances for the members of a supply chain to the minimization of the variability in total costs. The authors employed an MINLP model, and the robustness of the solutions was improved by incorporating the Taguchi Method of Tolerance Design (TMTD). They tested their model and showed the effectiveness of it.

The studies reviewed here highlight the importance of considering uncertainty in handling SSOA. Table 1 presents a summary of the reviewed studies, taking into account the uncertainty associated with SSOA can lead to more robust decisions in supply chain management. The studies also reveal that SSOA is a complex process that requires the integration of various criteria, including economic, social, and environmental considerations. Additionally, the studies demonstrate the need to address the challenges of sustainable SSOA, which has received limited attention in prior literature. Therefore, the current study focused on developing a new methodology to deal with sustainable SSOA problems under uncertainty.

Table 1. Summary of the reviewed studies.

No.	Author(s) and Reference	Year of Publication	Description of the Approach for SSOA
1	Esmaeili-Najafabadi et al. [40]	2019	An MINLP model that incorporates two precautionary measures aimed at mitigating disruption risks
2	Moheb-Alizadeh and Handfield [41]	2019	An MODM model considering multiple periods, products, and transportation modes
3	Hosseini et al. [42]	2019	A graphical model to obtain the likelihood of disruption scenarios for SSOA
4	Kellner and Utz [43]	2019	An MODM model for evaluation of supplier sustainability based on costs and supply risk.
5	Duan et al. [44]	2019	An integrated model for SSOA by combining AQM, linguistic Z-numbers, and an MODM model
6	Mohammed et al. [45]	2019	A hybrid approach based on AHP, fuzzy TOPSIS and an MODM model
7	Safaeian et al. [46]	2019	An MODM model based on the Zimmermann fuzzy approach and NSGA
8	Alegoz and Yapicioglu [47]	2019	A hybrid approach based on trapezoidal type-2 fuzzy AHP, fuzzy TOPSIS and goal programming
9	Mari et al. [48]	2019	A possibilistic fuzzy MODM model and an interactive fuzzy optimization methodology
10	Laosirihongthong et al. [49]	2019	An approach based on the fuzzy AHP and a cost-minimization model
11	Feng and Gong [50]	2020	An integrated approach using MODM and the linguistic entropy weight method
12	Khoshfetrat et al. [51]	2020	An integrated approach based on AHP and MODM model in a fuzzy environment
13	Jia et al. [52]	2020	A robust MODM model based on four conflicting objectives
14	Wong [53]	2020	A fuzzy goal programming model that considered various factors like suppliers' dynamic risk
15	You et al. [54]	2020	A framework that employed Double Hierarchy Hesitant Linguistic Term Sets
16	Rezaei et al. [55]	2020	An integrated approach using the AHP method and MODM model
17	Kaviani et al. [56]	2020	An approach that combined fuzzy multi-objective optimization and intuitionistic fuzzy AHP
18	Rezaei et al. [57]	2020	A two-stage model based on stochastic programming that uses a conditional value-at-risk
19	Wang et al. [58]	2020	A model based on ANP and integer programming
20	Çalık [59]	2020	An approach based on the AHP method, interval type-2 fuzzy sets and MODM model

Table 1. *Cont.*

No.	Author(s) and Reference	Year of Publication	Description of the Approach for SSOA
21	Khalili Nasr et al. [60]	2021	A two-stage model based on a fuzzy BWM and a linear MODM model
22	Kaur and Prakash Singh [61]	2021	An integrated approach based on DEA, fuzzy AHP, and TOPSIS
23	Islam et al. [62]	2021	A new two-stage approach based on a Relational Regressor Chain, ε-constraint and weighted-sum methods
24	Rezaei et al. [63]	2021	A framework based on MINLP models, risk reduction strategies and FMEA technique
25	Firouzi and Jadidi [64]	2021	A fuzzy MODM model that could manage the uncertainties brought about by disasters
26	Li et al. [65]	2021	An approach based on the risk value assessed and an MODM model
27	Yousefi et al. [66]	2021	A two-stage hybrid approach that employed DEA and an MODM model
28	Beiki et al. [67]	2021	A new approach by combining an MODM model with the language entropy weight method
29	Esmaeili-Najafabadi et al. [68]	2021	A multi-objective approach based on VaR, CVaR, and PSO
30	Mohammed et al. [69]	2021	An integrated approach based on AHP, TOPSIS, and the ε-constraint method
31	Hosseini et al. [70]	2022	An approach based on the evidential reasoning, BWM, stochastic programming and dynamic programming
32	Ali et al. [71]	2022	A hybrid approach using fuzzy AHP, fuzzy TOPSIS and an MODM model
33	Goodarzi et al. [72]	2022	A framework based on fuzzy Delphi, Gray Correlation method, TOPSIS and MODM models
34	Liaqait et al. [73]	2022	Fuzzy MCDM techniques, demand forecasting, and MODM mathematical models
35	Gai et al. [74]	2022	A two-stage approach that incorporated linguistic Z-Numbers, MULTIMOORA, and an MODM model
36	Aouadni and Euchi [75]	2022	A hybrid methodology based on BWM, TOPSIS and bi-objective programming
37	Galankashi et al. [76]	2022	An integrated approach based on a fuzzy AHP and an MODM model with multiple periods
38	Liu et al. [77]	2022	An approach based on a modified BWM method and goal programming
39	Bai et al. [78]	2022	An MODM mathematical model that can assess various procurement policies
40	Ahmad et al. [79]	2022	An integrated approach based on an MINLP model and the Taguchi Method of Tolerance Design

3. Methodology

In this section, a new decision-making approach is presented based on interval type-2 fuzzy sets, ROG of fuzzy sets, SMART, MEREC and WASPAS. Then a model is described to deal with the SSOA problem. The preliminaries and different components of the decision-making approach are delineated in the following subsections.

3.1. Interval Type-2 Fuzzy Sets

Interval type-2 fuzzy sets are a type of fuzzy set that allows for a more precise representation of uncertainty in data. While traditional fuzzy sets assign each element a membership value between 0 and 1, IT2FS assign each element a membership function that is itself a fuzzy set. This allows for a more nuanced representation of uncertainty, as the membership function can vary within a given interval. IT2FS can also be used to construct fuzzy preference relations, which provide a way to model the preferences of decision-makers. This can be useful in group decision-making scenarios, where there may be multiple decision-makers with different preferences. The use of IT2FS in constructing these relations can help to improve the efficiency of the decision-making process. Furthermore, IT2FS can be used to rank alternatives and criteria weights, which can be useful in determining the most appropriate course of action [80,81].

The study uses a trapezoidal form of IT2FSs that is defined by a two-level membership function denoted by $\mu_F(x)$. This function includes an Upper Membership Function (UMF) and a Lower Membership Function (LMF) that form the Footprint of Uncertainty (FOU) for an interval type-2 fuzzy set. The trapezoidal IT2FS is formed by UMF and LMF, which have trapezoidal shapes. The trapezoidal membership function has commonly been used in fuzzy sets and fuzzy logic systems. It has a simple shape that is easy to understand and interpret. The trapezoidal membership function is more flexible than the triangular membership function, as they allow for a wider range of uncertainty to be represented. This type of membership function is more precise, as they do not assume a normal distribution for the uncertainty. The trapezoidal membership function can represent both symmetric and asymmetric uncertainty, making them a more versatile choice for many applications. Moreover, it can be easily combined with other types using standard fuzzy set operations, such as union and intersection. It can also be easily converted to crisp numbers, which makes it more useful in practical applications. This type of membership function can be used in a wide range of applications, including control systems, decision-making, and data analysis, making it a popular choice for many researchers and practitioners in the field of fuzzy logic. Figure 1 depicts a trapezoidal IT2FS.

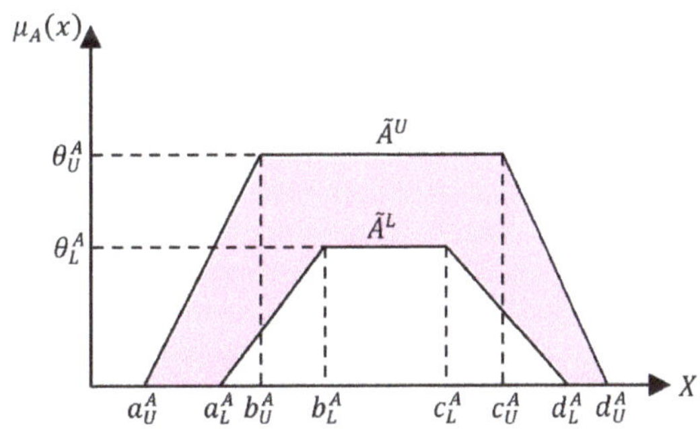

Figure 1. Trapezoidal IT2FS representation.

The mathematical expression for defining this trapezoidal IT2FS is as follows [82,83].

$$\tilde{\tilde{A}} = (\tilde{A}_i |\, i \in \{L, U\}) = (a_i^A, b_i^A, c_i^A, d_i^A; \theta_i^A |\, i \in \{L, U\}) \tag{1}$$

Assuming $\tilde{\tilde{B}}$ is another trapezoidal IT2FS with the same definition and k is a definite number, we can define some fundamental mathematical operations of trapezoidal IT2FSs as follows.

Addition: $\oplus$

$$\tilde{\tilde{A}} \oplus \tilde{\tilde{B}} = (a_i^A + a_i^B, b_i^A + b_i^B, c_i^A + c_i^B, d_i^A + d_i^B; \min(\theta_i^A, \theta_i^B) |\, i \in \{L, U\}) \tag{2}$$

$$\tilde{\tilde{A}} \oplus k = (a_i^A + k, b_i^A + k, c_i^A + k, d_i^A + k; \theta_i^A |\, i \in \{L, U\}) \tag{3}$$

Subtraction: $\ominus$

$$\tilde{\tilde{A}} \ominus \tilde{\tilde{B}} = (a_i^A - d_i^B, b_i^A - c_i^B, c_i^A - b_i^B, d_i^A - a_i^B; \min(\theta_i^A, \theta_i^B) |\, i \in \{L, U\}) \tag{4}$$

$$\tilde{\tilde{A}} \ominus k = (a_i^A - k, b_i^A - k, c_i^A - k, d_i^A - k; \theta_i^A |\, i \in \{L, U\}) \tag{5}$$

Multiplication: ⊗

$$\tilde{\tilde{A}} \otimes \tilde{\tilde{B}} = (min\ I_1, min\ I_2, max\ I_2, max\ I_1; \min(\theta_i^A, \theta_i^B) |\ i \in \{L, U\}) \tag{6}$$

where $I_1 = \{a_i^A a_i^B, a_i^A d_i^B, d_i^A a_i^B, d_i^A d_i^B\}$ and $I_2 = \{b_i^A b_i^B, b_i^A c_i^B, c_i^A b_i^B, c_i^A c_i^B\}$.

$$\tilde{\tilde{A}} \otimes k = \begin{cases} (a_i^A k, b_i^A k, c_i^A k, d_i^A k; \theta_i^A |\ i \in \{L, U\})\ if\ k \geq 0 \\ (d_i^A k, c_i^A k, b_i^A k, a_i^A k; \theta_i^A |\ i \in \{L, U\})\ if\ k < 0 \end{cases} \tag{7}$$

Exponentiation: ∧

$$\tilde{\tilde{A}} \wedge k = ((a_i^A)^k, (b_i^A)^k, (c_i^A)^k, (d_i^A)^k; \theta_i^A |\ i \in \{L, U\}) \tag{8}$$

Defuzzified crisp score: Γ

$$\Gamma(\tilde{\tilde{A}}) = \frac{1}{2}\left(\sum_{i \in \{L, U\}} \frac{a_i^A + (1 + \theta_i^A)(b_i^A + c_i^A) + d_i^A}{4 + 2\theta_i^A}\right) \tag{9}$$

3.2. Comparative Ranking of Trapezoidal IT2FSs Based on ROG

The section puts forward a technique for comparative ranking of trapezoidal interval type-2 fuzzy sets. Several methods have been proposed for ranking fuzzy numbers. Lee and Li [84] developed a ranking approach to sort fuzzy numbers based on the fuzzy mean and spread of these numbers. However, the method becomes challenging to compare when fuzzy numbers have a high mean value with a high spread or a low mean value with a low spread. Cheng [85] proposed the coefficient of variance (CV index) to address this limitation, which ranks fuzzy numbers by their smaller CV index. Additionally, Cheng [85] proposed the distance-based method to rank fuzzy numbers. However, both the distance-based method and the CV index have limitations, with the distance method contradicting the CV index in some cases. Chu [86] proposed a ranking approach that uses the area between the centroid and the original point to address these limitations, but it fails to rank fuzzy numbers with the same centroid point. As an improvement over previous methods Deng et al. [87] proposed a ranking method that was free from the limitations of the mentioned methods. The technique of ranking proposed in this section is derived from the modified area method suggested by Deng et al. [87]. This method assesses the ranking of a fuzzy set by examining the area between the original point and the Radius of Gyration (ROG) point. Deng et al. [87] initially introduced this ranking technique for generalized trapezoidal fuzzy numbers, using the moment of inertia concerning the x and y axes.

This research utilizes the idea of ranking based on ROG and applies it to present a comparative ranking method for trapezoidal IT2FSs, which is a novel approach. Suppose we are working with a collection of n trapezoidal IT2FSs, which we will refer to as $\tilde{\tilde{E}}_1$, $\tilde{\tilde{E}}_2$, …, $\tilde{\tilde{E}}_n$. Below are the steps that describe the ROG-based ranking method that is proposed for ranking trapezoidal IT2FSs.

Step 1. Compute the moment of inertia for the upper and lower membership functions of each set with respect to both the x and y axes (Ix and Iy), using the equations provided below.

$$Ix_i^{E_k} = Ix_{1i}^{E_k} + Ix_{2i}^{E_k} + Ix_{3i}^{E_k}, \quad i \in \{L, U\}, k \in \{1, 2, \ldots, n\} \tag{10}$$

$$Iy_i^{E_k} = Iy_{1i}^{E_k} + Iy_{2i}^{E_k} + Iy_{3i}^{E_k}, \quad i \in \{L, U\}, k \in \{1, 2, \ldots, n\} \tag{11}$$

where

$$Ix_{1i}^{E_k} = \frac{(b_i^{E_k} - a_i^{E_k})(\theta_i^{E_k})^3}{12}, \quad i \in \{L, U\}, k \in \{1, 2, \ldots, n\} \tag{12}$$

$$Ix_{2i}^{E_k} = \frac{(c_i^{E_k} - b_i^{E_k})(\theta_i^{E_k})^3}{3}, \quad i \in \{L, U\},\ k \in \{1, 2, \ldots, n\} \quad (13)$$

$$Ix_{3i}^{E_k} = \frac{(d_i^{E_k} - c_i^{E_k})(\theta_i^{E_k})^3}{12}, \quad i \in \{L, U\}, k \in \{1, 2, \ldots, n\} \quad (14)$$

$$Iy_{1i}^{E_k} = \frac{(b_i^{E_k} - a_i^{E_k})^3 \theta_i^{E_k}}{4} + \frac{(b_i^{E_k} - a_i^{E_k})(a_i^{E_k})^2 \theta_i^{E_k}}{2} + \frac{2(b_i^{E_k} - a_i^{E_k})^2 a_i^{E_k} \theta_i^{E_k}}{3}, \quad i \in \{L, U\},\ k \in \{1, 2, \ldots, n\} \quad (15)$$

$$Iy_{2i}^{E_k} = \frac{(c_i^{E_k} - b_i^{E_k})^3 \theta_i^{E_k}}{3} + (c_i^{E_k} - b_i^{E_k})(b_i^{E_k})^2 \theta_i^{E_k} + (c_i^{E_k} - b_i^{E_k})^2 b_i^{E_k} \theta_i^{E_k}, \quad i \in \{L, U\}, k \in \{1, 2, \ldots, n\} \quad (16)$$

$$Iy_{3i}^{E_k} = \frac{(d_i^{E_k} - c_i^{E_k})^3 \theta_i^{E_k}}{12} + \frac{(d_i^{E_k} - c_i^{E_k})(c_i^{E_k})^2 \theta_i^{E_k}}{2} + \frac{(d_i^{E_k} - c_i^{E_k})^2 c_i^{E_k} \theta_i^{E_k}}{3}, \quad i \in \{L, U\},\ k \in \{1, 2, \ldots, n\} \quad (17)$$

Step 2. Use the following equations to compute the ROG point for the UMF and LMF of each trapezoidal IT2FS.

$$Rx_i^{E_k} = \sqrt{\frac{Ix_i^{E_k}}{(((c_i^{E_k} - b_i^{E_k}) + (d_i^{E_k} - a_i^{E_k})).\theta_i^{E_k})/2}}, \quad i \in \{L, U\},\ k \in \{1, 2, \ldots, n\} \quad (18)$$

$$Ry_i^{E_k} = \sqrt{\frac{Iy_i^{E_k}}{(((c_i^{E_k} - b_i^{E_k}) + (d_i^{E_k} - a_i^{E_k})).\theta_i^{E_k})/2}}, \quad i \in \{L, U\},\ k \in \{1, 2, \ldots, n\} \quad (19)$$

Step 3. Employ the following formula to compute interval areas based on both the obtained ROG point and the original point.

$$LR_{E_k} = min(Rx_L^{E_k} Ry_L^{E_k}, Rx_U^{E_k} Ry_U^{E_k}), \quad k \in \{1, 2, \ldots, n\} \quad (20)$$

$$UR_{E_k} = max(Rx_L^{E_k} Ry_L^{E_k}, Rx_U^{E_k} Ry_U^{E_k}), \quad k \in \{1, 2, \ldots, n\} \quad (21)$$

Step 4. Compute the degree of possibility for each pair of fuzzy sets in relation to one another using the following equation. Let $\tilde{\tilde{E}}_s$ and $\tilde{\tilde{E}}_m$ be two unequal trapezoidal IT2FSs.

$$Pos(\tilde{\tilde{E}}_s \geq \tilde{\tilde{E}}_m) = \begin{cases} 1 & if \quad \Delta_N \geq 0\ and\ \Delta_P \geq 0 \\ \frac{\Delta_P}{\Delta_P - \Delta_N} & if \quad \Delta_N \leq 0\ and\ \Delta_P \geq 0 \\ 0 & if \quad \Delta_N \leq 0\ and\ \Delta_P \leq 0 \end{cases} \quad (22)$$

where $\Delta_N = LR_{E_s} - UR_{E_m}$, $\Delta_P = UR_{E_s} - LR_{E_m}$, and the degree of possibility for $\tilde{\tilde{E}}_s$ over $\tilde{\tilde{E}}_m$ is denoted by $Pos(\tilde{\tilde{E}}_s \geq \tilde{\tilde{E}}_m)$.

Step 5. Compute the comparative ranking values for the trapezoidal IT2FSs using the following equation [30,88,89].

$$CR(\tilde{\tilde{E}}_k) = \frac{1}{n(n-1)} \left(\sum_{l=1}^{n} Pos(\tilde{\tilde{E}}_k \geq \tilde{\tilde{E}}_l) + \frac{n}{2} - 1 \right), \quad k \in \{1, 2, \ldots, n\} \quad (23)$$

3.3. *The Proposed MCDM Approach*

This section presents a new method for multi-criteria decision-making when the experts' judgments are expressed as trapezoidal interval type-2 fuzzy sets. The proposed method combines SMART, MEREC, and the ROG-based ranking technique to provide a comprehensive approach to decision-making. To use this approach, decision-makers must first define the problem and provide subjective assessments. SMART provides subjective weights for the criteria, MEREC determines the objective weights, and WASPAS is used for the fuzzy evaluation of the suppliers. The SMART method has several advantages that make it a popular approach to multi-criteria decision-making. It is a straightforward and easy-to-understand approach to decision-making. It requires minimal training and can be applied by individuals with different levels of expertise. The SMART method is a flexible approach that can be applied in a wide range of decision-making contexts. It allows decision-makers to incorporate both qualitative and quantitative criteria. Moreover, it is a relatively quick approach to decision-making that can help decision-makers save time and resources. In the MEREC method, the determination of objective weights takes a unique approach compared to other objective weighting methods. Instead of using variations in criteria to calculate weights, this method utilizes the removal effects of criteria on the performances of alternatives as a measure for weighting. Such a perspective is new and distinct from other approaches to determining objective criteria weights. The MEREC method provides insights into the relative importance of each criterion to the decision. In addition, the efficiency of the WASPAS method for the evaluation process has been demonstrated through numerous studies, making it an effective MCDM method [90]. The subjective weights and objective weights are combined to provide more realistic decisions. The ROG-based technique is used to calculate ranking values for the suppliers based on their aggregated WASPAS measures. The following steps represent the procedure of determination of relative scores of suppliers using the proposed MCDM approach.

Step 1. Form a team of decision-makers (DMs). This step involves assigning a group of experts who will carry out the decision-making process. Typically, these experts are chosen from senior-level executives or other positions of high responsibility within an organization. Let us assume that there is a group of q decision-makers ($\mathcal{D}_1$ to $\mathcal{D}_q$).

Step 2. Collect information about the potential suppliers and evaluation criteria. Gather data on the issue and extract the options that require assessment as well as the standards that can account for various aspects of the choices. Suppose that there are n alternatives ($\mathcal{A}_1$ to $\mathcal{A}_n$) and m criteria ($\mathcal{C}_1$ to $\mathcal{C}_m$) involved in the MCDM problem.

Step 3. Gather the preliminary evaluations of the criteria from each decision-maker. Ask each member of the decision-making group to provide their initial evaluations of the criteria. Different techniques, such as linguistic variables or the Likert scale can be employed to gather their opinions. As per the proposed approach's framework, a scale ranging from 0 to 100 is utilized for evaluations, with 0 denoting the least important and 100 the most important criteria.

Step 4. Obtain the initial evaluations of the alternatives' performances on each criterion from all experts. To capture the uncertainty of the evaluation process, linguistic variables are employed in this stage to gather the opinions of the decision-makers. The primary benefit of linguistic variables is their ability to be converted into trapezoidal IT2FSs. The range of linguistic variables encompasses "Very Poor" (VP) to "Very Good" (VG), and the full list of these variables is presented in Table 2 [82]. Let $\tilde{\tilde{x}}_{ijk}$ indicate the evaluation of jth criterion for ith alternative based on the perspective of kth decision-maker.

Table 2. The linguistic variables and related fuzzy numbers.

Linguistic Variables	Trapezoidal IT2FSs
Very Poor (VP)	((0, 0, 0, 1; 1), (0, 0, 0, 0.5; 0.9))
Poor (P)	((0, 1, 1.5, 3; 1), (0.5, 1, 1.5, 2; 0.9))
Medium Poor (MP)	((1, 3, 3.5, 5; 1), (2, 3, 3.5, 4; 0.9))
Fair (F)	((3, 5, 5.5, 7; 1), (4, 5, 5.5, 6; 0.9))
Medium Good (MG)	((5, 7, 7.5, 9; 1), (6, 7, 7.5, 8; 0.9))
Good (G)	((7, 8.5, 9, 10; 1), (8, 8.5, 9, 9.5; 0.9))
Very Good (VG)	((9, 10, 10, 10; 1), (9.5, 10, 10, 10; 0.9))

Step 5. Calculate the subjective weight of each criterion using the SMART technique. Use the evaluations obtained from the experts in Step 3 for this step. Let IP_{jk} denote the

importance or points assigned by the kth decision-maker to the jth criterion. Then, apply the following equation to determine the subjective weight of each criterion (w_j^s) [37].

$$w_j^s = \frac{\sum_k IP_{jk}}{\sum_k \sum_j IP_{jk}} \tag{24}$$

Step 6. Create an interval type-2 fuzzy decision-matrix by combining the alternatives' performances. Utilize Table 2, apply arithmetic operations of IT2FSs, and use the initial evaluations gathered in Step 4 to consolidate the alternatives' performances and turn them into trapezoidal interval type-2 fuzzy numbers. It is worth noting that this type of fuzzy set is an effective method for capturing decision-making information uncertainty. The results of this phase are the elements of the interval type-2 fuzzy decision-matrix ($\tilde{\tilde{x}}_{ij}$). These elements are calculated as follows.

$$\tilde{\tilde{x}}_{ij} = \frac{1}{q} \overset{q}{\underset{k=1}{\oplus}} \tilde{\tilde{x}}_{ijk} \tag{25}$$

Step 7. Defuzzify the decision-matrix and calculate the objective weights of criteria using the MEREC method [38]. In order to calculate the objective criteria weights (w_j^o) using MEREC, the defuzzified decision-matrix needs to be obtained first. The elements of the crisp matrix (x_{ij}^d) can be computed based on the results of Step 6 and Equation (9) as follows.

$$x_{ij}^d = \Gamma(\tilde{\tilde{x}}_{ij}) \tag{26}$$

Step 8. Combine the subjective and objective criteria weights to obtain more realistic weights for the criteria. By fusing the subjective criteria weights obtained using SMART in Step 5 with the objective weights determined by MEREC in Step 7, the combined weights of criteria (w_j^c) can be computed using the following formula with a combination parameter ω.

$$w_j^c = \omega w_j^s + (1-\omega) w_j^o \tag{27}$$

Step 9. Normalize the interval type-2 fuzzy decision-matrix. The WASPAS method typically utilizes a linear normalization approach, but given the utilization of trapezoidal IT2FSs, we need to adapt the normalization approach in this stage. The beneficial criteria are represented by BC while NC is used to represent non-beneficial criteria. Utilize the subsequent equations to normalize the interval type-2 fuzzy decision-matrix. Keep in mind that the calculations use Equations (2) to (9).

$$\tilde{\tilde{x}}_{ij}^n = \begin{cases} \tilde{\tilde{x}}_{ij} \otimes \frac{1}{\max_i x_{ij}^d} & if\ j \in BC \\ 1 \ominus (\tilde{\tilde{x}}_{ij} \otimes \frac{1}{\max_i x_{ij}^d}) & if\ j \in NC \end{cases} \tag{28}$$

Step 10. Determine the WSM ($\tilde{\tilde{Q}}_i^S$) and WPM ($\tilde{\tilde{Q}}_i^P$) measures of the WASPAS method by applying the following equations. As a result of using trapezoidal IT2FSs, this step differs slightly from the classic WASPAS method and has been modified for more efficient handling.

$$\tilde{\tilde{Q}}_i^S = \sum_{j=1}^{m} \oplus (\tilde{\tilde{x}}_{ij}^n \otimes w_j^c) \tag{29}$$

$$\tilde{\tilde{Q}}_i^P = \prod_{j=1}^{m} \otimes ((1 \oplus \tilde{\tilde{x}}_{ij}^n) \wedge w_j^c) \tag{30}$$

Step 11. Calculate the composite WASPAS measure. Utilizing the normalized WSM and WPM measures and the combination parameter γ, the composite WASPAS measure is computed in this step.

$$\tilde{\tilde{Q}}_i = (\tilde{\tilde{Q}}_i^{SN} \otimes \gamma) \oplus (\tilde{\tilde{Q}}_i^{PN} \otimes (1-\gamma)) \tag{31}$$

where

$$\tilde{\tilde{Q}}_i^{SN} = \tilde{\tilde{Q}}_i^{S} \otimes \frac{1}{\max_l \Gamma(\tilde{\tilde{Q}}_l^{S})} \tag{32}$$

$$\tilde{\tilde{Q}}_i^{PN} = \tilde{\tilde{Q}}_i^{P} \otimes \frac{1}{\max_l \Gamma(\tilde{\tilde{Q}}_l^{P})} \tag{33}$$

Step 12. Determine the ranking values of the composite WASPAS measures. This step employs Equations (10) to (23) (the proposed ROG-based ranking technique) to determine ranking values (S_i) for the composite WASPAS measurements of the alternatives (suppliers). These ranking values will be utilized as relative scores for suppliers in the evaluation and order allocation model.

$$S_i = CR(\tilde{\tilde{Q}}_i) \tag{34}$$

3.4. A Mathematical Model for the SSOA Problem

In this sub-section, a multi-objective mathematical model is presented for the supplier selection and order allocation problem. The model is based on the minimization of the total purchasing cost and total distance, and the maximization of the total scores of the selected suppliers. The aim of this model is to select the most suitable supplier for some production centers and determine the number of orders for the selected suppliers. Notations of the model are represented in Table 3.

Table 3. Notations of the SSOA model.

Parameters/Variables	Description
c_{ij}^{o}	Unit purchasing cost of the ith supplier for jth production center
d_{ij}	Distance between ith supplier and jth production center
S_i	The relative score of ith supplier
O_i^{min}	Minimum quantity to be ordered from ith supplier
CP_i	Supply capacity of ith supplier
DEM_j	Demand of jth production center
K^{min}	Minimum number of suppliers that need to be selected
x_{ij}^{o}	Variable: order quantity of the ith supplier for jth production center
y_i	Binary variable: = 1 if ith supplier is selected; = 0 otherwise
Z_1	Total purchasing cost
Z_1^{min}	Minimum value of Z_1
Z_1^{max}	Maximum value of Z_1
Z_2	Total distance-based measure
Z_2^{min}	Minimum value of Z_2
Z_2^{max}	Maximum value of Z_2
Z_3	Total relative score of selected suppliers
Z_3^{min}	Minimum value of Z_3
Z_3^{max}	Maximum value of Z_3

Using the fuzzy multi-objective programming approach [91], the SSOA problem considered in this study is defined as follows.

$$Max\ \lambda \tag{35a}$$

$$Z_1 = \sum_i \sum_j c_{ij}^o x_{ij}^o \tag{35b}$$

$$Z_2 = \sum_i \sum_j d_{ij} x_{ij}^o \tag{35c}$$

$$Z_3 = \sum_i \sum_j S_i x_{ij}^o \tag{35d}$$

$$\lambda \leq 1 - \frac{Z_1 - Z_1^{min}}{Z_1^{max} - Z_1^{min}} \tag{35e}$$

$$\lambda \leq 1 - \frac{Z_2 - Z_2^{min}}{Z_2^{max} - Z_2^{min}} \tag{35f}$$

$$\lambda \leq 1 + \frac{Z_3 - Z_3^{max}}{Z_3^{max} - Z_3^{min}} \tag{35g}$$

$$\sum_j x_{ij}^o \geq y_i O_i^{min} \quad \forall i \tag{35h}$$

$$\sum_j x_{ij}^o \leq y_i CP_i \quad \forall i \tag{35i}$$

$$\sum_i x_{ij}^o = DEM_j \quad \forall j \tag{35j}$$

$$\sum_i y_i \geq K^{min} \tag{35k}$$

$$x_{ij}^o \geq 0 \text{ and } y_i \in \{0,1\} \tag{35l}$$

To obtain the minimum and maximum values of Z_1, Z_2 and Z_3, the model needs to be solved separately by considering the related objective functions and constraints (35h) to (35l). By defining the objective functions in Equations (35b) to (35d) and maximizing λ through the constraints in (35e) to (35g), these functions can achieve values that approach their ideal states. The relative score of the ith supplier in Equation (35d) is determined using the MCDM approach presented in the previous subsection. The minimum order allocated to selected suppliers is defined by constraint (35h), while constraint (35i) requires that the order quantity of the selected suppliers does not exceed their maximum supply capacity. The purpose of constraint (35j) is to ensure that the selected suppliers' order quantity satisfies the minimum demand requirement of each production center. To have a more reliable procurement process, it is common to select multiple suppliers for a production center. Therefore, constraint (35k) sets a minimum number of suppliers that must be selected. Constraint (35l) shows the type of variables.

The procedure for using the methodology is presented in Figure 2.

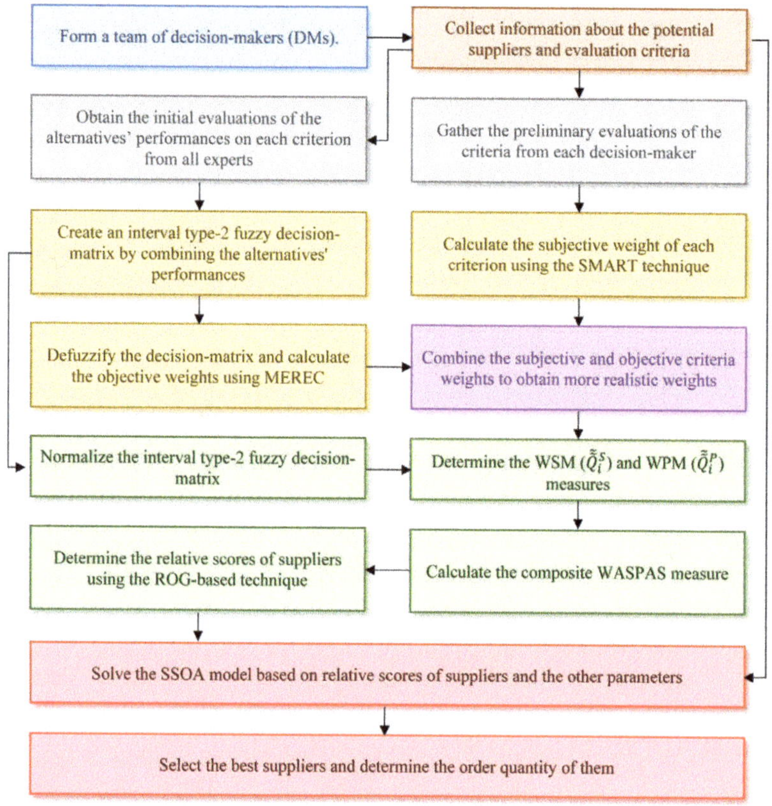

Figure 2. The framework of the methodology.

4. Results and Discussion

In this section, firstly the proposed methodology is applied to deal with a sustainable SSOA problem. Then the results are discussed through a sensitivity analysis.

4.1. The Application of the Methodology in Sustainable SSOA

The proposed methodology was used to deal with an example of the SSOA problem in a company. The company operates in the food industry with more than 200 employees in Golestan, Iran. It has two production centers (PC_1 and PC_2) and needs to purchase raw materials from some potential suppliers. The company's management team is highly qualified and has extensive experience in the industry. They must decide which suppliers to purchase from and determine the quantity of raw materials to order. The company has implemented various measures to reduce waste, use renewable energy sources, and optimize its logistics operations. It has a strong commitment to sustainability and is actively involved in various initiatives to reduce its carbon footprint. The following is the description of using different steps of the proposed MCDM approach and using the mathematical model in selecting suppliers and allocating orders to them.

Step 1. In this step, a group of experts was formed. This group consists of two experts from the procurement department ($\mathcal{D}_1$ and $\mathcal{D}_2$), two experts from the operations department ($\mathcal{D}_3$ and $\mathcal{D}_4$), two experts from the finance department ($\mathcal{D}_5$ and $\mathcal{D}_6$), one expert from the marketing department ($\mathcal{D}_7$), and one expert from the research and development (R&D) department ($\mathcal{D}_8$). The experts have a good knowledge of fuzzy sets, decision-making techniques and supply chain management practices and principles. Table 4 presents some details about the experts.

Table 4. Information about experts.

Expert	Department	Job Title	Years of Experience	Gender	Academic Degree
$\mathcal{D}_1$	Procurement department	Purchasing Director	8	Male	PhD in Management
$\mathcal{D}_2$	Procurement department	Sourcing Specialist	6	Female	MA in Business Management
$\mathcal{D}_3$	Operations department	Operations Manager	7	Male	PhD in Operations Research
$\mathcal{D}_4$	Operations department	Supply Chain Analyst	2	Female	BA in Industrial Engineering
$\mathcal{D}_5$	Finance department	Finance manager	8	Female	MA in Accounting & Finance
$\mathcal{D}_6$	Finance department	Risk analyst	4	Male	BA in Accounting & Finance
$\mathcal{D}_7$	Marketing department	Chief marketing officer	7	Male	MA in Marketing
$\mathcal{D}_8$	R&D department	Project manager	10	Male	PhD in Industrial Engineering

By bringing together decision-makers from these different departments, the company can ensure that all relevant factors are considered and that the selected suppliers meet the company's requirements and standards. Each department brings its unique expertise and perspective to the decision-making process, resulting in a well-informed and comprehensive decision.

Step 2. The potential suppliers and evaluation criteria should be identified in this step. The decision-making group has identified eight potential suppliers (Sup_1 to Sup_8) which can be seen in Figure 3. These alternatives need to be evaluated with respect to sustainability criteria. According to the literature, the decision-making group agreed on fifteen criteria within three dimensions of sustainability [92–95]. The criteria and their descriptions are presented in Table 5.

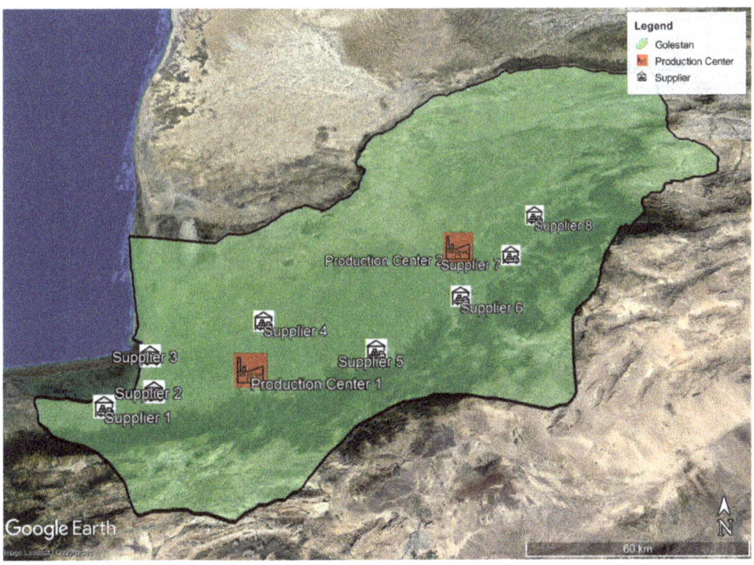

Figure 3. The geographical representation of the suppliers and production centers.

Table 5. The evaluation criteria and their descriptions.

Dimension	Criteria	Description
Environmental sustainability	Climate change mitigation (C_{11})	This involves reducing greenhouse gas emissions and implementing measures to mitigate the effects of climate change, such as investing in renewable energy, improving energy efficiency, and adopting low-carbon transportation options.
	Resource conservation (C_{12})	This involves reducing the consumption of non-renewable resources and conserving natural resources such as water, land, and forests. Companies can achieve this by implementing sustainable sourcing practices, using recycled materials, and minimizing waste.
	Pollution prevention (C_{13})	This involves minimizing or eliminating the release of harmful substances into the environment, such as toxic chemicals or air pollutants. Companies can achieve this by implementing pollution prevention measures, such as using clean production processes, reducing emissions, and properly disposing of hazardous waste.
	Biodiversity conservation (C_{14})	This involves protecting and conserving biodiversity and ecosystem services, such as pollination, soil fertility, and water quality. Companies can achieve this by using sustainable land management practices, protecting endangered species and habitats, and reducing deforestation.
	Adoption of circular economy principles (C_{15})	This involves moving away from the traditional linear model of "take-make-dispose" and instead adopting a circular economy model where waste is minimized and resources are kept in use for as long as possible. This can be achieved by implementing recycling programs, designing products for reuse, and finding ways to extend the lifespan of products.
Social sustainability	Labor standards (C_{21})	This involves ensuring fair wages, safe working conditions, and other labor standards throughout the supply chain. Companies can achieve this by implementing codes of conduct for suppliers, auditing their supply chains for compliance, and providing training and support to suppliers to help them meet these standards.
	Human rights (C_{22})	This involves respecting and promoting human rights, including freedom from discrimination, the right to privacy, and the right to freedom of association. Companies can achieve this by implementing human rights policies, engaging with stakeholders to understand their concerns, and monitoring their supply chains to identify and address human rights abuses.
	Community engagement (C_{23})	This involves engaging with local communities in a respectful and transparent manner, and taking their concerns into account in decision-making processes. Companies can achieve this by implementing community engagement strategies, conducting impact assessments to understand the potential impacts of their operations on local communities, and providing support to local communities to help build their capacity and improve their well-being.
	Health and safety (C_{24})	This involves ensuring that the health and safety of workers and local communities are protected from harm. Companies can achieve this by implementing health and safety policies and procedures, providing training and support to workers and suppliers, and conducting risk assessments to identify and address potential health and safety hazards.

Table 5. *Cont.*

Dimension	Criteria	Description
	Diversity and inclusion (C_{25})	This involves promoting diversity and inclusion throughout the supply chain, including ensuring that women and other underrepresented groups have equal opportunities to participate in economic activities. Companies can achieve this by implementing diversity and inclusion policies and programs, providing training and support to suppliers, and monitoring their supply chains for compliance.
Economic sustainability	Cost-efficiency (C_{31})	This involves reducing costs while maintaining or improving the quality of products and services. Companies can achieve this by implementing efficiency measures, such as improving production processes, reducing waste, and optimizing logistics and transportation.
	Innovation (C_{32})	This involves developing and implementing new products, services, or business models that create value for the company and its stakeholders. Companies can achieve this by investing in research and development, collaborating with other organizations to share knowledge and expertise, and exploring new markets or opportunities.
	Resilience (C_{33})	This involves building resilience into the supply chain to ensure that it can withstand disruptions, such as natural disasters, political instability, or economic downturns. Companies can achieve this by diversifying their suppliers, implementing risk management strategies, and maintaining adequate inventory levels.
	Responsible investment (C_{34})	This involves investing in companies or projects that have a positive impact on the environment, society, or economy. Companies can achieve this by implementing responsible investment policies, conducting due diligence on potential investments, and engaging with stakeholders to understand their concerns.
	Long-term perspective (C_{35})	This involves taking a long-term perspective when making business decisions, and considering the potential impacts of those decisions on future generations. Companies can achieve this by implementing sustainability strategies that consider the environmental, social, and economic impacts of their operations over the long term.

Steps 3 to 5. In these steps, the experts expressed their evaluations on each criterion based on a scale ranging from 0 to 100, then they were asked to evaluate each supplier with respect to the criteria using linguistic variables. The data from these steps are presented in Tables 6 and 7. It should be noted that the experts' evaluations of the suppliers are partially provided in Table 7 due to limitations in space. The detailed data can be found in Reference [96], named Evaluation Data. According to the evaluations of the experts on each criterion, the subjective criteria weights can be determined using the SMART method and Equation (24). The subjective weights are shown in the last column of Table 6.

Table 6. The evaluations of the experts on each criterion.

	$\mathcal{D}_1$	$\mathcal{D}_2$	$\mathcal{D}_3$	$\mathcal{D}_4$	$\mathcal{D}_5$	$\mathcal{D}_6$	$\mathcal{D}_7$	$\mathcal{D}_8$	Sum	w_j^s
C_{11}	40	35	45	50	30	30	40	40	310	0.0665
C_{12}	40	45	50	40	35	40	30	30	310	0.0665
C_{13}	30	40	40	40	45	45	50	30	320	0.0687
C_{14}	20	25	25	20	10	20	30	20	170	0.0365
C_{15}	30	20	30	35	40	20	30	20	225	0.0483
C_{21}	40	50	60	30	30	40	45	50	345	0.0740
C_{22}	25	20	20	30	35	25	20	25	200	0.0429
C_{23}	30	30	40	40	30	40	30	30	270	0.0579
C_{24}	30	40	40	45	35	40	20	25	275	0.0590
C_{25}	15	10	10	15	20	10	10	20	110	0.0236
C_{31}	60	70	60	80	80	70	75	70	565	0.1212
C_{32}	40	50	50	40	30	40	45	50	345	0.0740
C_{33}	40	30	50	45	35	50	45	45	340	0.0730
C_{34}	50	60	60	50	70	60	70	60	480	0.1030
C_{35}	45	50	55	60	45	40	60	40	395	0.0848

Table 7. Experts' evaluation of suppliers on each criterion.

		C_{11}	C_{12}	C_{13}	C_{14}	C_{15}	C_{21}	C_{22}	C_{23}	C_{24}	C_{25}	C_{31}	C_{32}	C_{33}	C_{34}	C_{35}
$\mathcal{D}_1$	Sup_1	VG	G	P	MG	P	G	MP	MG	VG	F	G	G	VG	MG	VG
	Sup_2	MP	MG	F	MP	P	MG	P	MP	P	MG	G	MG	MP	P	VP
	Sup_3	MP	P	MP	P	VP	MG	F	VP	MP	P	F	F	MP	F	F
	Sup_4	P	MG	P	F	MG	VP	MP	MP	P	F	P	MP	MP	F	VP
	Sup_5	MG	P	MP	F	MP	MG	MG	F	P	MP	MG	P	MG	F	F
	Sup_6	G	MP	F	MP	F	G	G	P	G	MG	MG	VG	VG	MG	G
	Sup_7	VP	P	P	MG	MP	MP	MP	F	VP	MP	VP	F	P	F	VP
	Sup_8	MG	P	F	MP	F	P	MP	VP	MP	P	MG	MG	MP	P	P
$\mathcal{D}_2$	Sup_1	MG	G	P	MG	MP	VG	MP	MG	MG	F	G	G	MG	F	MG
	Sup_2	MP	G	MP	P	VP	G	MP	P	P	F	F	MP	P	VP	VP
	Sup_3	P	P	MG	F	P	F	MP	MP	P	P	MG	F	VP	MP	F
	Sup_4	VP	G	MP	MP	MP	VP	P	F	P	F	VP	P	F	F	P
	Sup_5	F	P	F	P	MP	MP	MG	MG	F	MP	MG	F	F	MG	MP
	Sup_6	G	F	MG	F	MG	G	MG	MP	G	F	VG	VG	G	MG	G
	Sup_7	VP	P	P	MP	MP	MP	P	MP	VP	MP	VP	F	MP	MP	MP
	Sup_8	G	MP	MP	VP	MP	P	MP	P	MP	P	MP	F	VP	MP	P
⋮	⋮	⋮	⋮	⋮	⋮	⋮	⋮	⋮	⋮	⋮	⋮	⋮	⋮	⋮	⋮	⋮
$\mathcal{D}_8$	Sup_1	MG	G	P	MG	MP	VG	MP	MG	MG	F	G	G	MG	F	MG
	Sup_2	MP	G	MP	P	VP	G	MP	P	P	F	F	MP	P	VP	VP
	Sup_3	P	P	MG	F	P	F	MP	MP	P	P	MG	F	VP	MP	F
	Sup_4	VP	G	MP	MP	MP	VP	P	F	P	F	VP	P	F	F	P
	Sup_5	F	P	F	P	MP	MP	MG	MG	F	MP	MG	F	F	MG	MP
	Sup_6	G	F	MG	F	MG	G	MG	MP	G	F	VG	VG	G	MG	G
	Sup_7	VP	P	P	MP	MP	MP	P	MP	VP	MP	VP	F	MP	MP	MP
	Sup_8	G	MP	MP	VP	MP	P	MP	P	MP	P	MP	F	VP	MP	P

Step 6. The interval type-2 fuzzy decision-matrix can be calculated based on Table 7 and Equation (25). According to the number of suppliers and criteria in this case, the matrix has 120 (8×15) elements which are defined as trapezoidal IT2FSs. Due to limitations in space, it is not possible to present all of the elements of the decision-matrix in this paper. The decision-matrix is partially provided in Table 8, and the detailed matrix can be found in Reference [96], named Decision Matrix.

Table 8. Interval type-2 fuzzy decision-matrix.

	a_U	b_U	c_U	d_U	θ_U	a_L	b_L	c_L	d_L	θ_L
$\tilde{\tilde{x}}_{1,11}$	6.25	7.94	8.31	9.38	1	7.13	7.94	8.31	8.69	0.9
$\tilde{\tilde{x}}_{1,12}$	5.75	7.5	8	9.25	1	6.75	7.5	8	8.5	0.9
$\tilde{\tilde{x}}_{1,13}$	1.25	2.75	3.25	4.75	1	2	2.75	3.25	3.75	0.9
$\tilde{\tilde{x}}_{1,14}$	3.75	5.75	6.25	7.75	1	4.75	5.75	6.25	6.75	0.9
$\tilde{\tilde{x}}_{1,15}$	0.38	1.5	1.88	3.25	1	0.94	1.5	1.88	2.38	0.9
⋮	⋮	⋮	⋮	⋮	⋮	⋮	⋮	⋮	⋮	⋮
$\tilde{\tilde{x}}_{8,31}$	2.5	4.5	5	6.5	1	3.5	4.5	5	5.5	0.9
$\tilde{\tilde{x}}_{8,32}$	4.75	6.63	7.13	8.5	1	5.75	6.63	7.13	7.63	0.9
$\tilde{\tilde{x}}_{8,33}$	0.25	1.13	1.44	2.75	1	0.69	1.13	1.44	1.94	0.9
$\tilde{\tilde{x}}_{8,34}$	1.5	3.25	3.75	5.25	1	2.38	3.25	3.75	4.25	0.9
$\tilde{\tilde{x}}_{8,35}$	0	0.38	0.56	1.75	1	0.19	0.38	0.56	1.06	0.9

Steps 7 and 8. According to the decision-matrix obtained in the previous step, the defuzzified decision matrix was calculated. Then the MEREC method was used and the objective weights of the criteria were determined. The defuzzified decision matrix and the objective criteria weights are represented in Table 9. Based on the objective weights of the criteria and the subjective weights obtained in the previous steps, the combined weights can be calculated. The last column of Table 9 shows the combined weights of the criteria. It should be noted that ω = 0.5 was considered for the combination parameter.

Table 9. The defuzzified decision-matrix and the objective criteria weights.

	Sup_1	Sup_2	Sup_3	Sup_4	Sup_5	Sup_6	Sup_7	Sup_8	w_j^o	w_j^c
C_{11}	8.04	3.92	1.70	0.85	5.67	9.00	1.27	6.86	0.0811	0.0738
C_{12}	7.69	7.94	2.48	7.05	2.48	4.67	2.01	2.71	0.0418	0.0542
C_{13}	2.98	3.45	5.92	1.79	2.96	6.55	1.32	5.42	0.0556	0.0621
C_{14}	5.92	1.09	3.45	3.21	2.49	3.70	5.42	1.79	0.0608	0.0486
C_{15}	1.70	0.85	1.32	4.42	1.41	5.42	1.70	3.21	0.0544	0.0514
C_{21}	9.29	7.44	5.17	1.18	4.92	8.76	4.92	1.09	0.0795	0.0768
C_{22}	4.67	2.71	3.21	0.80	7.11	6.61	3.20	4.67	0.0883	0.0656
C_{23}	7.69	1.70	0.94	2.96	7.49	2.32	4.67	1.56	0.0660	0.0620
C_{24}	8.23	0.85	1.56	2.48	3.45	9.15	1.18	1.56	0.0596	0.0593
C_{25}	4.17	4.67	1.78	6.99	4.17	5.42	3.68	2.74	0.0482	0.0359
C_{31}	9.29	6.30	4.92	1.47	5.92	8.95	1.56	4.67	0.0645	0.0929
C_{32}	7.69	6.17	4.67	2.23	3.20	9.44	4.92	6.80	0.0496	0.0618
C_{33}	8.57	3.45	0.94	3.70	5.67	9.29	2.71	1.32	0.0748	0.0739
C_{34}	6.61	0.94	3.68	5.17	7.44	7.11	4.17	3.45	0.0894	0.0962
C_{35}	8.95	1.03	6.99	1.00	3.70	8.85	1.41	0.56	0.0862	0.0855

Steps 9 to 11. Based on the decision-matrix obtained in Step 6, and Equation (28), the normalized decision-matrix can be computed. Because of limitations in space, this matrix is not presented with details. The partial version of the normalized matrix is presented in Table 10, the detailed version can be found in Reference [96], named Normalized Decision Matrix.

Steps 10 and 11. According to the normalized decision-matrix and Equations (29) to (33), the values of WSM ($\tilde{\tilde{Q}}_i^S$), WPM ($\tilde{\tilde{Q}}_i^P$) and composite WASPAS measure ($\tilde{\tilde{Q}}_i$) were computed. The computations were carried out with γ = 0.5. These values are shown in Table 11.

Table 10. The normalized decision matrix.

	a_U	b_U	c_U	d_U	θ_U	a_L	b_L	c_L	d_L	θ_L
$\tilde{\tilde{x}}^n_{1,11}$	0.69	0.88	0.92	1.04	1	0.79	0.88	0.92	0.97	0.9
$\tilde{\tilde{x}}^n_{1,12}$	0.72	0.94	1.01	1.17	1	0.85	0.94	1.01	1.07	0.9
$\tilde{\tilde{x}}^n_{1,13}$	0.19	0.42	0.5	0.73	1	0.31	0.42	0.5	0.57	0.9
$\tilde{\tilde{x}}^n_{1,14}$	0.63	0.97	1.06	1.31	1	0.8	0.97	1.06	1.14	0.9
$\tilde{\tilde{x}}^n_{1,15}$	0.07	0.28	0.35	0.6	1	0.17	0.28	0.35	0.44	0.9
⋮	⋮	⋮	⋮	⋮	⋮	⋮	⋮	⋮	⋮	⋮
$\tilde{\tilde{x}}^n_{8,31}$	0.27	0.48	0.54	0.7	1	0.38	0.48	0.54	0.59	0.9
$\tilde{\tilde{x}}^n_{8,32}$	0.5	0.7	0.76	0.9	1	0.61	0.7	0.76	0.81	0.9
$\tilde{\tilde{x}}^n_{8,33}$	0.03	0.12	0.15	0.3	1	0.07	0.12	0.15	0.21	0.9
$\tilde{\tilde{x}}^n_{8,34}$	0.2	0.44	0.5	0.71	1	0.32	0.44	0.5	0.57	0.9
$\tilde{\tilde{x}}^n_{8,35}$	0	0.04	0.06	0.2	1	0.02	0.04	0.06	0.12	0.9

Table 11. Different measures of the WASPAS method.

	a_U	b_U	c_U	d_U	θ_U	a_L	b_L	c_L	d_L	θ_L
$\tilde{\tilde{Q}}_1^S$	0.63	0.83	0.89	1.03	1	0.74	0.83	0.89	0.94	0.9
$\tilde{\tilde{Q}}_2^S$	0.22	0.39	0.45	0.62	1	0.31	0.39	0.45	0.51	0.9
$\tilde{\tilde{Q}}_3^S$	0.21	0.4	0.46	0.64	1	0.31	0.4	0.46	0.52	0.9
$\tilde{\tilde{Q}}_4^S$	0.17	0.34	0.4	0.57	1	0.26	0.34	0.4	0.46	0.9
$\tilde{\tilde{Q}}_5^S$	0.35	0.57	0.63	0.81	1	0.46	0.57	0.63	0.69	0.9
$\tilde{\tilde{Q}}_6^S$	0.67	0.87	0.93	1.06	1	0.78	0.87	0.93	0.98	0.9
$\tilde{\tilde{Q}}_7^S$	0.16	0.35	0.4	0.58	1	0.25	0.35	0.4	0.46	0.9
$\tilde{\tilde{Q}}_8^S$	0.21	0.39	0.44	0.62	1	0.3	0.39	0.44	0.5	0.9
$\tilde{\tilde{Q}}_1^P$	1.61	1.82	1.87	2.02	1	1.72	1.82	1.87	1.93	0.9
$\tilde{\tilde{Q}}_2^P$	1.21	1.37	1.42	1.59	1	1.29	1.37	1.42	1.48	0.9
$\tilde{\tilde{Q}}_3^P$	1.2	1.39	1.44	1.62	1	1.29	1.39	1.44	1.5	0.9
$\tilde{\tilde{Q}}_4^P$	1.16	1.32	1.37	1.54	1	1.24	1.32	1.37	1.43	0.9
$\tilde{\tilde{Q}}_5^P$	1.33	1.55	1.61	1.8	1	1.44	1.55	1.61	1.68	0.9
$\tilde{\tilde{Q}}_6^P$	1.65	1.86	1.91	2.06	1	1.76	1.86	1.91	1.97	0.9
$\tilde{\tilde{Q}}_7^P$	1.15	1.33	1.38	1.56	1	1.24	1.33	1.38	1.44	0.9
$\tilde{\tilde{Q}}_8^P$	1.19	1.37	1.42	1.59	1	1.28	1.37	1.42	1.48	0.9
$\tilde{\tilde{Q}}_1$	0.78	0.95	1	1.12	1	0.87	0.95	1	1.04	0.9
$\tilde{\tilde{Q}}_2$	0.45	0.59	0.63	0.77	1	0.52	0.59	0.63	0.68	0.9
$\tilde{\tilde{Q}}_3$	0.44	0.6	0.64	0.79	1	0.52	0.6	0.64	0.69	0.9
$\tilde{\tilde{Q}}_4$	0.4	0.54	0.59	0.73	1	0.48	0.54	0.59	0.64	0.9
$\tilde{\tilde{Q}}_5$	0.55	0.73	0.78	0.93	1	0.64	0.73	0.78	0.84	0.9
$\tilde{\tilde{Q}}_6$	0.81	0.99	1.03	1.15	1	0.91	0.99	1.03	1.07	0.9
$\tilde{\tilde{Q}}_7$	0.4	0.55	0.59	0.74	1	0.47	0.55	0.59	0.64	0.9
$\tilde{\tilde{Q}}_8$	0.43	0.58	0.62	0.77	1	0.51	0.58	0.62	0.68	0.9

Step 12. Using the ROG-based method proposed for the comparative ranking of IT2FSs, the ranking values (S_i) or relative scores for suppliers were determined in this step. These values in addition to the other parameters related to the considered company are

presented in Table 12. Based on these parameters and Model (35) the SSOA problem was solved, and the quantity of the order from each supplier was obtained. The outcomes of solving the SSOA problem are shown in Table 13. It should be noted that the solver of LINGO 18 Software (Commercial Version) was used to handle the optimization problem.

Table 12. The parameters of the SSOA model.

Supplier	S_i	O_i^{min} (Tons)	CP_i (Tons)	c_{i1}^o ($\times 10^5$ IRR)	c_{i2}^o ($\times 10^5$ IRR)	d_{i1} (Km)	d_{i2} (Km)
Sup_1	0.1696	5000	40,000	200	260	45	150
Sup_2	0.1155	7000	25,000	260	270	30	120
Sup_3	0.1230	4500	26,000	250	260	35	110
Sup_4	0.0717	5500	26,000	260	280	20	80
Sup_5	0.1518	4000	20,000	280	260	40	60
Sup_6	0.1875	5800	35,000	290	200	80	20
Sup_7	0.0711	3900	15,000	280	230	100	25
Sup_8	0.1097	4200	25,000	270	240	130	35
$DEM_1 = 50{,}000$			$DEM_2 = 40{,}000$			$K^{min} = 2$	

Table 13. The results of solving the SSOA problem.

Supplier	x_{ij}^o PC_1	x_{ij}^o PC_2	y_i	Objective Functions	
Sup_1	34,079.12	0	1	$Z_1^{min} = 0.1865 \times 10^8$	$Z_1 = 0.1927956 \times 10^8$
Sup_2	9705.485	0	1	$Z_1^{max} = 0.2541 \times 10^8$	
Sup_3	0	0	0	$Z_2^{min} = 2{,}065{,}000$	$Z_2 = 2{,}948{,}341$
Sup_4	0	0	0	$Z_2^{max} = 0.1155 \times 10^8$	
Sup_5	6215.396	0	1	$Z_3^{min} = 8445.527$	$Z_3 = 14{,}956.37$
Sup_6	0	35,000	1	$Z_3^{max} = 15{,}625$	
Sup_7	0	0	0		$\lambda = 0.9069$
Sup_8	0	5000	1		

The information presented in Table 13 suggests that for meeting the demand of Production Center 1, Suppliers 1, 2, and 5 are identified as suitable suppliers, while for Production Center 2, Suppliers 6 and 8 are considered as the optimal options. It is important to note that the selection of suppliers for each production center was not solely based on their relative scores or performance ratings. The proposed methodology also considered the order quantities that each supplier could provide to ensure that the total demand of each production center could be met.

4.2. Sensitivity Analysis

Performing a sensitivity analysis on the weights of criteria used for supplier selection and order allocation can provide valuable insights into the decision-making process and help companies to make more informed choices. It is essential to conduct such an analysis because the weights assigned to each criterion can significantly impact the relative scores of the suppliers. Since one of the objective functions of the mathematical model is related to the relative scores the changes in the weights can affect the quantity of orders allocated to suppliers. By conducting a sensitivity analysis, companies can test different weight combinations and observe the effect of each change on the final scores of suppliers and order allocation. This analysis can help companies understand the trade-offs between different criteria and make a more balanced decision. Additionally, performing a sensitivity analysis can help companies identify which criteria have the most significant impact on supplier selection and order allocation decisions. To make this analysis, a pattern of changing weights has been used in this study. The pattern of changing weights involves defining m sets, where m is the total number of criteria being considered. In each set, one criterion is assigned the highest weight, while another is assigned the lowest weight. The remaining criteria are assigned weights that lie between these two extremes. These weights are presented in Table 14 and graphically shown in Figure 4.

Table 14. The weights used for the sensitivity analysis.

	C_{11}	C_{12}	C_{13}	C_{14}	C_{15}	C_{21}	C_{22}	C_{23}	C_{24}	C_{25}	C_{31}	C_{32}	C_{33}	C_{34}	C_{35}
Set 1	0.008	0.017	0.025	0.033	0.042	0.050	0.058	0.067	0.075	0.083	0.092	0.100	0.108	0.117	0.125
Set 2	0.017	0.025	0.033	0.042	0.050	0.058	0.067	0.075	0.083	0.092	0.100	0.108	0.117	0.125	0.008
Set 3	0.025	0.033	0.042	0.050	0.058	0.067	0.075	0.083	0.092	0.100	0.108	0.117	0.125	0.008	0.017
Set 4	0.033	0.042	0.050	0.058	0.067	0.075	0.083	0.092	0.100	0.108	0.117	0.125	0.008	0.017	0.025
Set 5	0.042	0.050	0.058	0.067	0.075	0.083	0.092	0.100	0.108	0.117	0.125	0.008	0.017	0.025	0.033
Set 6	0.050	0.058	0.067	0.075	0.083	0.092	0.100	0.108	0.117	0.125	0.008	0.017	0.025	0.033	0.042
Set 7	0.058	0.067	0.075	0.083	0.092	0.100	0.108	0.117	0.125	0.008	0.017	0.025	0.033	0.042	0.050
Set 8	0.067	0.075	0.083	0.092	0.100	0.108	0.117	0.125	0.008	0.017	0.025	0.033	0.042	0.050	0.058
Set 9	0.075	0.083	0.092	0.100	0.108	0.117	0.125	0.008	0.017	0.025	0.033	0.042	0.050	0.058	0.067
Set 10	0.083	0.092	0.100	0.108	0.117	0.125	0.008	0.017	0.025	0.033	0.042	0.050	0.058	0.067	0.075
Set 11	0.092	0.100	0.108	0.117	0.125	0.008	0.017	0.025	0.033	0.042	0.050	0.058	0.067	0.075	0.083
Set 12	0.100	0.108	0.117	0.125	0.008	0.017	0.025	0.033	0.042	0.050	0.058	0.067	0.075	0.083	0.092
Set 13	0.108	0.117	0.125	0.008	0.017	0.025	0.033	0.042	0.050	0.058	0.067	0.075	0.083	0.092	0.100
Set 14	0.117	0.125	0.008	0.017	0.025	0.033	0.042	0.050	0.058	0.067	0.075	0.083	0.092	0.100	0.108
Set 15	0.125	0.008	0.017	0.025	0.033	0.042	0.050	0.058	0.067	0.075	0.083	0.092	0.100	0.108	0.117

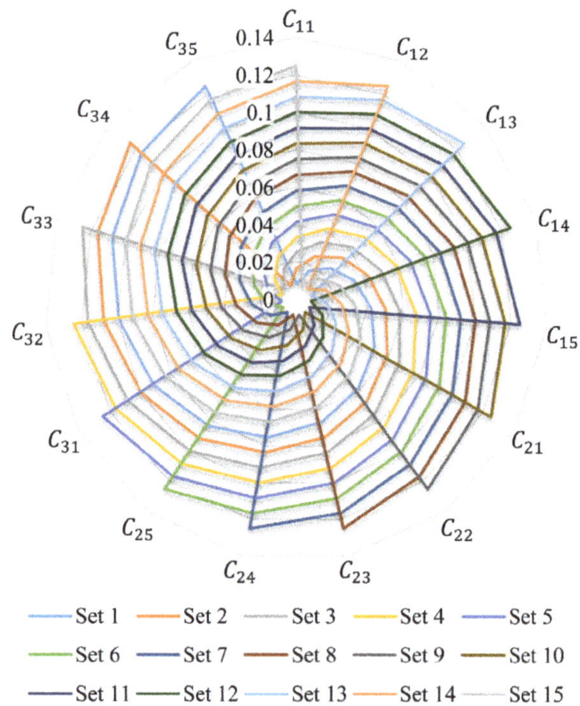

Figure 4. The pattern of changing criteria weights.

The weights provided in Table 14 were used instead of w_j^c in Equation (27) to see the changes in the relative scores of the suppliers. The variations in the relative scores of each supplier can be seen in Figure 5. Based on the data presented in Figure 5, it appears that the relative score of Supplier 5 remains consistently stable across all sets, suggesting a high degree of reliability in terms of meeting the defined criteria. Suppliers 1 and 6 also demonstrate a relatively stable score, albeit with some minor variations. Conversely, the scores of the remaining suppliers appear to vary considerably with changes in the criteria weights across the different sets. It is worth noting that a stable relative score can be interpreted as a higher degree of consistency in meeting the defined criteria, and therefore, may be indicative of a more reliable supplier. This pattern of changing weights allows companies to test the impact of assigning different levels of importance to each criterion and observe the resulting effect on supplier selection and order allocation.

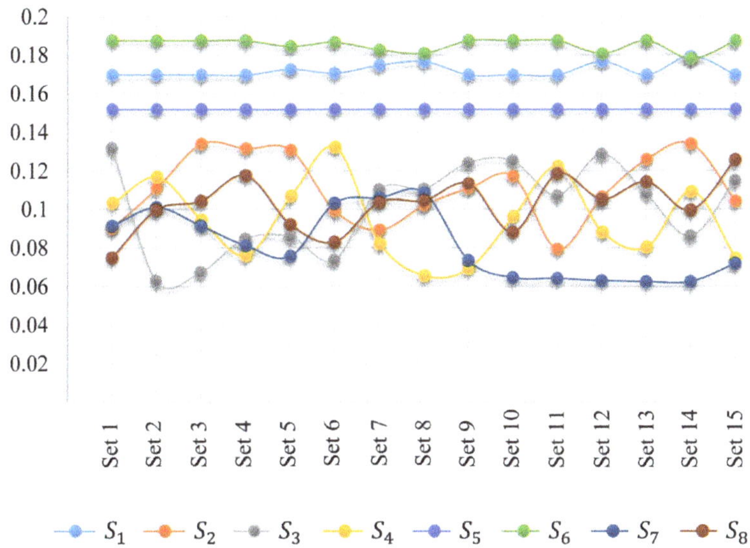

Figure 5. The effect of changing criteria weights on the relative scores of the suppliers.

The results presented in Figure 6 reveal the effects of changing the criteria weights on the resulting order quantities for two production centers. Specifically, the order quantity for Production Center 1 varies for Suppliers 1, 2, 4, and 5, while Supplier 1 consistently receives the highest order quantity allocation. For Production Center 2, the order quantity varies for Suppliers 5, 7, and 8, but a fixed quantity of orders is allocated to Supplier 6. These findings have important implications for supplier selection and order allocation decisions. Firstly, they highlight the importance of considering the relative scores of the suppliers when allocating orders, as Supplier 1 consistently receives the highest order quantity allocation for Production Center 1, indicating that it is the most reliable supplier for this production center. Similarly, Supplier 6 receives a fixed quantity of orders for Production Center 2, suggesting that it is the most reliable supplier for this production center. Furthermore, the analysis underscores the need to consider the effects of the criteria weights on the relative scores of the suppliers. The relative scores of Suppliers 1, 5, and 6 remain relatively stable across the different sets of criteria weights, indicating that they are more robust and reliable suppliers. In contrast, the relative scores of Suppliers 2, 3, 4, 7, and 8 vary significantly across the different sets of criteria weights, highlighting their sensitivity to changes in the criteria weights. Therefore, the results of the sensitivity analysis suggest that suppliers with more stable relative scores are generally more reliable, as they are less affected by changes in the criteria weights. Overall, the findings demonstrate the importance of conducting sensitivity analysis on the effects of criteria weights on supplier selection and order allocation decisions. This can help decision-makers identify the most reliable suppliers and allocate orders in a way that maximizes efficiency and sustainability criteria.

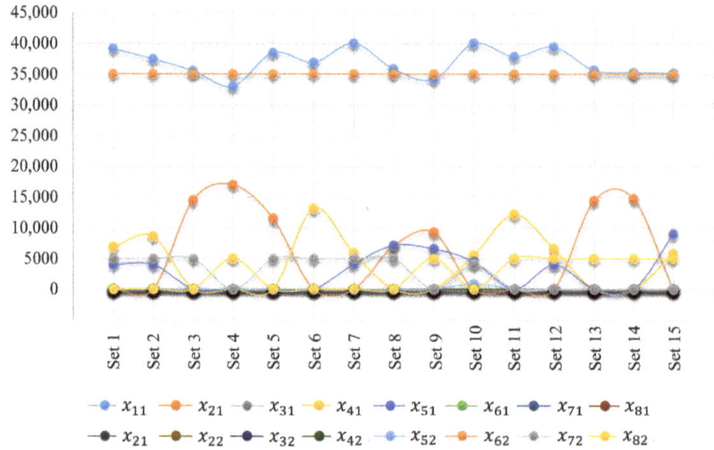

Figure 6. The effect of changing criteria weights on the order quantity.

4.3. Comparative Analysis

The subsection in your paper presents a comprehensive comparison between the results of the proposed method and those of six other well-established methods, namely Simple Additive Weighting (SAW), Complex Proportional Assessment (COPRAS), TOPSIS, VIKOR (stands for "VIekriterijumsko KOmpromisno Rangiranje"), Evaluation Based on Distance from Average Solution (EDAS), and MULTIMOORA. This comparison aims to validate the accuracy and efficiency of the proposed method while identifying its strengths and weaknesses relative to other methods. To measure the strength of the relationship between the results, the study uses Spearman's rank correlation coefficient (ρ), which is a robust measure of the correlation between the rankings obtained from the proposed method and the rankings from the other methods. Table 15 presents the ranking results of different methods and the correlation coefficient.

Table 15. The results of the comparison.

Supplier	SAW	COPRAS	TOPSIS	VIKOR	EDAS	MULTIMOORA	Proposed Approach
Sup_1	2	2	2	1	2	2	2
Sup_2	5	5	5	6	5	5	5
Sup_3	4	4	4	4	4	4	4
Sup_4	7	7	7	8	8	7	7
Sup_5	3	3	3	3	3	3	3
Sup_6	1	1	1	2	1	1	1
Sup_7	8	8	8	7	7	8	8
Sup_8	6	6	6	5	6	6	6
ρ	1	1	1	0.929	0976	1	—

The results show that Supplier 6 has the first rank in the results of all methods except VIKOR, where it ranks second. Meanwhile, Supplier 1 has the second rank in the results of all methods except VIKOR, where it ranks first. Additionally, Supplier 5 ranks third in all of the six methods. Based on the interpretation of correlation values presented by Walters [97], the values of Spearman's rank correlation coefficient demonstrate a very strong relationship between the results of the proposed method and those of the other methods. This confirms the validity of the results obtained from the proposed method and suggests that it is a reliable and effective tool for supplier selection.

5. Conclusions

The SSOA problem is a critical aspect of supply chain management. Efficient supplier selection and order allocation can significantly impact the overall sustainability of the supply chain. The SSOA problem becomes more complex when considering sustainability criteria, as these criteria are often uncertain and subjective. Therefore, the development of effective methodologies for sustainable SSOA is crucial for achieving sustainability goals in the supply chain. The proposed methodology in this study integrates multiple techniques to address the sustainable SSOA problem. The methodology utilizes a new ranking method based on the concept of Radius of Gyration for interval type-2 fuzzy sets, which can handle the uncertainty in supplier evaluation. To determine the weights of evaluation criteria, both subjective weights obtained using the SMART and expert preferences, and objective weights calculated using the MEREC method are combined. The proposed methodology also incorporates sustainability criteria and uses the WASPAS method to evaluate supplier performance as type-2 fuzzy sets. The ROG-based ranking method is then employed to calculate the relative scores of suppliers, and an MODM linear mathematical model is presented to identify suitable suppliers and allocate their order quantities. The proposed methodology was applied to a sustainable SSOA problem in Golestan, Iran. The results demonstrated that the proposed approach was effective in selecting suitable suppliers and allocating their order quantities based on sustainability criteria. The application of the proposed methodology resulted in the selection of five suitable suppliers and the allocation of orders among them. The sensitivity analysis also showed that the proposed methodology was robust and could handle changes in the weight of evaluation criteria. Future research can be conducted in various directions to further enhance the proposed methodology. To further improve the understanding of the advantages of a ROG-based approach, a comprehensive comparison with other ranking approaches could be conducted in future research. Other types of fuzzy sets and membership functions, such as symmetric IT2FSs, Fermatean fuzzy sets and Pythagorean fuzzy sets, can be explored to evaluate supplier performance. Additionally, other weighting methods, such as SWARA (Stepwise Weight Assessment Ratio Analysis) and SECA (Simultaneous Evaluation of Criteria and Alternatives), can be used to determine the weights of evaluation criteria. Furthermore, other MCDM methods, such as CoCoSo (Combined Compromise Solution) and MARCOS (Measurement of Alternatives and Ranking according to COmpromise Solution), can be investigated to compare their performance with the proposed approach. The proposed method can also be extended to a dynamic decision-making approach. This could involve setting up rules or algorithms to adjust the decision based on predefined criteria or using machine learning techniques to learn from past decisions and adjust the decision-making process accordingly. Overall, the proposed methodology provides a solid foundation for future research and can be further enhanced to tackle more complex sustainability challenges in the supply chain.

Funding: This research received no external funding.

Data Availability Statement: Not applicable.

Conflicts of Interest: The author declares no conflict of interest.

References

1. Vahidi, F.; Torabi, S.A.; Ramezankhani, M.J. Sustainable supplier selection and order allocation under operational and disruption risks. *J. Clean. Prod.* **2018**, *174*, 1351–1365. [CrossRef]
2. Ghadimi, P.; Ghassemi Toosi, F.; Heavey, C. A multi-agent systems approach for sustainable supplier selection and order allocation in a partnership supply chain. *Eur. J. Oper. Res.* **2018**, *269*, 286–301. [CrossRef]
3. Ashby, A.; Leat, M.; Hudson-Smith, M. Making connections: A review of supply chain management and sustainability literature. *Supply Chain. Manag. Int. J.* **2012**, *17*, 497–516. [CrossRef]
4. Giri, B.C.; Molla, M.U.; Biswas, P. Pythagorean fuzzy DEMATEL method for supplier selection in sustainable supply chain management. *Expert Syst. Appl.* **2022**, *193*, 116396. [CrossRef]

5. Yadav, S.; Choi, T.-M.; Kumar, A.; Luthra, S.; Naz, F. A meta-analysis of sustainable supply chain practices and performance: The moderating roles of type of economy and innovation. *Int. J. Oper. Prod. Manag.* **2023**, *43*, 802–845. [CrossRef]
6. Acerbi, F.; Rocca, R.; Fumagalli, L.; Taisch, M. Enhancing the cosmetics industry sustainability through a renewed sustainable supplier selection model. *Prod. Manuf. Res.* **2023**, *11*, 2161021. [CrossRef]
7. Wilhelm, M.M.; Blome, C.; Bhakoo, V.; Paulraj, A. Sustainability in multi-tier supply chains: Understanding the double agency role of the first-tier supplier. *J. Oper. Manag.* **2016**, *41*, 42–60. [CrossRef]
8. Manupati, V.K.; Schoenherr, T.; Ramkumar, M.; Wagner, S.M.; Pabba, S.K.; Inder Raj Singh, R. A blockchain-based approach for a multi-echelon sustainable supply chain. *Int. J. Prod. Res.* **2020**, *58*, 2222–2241. [CrossRef]
9. Zeballos, L.J.; Méndez, C.A.; Barbosa-Povoa, A.P.; Novais, A.Q. Multi-period design and planning of closed-loop supply chains with uncertain supply and demand. *Comput. Chem. Eng.* **2014**, *66*, 151–164. [CrossRef]
10. Rösner, T. Supplier selection and order allocation in purchasing and supply management: A systematic literature review. *Int. J. Procure. Manag.* **2022**, *16*, 31–74. [CrossRef]
11. Gören, H.G. A decision framework for sustainable supplier selection and order allocation with lost sales. *J. Clean. Prod.* **2018**, *183*, 1156–1169. [CrossRef]
12. Tronnebati, I.; El Yadari, M.; Jawab, F. A Review of Green Supplier Evaluation and Selection Issues Using MCDM, MP and AI Models. *Sustainability* **2022**, *14*, 16714. [CrossRef]
13. Keshavarz-Ghorabaee, M.; Amiri, M.; Zavadskas, E.K.; Turskis, Z.; Antuchevičienė, J. Ranking of Bridge Design Alternatives: A TOPSIS-FADR Method. *Balt. J. Road Bridge Eng.* **2018**, *13*, 209–237. [CrossRef]
14. Resende, C.H.L.; Geraldes, C.A.S.; Lima, F.R. Decision Models for Supplier Selection in Industry 4.0 Era: A Systematic Literature Review. *Procedia Manuf.* **2021**, *55*, 492–499. [CrossRef]
15. Paul, A.; Shukla, N.; Paul, S.K.; Trianni, A. Sustainable Supply Chain Management and Multi-Criteria Decision-Making Methods: A Systematic Review. *Sustainability* **2021**, *13*, 7104. [CrossRef]
16. Gupta, P.; Govindan, K.; Mehlawat, M.K.; Kumar, S. A weighted possibilistic programming approach for sustainable vendor selection and order allocation in fuzzy environment. *Int. J. Adv. Manuf. Technol.* **2016**, *86*, 1785–1804. [CrossRef]
17. De Capua, C.; Morello, R.; Carbone, R. Measurement uncertainty in decision-making: How to take reliable decisions under uncertainty. *Int. J. Meas. Technol. Instrum. Eng.* **2011**, *1*, 40–52. [CrossRef]
18. Chen, K.-S.; Chang, T.-C.; Huang, C.-C. Supplier Selection by Fuzzy Assessment and Testing for Process Quality under Consideration with Data Imprecision. *Mathematics* **2020**, *8*, 1420. [CrossRef]
19. Xie, Y.; He, L.; Xiang, W.; Peng, Z.; Ming, X.; Goh, M. Prioritizing risk factors in sustainable supply chain using fuzzy Kano and interval-valued intuitionistic fuzzy QFD. *Kybernetes* **2022**. [CrossRef]
20. Smets, P. Imperfect Information: Imprecision and Uncertainty. In *Uncertainty Management in Information Systems: From Needs to Solutions*; Motro, A., Smets, P., Eds.; Springer: Boston, MA, USA, 1997; pp. 225–254. [CrossRef]
21. Gonzalez, A.; Pons, O.; Vila, M.A. Dealing with uncertainty and imprecision by means of fuzzy numbers. *Int. J. Approx. Reason.* **1999**, *21*, 233–256. [CrossRef]
22. Kaya, I.; Çolak, M.; Terzi, F. A comprehensive review of fuzzy multi criteria decision making methodologies for energy policy making. *Energy Strategy Rev.* **2019**, *24*, 207–228. [CrossRef]
23. Farid, H.M.A.; Bouye, M.; Riaz, M.; Jamil, N. Fermatean Fuzzy CODAS Approach with Topology and Its Application to Sustainable Supplier Selection. *Symmetry* **2023**, *15*, 433. [CrossRef]
24. Hailiang, Z.; Khokhar, M.; Islam, T.; Sharma, A. A model for green-resilient supplier selection: Fuzzy best–worst multi-criteria decision-making method and its applications. *Environ. Sci. Pollut. Res.* **2023**, *30*, 54035–54058. [CrossRef] [PubMed]
25. Kaya, S.K. A novel two-phase group decision-making model for circular supplier selection under picture fuzzy environment. *Environ. Sci. Pollut. Res.* **2023**, *30*, 34135–34157. [CrossRef] [PubMed]
26. Keshavarz Ghorabaee, M.; Amiri, M.; Zavadskas, E.K.; Antucheviciene, J. Supplier evaluation and selection in fuzzy environments: A review of MADM approaches. *Econ. Res.-Ekon. Istraz.* **2017**, *30*, 1073–1118. [CrossRef]
27. Güneri, B.; Deveci, M. Evaluation of supplier selection in the defense industry using q-rung orthopair fuzzy set based EDAS approach. *Expert Syst. Appl.* **2023**, *222*, 119846. [CrossRef]
28. Celik, E.; Gul, M. Hazard identification, risk assessment and control for dam construction safety using an integrated BWM and MARCOS approach under interval type-2 fuzzy sets environment. *Autom. Constr.* **2021**, *127*, 103699. [CrossRef]
29. Liu, P.; Gao, H.; Ma, J. Novel green supplier selection method by combining quality function deployment with partitioned Bonferroni mean operator in interval type-2 fuzzy environment. *Inf. Sci.* **2019**, *490*, 292–316. [CrossRef]
30. Keshavarz Ghorabaee, M.; Zavadskas, E.K.; Amiri, M.; Antucheviciene, J. Evaluation by an area-based method of ranking interval type-2 fuzzy sets (EAMRIT-2F) for multi-criteria group decision-making. *Transform. Bus. Econ.* **2016**, *15*, 76–95.
31. Ecer, F. Multi-criteria decision making for green supplier selection using interval type-2 fuzzy AHP: A case study of a home appliance manufacturer. *Oper. Res.* **2022**, *22*, 199–233. [CrossRef]
32. Yucesan, M.; Mete, S.; Serin, F.; Celik, E.; Gul, M. An Integrated Best-Worst and Interval Type-2 Fuzzy TOPSIS Methodology for Green Supplier Selection. *Mathematics* **2019**, *7*, 182. [CrossRef]
33. Hoseini, S.A.; Hashemkhani Zolfani, S.; Skačkauskas, P.; Fallahpour, A.; Saberi, S. A Combined Interval Type-2 Fuzzy MCDM Framework for the Resilient Supplier Selection Problem. *Mathematics* **2022**, *10*, 44. [CrossRef]

34. Boral, S.; Chaturvedi, S.K.; Howard, I.; Naikan, V.N.A.; McKee, K. An integrated interval type-2 fuzzy sets and multiplicative half quadratic programming-based MCDM framework for calculating aggregated risk ranking results of failure modes in FMECA. *Process Saf. Environ. Prot.* **2021**, *150*, 194–222. [CrossRef]
35. Boral, S.; Chakraborty, S. Failure analysis of CNC machines due to human errors: An integrated IT2F-MCDM-based FMEA approach. *Eng. Fail. Anal.* **2021**, *130*, 105768. [CrossRef]
36. Celik, E.; Gul, M.; Aydin, N.; Gumus, A.T.; Guneri, A.F. A comprehensive review of multi criteria decision making approaches based on interval type-2 fuzzy sets. *Knowl. Based Syst.* **2015**, *85*, 329–341. [CrossRef]
37. Zardari, N.H.; Ahmed, K.; Shirazi, S.M.; Yusop, Z.B. *Weighting Methods and their Effects on Multi-Criteria Decision Making Model Outcomes in Water Resources Management*; Springer: New York, NY, USA, 2014.
38. Keshavarz-Ghorabaee, M.; Amiri, M.; Zavadskas, E.K.; Turskis, Z.; Antucheviciene, J. Determination of Objective Weights Using a New Method Based on the Removal Effects of Criteria (MEREC). *Symmetry* **2021**, *13*, 525. [CrossRef]
39. Zavadskas, E.K.; Turskis, Z.; Antucheviciene, J.; Zakarevicius, A. Optimization of weighted aggregated sum product assessment. *Elektron. Ir. Elektrotechnika* **2012**, *122*, 3–6. [CrossRef]
40. Esmaeili-Najafabadi, E.; Fallah Nezhad, M.S.; Pourmohammadi, H.; Honarvar, M.; Vahdatzad, M.A. A joint supplier selection and order allocation model with disruption risks in centralized supply chain. *Comput. Ind. Eng.* **2019**, *127*, 734–748. [CrossRef]
41. Moheb-Alizadeh, H.; Handfield, R. Sustainable supplier selection and order allocation: A novel multi-objective programming model with a hybrid solution approach. *Comput. Ind. Eng.* **2019**, *129*, 192–209. [CrossRef]
42. Hosseini, S.; Morshedlou, N.; Ivanov, D.; Sarder, M.D.; Barker, K.; Khaled, A.A. Resilient supplier selection and optimal order allocation under disruption risks. *Int. J. Prod. Econ.* **2019**, *213*, 124–137. [CrossRef]
43. Kellner, F.; Utz, S. Sustainability in supplier selection and order allocation: Combining integer variables with Markowitz portfolio theory. *J. Clean. Prod.* **2019**, *214*, 462–474. [CrossRef]
44. Duan, C.-Y.; Liu, H.-C.; Zhang, L.-J.; Shi, H. An Extended Alternative Queuing Method with Linguistic Z-numbers and Its Application for Green Supplier Selection and Order Allocation. *Int. J. Fuzzy Syst.* **2019**, *21*, 2510–2523. [CrossRef]
45. Mohammed, A.; Harris, I.; Govindan, K. A hybrid MCDM-FMOO approach for sustainable supplier selection and order allocation. *Int. J. Prod. Econ.* **2019**, *217*, 171–184. [CrossRef]
46. Safaeian, M.; Fathollahi-Fard, A.M.; Tian, G.; Li, Z.; Ke, H. A multi-objective supplier selection and order allocation through incremental discount in a fuzzy environment. *J. Intell. Fuzzy Syst.* **2019**, *37*, 1435–1455. [CrossRef]
47. Alegoz, M.; Yapicioglu, H. Supplier selection and order allocation decisions under quantity discount and fast service options. *Sustain. Prod. Consum.* **2019**, *18*, 179–189. [CrossRef]
48. Mari, S.I.; Memon, M.S.; Ramzan, M.B.; Qureshi, S.M.; Iqbal, M.W. Interactive Fuzzy Multi Criteria Decision Making Approach for Supplier Selection and Order Allocation in a Resilient Supply Chain. *Mathematics* **2019**, *7*, 137. [CrossRef]
49. Laosirihongthong, T.; Samaranayake, P.; Nagalingam, S. A holistic approach to supplier evaluation and order allocation towards sustainable procurement. *Benchmarking Int. J.* **2019**, *26*, 2543–2573. [CrossRef]
50. Feng, J.; Gong, Z. Integrated linguistic entropy weight method and multi-objective programming model for supplier selection and order allocation in a circular economy: A case study. *J. Clean. Prod.* **2020**, *277*, 122597. [CrossRef]
51. Khoshfetrat, S.; Rahiminezhad Galankashi, M.; Almasi, M. Sustainable supplier selection and order allocation: A fuzzy approach. *Eng. Optim.* **2020**, *52*, 1494–1507. [CrossRef]
52. Jia, R.; Liu, Y.; Bai, X. Sustainable supplier selection and order allocation: Distributionally robust goal programming model and tractable approximation. *Comput. Ind. Eng.* **2020**, *140*, 106267. [CrossRef]
53. Wong, J.-T. Dynamic procurement risk management with supplier portfolio selection and order allocation under green market segmentation. *J. Clean. Prod.* **2020**, *253*, 119835. [CrossRef]
54. You, S.-Y.; Zhang, L.-J.; Xu, X.-G.; Liu, H.-C. A New Integrated Multi-Criteria Decision Making and Multi-Objective Programming Model for Sustainable Supplier Selection and Order Allocation. *Symmetry* **2020**, *12*, 302. [CrossRef]
55. Rezaei, A.; Rahiminezhad Galankashi, M.; Mansoorzadeh, S.; Mokhatab Rafiei, F. Supplier Selection and Order Allocation with Lean Manufacturing Criteria: An Integrated MCDM and Bi-objective Modelling Approach. *Eng. Manag. J.* **2020**, *32*, 253–271. [CrossRef]
56. Kaviani, M.A.; Peykam, A.; Khan, S.A.; Brahimi, N.; Niknam, R. A new weighted fuzzy programming model for supplier selection and order allocation in the food industry. *J. Model. Manag.* **2020**, *15*, 381–406. [CrossRef]
57. Rezaei, S.; Ghalehkhondabi, I.; Rafiee, M.; Namdar Zanganeh, S. Supplier selection and order allocation in CLSC configuration with various supply strategies under disruption risk. *OPSEARCH* **2020**, *57*, 908–934. [CrossRef]
58. Wang, C.; Yang, Q.; Dai, S. Supplier Selection and Order Allocation under a Carbon Emission Trading Scheme: A Case Study from China. *Int. J. Environ. Res. Public Health* **2020**, *17*, 111. [CrossRef] [PubMed]
59. Çalık, A. A hybrid approach for selecting sustainable suppliers and determining order allocation based on interval type-2 fuzzy sets. *J. Enterp. Inf. Manag.* **2020**, *33*, 923–945. [CrossRef]
60. Khalili Nasr, A.; Tavana, M.; Alavi, B.; Mina, H. A novel fuzzy multi-objective circular supplier selection and order allocation model for sustainable closed-loop supply chains. *J. Clean. Prod.* **2021**, *287*, 124994. [CrossRef]
61. Kaur, H.; Prakash Singh, S. Multi-stage hybrid model for supplier selection and order allocation considering disruption risks and disruptive technologies. *Int. J. Prod. Econ.* **2021**, *231*, 107830. [CrossRef]

62. Islam, S.; Amin, S.H.; Wardley, L.J. Machine learning and optimization models for supplier selection and order allocation planning. *Int. J. Prod. Econ.* **2021**, *242*, 108315. [CrossRef]
63. Rezaei, A.; Aghsami, A.; Rabbani, M. Supplier selection and order allocation model with disruption and environmental risks in centralized supply chain. *Int. J. Syst. Assur. Eng. Manag.* **2021**, *12*, 1036–1072. [CrossRef]
64. Firouzi, F.; Jadidi, O. Multi-objective model for supplier selection and order allocation problem with fuzzy parameters. *Expert Syst. Appl.* **2021**, *180*, 115129. [CrossRef]
65. Li, F.; Wu, C.-H.; Zhou, L.; Xu, G.; Liu, Y.; Tsai, S.-B. A model integrating environmental concerns and supply risks for dynamic sustainable supplier selection and order allocation. *Soft. Comput.* **2021**, *25*, 535–549. [CrossRef]
66. Yousefi, S.; Jahangoshai Rezaee, M.; Solimanpur, M. Supplier selection and order allocation using two-stage hybrid supply chain model and game-based order price. *Oper. Res.* **2021**, *21*, 553–588. [CrossRef]
67. Beiki, H.; Mohammad Seyedhosseini, S.; Ponkratov, V.; Olegovna Zekiy, A.; Ivanov, S.A. Addressing a sustainable supplier selection and order allocation problem by an integrated approach: A case of automobile manufacturing. *J. Ind. Prod. Eng.* **2021**, *38*, 239–253. [CrossRef]
68. Esmaeili-Najafabadi, E.; Azad, N.; Saber Fallah Nezhad, M. Risk-averse supplier selection and order allocation in the centralized supply chains under disruption risks. *Expert Syst. Appl.* **2021**, *175*, 114691. [CrossRef]
69. Mohammed, A.; Harris, I.; Soroka, A.; Naim, M.; Ramjaun, T.; Yazdani, M. Gresilient supplier assessment and order allocation planning. *Ann. Oper. Res.* **2021**, *296*, 335–362. [CrossRef]
70. Hosseini, Z.S.; Flapper, S.D.; Pirayesh, M. Sustainable supplier selection and order allocation under demand, supplier availability and supplier grading uncertainties. *Comput. Ind. Eng.* **2022**, *165*, 107811. [CrossRef]
71. Ali, H.; Zhang, J.; Liu, S.; Shoaib, M. An integrated decision-making approach for global supplier selection and order allocation to create an environment-friendly supply chain. *Kybernetes* **2022**. *ahead-of-print*. [CrossRef]
72. Goodarzi, F.; Abdollahzadeh, V.; Zeinalnezhad, M. An integrated multi-criteria decision-making and multi-objective optimization framework for green supplier evaluation and optimal order allocation under uncertainty. *Decis. Anal. J.* **2022**, *4*, 100087. [CrossRef]
73. Liaqait, R.A.; Warsi, S.S.; Agha, M.H.; Zahid, T.; Becker, T. A multi-criteria decision framework for sustainable supplier selection and order allocation using multi-objective optimization and fuzzy approach. *Eng. Optim.* **2022**, *54*, 928–948. [CrossRef]
74. Gai, L.; Liu, H.-C.; Wang, Y.; Xing, Y. Green supplier selection and order allocation using linguistic Z-numbers MULTIMOORA method and bi-objective non-linear programming. *Fuzzy Optim. Decis. Mak.* **2022**. [CrossRef]
75. Aouadni, S.; Euchi, J. Using Integrated MMD-TOPSIS to Solve the Supplier Selection and Fair Order Allocation Problem: A Tunisian Case Study. *Logistics* **2022**, *6*, 8. [CrossRef]
76. Galankashi, M.R.; Rezaei, A.; Keyvanpazhooh, M.; Rafiei, F.M. Assessment of suppliers and optimum order allocation in agile automotive manufacturing companies. *Int. J. Ind. Syst. Eng.* **2022**, *42*, 245–276. [CrossRef]
77. Liu, P.; Hendalianpour, A.; Fakhrabadi, M.; Feylizadeh, M. Integrating IVFRN-BWM and Goal Programming to Allocate the Order Quantity Considering Discount for Green Supplier. *Int. J. Fuzzy Syst.* **2022**, *24*, 989–1011. [CrossRef]
78. Bai, C.; Zhu, Q.; Sarkis, J. Supplier portfolio selection and order allocation under carbon neutrality: Introducing a "Cool"ing model. *Comput. Ind. Eng.* **2022**, *170*, 108335. [CrossRef]
79. Ahmad, M.T.; Firouz, M.; Mondal, S. Robust supplier-selection and order-allocation in two-echelon supply networks: A parametric tolerance design approach. *Comput. Ind. Eng.* **2022**, *171*, 108394. [CrossRef]
80. Chen, S.-M.; Lee, L.-W. Fuzzy multiple attributes group decision-making based on the interval type-2 TOPSIS method. *Expert Syst. Appl.* **2010**, *37*, 2790–2798. [CrossRef]
81. Celik, E.; Bilisik, O.N.; Erdogan, M.; Gumus, A.T.; Baracli, H. An integrated novel interval type-2 fuzzy MCDM method to improve customer satisfaction in public transportation for Istanbul. *Transp. Res. Part E Logist. Transp. Rev.* **2013**, *58*, 28–51. [CrossRef]
82. Keshavarz Ghorabaee, M.; Amiri, M.; Salehi Sadaghiani, J.; Zavadskas, E.K. Multi-criteria project selection using an extended VIKOR method with interval type-2 fuzzy sets. *Int. J. Inf. Technol. Decis. Mak.* **2015**, *14*, 993–1016. [CrossRef]
83. Kahraman, C.; Öztayşi, B.; Uçal Sarı, İ.; Turanoğlu, E. Fuzzy analytic hierarchy process with interval type-2 fuzzy sets. *Knowl. Based Syst.* **2014**, *59*, 48–57. [CrossRef]
84. Lee, E.S.; Li, R.J. Comparison of fuzzy numbers based on the probability measure of fuzzy events. *Comput. Math. Appl.* **1988**, *15*, 887–896. [CrossRef]
85. Cheng, C.-H. A new approach for ranking fuzzy numbers by distance method. *Fuzzy Sets Syst.* **1998**, *95*, 307–317. [CrossRef]
86. Chu, T.-C.; Tsao, C.-T. Ranking fuzzy numbers with an area between the centroid point and original point. *Comput. Math. Appl.* **2002**, *43*, 111–117. [CrossRef]
87. Deng, Y.; Zhenfu, Z.; Qi, L. Ranking fuzzy numbers with an area method using radius of gyration. *Comput. Math. Appl.* **2006**, *51*, 1127–1136. [CrossRef]
88. Xu, Z. Algorithm for priority of fuzzy complementary judgment matrix. *J. Syst. Eng.* **2001**, *16*, 311–314.
89. Chen, S.-M.; Lee, L.-W. Fuzzy multiple attributes group decision-making based on the ranking values and the arithmetic operations of interval type-2 fuzzy sets. *Expert Syst. Appl.* **2010**, *37*, 824–833. [CrossRef]
90. Mardani, A.; Nilashi, M.; Zakuan, N.; Loganathan, N.; Soheilirad, S.; Saman, M.Z.M.; Ibrahim, O. A systematic review and meta-Analysis of SWARA and WASPAS methods: Theory and applications with recent fuzzy developments. *Appl. Soft. Comput.* **2017**, *57*, 265–292. [CrossRef]

91. Zimmermann, H.J. Fuzzy programming and linear programming with several objective functions. *Fuzzy Sets Syst.* **1978**, *1*, 45–55. [CrossRef]
92. Schramm, V.B.; Cabral, L.P.B.; Schramm, F. Approaches for supporting sustainable supplier selection—A literature review. *J. Clean. Prod.* **2020**, *273*, 123089. [CrossRef]
93. Zimmer, K.; Fröhling, M.; Schultmann, F. Sustainable supplier management—A review of models supporting sustainable supplier selection, monitoring and development. *Int. J. Prod. Res.* **2016**, *54*, 1412–1442. [CrossRef]
94. Rashidi, K.; Noorizadeh, A.; Kannan, D.; Cullinane, K. Applying the triple bottom line in sustainable supplier selection: A meta-review of the state-of-the-art. *J. Clean. Prod.* **2020**, *269*, 122001. [CrossRef]
95. Koberg, E.; Longoni, A. A systematic review of sustainable supply chain management in global supply chains. *J. Clean. Prod.* **2019**, *207*, 1084–1098. [CrossRef]
96. Keshavarz-Ghorabaee, M. Supplementary data for "Sustainable supplier selection and order allocation using an integrated ROG-based type-2 fuzzy decision-making approach". *Figshare* **2023**. [CrossRef]
97. Walters, S.J. *Quality of Life Outcomes in Clinical Trials and Health-Care Evaluation: A Practical Guide to Analysis and Interpretation*; Wiley: New York, NY, USA, 2009.

mathematics

MDPI

Article

Fuzzy Evaluation Model of Machining Process Loss

Kuen-Suan Chen [1,2,3], Tsun-Hung Huang [1], Jin-Shyong Lin [4,*], Chun-Min Yu [1,*] and Chun-Ming Yang [5]

1 Department of Industrial Engineering and Management, National Chin-Yi University of Technology, Taichung 411030, Taiwan; kschen@ncut.edu.tw (K.-S.C.); toby@ncut.edu.tw (T.-H.H.)
2 Department of Business Administration, Chaoyang University of Technology, Taichung 413310, Taiwan
3 Department of Business Administration, Asia University, Taichung 413305, Taiwan
4 Department of Mechanical Engineering, National Chin-Yi University of Technology, Taichung 411030, Taiwan
5 School of Economics and Management, Dongguan University of Technology, Dongguan 523808, China; 2020812@dgut.edu.cn
* Correspondence: linjsh@ncut.edu.tw (J.-S.L.); march@ncut.edu.tw (C.-M.Y.)

Abstract: In facing the many negative impacts of global warming on the earth's environment, the machining industry must reduce the rates of product rework and scrap in the manufacturing process by enhancing the process quality of the processed product. According to the concept of the Taguchi loss function, the closer the measured value of the processed product is to the target value T, then the longer the mean time between failures (*MTBF*) of the product. Clearly, raising the process quality of the processed product can effect energy saving and waste reduction during production and sales, which can help enterprises fulfill their corporate social responsibilities. On the basis of the Taguchi loss function, this study used the process expected loss to evaluate the process loss. Next, the process expected loss was used as an evaluation index, in which the accuracy index and the precision index can help the machining industry find the direction for improvement. Additionally, this study first derived a confidence interval of the process expected loss. Then, it was built on the confidence interval, and a confidence interval-based fuzzy test was developed for the process expected loss. Finally, an empirical example was adopted to explain the application of the fuzzy evaluation model of the machining process proposed in this paper.

Keywords: Taguchi loss function; process expected loss; confidence interval; confidence interval-based fuzzy test; corporate social responsibility

MSC: 62C05; 62C86

Citation: Chen, K.-S.; Huang, T.-H.; Lin, J.-S.; Yu, C.-M.; Yang, C.-M. Fuzzy Evaluation Model of Machining Process Loss. *Mathematics* **2023**, *11*, 4596. https://doi.org/10.3390/math11224596

Academic Editor: Aleksandar Aleksić

Received: 14 October 2023
Revised: 4 November 2023
Accepted: 8 November 2023
Published: 9 November 2023

1. Introduction

As global warming has impacted the earth's environment, how to coexist with the natural environment has become an important issue. How to balance economic growth and sustainable development is an issue that governments and enterprises in various countries must face together [1,2]. Given this line of thinking, we must increase the extent of our responsibility for the social and ecological environment [3,4]. Therefore, corporate social responsibility (CSR) has now become a common business concept that is being promoted across the globe [5]. It is clear that products must be produced with circular economy green thinking—reduce, reuse, and recycle—from development and design to production and processing [6]. Only in this way can enterprises fulfill their social responsibilities and ease their burdens on society and the ecological environment.

Taiwan is the fifth largest exporter and the seventh largest producer of machinery and machine tools in the world [7,8]. For the machining industry, elevating the process quality of processed products can decrease the proportion of rework and scrap of processed products in the production stage. In addition, based on the concept of the Taguchi loss function, when the measured value of the processed product is closer to the target value

T, the mean time between failures (MTBF) of the product is longer [9]. According to Kethley [10], the Taguchi loss function is expressed as follows:

$$L(X) = k(X - T)^2, \tag{1}$$

where k is the multiplier and T is the target value. In order to not lose generality, it was assumed that the tolerance of the product quality characteristics is $T \pm d$. For individual processed products, as the size of the processed product is closer to the target value T, the process loss is lower. For the entire machining process, the expected process loss can be used as a tool through which to evaluate the process loss [8]. In a normal manufacturing process, the expected value of the Taguchi loss function with $k/d = 1$ is displayed as follows:

$$\theta_L = E[L(X)] = \delta^2 + \gamma^2, \tag{2}$$

where $\delta = (\mu - T)/d$ and $\gamma = \sigma/d$. It is clear that δ and γ are two important factors that affect the expected value of process loss. δ refers to the accuracy index, and γ refers to the precision index [8]. When the process mean μ is closer to the target value T, then the value of δ is closer to 0, and the process expected loss is lower. Similarly, when the value of process variance σ^2 is smaller, then the value of γ is also smaller, and the process expected loss is lower further still. In addition, the process yield of the product is also the function of δ and γ as per the following:

$$\begin{aligned} p &= P(T - d \leq X \leq T + d) \\ &= P\left(\frac{T-\mu-d}{\sigma} \leq Z \leq \frac{T-\mu+d}{\sigma}\right) \\ &= \Phi\left(\frac{1-\delta}{\gamma}\right) + \Phi\left(\frac{1+\delta}{\gamma}\right) - 1 \end{aligned} \tag{3}$$

Based on the abovementioned, as the process mean μ is closer to the target value T, the value of δ is closer to 0. Except for the lower process expected loss, the process yield is also higher. Likewise, when the value of process variance σ^2 is smaller, the process expected loss is lower, whereas the process yield is higher. Clearly, the decrease in the process expected loss can ensure an improvement in the process yield. Thus, the rates of rework and scrap for the processed product can be lowered in the manufacturing stage. In the meantime, *MTBF* can also be extended after the product is sold, thereby achieving the effect of energy saving and carbon reduction [11]. Based on the abovementioned, through aiming to assist all machining manufacturers, we examine whether the accuracy (δ) and precision (γ) of the process can meet the requirements of the Six Sigma quality level, whether improving the process allows it to reach the required level, and whether establishing a complete process loss evaluation and improvement model aids with this aim (which is an important issue).

Therefore, this paper uses the expected value of the Taguchi loss function as the evaluation index to develop a complete process loss evaluation and improvement model, which can not only assist machining manufacturers in enhancing process quality, but can also achieve the effect of energy saving and carbon reduction.

Since the process expected loss θ_L has unknown parameters, the misjudgment led by sampling error may take place if the evaluation index θ_L of the machining process loss is only evaluated by point estimation [12]. In addition, companies emphasize the mechanism of quick responses, and—given costs and timeliness—usually the sample size is not big. Accordingly, the sample size will make statistical tests vary, and then inconsistent decisions will be made [11,13–15]. In addition, many studies have pointed out that the confidence interval-based fuzzy testing model can incorporate past accumulated data with experts' experiences, so the precision of the test can be maintained in the state of small samples [10]. Furthermore, according to the studies mentioned in the above literature [8,9,13], fuzzy tests based on confidence intervals collect data with real numbers instead of fuzzy numbers, which are relatively simple and easy to collect. Therefore, in this paper, we propose a confidence interval-based fuzzy testing model for index θ_L.

The remainder of this paper is organized as follows. In Section 2, we derive a 100% confidence interval of δ and develop a confidence interval-based fuzzy test for this index. In Section 3, we derive a 100% confidence interval of the process expected loss and construct a confidence interval-based fuzzy test for the process expected loss. Next, in Section 4, an application example is used to illustrate the application of the fuzzy evaluation model of the machining process that is proposed by this study. Lastly, conclusions are made in Section 5.

2. Confidence Interval-Based Fuzzy Test of δ

According to Equation (2), the process expected loss is denoted as $\theta_L = \delta^2 + \gamma^2$, where δ and γ are two unknown parameters. As mentioned earlier, $\delta = (\mu - T)/d$ is the accuracy index. As noted by Kethley [11], accuracy can be leveled up by adjusting machine parameters; as such, the cost of improvement will be relatively low, and it will be relatively easy to succeed. Hence, this paper first develops a fuzzy test of the accuracy index δ, as well as decides whether to improve it. In the era of IE 4.0 smart manufacturing, the offset that can stably control the process is viewed as quantity, even if the value of the accuracy index δ is quite close to 0.

Let $X_1, X_2, \ldots, X_n$ be a random sample of random variable X; then, the maximum likelihood estimators of δ and γ are respectively expressed as follows:

$$\hat{\delta} = \frac{1}{n}\sum_{j=1}^{n}\left(\frac{X_j - T}{d}\right) = \frac{\overline{X} - T}{d} \tag{4}$$

and

$$\hat{\gamma} = \frac{\sqrt{\sum_{j=1}^{n}\left(X_j - \overline{X}\right)^2}}{\sqrt{n}d} = \frac{S}{d}, \tag{5}$$

where $S = \sqrt{n^{-1}\sum_{j=1}^{n}\left(X_j - \overline{X}\right)^2}$ is the sample standard deviation. Thus, the estimator of the process expected loss θ_L is denoted as follows:

$$\hat{\theta}_L = \frac{1}{n}\sum_{j=1}^{n}\left(\frac{X_j - T}{d}\right)^2 = \hat{\delta}^2 + \hat{\gamma}^2. \tag{6}$$

In the normal manufacturing process, the expected value of the estimator is derived as follows:

$$E\left(\hat{\theta}_L\right) = \frac{1}{n}\sum_{j=1}^{n} E\left(\left(\frac{X_j - T}{d}\right)^2\right) = \frac{1}{n}\sum_{j=1}^{n}\left\{Var\left(\frac{X_j - T}{d}\right) + \left(E\left(\frac{X_j - T}{d}\right)\right)^2\right\} = \delta^2 + \gamma^2. \tag{7}$$

It is clear that the expected value of the estimator $\hat{\theta}_L$ is equal to θ_L. Thus, $\hat{\theta}_L$ is an unbiased estimator of the process expected loss θ_L.

According to Chen et al. [16], when the process quality reaches the k-sigma quality level, then we have $|\delta| \leq 1.5/k$ and $\gamma \leq 1/k$. However, since the German government introduced Industry 4.0 in 2011, a number of companies have integrated information, communication technologies, and digital manufacturing technologies to promote a fully networked production environment of smart manufacturing [17,18]. Based on the studies of Liu et al. [19], with the development and rapid evolution of emerging technologies such as the Internet of Things (IoT) and Big Data analysis, the manufacturing industry has also integrated and applied related technologies to move toward the goal of smart manufacturing. In addition, according to the research conducted by Askr et al. [20], process offsets can be easily reduced by controlling optimal machine parameters. Clearly, we can reduce process losses by starting from lowering process offsets, as well as by striving to diminish the process variations to achieve the purpose of lowering process losses. In order to quickly control process shifts, process monitoring and adjustment must be carried out in the state of small samples and the non-mass production of defective sizes. This study first proposes a confidence interval-based fuzzy test of δ. As mentioned earlier, the machining industry has been moving toward smart manufacturing since Industry 4.0. As such, the process mean, which is required to deviate from the target value T, is quite small. Thus, the null hypothesis H_0 is $\mu = T$, and the alternative hypothesis H_1 is $\mu \neq T$, which is equivalent to

$$H_0 : \delta = 0; \tag{8}$$

$$H_1 : \delta \neq 0. \tag{9}$$

Let the random variable $T = \sqrt{n}(\hat{\delta} - \delta)/\hat{\gamma}$, then T is distributed as a t distribution with an $n-1$ degree of freedom, which is denoted as $T\sim t_{n-1}$. This study first adopts a confidence interval-based fuzzy test of δ. Next, this study then derives the $100 \times (1-\alpha)\%$ confidence limit of δ as follows:

$$\begin{aligned} & p\left(-t_{\alpha/2;n-1} \leq \frac{\sqrt{n}(\hat{\delta}-\delta)}{\hat{\gamma}} \leq t_{\alpha/2;n-1}\right) = 1-\alpha \\ \Rightarrow\ & p\left(\hat{\delta} - \frac{t_{\alpha/2;n-1}}{\sqrt{n}}\hat{\gamma} \leq \delta \leq \hat{\delta} + \frac{t_{\alpha/2;n-1}}{\sqrt{n}}\hat{\gamma}\right) = 1-\alpha. \end{aligned} \tag{10}$$

Therefore, $[L\delta,\ U\delta]$ is the $100 \times (1-\alpha)\%$ confidence interval of δ, as shown below:

$$[L\delta,\ U\delta] = \left[\hat{\delta} - \frac{t_{\alpha/2;n-1}}{\sqrt{n}}\hat{\gamma},\ \hat{\delta} + \frac{t_{\alpha/2;n-1}}{\sqrt{n}}\hat{\gamma}\right]. \tag{11}$$

As noted by some studies, the α-cuts of the triangular fuzzy number $\tilde{\hat{\delta}}$ is presented as follows:

$$\hat{\delta}[\alpha] = \left[\hat{\delta}_1(\alpha), \hat{\delta}_2(\alpha)\right] \text{ for } 0.01 \leq \alpha \leq 1. \tag{12}$$

When $0 \leq \alpha \leq 0.01$, then the set is $\hat{\delta}[\alpha] = \left[\hat{\delta}_1(0.01), \hat{\delta}_2(0.01)\right]$, where

$$\hat{\delta}_1(\alpha) = \hat{\delta} - \frac{t_{\alpha/2;n-1}}{\sqrt{n}}\hat{\gamma} \text{ and } \hat{\delta}_2(\alpha) = \hat{\delta} + \frac{t_{\alpha/2;n-1}}{\sqrt{n}}\hat{\gamma}. \tag{13}$$

It is clear that when $\alpha = 1$, then $\hat{\delta}_1(1) = \hat{\delta}_2(1) = \hat{\delta}$, and the triangular fuzzy number of $\tilde{\hat{\delta}}$ is $\tilde{\hat{\delta}} = (\hat{\delta}_L, \hat{\delta}_M, \hat{\delta}_R)$, where $\hat{\delta}_M = \hat{\delta}$,

$$\hat{\delta}_L = \hat{\delta} - \frac{t_{0.005;n-1}}{\sqrt{n}}\hat{\gamma} \text{ and } \hat{\delta}_R = \hat{\delta} + \frac{t_{0.005;n-1}}{\sqrt{n}}\hat{\gamma}. \tag{14}$$

Therefore, the membership function of fuzzy number $\tilde{\hat{\delta}}$ is

$$\Im(x) = \begin{cases} 0 & if\ x < \hat{\delta} - \frac{t_{0.005;n-1}}{\sqrt{n}}\hat{\gamma} \\ \beta_1 & if\ \hat{\delta} - \frac{t_{0.005;n-1}}{\sqrt{n}}\hat{\gamma} \leq x < \hat{\delta} \\ 1 & if\ x = \hat{\delta} \\ \beta_2 & if\ \hat{\delta} < x \leq \hat{\delta} + \frac{t_{0.005;n-1}}{\sqrt{n}}\hat{\gamma} \\ 0 & if\ \hat{\delta} + \frac{t_{0.005;n-1}}{\sqrt{n}}\hat{\gamma} < x \end{cases}, \tag{15}$$

where β_1 is determined by

$$x = \hat{\delta} - \frac{t_{\beta_1/2;n-1}}{\sqrt{n}}\hat{\gamma}; \tag{16}$$

and β_2 is determined by

$$x = \hat{\delta} + \frac{t_{\beta_2/2;n-1}}{\sqrt{n}}\hat{\gamma}. \tag{17}$$

Let set G_T be the area in the graph of membership function $\Im(x)$ as follows:

$$G_T = \{ (x,\alpha)|\hat{\delta}_1(\alpha) \le x \le \hat{\delta}_2(\alpha), 0 \le \alpha \le 1 \}. \quad (18)$$

In addition, let the set G_R be the area in the graph of membership function $\Im(x)$ but to the right of the vertical line x = 0, as depicted below:

$$G_R = \{ (x,\alpha)|0 \le x \le \hat{\delta}_2(\alpha), 0 \le \alpha \le a \}, \quad (19)$$

where $\alpha = a$ such that $\hat{\delta}_2(\alpha) = 0$. As noted by Chen et al. [21], let $\Delta_T = \hat{\delta}_R - \hat{\delta}_L$, then

$$\Delta_T = \hat{\delta}_R - \hat{\delta}_L = 2\frac{t_{0.005;n-1}}{\sqrt{n}}\hat{\gamma}. \quad (20)$$

Let $\Delta_R = \hat{\delta}_R - 0$, then

$$\Delta_R = \hat{\delta}_R - 0 = \hat{\delta} + \frac{t_{0.005;n-1}}{\sqrt{n}}\hat{\gamma}. \quad (21)$$

The membership function $\Im(x)$ relative to Δ_T and Δ_R is shown as follows (Figure 1):

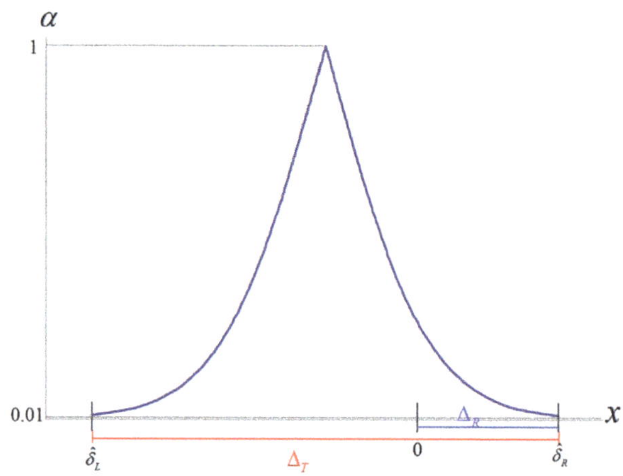

Figure 1. Membership function $\Im(x)$ relative to Δ_T and Δ_R.

According to Chen and Yu [22], let decision variable $D = \Delta_R/\Delta_T$ be expressed as follows:

$$D = \frac{\Delta_R}{\Delta_T} = \frac{\hat{\delta} + \frac{t_{0.005;n-1}}{\sqrt{n}}\hat{\gamma}}{2\frac{t_{0.005;n-1}}{\sqrt{n}}\hat{\gamma}}. \quad (22)$$

In this paper, we proposed fuzzy test rules and improvements in the measuring of the accuracy index δ based on the above decision variable D. Let the value of ϕ be between zero and 0.5, then the fuzzy test rules and improvement measures are made as follows:

1. When $D \le \phi$, then reject H_0 and conclude $\delta > 0$, thus indicating that the process mean μ is shifted to the right. Thus, the process must be adjusted, and the mean must be moved to the left to lift the accuracy of the process.
2. When $\phi \le D \le 1 - \phi$, then do not reject H_0 and conclude $\delta = 0$, which means that the process mean μ is not deviated from the target value T; as such, the process does not need to be adjusted.

3. When $D > 1 - \phi$, then reject H_0 and conclude $\delta < 0$, thus showing that the process mean μ is shifted to the left. Therefore, the process must be adjusted, and the mean must be moved to the left to increase the accuracy of the process.

According to the above fuzzy test rules, the machining industry can deviate the process mean μ from the target value T to a relatively small degree in order to cut down process losses.

3. Confidence Interval-Based Fuzzy Test of Process Expected Loss

As mentioned in the previous section, in the era of industrial smart manufacturing, the Internet of Things and Big Data analysis technology have gradually matured, and the process offset that can be stably controlled is seen as a quantity, even if the value of accuracy index δ is quite close to 0. When the value of accuracy index δ is close to 0, the process expected loss (θ_L) is the only remaining process variation that needs to be controlled. Therefore, this section develops the fuzzy test of the process expected loss based on this premise.

As mentioned before, the machining industry can make the process mean μ deviate from the target value T to a minimal degree in the environment of smart manufacturing. In addition, according to the abovementioned fuzzy test rules, the industry can tell whether the degree to which the process mean μ is shifted from the target value T is relatively small [9]. At the same time, through the mechanism of improvement and adjustment, the deviation of the process mean μ from the target value T can be minimized. Based on this condition, we proposed a confidence interval-based fuzzy test of the process expected loss that is investigated in this paper. As noted above, when the process quality reaches the k-sigma quality level, then the required value of γ is smaller than or equal to $1/k$. When the process mean μ deviates from the target value T to a relatively low degree, i.e., $\delta = 0$, then the required value of the process expected loss is smaller than or equal to $1/k^2$. Then, the null hypothesis and alternative hypothesis can be shown as below:

$$H_0' : \theta_L \le 1/k^2; \tag{23}$$

$$H_1' : \theta_L > 1/k^2. \tag{24}$$

Let a random variable be $K = n\hat{\theta}_L^2/\theta_L^2$ as follows:

$$K = \frac{1}{n}\sum_{j=1}^{n} Y_j^2/\theta_L^2, \tag{25}$$

where $Y_j = (X_j - T)/d$ is distributed as a normal distribution with a mean δ and standard deviation γ. Then, K is distributed as a chi-square distribution with an $n-1$ degree of freedom with $\delta = 0$, which is denoted as $K \sim \chi^2_{n-1}$. Similar to δ, this study first uses a confidence interval-based fuzzy testing model for the process expected loss θ_L. Then, this study derives the $100 \times (1-\alpha)\%$ confidence limits of the process expected loss θ_L as follows:

$$\begin{aligned} & p\left(\chi^2_{\alpha/2;n-1} \le \frac{1}{n\theta_L^2}\sum_{j=1}^{n} Y_j^2 \le \chi^2_{1-\alpha/2;n-1}\right) = 1-\alpha \\ \Rightarrow & p\left(\frac{\sqrt{\sum_{j=1}^{n} Y_j^2}}{\sqrt{n\chi^2_{1-\alpha/2;n-1}}} \le \theta_L \le \frac{\sqrt{\sum_{j=1}^{n} Y_j^2}}{\sqrt{n\chi^2_{\alpha/2;n-1}}}\right) = 1-\alpha. \end{aligned} \tag{26}$$

Therefore, $[L\theta_L,\ U\theta_L]$ represents the $100 \times (1-\alpha)\%$ confidence interval of the process expected loss θ_L as follows:

$$[L\theta_L,\ U\theta_L] = \left[\frac{\sqrt{\sum_{j=1}^{n} Y_j^2}}{\sqrt{n\chi^2_{1-\alpha/2;n-1}}}, \frac{\sqrt{\sum_{j=1}^{n} Y_j^2}}{\sqrt{n\chi^2_{\alpha/2;n-1}}}\right]. \tag{27}$$

As noted by some studies, the α-cuts of the triangular fuzzy number $\widetilde{\hat{\theta}}_L$ is illustrated as follows [10,13]:

$$\hat{\theta}_L[\alpha] = \left[\hat{\theta}_{L1}(\alpha), \hat{\theta}_{L2}(\alpha)\right] \text{ for } 0.01 \le \alpha \le 1. \tag{28}$$

When $0 \le \alpha \le 0.01$, then set $\hat{\theta}_L[\alpha] = \left[\hat{\theta}_{L1}(0.01), \hat{\theta}_{L2}(0.01)\right]$, where

$$\hat{\theta}_{L1}(\alpha) = \frac{\sqrt{\sum_{j=1}^{n} Y_j^2}}{\sqrt{n\chi^2_{1-\alpha/2;n-1}}} \text{ and } \hat{\theta}_{L2}(\alpha) = \frac{\sqrt{\sum_{j=1}^{n} Y_j^2}}{\sqrt{n\chi^2_{\alpha/2;n-1}}}. \tag{29}$$

Based on the above, the triangular fuzzy number of $\widetilde{\hat{\theta}}_L$ is $\widetilde{\hat{\theta}}_L = (L\theta, M\theta, R\theta)$, where

$$L\theta = \frac{\sqrt{\sum_{j=1}^{n} Y_j^2}}{\sqrt{n\chi^2_{0.995;n-1}}}; \tag{30}$$

$$M\theta = \frac{\sqrt{\sum_{j=1}^{n} Y_j^2}}{\sqrt{n\chi^2_{0.5;n-1}}}; \tag{31}$$

$$R\theta = \frac{\sqrt{\sum_{j=1}^{n} Y_j^2}}{\sqrt{n\chi^2_{0.05;n-1}}}. \tag{32}$$

Therefore, the membership function of fuzzy number $\widetilde{\hat{\delta}}$ is

$$\Re(y) = \begin{cases} 0 & if\ y < L\theta \\ \chi_1 & if\ L\theta \le y < M\theta \\ 1 & if\ y = M\theta \\ \chi_2 & if\ M\theta < x \le R\theta \\ 0 & if\ R\theta < x \end{cases}, \tag{33}$$

where χ_1 is determined by

$$y = \frac{\sqrt{\sum_{j=1}^{n} Y_j^2}}{\sqrt{n\chi^2_{1-\chi_1/2;n-1}}}, \tag{34}$$

and χ_2 is determined by

$$y = \frac{\sqrt{\sum_{j=1}^{n} Y_j^2}}{\sqrt{n\chi^2_{\chi_2/2;n-1}}}. \tag{35}$$

Let set G'_T be the area in the graph of membership function $\Re(y)$, as displayed below:

$$G'_T = \left\{ (y,\alpha) \middle| \hat{\theta}_{L1}(\alpha) \le y \le \hat{\theta}_{L2}(\alpha), 0 \le \alpha \le 1 \right\}. \tag{36}$$

In addition, let set G'_R be the area in the graph of membership function $\Re(y)$ but to the right of the vertical line y $1/k$, as expressed below:

$$G'_R = \left\{ (y,\alpha) \middle| \hat{\theta}_{L1}(\alpha) \le y \le 1/k^2, 0 \le \alpha \le b \right\}, \tag{37}$$

where $\alpha = b$ such that $\hat{\theta}_{L1}(\alpha) = 1/k^2$. As noted by Chen et al. [21], let $\Delta'_T = R\theta - L\theta$, then

$$\Delta'_T = R\theta - L\theta = \frac{\sqrt{\sum_{j=1}^{n} Y_j^2}}{\sqrt{n\chi^2_{0.005;n-1}}} - \frac{\sqrt{\sum_{j=1}^{n} Y_j^2}}{\sqrt{n\chi^2_{0.995;n-1}}}. \tag{38}$$

Let $\Delta'_R = 1/k^2 - L\theta$, then

$$\Delta'_R = \frac{1}{k^2} - \frac{\sqrt{\sum_{j=1}^{n} Y_j^2}}{\sqrt{n\chi^2_{0.995;n-1}}}. \tag{39}$$

The membership function $\Re(y)$ relative to Δ'_T and Δ'_R is shown as follows (Figure 2):

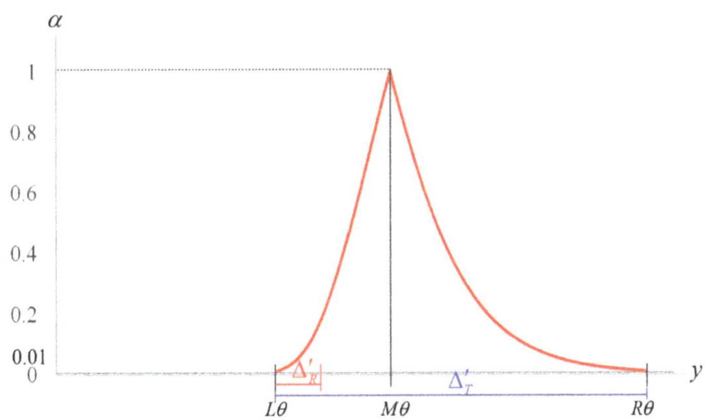

Figure 2. Membership function $\Re(y)$ relative to Δ'_T and Δ'_R.

According to Chen and Yu [22], let the decision variable be $D' = \Delta'_R/\Delta'_T$, which is depicted as follows:

$$D' = \frac{\Delta'_R}{\Delta'_T} = \frac{\frac{1}{k^2} - \frac{\sqrt{\sum_{j=1}^{n} Y_j^2}}{\sqrt{n\chi^2_{0.995;n-1}}}}{\frac{\sqrt{\sum_{j=1}^{n} Y_j^2}}{\sqrt{n\chi^2_{0.005;n-1}}} - \frac{\sqrt{\sum_{j=1}^{n} Y_j^2}}{\sqrt{n\chi^2_{0.995;n-1}}}}.$$

Then, in this study, test rules and improvement measures for the process expected loss θ_L were proposed based on the above decision variable D'. Let the value of ϕ fall between zero and 0.5. Then, the fuzzy test rules and improvement measures are to be made as follows:

1. When $D' \leq \phi$, then reject H_0 and conclude that the process expected loss does not meet the required level ($\theta_L > 1/k^2$), thus indicating that process variation must be reduced to cut down process losses.
2. When $D' > \phi$, then do not reject H_0 and conclude that the process expected loss meets the required level ($\theta_L \leq 1/k^2$), thus showing that the process does not require any improvement or adjustment.

According to the above two fuzzy test rules, the machining industry can make the process mean μ deviate from the target value T to a relatively low degree. Meanwhile, the process variation can be lowered to reduce the process loss.

4. Practical Application

As mentioned earlier, Taiwan is ranked fifth in the world's export and seventh in global production of machinery and machine tools. For the machining industry, boosting the process quality of processed products can decrease the rework and scrap rates of processed products in the stage of production [23]. In addition, according to the concept of the Taguchi loss function, the mean time between failures of the product is longer as the measured value of the processed product is closer to the target value T. Since central Taiwan is a stronghold of the machinery and machine tool industry, it has not only the manufacturers of machinery and machine tools, but also many component processing factories in its surrounding area. In order to diminish the process losses of these components, this study takes the inner diameter of a gear processed by a machining factory in central Taiwan as an example through which to demonstrate the application of the two fuzzy testing models proposed in this paper. First, this study uses the confidence interval-based fuzzy test proposed in Section 2 to test and monitor the process mean so as to ensure that the quantity of the process mean deviating from the target value T is relatively small. Next, when the process mean is shifted from the target value T to a relatively low degree, then the confidence interval-based fuzzy test of the process expected loss proposed in Section 3 is employed to test and evaluate whether the process expected loss can meet the required level.

4.1. Fuzzy Test of δ

When aiming to minimize the process expected loss, engineers require that the deviation of the mean μ from the target value T must be small in the gear inner diameter machining process. Thus, the null hypothesis and alternative hypothesis H_1 are defined as below:

$$H_0 : \delta = 0;$$

$$H_1 : \delta \neq 0.$$

In order to pursue a mechanism of rapid responses, the Taiwanese industry usually takes 15 or 16 samples from the production line for measurement and sampling. In order to perform the above fuzzy test, 16 samples were randomly selected from the processed product of a certain type of gear. The data of these 16 samples are listed as follows:

$x_1 = 3.508$,	$x_2 = 3.506$,	$x_3 = 3.533$,	$x_4 = 3.506$,
$x_5 = 3.499$,	$x_6 = 3.467$,	$x_7 = 3.500$,	$x_8 = 3.515$
$x_9 = 3.477$,	$x_{10} = 3.501$,	$x_{11} = 3.505$,	$x_{12} = 3.516$,
$x_{13} = 3.478$,	$x_{14} = 3.500$,	$x_{15} = 3.490$,	$x_{16} = 3.512$

The tolerance of the inner diameter for this type of gear was found to be 3.5 ± 0.05, that is, target $T = 3.5$ and $d = 0.05$. Then, the maximum likelihood estimators of δ and γ are respectively displayed as follows:

$$\hat{\delta} = \frac{\overline{X} - T}{d} = \frac{3.501 - 3.5}{0.05} = 0.02$$

and

$$\hat{\gamma} = \frac{S}{d} = \frac{0.017}{0.05} = 0.34.$$

Therefore,

$$\hat{\delta}_L = \hat{\delta} - \frac{t_{0.005;15}}{\sqrt{16}}\hat{\gamma} = 0.02 - \frac{2.947}{4} \times 0.34 = 0.233,$$

$$\hat{\delta}_R = \hat{\delta} + \frac{t_{0.005;15}}{\sqrt{16}}\hat{\gamma} = 0.02 + \frac{2.947}{4} \times 0.34 = 0.273,$$

and the triangular fuzzy number is presented as $\widetilde{\delta} = (-0.233, 0.02, 0.273)$. According to Equations (20) and (21), Δ_R and Δ_T are calculated as follows:

$$\Delta_R = \hat{\delta}_R - 0 = 0.273;$$

$$\Delta_T = \hat{\delta}_R - \hat{\delta}_L = 0.273 - (-0.233) = 0.506.$$

Thus, $D = \Delta_R/\Delta_T = 0.273/0.506 = 0.540$. Let $\phi = 0.2$, and, according to the fuzzy test rule (2), when $0.2 \leq \mathrm{D} \leq 0.8$, then do not reject H_0 and conclude $\delta = 0$. This represents that the process mean μ does not deviate from the target value T; as such, the process does not need to make any adjustment.

4.2. Fuzzy Test of Process Expected Loss

According to the abovementioned fuzzy test results, it was revealed that the process mean μ that shifted from the target value T was extremely small, i.e., $\delta = 0$. As noted above, when the process quality reaches the Six Sigma quality level, then the required value of γ is smaller than or equal to $1/6$. Thus, the required value of the process expected loss θ_L is smaller than or equal to $1/36$. As such, the null hypothesis and alternative hypothesis are shown below:

$$H_0' : \theta_L \leq 1/36;$$

$$H_1' : \theta_L > 1/36.$$

In order to pursue a mechanism of rapid responses, the Taiwanese industry usually takes 15 or 16 samples from the production line for measurement and sampling. Let $Y_j = (X_j - 3.5)/0.05$, then 16 samples, after variable transformation is applied, are listed as follows:

$y_1 = 0.154$	$y_2 = 0.129$	$y_3 = 0.656$	$y_4 = 0.127$
$y_5 = -0.018$	$y_6 = -0.664$	$y_7 = 0.008$	$y_8 = 0.299$
$y_9 = -0.460$	$y_{10} = 0.017,$	$y_{11} = 0.106$	$y_{12} = 0.322$
$y_{13} = -0.443$	$y_{14} = -0.007$	$y_{15} = -0.209$	$y_{16} = 0.246$

Therefore, the sum of the squares of these 16 samples is denoted as $\sum_{j=1}^{16} y_j^2 = 1.646$, and the values of $L\theta$, $M\theta$, and $R\theta$ are calculated as follows:

$$L\theta = \frac{\sqrt{\sum_{j=1}^{16} y_j^2}}{\sqrt{16 \times \chi^2_{0.995;15}}} = \frac{\sqrt{1.646}}{\sqrt{16 \times 32.801}} = 0.003;$$

$$M\theta = \frac{\sqrt{\sum_{j=1}^{16} y_j^2}}{\sqrt{16 \times \chi^2_{0.5;15}}} = \frac{\sqrt{1.646}}{\sqrt{16 \times 14.339}} = 0.085;$$

$$R\theta = \frac{\sqrt{\sum_{j=1}^{16} y_j^2}}{\sqrt{16 \times \chi^2_{0.005;15}}} = \frac{\sqrt{1.646}}{\sqrt{16 \times 4.601}} = 0.150.$$

Based on the above, the triangular fuzzy number of $\widetilde{\hat{\theta}}_L$ is denoted as $\widetilde{\hat{\theta}}_L = (0.003, 0.085,$ and $0.150)$. According to Equations (38) and (39), Δ_R' and Δ_T' are calculated as follows:

$$\Delta_R' = 1/36 - L\theta = 0.028 - 0.003 = 0.025;$$

$$\Delta_T' = R\theta - L\theta = 0.150 - 0.003 = 0.147.$$

Thus, $D' = \Delta_R'/\Delta_T' = 0.025/0.147 = 0.170$. Let $\phi = 0.2$, and, according to the fuzzy test rule (1) of the process expected loss, when $D' \leq 0.2$, then reject H_0 and conclude that the process expected loss does not meet the required level ($\theta_L > 1/36$). This means that the process variation must be dwindled so as to lessen the process loss.

Based on the above evaluation results, the process engineers carried out improvement measures to lower the process variation. After collecting the improved 16 variables, the transformed sample data were written as follows:

$y'_1 = 0.114$	$y'_2 = -0.027$	$y'_3 = -0.168$	$y'_4 = -0.175$
$y'_5 = -0.064$	$y'_6 = 0.358$	$y'_7 = -0.246$	$y'_8 = 0.275$
$y'_9 = -0.299$	$y'_{10} = 0.046$	$y'_{11} = 0.021$	$y'_{12} = -0.152$
$y'_{13} = -0.109$	$y'_{14} = 0.130$	$y'_{15} = 0.156$	$y'_{16} = 0.368$

Therefore, the sum of the squares of these 16 improved pieces of sample data was denoted as $\sum_{j=1}^{16} y_j'^2 = 0.644$, and the values of $L\theta'$, $M\theta'$, and $R\theta'$ were computed as follows:

$$L\theta' = \frac{\sqrt{\sum_{j=1}^{16} y_j'^2}}{\sqrt{16\chi^2_{0.995;15}}} = \frac{\sqrt{0.644}}{\sqrt{16 \times 32.801}} = 0.004;$$

$$M\theta' = \frac{\sqrt{\sum_{j=1}^{16} y_j'^2}}{\sqrt{16 \times \chi^2_{0.5;15}}} = \frac{\sqrt{0.644}}{\sqrt{16 \times 14.339}} = 0.053;$$

$$R\theta' = \frac{\sqrt{\sum_{j=1}^{16} y_j'^2}}{\sqrt{16 \times \chi^2_{0.005;15}}} = \frac{\sqrt{0.644}}{\sqrt{16 \times 4.601}} = 0.094.$$

Based on the above, the triangular fuzzy number of $\tilde{\hat{\theta}}_L$ was expressed as $\tilde{\hat{\theta}}_L$ = (0.004, 0.053, and 0.094). According to Equations (38) and (39), Δ'_R and Δ'_T were calculated as follows:

$$\Delta'_R = 1/36 - L\theta = 0.028 - 0.004 = 0.024$$

$$\Delta'_T = R\theta - L\theta = 0.094 - 0.004 = 0.090$$

Thus, $D' = \Delta'_R/\Delta'_T = 0.024/0.090 = 0.267$. According to the fuzzy test rule (2) of the process expected loss with $\phi = 0.2$, when $D' > 0.2$, then do not reject H_0 and conclude that the process expected loss meets the required level ($\theta_L \leq 1/36$). This indicates the process improvement has a remarkable effect.

5. Conclusions

Governments and enterprises must take into account economic growth, as well as the natural environment, when dealing with the issue of global warming [2,3]. Under this thinking, we must shoulder the responsibility for the social and ecological environment; as such, corporate social responsibility (CSR) has now become a common business philosophy urged on by the state of the world [5]. The expected value of the Taguchi loss function is $\delta^2 + \gamma^2$, and δ and γ are two key factors that affect the expected values of process losses. In addition, the product process yield is a function of δ and γ [9]. As the process mean μ is closer to the target value T (the value of δ is closer to 0), or as the value of the process variation is smaller (the value of γ is smaller), then the process expected loss is lower, whereas the process yield is higher. It is clear that decreasing the process expected loss can ensure an increase in process yield. In addition to a decrease in the rework and scrap of the processed product in the stage of production, *MTBF* can also be extended after the product is sold, thereby achieving the effect of energy saving and carbon reduction [9]. Therefore, in this paper, the expected value of the Taguchi loss function was adopted as an evaluation index of the machining process loss. Since the process expected loss θ_L contained unknown parameters, the misjudgment caused by sampling errors may be incurred if only the point estimates are used to assess the evaluation index θ_L of the machining process loss [16]. Furthermore, enterprises emphasize the mechanism of quick responses, and—given costs and timeliness—the sample size is usually not large. Consequently, statistical tests will vary due to the sample size, such that inconsistent decisions will be generated. As a result, this study first derived a $100 \times (1-\alpha)\%$ confidence interval of δ, as well as developed a confidence interval-based fuzzy test for this index [13]. Next, this study derived a 100% confidence interval of the process expected loss, as well as established a confidence interval-based fuzzy test for the process expected loss. Numerous studies have suggested that the confidence interval-based fuzzy testing model can integrate past accumulated data with experts' experiences, such that the accuracy of the test can be maintained in the case of small samples [11,13]. It is clear that the model proposed in this paper has the following advantages:

1. The expected value of the Taguchi loss function is used as an evaluation index, in which the accuracy index and the precision index can help the machining industry find the correct direction for improvement.
2. In addition to reflecting the process expected loss, this index can also reflect the process yield.
3. Since the fuzzy evaluation model based on the confidence interval can integrate experts' experiences with past data, the evaluation accuracy can still be maintained in small samples in order to meet the requirements of enterprises for the purpose of rapid responses.
4. Apart from assisting machining manufacturers in boosting the quality of the machining process, the model can also effect energy saving and carbon reduction at the same time, such that machining manufacturers can reach their goals of fulfilling their corporate social responsibilities.

Overall, the evaluation model of the process loss built in this paper can help machining manufacturers review and enhance their own machining process capabilities through accuracy index δ and the fuzzy test of the process expected loss θ_L, thus achieving the effects of cost reduction, energy saving, and carbon reduction. The benefits of cost reduction and energy saving, as well as carbon reduction, are concepts; as such, they can be further explored in future research. In addition, an evaluation model could be established for the benefits of cost and carbon reduction so as to improve process capabilities.

Author Contributions: Conceptualization, K.-S.C., J.-S.L. and C.-M.Y. (Chun-Min Yu); methodology, K.-S.C. and J.-S.L.; software, T.-H.H.; validation, C.-M.Y. (Chun-Min Yu) and C.-M.Y. (Chun-Ming Yang); formal analysis, K.-S.C., J.-S.L. and C.-M.Y. (Chun-Min Yu); resources, J.-S.L.; data curation, T.-H.H.; writing—original draft preparation, K.-S.C., T.-H.H., J.-S.L., C.-M.Y. (Chun-Min Yu) and C.-M.Y. (Chun-Ming Yang); writing—review and editing, K.-S.C., J.-S.L. and C.-M.Y. (Chun-Min Yu); visualization, C.-M.Y. (Chun-Min Yu); supervision, K.-S.C.; project administration, J.-S.L. All authors have read and agreed to the published version of the manuscript.

Funding: This research received no external funding.

Data Availability Statement: All data generated or analyzed during this study are included in the article.

Conflicts of Interest: The authors declare no conflict of interest.

References

1. Niklewicz-Pijaczyńska, M.; Stańczyk, E.; Gardocka-Jałowiec, A.; Gródek-Szostak, Z.; Niemczyk, A.; Szalonka, K.; Homa, M. A strategy for planned product aging in view of sustainable development challenges. *Energies* **2021**, *14*, 7793. [CrossRef]
2. He, Q.R.; Chen, P.K. Developing a green supplier evaluation system for the chinese semiconductor manufacturing industry based on supplier willingness. *Oper. Manag. Res.* **2022**, *16*, 227–244. [CrossRef]
3. Simmou, W.; Govindan, K.; Sameer, I.; Hussainey, K.; Simmou, S. Doing good to be green and live clean!—Linking corporate social responsibility strategy, green innovation, and environmental performance: Evidence from maldivian and moroccan small and medium-sized enterprises. *J. Clean. Prod.* **2023**, *384*, 135265. [CrossRef]

4. Yang, Y.; Jiang, Y. Buyer-supplier CSR alignment and firm performance: A contingency theory perspective. *J. Bus. Res* **2023**, *154*, 113340. [CrossRef]
5. Sütoová, A.; Kóča, F. Corporate social responsibility standards: Is it possible to meet diverse customer requirements? *Manag. Syst. Prod. Eng* **2022**, *30*, 54–61. [CrossRef]
6. Yu, D.; Chen, Y. Dynamic structure and knowledge diffusion trajectory research in green supply chain. *J. Intell. Fuzzy Syst.* **2021**, *40*, 4979–4991. [CrossRef]
7. Chien, C.F.; Hong, T.Y.; Guo, H.Z. An empirical study for smart production for TFT-LCD to empower Industry 3.5. *J. Chin. Inst. Eng.* **2017**, *40*, 552–561. [CrossRef]
8. Wu, M.F.; Chen, H.Y.; Chang, T.C.; Wu, C.F. Quality evaluation of internal cylindrical grinding process with multiple quality characteristics for gear products. *Int. J. Prod. Res.* **2019**, *57*, 6687–6701. [CrossRef]
9. Chen, H.Y.; Lin, K.P. Fuzzy supplier selection model based on lifetime performance index. *Expert Syst. Appl.* **2022**, *208*, 118135. [CrossRef]
10. Kethley, R. Using taguchi loss functions to develop a single objective function in a multi-criteria context: A scheduling example. *Int. J. Inf. Manag. Sci.* **2008**, *19*, 589.
11. Cheng, S.W. Practical implementation of the process capability indices. *Qual. Eng* **1994**, *7*, 239–259. [CrossRef]
12. Xu, Y.; Zhang, X.; Meng, P. A novel intelligent deep learning-based uncertainty-guided network training in market price. *IEEE Trans. Ind. Inform.* **2022**, *18*, 5705–5711. [CrossRef]
13. Chiou, K.C. Building Up of Fuzzy Evaluation Model of Life Performance Based on Type-II Censored Data. *Mathematics* **2023**, *11*, 3686. [CrossRef]
14. Blanco-Mesa, F.; Merigó, J.M.; Gil-Lafuente, A.M. Fuzzy decision making: A bibliometric-based review. *J. Intell. Fuzzy Syst.* **2017**, *32*, 2033–2050. [CrossRef]
15. Yu, D.; Chen, Y.; Xu, Z. The longitudinal research of type-2 fuzzy sets domain: From conceptual structure and knowledge diffusion perspectives. *Inf. Sci.* **2021**, *568*, 317–332. [CrossRef]
16. Chen, K.S.; Huang, C.T.; Chang, T.C. Decision-making for the selection of suppliers based on the process quality assessment. *Int. J. Reliab. Qual. Saf. Eng.* **2020**, *27*, 2050016. [CrossRef]
17. Xu, X.; Lu, Y.; Vogel-Heuser, B.; Wang, L. Industry 4.0 and Industry 5.0—Inception, conception and perception. *J. Manuf. Syst.* **2021**, *61*, 530–535. [CrossRef]
18. Bagherian, A.; Gershon, M.; Kumar, S.; Manoj Kumar, M. Analyzing the relationship between digitalization and energy sustainability: A comprehensive ISM-MICMAC and DEMATEL approach. *Expert Syst. Appl.* **2024**, *236*, 121193. [CrossRef]
19. Liu, J.; Ma, C.; Gui, H.; Wang, S. A four-terminal-architecture cloud-edge-based digital twin system for thermal error control of key machining equipment in production lines. *Mech. Syst. Signal Proc.* **2022**, *166*, 108488. [CrossRef]
20. Askr, H.; Abdel-Salam, M.; Hassanien, A.E. Copula entropy-based golden jackal optimization algorithm for high-dimensional feature selection problems. *Expert Syst. Appl.* **2024**, *238*, 121582. [CrossRef]
21. Chen, K.S.; Yu, C.M.; Tsaur, R.C. Fuzzy Evaluation Model for Lifetime Performance of Electronic Products with Redundant Backup Systems. *Ann. Oper. Res.* **2023**. [CrossRef]
22. Chen, K.S.; Yu, C.M. Confidence-Interval-Based Fuzzy Supplier Selection Model with Lifetime Performance Index. *Ann. Oper. Res.* **2023**. [CrossRef]
23. Nguyen, T.T.; Duong, Q.D.; Mia, M. Multi-response optimization of the actively driven rotary turning for energy efficiency, carbon emissions, and machining quality. *Proc. Inst. Mech. Eng. Part B-J. Eng. Manuf.* **2021**, *235*, 2155–2173. [CrossRef]

mathematics

Article

Frank Prioritized Aggregation Operators and WASPAS Method Based on Complex Intuitionistic Fuzzy Sets and Their Application in Multi-Attribute Decision-Making

Zeeshan Ali [1,2], Tahir Mahmood [1] and Miin-Shen Yang [3,*]

1 Department of Mathematics and Statistics, International Islamic University Islamabad, Islamabad 44000, Pakistan; tahirbakhat@iiu.edu.pk (T.M.)
2 KERMIT, Department of Mathematical Modeling, Statistics and Bioinformatics, Ghent University, Coupure Links 653, Ghent 9000, Belgium
3 Department of Applied Mathematics, Chung Yuan Christian University, Chung-Li, Taoyuan 32023, Taiwan
* Correspondence: msyang@math.cycu.edu.tw

Abstract: Complex intuitionistic fuzzy (CIF) information covers the degree of membership and the degree of non-membership in the form of polar coordinates with a valuable and dominant characteristic where the sum of the real parts (the same rule for the imaginary parts) of the pair must be contained in the unit interval. In this paper, we first derive the Frank operational laws for CIF information and then examine the prioritized aggregation operators based on Frank operational laws for managing the theory of CIF information. These are the CIF Frank prioritized averaging (CIFFPA) operator, the CIF Frank prioritized ordered averaging (CIFFPOA) operator, the CIF Frank prioritized geometric (CIFFPG) operator, and the CIF Frank prioritized ordered geometric (CIFFPOG) operator with properties of idempotency, monotonicity, and boundedness. Furthermore, we derive the WASPAS (weighted aggregates sum product assessment) under the consideration or presence of the CIF information and try to justify it with the help of a suitable example. Additionally, we illustrate some numerical examples in the presence of the MADM (multi-attribute decision-making) procedures for evaluating the comparison between the proposed operators with some well-known existing operators to show the validity and worth of the proposed approaches.

Keywords: fuzzy sets; intuitionistic fuzzy sets; complex intuitionistic fuzzy sets; frank prioritized aggregation operators; WASPAS techniques; multi-attribute decision-making

MSC: 03B52; 03E72; 47S40; 90C70

Citation: Ali, Z.; Mahmood, T.; Yang, M.-S. Frank Prioritized Aggregation Operators and WASPAS Method Based on Complex Intuitionistic Fuzzy Sets and Their Application in Multi-Attribute Decision-Making. *Mathematics* **2023**, *11*, 2058. https://doi.org/10.3390/math11092058

Academic Editor: Aleksandar Aleksic

Received: 30 March 2023
Revised: 24 April 2023
Accepted: 25 April 2023
Published: 26 April 2023

1. Introduction

Multi-attribute decision-making (MADM) procedures are some of the finest or best techniques for evaluating the valuable and dominant preference from the set of feasible ones under the consideration of the available data. Traditionally, the MADM problem is a part of the decision-making procedure which often needs experts to provide evaluation data about the attributes and the alternatives with fuzzy sets (FSs) [1] in which FSs have been applied in different fields [2–4]. Various attempts have been derived by the distinct individuals in proceeding the data values using different extensions such as hesitant soft fuzzy rough sets [5], and fuzzy Mandelbrot sets [6]. Furthermore, intuitionistic FSs (IFSs) are also one of the most valuable and dominant extensions of FSs which was performed by Atanassov [7]. IFSs cover the degree of membership and the degree of non-membership of a given element to the set of discourse with the characteristic in which the sum of the pair must be contained in the unit interval. FSs are the particular cases of IFSs if we remove the degree of non-membership with its applications [8–10]. Furthermore, the utilization of the second term in the grade of truth is very awkward, and so, in many situations, we may face a problem with two-dimensional information, where FSs and IFSs deal only with one-dimensional information. Therefore, Ramot, et al. [11] successfully utilized the second term in the grade of truth and gave their name in the form of complex FS (CFS), where the truth grade in CFS is computed in the form of complex numbers whose real and unreal (imaginary) parts are covered in the unit interval. Various attempts have been performed by various individuals using systems such as the Mamdani complex fuzzy inference system [12]. Additionally, Alkouri and Salleh [13] exposed the new theory of complex IFS (CIFS) with its applications [14], which is the modified version of the three different types of ideas such as FSs, IFSs, and CFSs.

Frank t-norm and t-conorm are used for computing any type of aggregation operators (AOs) which was derived by Frank [15] in 1979. Frank norms have a lot of benefits because the simple algebraic and Lukasiewicz's t-norm and t-conorm [16] are the special cases of Frank t-norm and t-conorm. Furthermore, the idea of prioritizing AOs for the first time was given by Yager [17], and then Yu and Xu [18] who considered prioritized intuitionistic fuzzy AOs. These AOs were computed based on algebraic operational laws. Moreover, the main idea of the weighted aggregated sum product assessment (WASPAS) technique was given by Zavadskas, et al. [19,20] with its applications [21,22], which is the generalization of two different techniques such as weighted sum assessment (WSA) and weighted product assessment (WPS). The WASPAS technique is very strong and valuable because this is the modified version of many techniques and many individuals have utilized it in numerous fields such as the computer sciences, pure mathematics, engineering sciences, artificial intelligence, and decision-making.

The theory of FSs, IFSs, CFSs, and CIFSs has gained a lot of attention from different fuzzy researchers because these structures are very beneficial and valuable for depicting awkward and unreliable information very easily. Various attempts have been derived by distinct individuals in proceeding with the data values using different extensions such as AOs for IFSs [23] and geometric AOs for IFSs [24]. Furthermore, the Frank power AOs based on IFSs [25] are also a combination of the Frank and power AOs which is a very awkward and complicated task. The complex fuzzy credibility of Frank AOs was derived by Yahya, et al. [26]. Under the consideration of hesitant fuzzy information, the theory of Frank AOs was invented by Qin, et al. [27]. In the presence of the dual hesitant set theory, the major theory of Frank AOs was evaluated by Tang, et al. [28]. The prioritized AOs for trapezoidal IFS were derived by Ye [29], and the simple prioritized AOs for IFS were evaluated by Yu and Xu [18]. Ali, et al. [30] derived the idea of prioritized AOs for CIF soft information with their application in decision-making procedures. Yu [31] examined the theory of generalized prioritized AOs for intuitionistic fuzzy environments, and Lin, et al. [32] derived the fuzzy number intuitionistic fuzzy prioritized AOs and their application in decision-making procedures. Furthermore, Garg and Rani [33] exposed the averaging operators for CIFSs. Garg and Rani [34] evaluated the geometric operators for CIFSs, and Mahmood, et al. [35] examined the Aczel–Alsina AOs for CIFSs. Sarfraz, et al. [36] examined the prioritized AOs for IFSs with IF-prioritized Aczel–Alsina averaging. Poryazov, et al. [37] applied AOs for IFSs to the estimation of service compositions in telecommunication systems. Dai [38] derived linguistic complex fuzzy sets with their properties.

Frank and prioritized AOs based on IFSs were derived by different researchers, however, the theory of Frank and prioritized AOs based on CIFSs has not yet been evaluated by researchers in the literature. The investigation of Frank and prioritized AOs based on CIFSs is a very challenging task. In this analysis, we have accepted this task and not only derive the theory of Frank and prioritized AOs based on CIFSs but also derive the combination of Frank and prioritized AOs based on CIFSs, where the simple Frank and prioritized AOs are the special case of the derived theory. Furthermore, we also invent the theory of WASPAS for CIFSs. Inspired by the above discussion, the major investigations of this analysis are listed below:

1. To discover Frank operational laws for managing the theory of CIF information;
2. To derive the CIF Frank prioritized averaging (CIFFPA) operator, the CIF Frank prioritized ordered averaging (CIFFPOA) operator, the CIF Frank prioritized geometric (CIFFPG) operator, and the CIF Frank prioritized ordered geometric (CIFFPOG) operator with their properties;
3. To expose the idea of the weighted aggregates sum product assessment (WASPAS) procedure under the consideration or presence of the CIF information and try to simplify it with the help of a suitable example;
4. To demonstrate an example in the presence of the MADM procedures for evaluating the comparison between the proposed operators with some well-known existing operators to show the validity and worth of the discovered approaches.

This article is arranged in the form as follows: in Section 2, we review the different types of norms, CIFS, and the WASPAS technique; in Section 3, we examine Frank operational laws, CIFFPA operator, CIFFPOA operator, CIFFPG operator, and CIFFPOG operator, and their properties of idempotency, monotonicity, and boundedness; in Section 4, we derive the WASPAS for CIF information and try to justify it with the help of a suitable example; and in Section 5, we illustrate some examples in the presence of the MADM procedures for evaluating CIF information. Furthermore, the comparisons between the proposed operators and some well-known existing operators, such as Xu [23], Xu and Yager [24], Yahya, et al. [26], Yu [31], Lin, et al. [32], Garg and Rani [33], Garg and Rani [34], and Mahmood, et al. [35], are used to show the validity and worth of the discovered approaches and are discussed in Section 6. The final concluding information is shown in Section 7.

2. Preliminaries

In this section, we describe the prevailing theory of Frank norms, algebraic norms, and Lukasiewicz's norms for positive numbers. Furthermore, we also explain the idea of the WASPAS method [19,20] for classical set theory. Moreover, the idea of CIFSs and their related work are also a part of this study. For a clear presentation, the meaning of the symbols used in this paper is shown in Table 1.

Table 1. Meanings of different symbols used in the paper.

Symbols	Meanings	Symbols	Meanings	Symbols	Meanings
$u_{\daleth}^{rp}(x)$	Real part of membership grade	$\upsilon_{\daleth}^{rp}(x)$	Real part of the non-membership grade	X	Universal set
$u_{\daleth}^{ip}(x)$	Imaginary part of membership grade	$\upsilon_{\daleth}^{ip}(x)$	Imaginary part of the non-membership grade	x	Element of the universal set
$r^{rp}(x)$	Real part of the refusal grade	$r^{ip}(x)$	Imaginary part of refusal grade	$r(x)$	Refusal grade
$\overline{\overline{\mathbf{I}\mathfrak{T}}}$	Complex intuitionistic fuzzy set	$\overline{\overline{I\mathfrak{T}}}_{\mathfrak{H}}$	Complex intuitionistic fuzzy value	$V_s\left(\overline{\overline{I\mathfrak{T}}}_{\mathfrak{H}}\right)$	Score value
$V_a(\overline{\overline{\mathbf{I}\mathfrak{T}}}_{\mathfrak{H}})$	Accuracy value	$W_{\mathcal{B}}$	Weighted vector	$^{\circ}\mathrm{F} \geq 0$	Scaler
$\mathfrak{Y}(\mho, \mho^*)$	t-norm	$\mathfrak{Y}^*(\mho, \mho^*)$	t-conorm	$\daleth\daleth \in (1, +\infty)$	Scaler

2.1. WASPAS Method for Classical Set Theory

The major influence of this section is to recall the theory of the WASPAS procedure for classical information. The main procedure of the WASPAS method contains various valuable and dominant steps. Before evaluating the normalization, we arrange a collection of classical data which may be of a benefit type or cost type. If the data are of a benefit type, then good, otherwise, using the below theory, we normalize the information, such as:

$$\mathsf{C}'_{\mathfrak{H}B} = \begin{cases} \frac{\tilde{\mathsf{C}}_{\mathfrak{H}B}}{\max_{\mathfrak{H}} \tilde{\mathsf{C}}_{\mathfrak{H}B}} & for\ benefit \\ \frac{\min_{\mathfrak{H}} \tilde{\mathsf{C}}_{\mathfrak{H}B}}{\tilde{\mathsf{C}}_{\mathfrak{H}B}} & for\ cost. \end{cases} \tag{1}$$

After performing the above evaluation, we calculate the WSA and WPA, such as:

$$T_{\mathfrak{H}}^{WSA} = \sum_{B=1}^{d} W_B \tilde{\mathsf{C}}'_{\mathfrak{H}B}; \tag{2}$$

$$T_{\mathfrak{H}}^{WPA} = \sum_{B=1}^{d} (\tilde{\mathsf{C}}'_{\mathfrak{H}B})^{W_B}. \tag{3}$$

Using the data in Equations (2) and (3), we calculate the aggregated measure based on convex theory, such as:

$$T_{\mathfrak{H}} = {}^{\circ}\mathrm{F} T_{\mathfrak{H}}^{WSA} + (1 - {}^{\circ}\mathrm{F}) T_{\mathfrak{H}}^{WPA}, {}^{\circ}\mathrm{F} \in [0, 1]. \tag{4}$$

Before ranking the alternatives, we discuss the special cases of the WASPAS technique such as: When $^{\circ}\mathrm{F} = 1$, we obtain the data in Equation (2):

1. When $^{\circ}\mathrm{F} = 1$, we obtain the data in Equation (2);
2. When $^{\circ}\mathrm{F} = 0$, we obtain the data in Equation (3).

Finally, we derive the ranking result for examining the best one from the family of finite preferences.

2.2. Existing Ideas

Definition 1 ([15]). *For any two positive numbers $\mho$ and $\mho^*$, we have the theory of Frank t-norm and t-conorm, such that:*

$$\mathfrak{Y}(\mho, \mho^*) = \log_{\daleth\daleth}\left(1 + \frac{\left(\daleth\daleth^{\mho} - 1\right)\left(\daleth\daleth^{\mho^*} - 1\right)}{\daleth\daleth - 1}\right), \daleth\daleth \in (1, +\infty). \tag{5}$$

$$\mathfrak{Y}^*(\mho, \mho^*) = 1 - \log_{\daleth\daleth}\left(1 + \frac{\left(\daleth\daleth^{1-\mho} - 1\right)\left(\daleth\daleth^{1-\mho^*} - 1\right)}{\daleth\daleth - 1}\right), \daleth\daleth \in (1, +\infty) \tag{6}$$

Definition 2 ([1]). *For any two positive numbers $\mho$ and $\mho^*$, we have the theory of algebraic t-norm and t-conorm if we put the value of $\daleth\daleth \rightarrow 1$ in Equations (5) and (6), such that:*

$$\mathfrak{Y}(\mho, \mho^*) = \mho * \mho^*. \tag{7}$$

$$\mathfrak{Y}^*(\mho, \mho^*) = \mho + \mho^* - \mho * \mho^*. \tag{8}$$

Definition 3 ([16]). *For any two positive numbers $\mho$ and $\mho^*$, we have the theory of Lukasiewicz t-norm and t-conorm if we put the value of $\daleth\daleth \rightarrow +\infty$ in Equations (5) and (6), such that:*

$$\mathfrak{Y}(\mho,\mho^*) = \max\{0, \mho + \mho^* - 1\}. \tag{9}$$

$$\mathfrak{Y}^*(\mho,\mho^*) = \min\{\mho + \mho^*, 1\} \tag{10}$$

Definition 4 ([13]). *A numerical or mathematical equation:*

$$\overline{\overline{I\mathfrak{T}}} = \left\{\left(\left(u_{\gimel}^{rp}(x), u_{\gimel}^{ip}(x)\right), \left(v_{\gimel}^{rp}(x), v_{\gimel}^{ip}(x)\right)\right) : x \in X\right\} \tag{11}$$

Stated the CIFS with a truth grade $\left(u_{\gimel}^{rp}(x), u_{\gimel}^{ip}(x)\right)$ and falsity grade $\left(v_{\gimel}^{rp}(x), v_{\gimel}^{ip}(x)\right)$ must be implementing the following rules, such that $0 \leq u_{\gimel}^{rp}(x) + v_{\gimel}^{rp}(x) \leq 1$ and $0 \leq u_{\gimel}^{ip}(x) + v_{\gimel}^{ip}(x) \leq 1$. The notion of neutral grade is stated by: $r(x) = \left(r^{rp}(x), r^{ip}(x)\right)$ $= \left(1 - \left(u_{\gimel}^{rp}(x) + v_{\gimel}^{rp}(x)\right), 1 - \left(u_{\gimel}^{ip}(x) + v_{\gimel}^{ip}(x)\right)\right)$ and the representation of the CIF values (CIFVs) is the following: $\overline{\overline{I\mathfrak{T}}}_{\mathfrak{H}} = \left(\left(u_{\gimel_{\mathfrak{H}}}^{rp}, u_{\gimel_{\mathfrak{H}}}^{ip}\right), \left(v_{\gimel_{\mathfrak{H}}}^{rp}, v_{\gimel_{\mathfrak{H}}}^{ip}\right)\right), \mathfrak{H} = 1, 2, \ldots, \gimel$. As noted in the presence of the above information, we recall the idea of score and accuracy function, such as:

$$V_s\left(\overline{\overline{I\mathfrak{T}}}_{\mathfrak{H}}\right) = \frac{1}{2}\left(u_{\gimel_{\mathfrak{H}}}^{rp} - v_{\gimel_{\mathfrak{H}}}^{rp} + u_{\gimel_{\mathfrak{H}}}^{ip} - v_{\gimel_{\mathfrak{H}}}^{ip}\right) \in [-1, 1]. \tag{12}$$

$$V_a\left(\overline{\overline{I\mathfrak{T}}}_{\mathfrak{H}}\right) = \frac{1}{2}\left(u_{\gimel_{\mathfrak{H}}}^{rp} + v_{\gimel_{\mathfrak{H}}}^{rp} + u_{\gimel_{\mathfrak{H}}}^{ip} + v_{\gimel_{\mathfrak{H}}}^{ip}\right) \in [0, 1] \tag{13}$$

To differentiate the above information, we recall some valuable characteristics: if $V_s\left(\overline{\overline{I\mathfrak{T}}}_1\right) > V_s\left(\overline{\overline{I\mathfrak{T}}}_2\right) \Rightarrow \overline{\overline{I\mathfrak{T}}}_1 > \overline{\overline{I\mathfrak{T}}}_2$; If $V_s\left(\overline{\overline{I\mathfrak{T}}}_1\right) < V_s\left(\overline{\overline{I\mathfrak{T}}}_2\right) \Rightarrow \overline{\overline{I\mathfrak{T}}}_1 < \overline{\overline{I\mathfrak{T}}}_2$; If $V_s(\overline{\overline{I\mathfrak{T}}}_1) = V_s(\overline{\overline{I\mathfrak{T}}}_2) \Rightarrow$ If $V_a\left(\overline{\overline{I\mathfrak{T}}}_1\right) > V_a\left(\overline{\overline{I\mathfrak{T}}}_2\right) \Rightarrow \overline{\overline{I\mathfrak{T}}}_1 > \overline{\overline{I\mathfrak{T}}}_2$; If $V_a\left(\overline{\overline{I\mathfrak{T}}}_1\right) < V_a\left(\overline{\overline{I\mathfrak{T}}}_2\right) \Rightarrow \overline{\overline{I\mathfrak{T}}}_1 < \overline{\overline{I\mathfrak{T}}}_2$.

3. CIF Frank Prioritized Aggregation Operators

In this section, we propose the idea of Frank operational laws for CIF information. Furthermore, we examine the theory of the CIFFPA operator, the CIFFPOA operator, the CIFFPG operator, and the CIFFPOG operator, and their properties (idempotency, monotonicity, and boundedness). From now on, we will be using the CIFVs $\overline{\overline{I\mathfrak{T}}}_{\mathfrak{H}} = \left(\left(u_{\gimel_{\mathfrak{H}}}^{rp}, u_{\gimel_{\mathfrak{H}}}^{ip}\right), \left(v_{\gimel_{\mathfrak{H}}}^{rp}, v_{\gimel_{\mathfrak{H}}}^{ip}\right)\right), \mathfrak{H} = 1, 2, \ldots, \gimel$ for constructing any ideas.

Definition 5. *The mathematical form of Frank operational laws is stated as follows: for $\daleth\daleth \in (1, +\infty)$,*

$$\overline{\overline{I\mathfrak{T}}}_1 \oplus \overline{\overline{I\mathfrak{T}}}_2 = \left(\begin{array}{c}\left(1 - \log_{\daleth\daleth}\left(1 + \frac{\left(\daleth\daleth^{1-u_{\gimel_1}^{rp}} - 1\right)\left(\daleth\daleth^{1-u_{\gimel_2}^{rp}} - 1\right)}{\daleth\daleth - 1}\right), 1 - \log_{\daleth\daleth}\left(1 + \frac{\left(\daleth\daleth^{1-u_{\gimel_1}^{ip}} - 1\right)\left(\daleth\daleth^{1-u_{\gimel_2}^{ip}} - 1\right)}{\daleth\daleth - 1}\right)\right), \\ \left(\log_{\daleth\daleth}\left(1 + \frac{\left(\daleth\daleth^{v_{\gimel_1}^{rp}} - 1\right)\left(\daleth\daleth^{v_{\gimel_2}^{rp}} - 1\right)}{\daleth\daleth - 1}\right), \log_{\daleth\daleth}\left(1 + \frac{\left(\daleth\daleth^{v_{\gimel_1}^{ip}} - 1\right)\left(\daleth\daleth^{v_{\gimel_2}^{ip}} - 1\right)}{\daleth\daleth - 1}\right)\right)\end{array}\right). \tag{14}$$

$$\overline{\overline{I\mathfrak{T}}}_1 \otimes \overline{\overline{I\mathfrak{T}}}_2 = \left(\begin{array}{c} \left(\log_{ℸℸ}\left(1+\frac{\left(ℸℸ^{u^{rp}_{⅃1}}-1\right)\left(ℸℸ^{u^{rp}_{⅃2}}-1\right)}{ℸℸ-1}\right), \log_{ℸℸ}\left(1+\frac{\left(ℸℸ^{u^{ip}_{⅃1}}-1\right)\left(ℸℸ^{u^{ip}_{⅃2}}-1\right)}{ℸℸ-1}\right)\right), \\ \left(1-\log_{ℸℸ}\left(1+\frac{\left(ℸℸ^{1-v^{rp}_{⅃1}}-1\right)\left(ℸℸ^{1-v^{rp}_{⅃2}}-1\right)}{ℸℸ-1}\right), 1-\log_{ℸℸ}\left(1+\frac{\left(ℸℸ^{1-v^{ip}_{⅃1}}-1\right)\left(ℸℸ^{1-v^{ip}_{⅃2}}-1\right)}{ℸℸ-1}\right)\right) \end{array} \right) \tag{15}$$

$$\dddot{\mathcal{P}}\overline{\overline{I\mathfrak{T}}}_1 = \left(\begin{array}{c} \left(1-\log_{ℸℸ}\left(1+\frac{\left(ℸℸ^{1-u^{rp}_{⅃1}}-1\right)^{\dddot{\mathcal{P}}}}{(ℸℸ-1)^{\dddot{\mathcal{P}}-1}}\right), 1-\log_{ℸℸ}\left(1+\frac{\left(ℸℸ^{1-u^{ip}_{⅃1}}-1\right)^{\dddot{\mathcal{P}}}}{(ℸℸ-1)^{\dddot{\mathcal{P}}-1}}\right)\right), \\ \left(\log_{ℸℸ}\left(1+\frac{\left(ℸℸ^{v^{rp}_{⅃1}}-1\right)^{\dddot{\mathcal{P}}}}{(ℸℸ-1)^{\dddot{\mathcal{P}}-1}}\right), \log_{ℸℸ}\left(1+\frac{\left(ℸℸ^{v^{ip}_{⅃1}}-1\right)^{\dddot{\mathcal{P}}}}{(ℸℸ-1)^{\dddot{\mathcal{P}}-1}}\right)\right) \end{array} \right). \tag{16}$$

$$\overline{\overline{I\mathfrak{T}}}_1^{\dddot{\mathcal{P}}} = \left(\begin{array}{c} \left(\log_{ℸℸ}\left(1+\frac{\left(ℸℸ^{u^{rp}_{⅃1}}-1\right)^{\dddot{\mathcal{P}}}}{(ℸℸ-1)^{\dddot{\mathcal{P}}-1}}\right), \log_{ℸℸ}\left(1+\frac{\left(ℸℸ^{u^{ip}_{⅃1}}-1\right)^{\dddot{\mathcal{P}}}}{(ℸℸ-1)^{\dddot{\mathcal{P}}-1}}\right)\right), \\ \left(1-\log_{ℸℸ}\left(1+\frac{\left(ℸℸ^{1-v^{rp}_{⅃1}}-1\right)^{\dddot{\mathcal{P}}}}{(ℸℸ-1)^{\dddot{\mathcal{P}}-1}}\right), 1-\log_{ℸℸ}\left(1+\frac{\left(ℸℸ^{1-v^{ip}_{⅃1}}-1\right)^{\dddot{\mathcal{P}}}}{(ℸℸ-1)^{\dddot{\mathcal{P}}-1}}\right)\right) \end{array} \right). \tag{17}$$

Definition 6. *The mathematical form of the CIFFPA operator is shown below:*

$$\begin{aligned} &CIFFPA\left(\overline{\overline{I\mathfrak{T}}}_1, \overline{\overline{I\mathfrak{T}}}_2, \ldots, \overline{\overline{I\mathfrak{T}}}_⅃\right) = \\ &\left(\frac{\overline{\overline{\mathring{A}}}_1}{\sum_{\mathfrak{H}=1}^{⅃}\overline{\overline{\mathring{A}}}_{\mathfrak{H}}}\right)\overline{\overline{I\mathfrak{T}}}_1 \oplus \left(\frac{\overline{\overline{\mathring{A}}}_2}{\sum_{\mathfrak{H}=1}^{⅃}\overline{\overline{\mathring{A}}}_{\mathfrak{H}}}\right)\overline{\overline{I\mathfrak{T}}}_2 \oplus \ldots \oplus \left(\frac{\overline{\overline{\mathring{A}}}_⅃}{\sum_{\mathfrak{H}=1}^{⅃}\overline{\overline{\mathring{A}}}_{\mathfrak{H}}}\right)\overline{\overline{I\mathfrak{T}}}_⅃ \\ &= \oplus_{\mathfrak{H}=1}^{⅃}\left(\frac{\overline{\overline{\mathring{A}}}_{\mathfrak{H}}}{\sum_{\mathfrak{H}=1}^{⅃}\overline{\overline{\mathring{A}}}_{\mathfrak{H}}}\right)\overline{\overline{I\mathfrak{T}}}_{\mathfrak{H}}. \end{aligned} \tag{18}$$

With the values of $\overline{\overline{\mathring{A}}}_1 = 1$ and $\overline{\overline{\mathring{A}}}_{\mathfrak{H}} = \prod_{\mathcal{B}=1}^{\mathfrak{H}-1} V_s\left(\overline{\overline{I\mathfrak{T}}}_{\mathcal{B}}\right)$.

Theorem 1. *With the help of the data in Equation (18), we show that the aggregated value of Equation (18) will again be in the form of CIFV, such as:*

$$CIFFPA\left(\overline{\overline{I\mathfrak{T}}}_1, \overline{\overline{I\mathfrak{T}}}_2, \ldots, \overline{\overline{I\mathfrak{T}}}_⅃\right) = \left(\begin{array}{c} \left(1-\log_{ℸℸ}\left(1+\frac{\prod_{\mathfrak{H}=1}^{⅃}\left(ℸℸ^{1-u^{rp}_{⅃\mathfrak{H}}}-1\right)^{\frac{\dddot{\overline{\overline{\mathring{A}}}}_{\mathfrak{H}}}{\sum_{\mathfrak{H}=1}^{⅃}\overline{\overline{\mathring{A}}}_{\mathfrak{H}}}}}{(ℸℸ-1)^{\sum_{\mathfrak{H}=1}^{⅃}\frac{\overline{\overline{\mathring{A}}}_{\mathfrak{H}}}{\sum_{\mathfrak{H}=1}^{⅃}\overline{\overline{\mathring{A}}}_{\mathfrak{H}}}-1}}\right), 1-\log_{ℸℸ}\left(1+\frac{\prod_{\mathfrak{H}=1}^{⅃}\left(ℸℸ^{1-u^{ip}_{⅃\mathfrak{H}}}-1\right)^{\frac{\dddot{\overline{\overline{\mathring{A}}}}_{\mathfrak{H}}}{\sum_{\mathfrak{H}=1}^{⅃}\overline{\overline{\mathring{A}}}_{\mathfrak{H}}}}}{(ℸℸ-1)^{\sum_{\mathfrak{H}=1}^{⅃}\frac{\overline{\overline{\mathring{A}}}_{\mathfrak{H}}}{\sum_{\mathfrak{H}=1}^{⅃}\overline{\overline{\mathring{A}}}_{\mathfrak{H}}}-1}}\right)\right), \\ \left(\log_{ℸℸ}\left(1+\frac{\prod_{\mathfrak{H}=1}^{⅃}\left(ℸℸ^{v^{rp}_{⅃\mathfrak{H}}}-1\right)^{\frac{\overline{\overline{\mathring{A}}}_{\mathfrak{H}}}{\sum_{\mathfrak{H}=1}^{⅃}\overline{\overline{\mathring{A}}}_{\mathfrak{H}}}}}{(ℸℸ-1)^{\sum_{\mathfrak{H}=1}^{⅃}\frac{\overline{\overline{\mathring{A}}}_{\mathfrak{H}}}{\sum_{\mathfrak{H}=1}^{⅃}\overline{\overline{\mathring{A}}}_{\mathfrak{H}}}-1}}\right), \log_{ℸℸ}\left(1+\frac{\prod_{\mathfrak{H}=1}^{⅃}\left(ℸℸ^{v^{ip}_{⅃\mathfrak{H}}}-1\right)^{\frac{\overline{\overline{\mathring{A}}}_{\mathfrak{H}}}{\sum_{\mathfrak{H}=1}^{⅃}\overline{\overline{\mathring{A}}}_{\mathfrak{H}}}}}{(ℸℸ-1)^{\sum_{\mathfrak{H}=1}^{⅃}\frac{\overline{\overline{\mathring{A}}}_{\mathfrak{H}}}{\sum_{\mathfrak{H}=1}^{⅃}\overline{\overline{\mathring{A}}}_{\mathfrak{H}}}-1}}\right)\right) \end{array} \right). \tag{19}$$

Proof. The procedure of Mathematical Induction is used in this proof as follows: if $ⅎ = 2$, then we obtain

$$\left(\frac{\overline{\overline{\mathring{A}}}_1}{\sum_{\mathfrak{H}=1}^{ⅎ}\overline{\overline{\mathring{A}}}_{\mathfrak{H}}}\right)\overline{\overline{I\mathfrak{T}}}_1=\left(\begin{array}{c}\left(1-\log_{\text{ᒣᒣ}}\left(1+\frac{\left(\text{ᒣᒣ}^{1-u^{rp}_{ⅎ_1}}-1\right)^{\frac{\dddot{\overline{\overline{\mathring{A}}}}_1}{\sum_{\mathfrak{H}=1}^{ⅎ}\overline{\overline{\mathring{A}}}_{\mathfrak{H}}}}}{(\text{ᒣᒣ}-1)^{\frac{\overline{\overline{\mathring{A}}}_1}{\sum_{\mathfrak{H}=1}^{ⅎ}\overline{\overline{\mathring{A}}}_{\mathfrak{H}}}-1}}\right),1-\log_{\text{ᒣᒣ}}\left(1+\frac{\left(\text{ᒣᒣ}^{1-u^{ip}_{ⅎ_1}}-1\right)^{\frac{\dddot{\overline{\overline{\mathring{A}}}}_1}{\sum_{\mathfrak{H}=1}^{ⅎ}\overline{\overline{\mathring{A}}}_{\mathfrak{H}}}}}{(\text{ᒣᒣ}-1)^{\frac{\overline{\overline{\mathring{A}}}_1}{\sum_{\mathfrak{H}=1}^{ⅎ}\overline{\overline{\mathring{A}}}_{\mathfrak{H}}}-1}}\right)\right),\\ \left(\log_{\text{ᒣᒣ}}\left(1+\frac{\left(\text{ᒣᒣ}^{v^{rp}_{ⅎ_1}}-1\right)^{\frac{\overline{\overline{\mathring{A}}}_1}{\sum_{\mathfrak{H}=1}^{ⅎ}\overline{\overline{\mathring{A}}}_{\mathfrak{H}}}}}{(\text{ᒣᒣ}-1)^{\frac{\overline{\overline{\mathring{A}}}_1}{\sum_{\mathfrak{H}=1}^{ⅎ}\overline{\overline{\mathring{A}}}_{\mathfrak{H}}}-1}}\right),\log_{\text{ᒣᒣ}}\left(1+\frac{\left(\text{ᒣᒣ}^{v^{ip}_{ⅎ_1}}-1\right)^{\frac{\overline{\overline{\mathring{A}}}_1}{\sum_{\mathfrak{H}=1}^{ⅎ}\overline{\overline{\mathring{A}}}_{\mathfrak{H}}}}}{(\text{ᒣᒣ}-1)^{\frac{\overline{\overline{\mathring{A}}}_1}{\sum_{\mathfrak{H}=1}^{ⅎ}\overline{\overline{\mathring{A}}}_{\mathfrak{H}}}-1}}\right)\right)\end{array}\right)$$

$$\left(\frac{\overline{\overline{\mathring{A}}}_2}{\sum_{\mathfrak{H}=1}^{ⅎ}\overline{\overline{\mathring{A}}}_{\mathfrak{H}}}\right)\overline{\overline{I\mathfrak{T}}}_2=\left(\begin{array}{c}\left(1-\log_{\text{ᒣᒣ}}\left(1+\frac{\left(\text{ᒣᒣ}^{1-u^{rp}_{ⅎ_2}}-1\right)^{\frac{\dddot{\overline{\overline{\mathring{A}}}}_2}{\sum_{\mathfrak{H}=1}^{ⅎ}\overline{\overline{\mathring{A}}}_{\mathfrak{H}}}}}{(\text{ᒣᒣ}-1)^{\frac{\overline{\overline{\mathring{A}}}_2}{\sum_{\mathfrak{H}=1}^{ⅎ}\overline{\overline{\mathring{A}}}_{\mathfrak{H}}}-1}}\right),1-\log_{\text{ᒣᒣ}}\left(1+\frac{\left(\text{ᒣᒣ}^{1-u^{ip}_{ⅎ_2}}-1\right)^{\frac{\dddot{\overline{\overline{\mathring{A}}}}_2}{\sum_{\mathfrak{H}=1}^{ⅎ}\overline{\overline{\mathring{A}}}_{\mathfrak{H}}}}}{(\text{ᒣᒣ}-1)^{\frac{\overline{\overline{\mathring{A}}}_2}{\sum_{\mathfrak{H}=1}^{ⅎ}\overline{\overline{\mathring{A}}}_{\mathfrak{H}}}-1}}\right)\right),\\ \left(\log_{\text{ᒣᒣ}}\left(1+\frac{\left(\text{ᒣᒣ}^{v^{rp}_{ⅎ_2}}-1\right)^{\frac{\overline{\overline{\mathring{A}}}_2}{\sum_{\mathfrak{H}=1}^{ⅎ}\overline{\overline{\mathring{A}}}_{\mathfrak{H}}}}}{(\text{ᒣᒣ}-1)^{\frac{\overline{\overline{\mathring{A}}}_2}{\sum_{\mathfrak{H}=1}^{ⅎ}\overline{\overline{\mathring{A}}}_{\mathfrak{H}}}-1}}\right),\log_{\text{ᒣᒣ}}\left(1+\frac{\left(\text{ᒣᒣ}^{v^{ip}_{ⅎ_2}}-1\right)^{\frac{\overline{\overline{\mathring{A}}}_2}{\sum_{\mathfrak{H}=1}^{ⅎ}\overline{\overline{\mathring{A}}}_{\mathfrak{H}}}}}{(\text{ᒣᒣ}-1)^{\frac{\overline{\overline{\mathring{A}}}_2}{\sum_{\mathfrak{H}=1}^{ⅎ}\overline{\overline{\mathring{A}}}_{\mathfrak{H}}}-1}}\right)\right)\end{array}\right)$$

Thus,

$$\left(\frac{\overline{\overline{\mathring{A}}}_1}{\sum_{\mathfrak{H}=1}^{ⅎ}\overline{\overline{\mathring{A}}}_{\mathfrak{H}}}\right)\overline{\overline{I\mathfrak{T}}}_1\oplus\left(\frac{\overline{\overline{\mathring{A}}}_2}{\sum_{\mathfrak{H}=1}^{ⅎ}\overline{\overline{\mathring{A}}}_{\mathfrak{H}}}\right)\overline{\overline{I\mathfrak{T}}}_2=\left(\begin{array}{c}\left(1-\log_{\text{ᒣᒣ}}\left(1+\frac{\left(\text{ᒣᒣ}^{1-u^{rp}_{ⅎ_1}}-1\right)^{\frac{\dddot{\overline{\overline{\mathring{A}}}}_1}{\sum_{\mathfrak{H}=1}^{ⅎ}\overline{\overline{\mathring{A}}}_{\mathfrak{H}}}}}{(\text{ᒣᒣ}-1)^{\frac{\overline{\overline{\mathring{A}}}_1}{\sum_{\mathfrak{H}=1}^{ⅎ}\overline{\overline{\mathring{A}}}_{\mathfrak{H}}}-1}}\right),1-\log_{\text{ᒣᒣ}}\left(1+\frac{\left(\text{ᒣᒣ}^{1-u^{ip}_{ⅎ_1}}-1\right)^{\frac{\dddot{\overline{\overline{\mathring{A}}}}_1}{\sum_{\mathfrak{H}=1}^{ⅎ}\overline{\overline{\mathring{A}}}_{\mathfrak{H}}}}}{(\text{ᒣᒣ}-1)^{\frac{\overline{\overline{\mathring{A}}}_1}{\sum_{\mathfrak{H}=1}^{ⅎ}\overline{\overline{\mathring{A}}}_{\mathfrak{H}}}-1}}\right)\right),\\ \left(\log_{\text{ᒣᒣ}}\left(1+\frac{\left(\text{ᒣᒣ}^{v^{rp}_{ⅎ_1}}-1\right)^{\frac{\overline{\overline{\mathring{A}}}_1}{\sum_{\mathfrak{H}=1}^{ⅎ}\overline{\overline{\mathring{A}}}_{\mathfrak{H}}}}}{(\text{ᒣᒣ}-1)^{\frac{\overline{\overline{\mathring{A}}}_1}{\sum_{\mathfrak{H}=1}^{ⅎ}\overline{\overline{\mathring{A}}}_{\mathfrak{H}}}-1}}\right),\log_{\text{ᒣᒣ}}\left(1+\frac{\left(\text{ᒣᒣ}^{v^{ip}_{ⅎ_1}}-1\right)^{\frac{\overline{\overline{\mathring{A}}}_1}{\sum_{\mathfrak{H}=1}^{ⅎ}\overline{\overline{\mathring{A}}}_{\mathfrak{H}}}}}{(\text{ᒣᒣ}-1)^{\frac{\overline{\overline{\mathring{A}}}_1}{\sum_{\mathfrak{H}=1}^{ⅎ}\overline{\overline{\mathring{A}}}_{\mathfrak{H}}}-1}}\right)\right)\end{array}\right)\oplus$$

$$\left(\begin{array}{c}\left(1-\log_{\text{ᒣᒣ}}\left(1+\frac{\left(\text{ᒣᒣ}^{1-u^{rp}_{ⅎ_2}}-1\right)^{\frac{\dddot{\overline{\overline{\mathring{A}}}}_2}{\sum_{\mathfrak{H}=1}^{ⅎ}\overline{\overline{\mathring{A}}}_{\mathfrak{H}}}}}{(\text{ᒣᒣ}-1)^{\frac{\overline{\overline{\mathring{A}}}_2}{\sum_{\mathfrak{H}=1}^{ⅎ}\overline{\overline{\mathring{A}}}_{\mathfrak{H}}}-1}}\right),1-\log_{\text{ᒣᒣ}}\left(1+\frac{\left(\text{ᒣᒣ}^{1-u^{ip}_{ⅎ_2}}-1\right)^{\frac{\dddot{\overline{\overline{\mathring{A}}}}_2}{\sum_{\mathfrak{H}=1}^{ⅎ}\overline{\overline{\mathring{A}}}_{\mathfrak{H}}}}}{(\text{ᒣᒣ}-1)^{\frac{\overline{\overline{\mathring{A}}}_2}{\sum_{\mathfrak{H}=1}^{ⅎ}\overline{\overline{\mathring{A}}}_{\mathfrak{H}}}-1}}\right)\right),\\ \left(\log_{\text{ᒣᒣ}}\left(1+\frac{\left(\text{ᒣᒣ}^{v^{rp}_{ⅎ_2}}-1\right)^{\frac{\overline{\overline{\mathring{A}}}_2}{\sum_{\mathfrak{H}=1}^{ⅎ}\overline{\overline{\mathring{A}}}_{\mathfrak{H}}}}}{(\text{ᒣᒣ}-1)^{\frac{\overline{\overline{\mathring{A}}}_2}{\sum_{\mathfrak{H}=1}^{ⅎ}\overline{\overline{\mathring{A}}}_{\mathfrak{H}}}-1}}\right),\log_{\text{ᒣᒣ}}\left(1+\frac{\left(\text{ᒣᒣ}^{v^{ip}_{ⅎ_2}}-1\right)^{\frac{\overline{\overline{\mathring{A}}}_2}{\sum_{\mathfrak{H}=1}^{ⅎ}\overline{\overline{\mathring{A}}}_{\mathfrak{H}}}}}{(\text{ᒣᒣ}-1)^{\frac{\overline{\overline{\mathring{A}}}_2}{\sum_{\mathfrak{H}=1}^{ⅎ}\overline{\overline{\mathring{A}}}_{\mathfrak{H}}}-1}}\right)\right)\end{array}\right)$$

$$= \left(\begin{array}{c}\left(1-\log_{\daleth\daleth}\left(1+\frac{\prod_{\mathfrak{H}=1}^{2}\left(\daleth\daleth^{1-u^{rp}_{\Finv\mathfrak{H}}}-1\right)^{\frac{\overline{\overline{\mathring{A}}}^{\cdots}_{\mathfrak{H}}}{\sum_{\mathfrak{H}=1}^{\Finv}\overline{\overline{\mathring{A}}}_{\mathfrak{H}}}}}{(\daleth\daleth-1)^{\sum_{\mathfrak{H}=1}^{\Finv}\frac{\overline{\overline{\mathring{A}}}_{\mathfrak{H}}}{\sum_{\mathfrak{H}=1}^{\Finv}\overline{\overline{\mathring{A}}}_{\mathfrak{H}}}-1}}\right),1-\log_{\daleth\daleth}\left(1+\frac{\prod_{\mathfrak{H}=1}^{2}\left(\daleth\daleth^{1-u^{ip}_{\Finv\mathfrak{H}}}-1\right)^{\frac{\overline{\overline{\mathring{A}}}^{\cdots}_{\mathfrak{H}}}{\sum_{\mathfrak{H}=1}^{\Finv}\overline{\overline{\mathring{A}}}_{\mathfrak{H}}}}}{(\daleth\daleth-1)^{\sum_{\mathfrak{H}=1}^{\Finv}\frac{\overline{\overline{\mathring{A}}}_{\mathfrak{H}}}{\sum_{\mathfrak{H}=1}^{\Finv}\overline{\overline{\mathring{A}}}_{\mathfrak{H}}}-1}}\right)\right),\\ \left(\log_{\daleth\daleth}\left(1+\frac{\prod_{\mathfrak{H}=1}^{2}\left(\daleth\daleth^{v^{rp}_{\Finv 1}}-1\right)^{\frac{\overline{\overline{\mathring{A}}}_{\mathfrak{H}}}{\sum_{\mathfrak{H}=1}^{\Finv}\overline{\overline{\mathring{A}}}_{\mathfrak{H}}}}}{(\daleth\daleth-1)^{\sum_{\mathfrak{H}=1}^{\Finv}\frac{\overline{\overline{\mathring{A}}}_{\mathfrak{H}}}{\sum_{\mathfrak{H}=1}^{\Finv}\overline{\overline{\mathring{A}}}_{\mathfrak{H}}}-1}}\right),\log_{\daleth\daleth}\left(1+\frac{\prod_{\mathfrak{H}=1}^{2}\left(\daleth\daleth^{v^{ip}_{\Finv 1}}-1\right)^{\frac{\overline{\overline{\mathring{A}}}_{\mathfrak{H}}}{\sum_{\mathfrak{H}=1}^{\Finv}\overline{\overline{\mathring{A}}}_{\mathfrak{H}}}}}{(\daleth\daleth-1)^{\sum_{\mathfrak{H}=1}^{\Finv}\frac{\overline{\overline{\mathring{A}}}_{\mathfrak{H}}}{\sum_{\mathfrak{H}=1}^{\Finv}\overline{\overline{\mathring{A}}}_{\mathfrak{H}}}-1}}\right)\right)\end{array}\right)$$

We obtain the correct theory. Furthermore, we assume that we also obtain the correct theory for $\Finv = \mathcal{B}$, such that:

$$CIFFPA\left(\overline{\overline{I\mathfrak{T}}}_1,\overline{\overline{I\mathfrak{T}}}_2,\ldots,\overline{\overline{I\mathfrak{T}}}_{\mathcal{B}}\right)= \left(\begin{array}{c}\left(1-\log_{\daleth\daleth}\left(1+\frac{\prod_{\mathfrak{H}=1}^{\mathcal{B}}\left(\daleth\daleth^{1-u^{rp}_{\Finv\mathfrak{H}}}-1\right)^{\frac{\overline{\overline{\mathring{A}}}^{\cdots}_{\mathfrak{H}}}{\sum_{\mathfrak{H}=1}^{\Finv}\overline{\overline{\mathring{A}}}_{\mathfrak{H}}}}}{(\daleth\daleth-1)^{\sum_{\mathfrak{H}=1}^{\Finv}\frac{\overline{\overline{\mathring{A}}}_{\mathfrak{H}}}{\sum_{\mathfrak{H}=1}^{\Finv}\overline{\overline{\mathring{A}}}_{\mathfrak{H}}}-1}}\right),1-\log_{\daleth\daleth}\left(1+\frac{\prod_{\mathfrak{H}=1}^{\mathcal{B}}\left(\daleth\daleth^{1-u^{ip}_{\Finv\mathfrak{H}}}-1\right)^{\frac{\overline{\overline{\mathring{A}}}^{\cdots}_{\mathfrak{H}}}{\sum_{\mathfrak{H}=1}^{\Finv}\overline{\overline{\mathring{A}}}_{\mathfrak{H}}}}}{(\daleth\daleth-1)^{\sum_{\mathfrak{H}=1}^{\Finv}\frac{\overline{\overline{\mathring{A}}}_{\mathfrak{H}}}{\sum_{\mathfrak{H}=1}^{\Finv}\overline{\overline{\mathring{A}}}_{\mathfrak{H}}}-1}}\right)\right),\\ \left(\log_{\daleth\daleth}\left(1+\frac{\prod_{\mathfrak{H}=1}^{\mathcal{B}}\left(\daleth\daleth^{v^{rp}_{\Finv 1}}-1\right)^{\frac{\overline{\overline{\mathring{A}}}_{\mathfrak{H}}}{\sum_{\mathfrak{H}=1}^{\Finv}\overline{\overline{\mathring{A}}}_{\mathfrak{H}}}}}{(\daleth\daleth-1)^{\sum_{\mathfrak{H}=1}^{\Finv}\frac{\overline{\overline{\mathring{A}}}_{\mathfrak{H}}}{\sum_{\mathfrak{H}=1}^{\Finv}\overline{\overline{\mathring{A}}}_{\mathfrak{H}}}-1}}\right),\log_{\daleth\daleth}\left(1+\frac{\prod_{\mathfrak{H}=1}^{\mathcal{B}}\left(\daleth\daleth^{v^{ip}_{\Finv 1}}-1\right)^{\frac{\overline{\overline{\mathring{A}}}_{\mathfrak{H}}}{\sum_{\mathfrak{H}=1}^{\Finv}\overline{\overline{\mathring{A}}}_{\mathfrak{H}}}}}{(\daleth\daleth-1)^{\sum_{\mathfrak{H}=1}^{\Finv}\frac{\overline{\overline{\mathring{A}}}_{\mathfrak{H}}}{\sum_{\mathfrak{H}=1}^{\Finv}\overline{\overline{\mathring{A}}}_{\mathfrak{H}}}-1}}\right)\right)\end{array}\right)$$

Furthermore, we prove it for $\Finv = \mathcal{B}+1$, such as:

$$\begin{aligned}CIFFPA\left(\overline{\overline{I\mathfrak{T}}}_1,\overline{\overline{I\mathfrak{T}}}_2,\ldots,\overline{\overline{I\mathfrak{T}}}_{\mathcal{B}+1}\right) &= \left(\frac{\overline{\overline{\mathring{A}}}_1}{\sum_{\mathfrak{H}=1}^{\Finv}\overline{\overline{\mathring{A}}}_{\mathfrak{H}}}\right)\overline{\overline{I\mathfrak{T}}}_1 \oplus \left(\frac{\overline{\overline{\mathring{A}}}_2}{\sum_{\mathfrak{H}=1}^{\Finv}\overline{\overline{\mathring{A}}}_{\mathfrak{H}}}\right)\overline{\overline{I\mathfrak{T}}}_2 \oplus \ldots \oplus \left(\frac{\overline{\overline{\mathring{A}}}_{\mathcal{B}}}{\sum_{\mathfrak{H}=1}^{\Finv}\overline{\overline{\mathring{A}}}_{\mathfrak{H}}}\right)\overline{\overline{I\mathfrak{T}}}_{\mathcal{B}} \oplus \left(\frac{\overline{\overline{\mathring{A}}}_{\mathcal{B}+1}}{\sum_{\mathfrak{H}=1}^{\Finv}\overline{\overline{\mathring{A}}}_{\mathfrak{H}}}\right)\overline{\overline{I\mathfrak{T}}}_{\mathcal{B}+1}\\ &= \oplus_{\mathfrak{H}=1}^{\mathcal{B}}\left(\frac{\overline{\overline{\mathring{A}}}_{\mathfrak{H}}}{\sum_{\mathfrak{H}=1}^{\Finv}\overline{\overline{\mathring{A}}}_{\mathfrak{H}}}\right)\overline{\overline{I\mathfrak{T}}}_{\mathfrak{H}} \oplus \left(\frac{\overline{\overline{\mathring{A}}}_{\mathcal{B}+1}}{\sum_{\mathfrak{H}=1}^{\mathcal{B}+1}\overline{\overline{\mathring{A}}}_{\mathfrak{H}}}\right)\overline{\overline{I\mathfrak{T}}}_{\mathcal{B}+1}\end{aligned}$$

$$= \left(\begin{array}{c}\left(1-\log_{\daleth\daleth}\left(1+\frac{\prod_{\mathfrak{H}=1}^{\mathcal{B}}\left(\daleth\daleth^{1-u^{rp}_{\Finv\mathfrak{H}}}-1\right)^{\frac{\overline{\overline{\mathring{A}}}^{\cdots}_{\mathfrak{H}}}{\sum_{\mathfrak{H}=1}^{\Finv}\overline{\overline{\mathring{A}}}_{\mathfrak{H}}}}}{(\daleth\daleth-1)^{\sum_{\mathfrak{H}=1}^{\Finv}\frac{\overline{\overline{\mathring{A}}}_{\mathfrak{H}}}{\sum_{\mathfrak{H}=1}^{\Finv}\overline{\overline{\mathring{A}}}_{\mathfrak{H}}}-1}}\right),1-\log_{\daleth\daleth}\left(1+\frac{\prod_{\mathfrak{H}=1}^{\mathcal{B}}\left(\daleth\daleth^{1-u^{ip}_{\Finv\mathfrak{H}}}-1\right)^{\frac{\overline{\overline{\mathring{A}}}^{\cdots}_{\mathfrak{H}}}{\sum_{\mathfrak{H}=1}^{\Finv}\overline{\overline{\mathring{A}}}_{\mathfrak{H}}}}}{(\daleth\daleth-1)^{\sum_{\mathfrak{H}=1}^{\Finv}\frac{\overline{\overline{\mathring{A}}}_{\mathfrak{H}}}{\sum_{\mathfrak{H}=1}^{\Finv}\overline{\overline{\mathring{A}}}_{\mathfrak{H}}}-1}}\right)\right),\\ \left(\log_{\daleth\daleth}\left(1+\frac{\prod_{\mathfrak{H}=1}^{\mathcal{B}}\left(\daleth\daleth^{v^{rp}_{\Finv 1}}-1\right)^{\frac{\overline{\overline{\mathring{A}}}_{\mathfrak{H}}}{\sum_{\mathfrak{H}=1}^{\Finv}\overline{\overline{\mathring{A}}}_{\mathfrak{H}}}}}{(\daleth\daleth-1)^{\sum_{\mathfrak{H}=1}^{\Finv}\frac{\overline{\overline{\mathring{A}}}_{\mathfrak{H}}}{\sum_{\mathfrak{H}=1}^{\Finv}\overline{\overline{\mathring{A}}}_{\mathfrak{H}}}-1}}\right),\log_{\daleth\daleth}\left(1+\frac{\prod_{\mathfrak{H}=1}^{\mathcal{B}}\left(\daleth\daleth^{v^{ip}_{\Finv 1}}-1\right)^{\frac{\overline{\overline{\mathring{A}}}_{\mathfrak{H}}}{\sum_{\mathfrak{H}=1}^{\Finv}\overline{\overline{\mathring{A}}}_{\mathfrak{H}}}}}{(\daleth\daleth-1)^{\sum_{\mathfrak{H}=1}^{\Finv}\frac{\overline{\overline{\mathring{A}}}_{\mathfrak{H}}}{\sum_{\mathfrak{H}=1}^{\Finv}\overline{\overline{\mathring{A}}}_{\mathfrak{H}}}-1}}\right)\right)\end{array}\right)$$

$$\oplus\left(\begin{array}{c}\left(1-\log_{⅂⅂}\left(1+\frac{\left(⅂⅂^{1-u^{rp}_{⅃_{\mathcal{B}+1}}}-1\right)^{\frac{\overline{\overline{\mathrm{A}}}^{\cdots}_{\mathcal{B}+1}}{\sum_{\mathfrak{H}=1}^{\mathcal{B}+1}\overline{\overline{\mathrm{A}}}_{\mathfrak{H}}}}}{(⅂⅂-1)^{\frac{\overline{\overline{\mathrm{A}}}_{\mathcal{B}+1}}{\sum_{\mathfrak{H}=1}^{\mathcal{B}+1}\overline{\overline{\mathrm{A}}}_{\mathfrak{H}}}-1}}\right),1-\log_{⅂⅂}\left(1+\frac{\left(⅂⅂^{1-u^{ip}_{⅃_{\mathcal{B}+1}}}-1\right)^{\frac{\overline{\overline{\mathrm{A}}}^{\cdots}_{\mathcal{B}+1}}{\sum_{\mathfrak{H}=1}^{\mathcal{B}+1}\overline{\overline{\mathrm{A}}}_{\mathfrak{H}}}}}{(⅂⅂-1)^{\frac{\overline{\overline{\mathrm{A}}}_{\mathcal{B}+1}}{\sum_{\mathfrak{H}=1}^{\mathcal{B}+1}\overline{\overline{\mathrm{A}}}_{\mathfrak{H}}}-1}}\right)\right),\\ \left(\log_{⅂⅂}\left(1+\frac{\left(⅂⅂^{v^{rp}_{⅃_{\mathcal{B}+1}}}-1\right)^{\frac{\overline{\overline{\mathrm{A}}}_{\mathcal{B}+1}}{\sum_{\mathfrak{H}=1}^{\mathcal{B}+1}\overline{\overline{\mathrm{A}}}_{\mathfrak{H}}}}}{(⅂⅂-1)^{\frac{\overline{\overline{\mathrm{A}}}_{\mathcal{B}+1}}{\sum_{\mathfrak{H}=1}^{\mathcal{B}+1}\overline{\overline{\mathrm{A}}}_{\mathfrak{H}}}-1}}\right),\log_{⅂⅂}\left(1+\frac{\left(⅂⅂^{v^{ip}_{⅃_{\mathcal{B}+1}}}-1\right)^{\frac{\overline{\overline{\mathrm{A}}}_{\mathcal{B}+1}}{\sum_{\mathfrak{H}=1}^{\mathcal{B}+1}\overline{\overline{\mathrm{A}}}_{\mathfrak{H}}}}}{(⅂⅂-1)^{\frac{\overline{\overline{\mathrm{A}}}_{\mathcal{B}+1}}{\sum_{\mathfrak{H}=1}^{\mathcal{B}+1}\overline{\overline{\mathrm{A}}}_{\mathfrak{H}}}-1}}\right)\right)\end{array}\right)$$

$$=\left(\begin{array}{c}\left(1-\log_{⅂⅂}\left(1+\frac{\prod_{\mathfrak{H}=1}^{\mathcal{B}+1}\left(⅂⅂^{1-u^{rp}_{⅃_{\mathfrak{H}}}}-1\right)^{\frac{\overline{\overline{\mathrm{A}}}^{\cdots}_{\mathfrak{H}}}{\sum_{\mathfrak{H}=1}^{⅃}\overline{\overline{\mathrm{A}}}_{\mathfrak{H}}}}}{(⅂⅂-1)^{\sum_{\mathfrak{H}=1}^{⅃}\frac{\overline{\overline{\mathrm{A}}}_{\mathfrak{H}}}{\sum_{\mathfrak{H}=1}^{⅃}\overline{\overline{\mathrm{A}}}_{\mathfrak{H}}}-1}}\right),1-\log_{⅂⅂}\left(1+\frac{\prod_{\mathfrak{H}=1}^{\mathcal{B}+1}\left(⅂⅂^{1-u^{ip}_{⅃_{\mathfrak{H}}}}-1\right)^{\frac{\overline{\overline{\mathrm{A}}}^{\cdots}_{\mathfrak{H}}}{\sum_{\mathfrak{H}=1}^{⅃}\overline{\overline{\mathrm{A}}}_{\mathfrak{H}}}}}{(⅂⅂-1)^{\sum_{\mathfrak{H}=1}^{⅃}\frac{\overline{\overline{\mathrm{A}}}_{\mathfrak{H}}}{\sum_{\mathfrak{H}=1}^{⅃}\overline{\overline{\mathrm{A}}}_{\mathfrak{H}}}-1}}\right)\right),\\ \left(\log_{⅂⅂}\left(1+\frac{\prod_{\mathfrak{H}=1}^{\mathcal{B}+1}\left(⅂⅂^{v^{rp}_{⅃_1}}-1\right)^{\frac{\overline{\overline{\mathrm{A}}}_{\mathfrak{H}}}{\sum_{\mathfrak{H}=1}^{⅃}\overline{\overline{\mathrm{A}}}_{\mathfrak{H}}}}}{(⅂⅂-1)^{\sum_{\mathfrak{H}=1}^{⅃}\frac{\overline{\overline{\mathrm{A}}}_{\mathfrak{H}}}{\sum_{\mathfrak{H}=1}^{⅃}\overline{\overline{\mathrm{A}}}_{\mathfrak{H}}}-1}}\right),\log_{⅂⅂}\left(1+\frac{\prod_{\mathfrak{H}=1}^{\mathcal{B}+1}\left(⅂⅂^{v^{ip}_{⅃_1}}-1\right)^{\frac{\overline{\overline{\mathrm{A}}}_{\mathfrak{H}}}{\sum_{\mathfrak{H}=1}^{⅃}\overline{\overline{\mathrm{A}}}_{\mathfrak{H}}}}}{(⅂⅂-1)^{\sum_{\mathfrak{H}=1}^{⅃}\frac{\overline{\overline{\mathrm{A}}}_{\mathfrak{H}}}{\sum_{\mathfrak{H}=1}^{⅃}\overline{\overline{\mathrm{A}}}_{\mathfrak{H}}}-1}}\right)\right)\end{array}\right)$$

This proves the theorem. □

Proposition 1 (Idempotency). *If we use* $\overline{\overline{I\mathfrak{T}}}_{\mathfrak{H}}=\overline{\overline{I\mathfrak{T}}}=\left(\left(u^{rp}_{⅃},u^{ip}_{⅃}\right),\left(v^{rp}_{⅃},v^{ip}_{⅃}\right)\right)$, *then*

$$CIFFPA\left(\overline{\overline{I\mathfrak{T}}}_1,\overline{\overline{I\mathfrak{T}}}_2,\ldots,\overline{\overline{I\mathfrak{T}}}_{⅃}\right)=\overline{\overline{I\mathfrak{T}}}. \tag{20}$$

Proof. Let

$$CIFFPA\left(\overline{\overline{I\mathfrak{T}}}_1,\overline{\overline{I\mathfrak{T}}}_2,\ldots,\overline{\overline{I\mathfrak{T}}}_{⅃}\right)=\left(\begin{array}{c}\left(1-\log_{⅂⅂}\left(1+\frac{\prod_{\mathfrak{H}=1}^{⅃}\left(⅂⅂^{1-u^{rp}_{⅃_{\mathfrak{H}}}}-1\right)^{\frac{\overline{\overline{\mathrm{A}}}^{\cdots}_{\mathfrak{H}}}{\sum_{\mathfrak{H}=1}^{⅃}\overline{\overline{\mathrm{A}}}_{\mathfrak{H}}}}}{(⅂⅂-1)^{\sum_{\mathfrak{H}=1}^{⅃}\frac{\overline{\overline{\mathrm{A}}}_{\mathfrak{H}}}{\sum_{\mathfrak{H}=1}^{⅃}\overline{\overline{\mathrm{A}}}_{\mathfrak{H}}}-1}}\right),1-\log_{⅂⅂}\left(1+\frac{\prod_{\mathfrak{H}=1}^{⅃}\left(⅂⅂^{1-u^{ip}_{⅃_{\mathfrak{H}}}}-1\right)^{\frac{\overline{\overline{\mathrm{A}}}^{\cdots}_{\mathfrak{H}}}{\sum_{\mathfrak{H}=1}^{⅃}\overline{\overline{\mathrm{A}}}_{\mathfrak{H}}}}}{(⅂⅂-1)^{\sum_{\mathfrak{H}=1}^{⅃}\frac{\overline{\overline{\mathrm{A}}}_{\mathfrak{H}}}{\sum_{\mathfrak{H}=1}^{⅃}\overline{\overline{\mathrm{A}}}_{\mathfrak{H}}}-1}}\right)\right),\\ \left(\log_{⅂⅂}\left(1+\frac{\prod_{\mathfrak{H}=1}^{⅃}\left(⅂⅂^{v^{rp}_{⅃_{\mathfrak{H}}}}-1\right)^{\frac{\overline{\overline{\mathrm{A}}}_{\mathfrak{H}}}{\sum_{\mathfrak{H}=1}^{⅃}\overline{\overline{\mathrm{A}}}_{\mathfrak{H}}}}}{(⅂⅂-1)^{\sum_{\mathfrak{H}=1}^{⅃}\frac{\overline{\overline{\mathrm{A}}}_{\mathfrak{H}}}{\sum_{\mathfrak{H}=1}^{⅃}\overline{\overline{\mathrm{A}}}_{\mathfrak{H}}}-1}}\right),\log_{⅂⅂}\left(1+\frac{\prod_{\mathfrak{H}=1}^{⅃}\left(⅂⅂^{v^{ip}_{⅃_{\mathfrak{H}}}}-1\right)^{\frac{\overline{\overline{\mathrm{A}}}_{\mathfrak{H}}}{\sum_{\mathfrak{H}=1}^{⅃}\overline{\overline{\mathrm{A}}}_{\mathfrak{H}}}}}{(⅂⅂-1)^{\sum_{\mathfrak{H}=1}^{⅃}\frac{\overline{\overline{\mathrm{A}}}_{\mathfrak{H}}}{\sum_{\mathfrak{H}=1}^{⅃}\overline{\overline{\mathrm{A}}}_{\mathfrak{H}}}-1}}\right)\right)\end{array}\right)$$

$$= \left(\begin{array}{c} \left(1 - \log_{\daleth\daleth}\left(1 + \frac{\prod_{\mathfrak{H}=1}^{\Finv}\left(\daleth\daleth^{1-u_{\Finv}^{rp}} - 1\right)^{\frac{\overline{\overline{\text{Å}}}_{\mathfrak{H}}}{\sum_{\mathfrak{H}=1}^{\Finv}\overline{\overline{\text{Å}}}_{\mathfrak{H}}}}}{(\daleth\daleth-1)^{\sum_{\mathfrak{H}=1}^{\Finv}\frac{\overline{\overline{\text{Å}}}_{\mathfrak{H}}}{\sum_{\mathfrak{H}=1}^{\Finv}\overline{\overline{\text{Å}}}_{\mathfrak{H}}}-1}} \right), 1 - \log_{\daleth\daleth}\left(1 + \frac{\prod_{\mathfrak{H}=1}^{\Finv}\left(\daleth\daleth^{1-u_{\Finv}^{ip}} - 1\right)^{\frac{\overline{\overline{\text{Å}}}_{\mathfrak{H}}}{\sum_{\mathfrak{H}=1}^{\Finv}\overline{\overline{\text{Å}}}_{\mathfrak{H}}}}}{(\daleth\daleth-1)^{\sum_{\mathfrak{H}=1}^{\Finv}\frac{\overline{\overline{\text{Å}}}_{\mathfrak{H}}}{\sum_{\mathfrak{H}=1}^{\Finv}\overline{\overline{\text{Å}}}_{\mathfrak{H}}}-1}} \right) \right), \\ \left(\log_{\daleth\daleth}\left(1 + \frac{\prod_{\mathfrak{H}=1}^{\Finv}\left(\daleth\daleth^{v_{\Finv}^{rp}} - 1\right)^{\frac{\overline{\overline{\text{Å}}}_{\mathfrak{H}}}{\sum_{\mathfrak{H}=1}^{\Finv}\overline{\overline{\text{Å}}}_{\mathfrak{H}}}}}{(\daleth\daleth-1)^{\sum_{\mathfrak{H}=1}^{\Finv}\frac{\overline{\overline{\text{Å}}}_{\mathfrak{H}}}{\sum_{\mathfrak{H}=1}^{\Finv}\overline{\overline{\text{Å}}}_{\mathfrak{H}}}-1}} \right), \log_{\daleth\daleth}\left(1 + \frac{\prod_{\mathfrak{H}=1}^{\Finv}\left(\daleth\daleth^{v_{\Finv}^{ip}} - 1\right)^{\frac{\overline{\overline{\text{Å}}}_{\mathfrak{H}}}{\sum_{\mathfrak{H}=1}^{\Finv}\overline{\overline{\text{Å}}}_{\mathfrak{H}}}}}{(\daleth\daleth-1)^{\sum_{\mathfrak{H}=1}^{\Finv}\frac{\overline{\overline{\text{Å}}}_{\mathfrak{H}}}{\sum_{\mathfrak{H}=1}^{\Finv}\overline{\overline{\text{Å}}}_{\mathfrak{H}}}-1}} \right) \right) \end{array} \right)$$

$$= \left(\begin{array}{c} \left(1 - \log_{\daleth\daleth}\left(1 + \frac{\left(\daleth\daleth^{1-u_{\Finv}^{rp}} - 1\right)^{\sum_{\mathfrak{H}=1}^{\Finv}\frac{\overline{\overline{\text{Å}}}_{\mathfrak{H}}}{\sum_{\mathfrak{H}=1}^{\Finv}\overline{\overline{\text{Å}}}_{\mathfrak{H}}}}}{(\daleth\daleth-1)^{\sum_{\mathfrak{H}=1}^{\Finv}\frac{\overline{\overline{\text{Å}}}_{\mathfrak{H}}}{\sum_{\mathfrak{H}=1}^{\Finv}\overline{\overline{\text{Å}}}_{\mathfrak{H}}}-1}} \right), 1 - \log_{\daleth\daleth}\left(1 + \frac{\left(\daleth\daleth^{1-u_{\Finv}^{ip}} - 1\right)^{\sum_{\mathfrak{H}=1}^{\Finv}\frac{\overline{\overline{\text{Å}}}_{\mathfrak{H}}}{\sum_{\mathfrak{H}=1}^{\Finv}\overline{\overline{\text{Å}}}_{\mathfrak{H}}}}}{(\daleth\daleth-1)^{\sum_{\mathfrak{H}=1}^{\Finv}\frac{\overline{\overline{\text{Å}}}_{\mathfrak{H}}}{\sum_{\mathfrak{H}=1}^{\Finv}\overline{\overline{\text{Å}}}_{\mathfrak{H}}}-1}} \right) \right), \\ \left(\log_{\daleth\daleth}\left(1 + \frac{\left(\daleth\daleth^{v_{\Finv}^{rp}} - 1\right)^{\sum_{\mathfrak{H}=1}^{\Finv}\frac{\overline{\overline{\text{Å}}}_{\mathfrak{H}}}{\sum_{\mathfrak{H}=1}^{\Finv}\overline{\overline{\text{Å}}}_{\mathfrak{H}}}}}{(\daleth\daleth-1)^{\sum_{\mathfrak{H}=1}^{\Finv}\frac{\overline{\overline{\text{Å}}}_{\mathfrak{H}}}{\sum_{\mathfrak{H}=1}^{\Finv}\overline{\overline{\text{Å}}}_{\mathfrak{H}}}-1}} \right), \log_{\daleth\daleth}\left(1 + \frac{\left(\daleth\daleth^{v_{\Finv}^{ip}} - 1\right)^{\sum_{\mathfrak{H}=1}^{\Finv}\frac{\overline{\overline{\text{Å}}}_{\mathfrak{H}}}{\sum_{\mathfrak{H}=1}^{\Finv}\overline{\overline{\text{Å}}}_{\mathfrak{H}}}}}{(\daleth\daleth-1)^{\sum_{\mathfrak{H}=1}^{\Finv}\frac{\overline{\overline{\text{Å}}}_{\mathfrak{H}}}{\sum_{\mathfrak{H}=1}^{\Finv}\overline{\overline{\text{Å}}}_{\mathfrak{H}}}-1}} \right) \right) \end{array} \right)$$

$$= \left(\begin{array}{c} \left(1 - \log_{\daleth\daleth}\left(1 + \frac{\left(\daleth\daleth^{1-u_{\Finv}^{rp}} - 1\right)}{(\daleth\daleth-1)^{1-1}} \right), 1 - \log_{\daleth\daleth}\left(1 + \frac{\left(\daleth\daleth^{1-u_{\Finv}^{ip}} - 1\right)}{(\daleth\daleth-1)^{1-1}} \right) \right), \\ \left(\log_{\daleth\daleth}\left(1 + \frac{\left(\daleth\daleth^{v_{\Finv}^{rp}} - 1\right)}{(\daleth\daleth-1)^{1-1}} \right), \log_{\daleth\daleth}\left(1 + \frac{\left(\daleth\daleth^{v_{\Finv}^{ip}} - 1\right)}{(\daleth\daleth-1)^{1-1}} \right) \right) \end{array} \right), \sum_{\mathfrak{H}=1}^{\Finv}\frac{\overline{\overline{\text{Å}}}_{\mathfrak{H}}}{\sum_{\mathfrak{H}=1}^{\Finv}\overline{\overline{\text{Å}}}_{\mathfrak{H}}} = 1$$

$$= \left(\begin{array}{c} \left(1 - \log_{\daleth\daleth}\left(1 + \left(\daleth\daleth^{1-u_{\Finv}^{rp}} - 1\right)\right), 1 - \log_{\daleth\daleth}\left(1 + \left(\daleth\daleth^{1-u_{\Finv}^{ip}} - 1\right)\right) \right), \\ \left(\log_{\daleth\daleth}\left(1 + \left(\daleth\daleth^{v_{\Finv}^{rp}} - 1\right)\right), \log_{\daleth\daleth}\left(1 + \left(\daleth\daleth^{v_{\Finv}^{ip}} - 1\right)\right) \right) \end{array} \right)$$

$$= \left(\left(1 - \log_{\daleth\daleth}\left(\daleth\daleth^{1-u_{\Finv}^{rp}}\right), 1 - \log_{\daleth\daleth}\left(\daleth\daleth^{1-u_{\Finv}^{ip}}\right) \right), \left(\log_{\daleth\daleth}\left(\daleth\daleth^{v_{\Finv}^{rp}}\right), \log_{\daleth\daleth}\left(\daleth\daleth^{v_{\Finv}^{ip}}\right) \right) \right)$$

$$= \left(\left(1 - \left(1 - u_{\Finv}^{rp}\right), 1 - \left(1 - u_{\Finv}^{ip}\right) \right), \left(v_{\Finv}^{rp}, v_{\Finv}^{ip} \right) \right)$$

$$= \left(\left(u_{\Finv}^{rp}, u_{\Finv}^{ip} \right), \left(v_{\Finv}^{rp}, v_{\Finv}^{ip} \right) \right) = \overline{\overline{I\mathfrak{T}}}.$$

This proves the proposition. □

Proposition 2 (Monotonicity). *If* $\overline{\overline{I\mathfrak{T}}}_{\mathfrak{H}} = \left(\left(u_{\Finv_{\mathfrak{H}}}^{rp}, u_{\Finv_{\mathfrak{H}}}^{ip} \right), \left(v_{\Finv_{\mathfrak{H}}}^{rp}, v_{\Finv_{\mathfrak{H}}}^{ip} \right) \right) \leq \overline{\overline{I\mathfrak{T}}}_{\mathfrak{H}}^{*} = \left(\left(u_{\Finv_{\mathfrak{H}}}^{rp\,*}, u_{\Finv_{\mathfrak{H}}}^{ip\,*} \right), \left(v_{\Finv_{\mathfrak{H}}}^{rp\,*}, v_{\Finv_{\mathfrak{H}}}^{ip\,*} \right) \right)$, *then*

$$CIFFPA\left(\overline{\overline{I\mathfrak{T}}}_1, \overline{\overline{I\mathfrak{T}}}_2, \ldots, \overline{\overline{I\mathfrak{T}}}_{\Finv}\right) \leq CIFFPA\left(\overline{\overline{I\mathfrak{T}}}_1^{*}, \overline{\overline{I\mathfrak{T}}}_2^{*}, \ldots, \overline{\overline{I\mathfrak{T}}}_{\Finv}^{*}\right) \tag{21}$$

Proof. Consider $\overline{\overline{I\mathfrak{T}}}_{\mathfrak{H}} = ((u_{\Finv_{\mathfrak{H}}}^{rp}, u_{\Finv_{\mathfrak{H}}}^{ip}), (v_{\Finv_{\mathfrak{H}}}^{rp}, v_{\Finv_{\mathfrak{H}}}^{ip})) \leq \overline{\overline{I\mathfrak{T}}}_{\mathfrak{H}}^{*} = ((u_{\Finv_{\mathfrak{H}}}^{rp\,*}, u_{\Finv_{\mathfrak{H}}}^{ip\,*}), (v_{\Finv_{\mathfrak{H}}}^{rp\,*}, v_{\Finv_{\mathfrak{H}}}^{ip\,*}))$. Notice that $u_{\Finv_{\mathfrak{H}}}^{rp} \leq u_{\Finv_{\mathfrak{H}}}^{rp\,*}, u_{\Finv_{\mathfrak{H}}}^{ip} \leq u_{\Finv_{\mathfrak{H}}}^{ip\,*}$ and $v_{\Finv_{\mathfrak{H}}}^{rp} \geq v_{\Finv_{\mathfrak{H}}}^{rp\,*}, v_{\Finv_{\mathfrak{H}}}^{ip} \geq v_{\Finv_{\mathfrak{H}}}^{ip\,*}$, then we have $u_{\Finv_{\mathfrak{H}}}^{rp} \leq u_{\Finv_{\mathfrak{H}}}^{rp\,*}$

$$\Rightarrow 1 - u_{\Finv_{\mathfrak{H}}}^{rp} \geq 1 - u_{\Finv_{\mathfrak{H}}}^{rp\,*} \Rightarrow \daleth\daleth^{1-u_{\Finv_{\mathfrak{H}}}^{rp}} \geq \daleth\daleth^{1-u_{\Finv_{\mathfrak{H}}}^{rp\,*}} \Rightarrow \daleth\daleth^{1-u_{\Finv_{\mathfrak{H}}}^{rp}} - 1 \geq \daleth\daleth^{1-u_{\Finv_{\mathfrak{H}}}^{rp\,*}} - 1$$

$$\Rightarrow \left(\daleth\daleth^{1-u^{rp}_{\Finv_{\mathfrak{H}}}}-1\right)^{\frac{\overset{\cdots}{\overline{\overline{\mathrm{A}}}}_{\mathfrak{H}}}{\sum_{\mathfrak{H}=1}^{\Finv}\overline{\overline{\mathrm{A}}}_{\mathfrak{H}}}} \geq \left(\daleth\daleth^{1-u^{rp\,*}_{\Finv_{\mathfrak{H}}}}-1\right)^{\frac{\overset{\cdots}{\overline{\overline{\mathrm{A}}}}_{\mathfrak{H}}}{\sum_{\mathfrak{H}=1}^{\Finv}\overline{\overline{\mathrm{A}}}_{\mathfrak{H}}}} \Rightarrow \prod_{\mathfrak{H}=1}^{\Finv}\left(\daleth\daleth^{1-u^{rp}_{\Finv_{\mathfrak{H}}}}-1\right)^{\frac{\overset{\cdots}{\overline{\overline{\mathrm{A}}}}_{\mathfrak{H}}}{\sum_{\mathfrak{H}=1}^{\Finv}\overline{\overline{\mathrm{A}}}_{\mathfrak{H}}}}$$

$$\geq \prod_{\mathfrak{H}=1}^{\Finv}\left(\daleth\daleth^{1-u^{rp\,*}_{\Finv_{\mathfrak{H}}}}-1\right)^{\frac{\overset{\cdots}{\overline{\overline{\mathrm{A}}}}_{\mathfrak{H}}}{\sum_{\mathfrak{H}=1}^{\Finv}\overline{\overline{\mathrm{A}}}_{\mathfrak{H}}}} \Rightarrow \frac{\prod_{\mathfrak{H}=1}^{\Finv}\left(\daleth\daleth^{1-u^{rp}_{\Finv_{\mathfrak{H}}}}-1\right)^{\frac{\overset{\cdots}{\overline{\overline{\mathrm{A}}}}_{\mathfrak{H}}}{\sum_{\mathfrak{H}=1}^{\Finv}\overline{\overline{\mathrm{A}}}_{\mathfrak{H}}}}}{(\daleth\daleth-1)^{\sum_{\mathfrak{H}=1}^{\Finv}\frac{\overline{\overline{\mathrm{A}}}_{\mathfrak{H}}}{\sum_{\mathfrak{H}=1}^{\Finv}\overline{\overline{\mathrm{A}}}_{\mathfrak{H}}}-1}} \geq \frac{\prod_{\mathfrak{H}=1}^{\Finv}\left(\daleth\daleth^{1-u^{rp\,*}_{\Finv_{\mathfrak{H}}}}-1\right)^{\frac{\overset{\cdots}{\overline{\overline{\mathrm{A}}}}_{\mathfrak{H}}}{\sum_{\mathfrak{H}=1}^{\Finv}\overline{\overline{\mathrm{A}}}_{\mathfrak{H}}}}}{(\daleth\daleth-1)^{\sum_{\mathfrak{H}=1}^{\Finv}\frac{\overline{\overline{\mathrm{A}}}_{\mathfrak{H}}}{\sum_{\mathfrak{H}=1}^{\Finv}\overline{\overline{\mathrm{A}}}_{\mathfrak{H}}}-1}}$$

$$\Rightarrow \log_{\daleth\daleth}\left(1+\frac{\prod_{\mathfrak{H}=1}^{\Finv}\left(\daleth\daleth^{1-u^{rp}_{\Finv_{\mathfrak{H}}}}-1\right)^{\frac{\overset{\cdots}{\overline{\overline{\mathrm{A}}}}_{\mathfrak{H}}}{\sum_{\mathfrak{H}=1}^{\Finv}\overline{\overline{\mathrm{A}}}_{\mathfrak{H}}}}}{(\daleth\daleth-1)^{\sum_{\mathfrak{H}=1}^{\Finv}\frac{\overline{\overline{\mathrm{A}}}_{\mathfrak{H}}}{\sum_{\mathfrak{H}=1}^{\Finv}\overline{\overline{\mathrm{A}}}_{\mathfrak{H}}}-1}}\right) \geq \log_{\daleth\daleth}\left(1+\frac{\prod_{\mathfrak{H}=1}^{\Finv}\left(\daleth\daleth^{1-u^{rp\,*}_{\Finv_{\mathfrak{H}}}}-1\right)^{\frac{\overset{\cdots}{\overline{\overline{\mathrm{A}}}}_{\mathfrak{H}}}{\sum_{\mathfrak{H}=1}^{\Finv}\overline{\overline{\mathrm{A}}}_{\mathfrak{H}}}}}{(\daleth\daleth-1)^{\sum_{\mathfrak{H}=1}^{\Finv}\frac{\overline{\overline{\mathrm{A}}}_{\mathfrak{H}}}{\sum_{\mathfrak{H}=1}^{\Finv}\overline{\overline{\mathrm{A}}}_{\mathfrak{H}}}-1}}\right) \Rightarrow 1-$$

$$\log_{\daleth\daleth}\left(1+\frac{\prod_{\mathfrak{H}=1}^{\Finv}\left(\daleth\daleth^{1-u^{rp}_{\Finv_{\mathfrak{H}}}}-1\right)^{\frac{\overset{\cdots}{\overline{\overline{\mathrm{A}}}}_{\mathfrak{H}}}{\sum_{\mathfrak{H}=1}^{\Finv}\overline{\overline{\mathrm{A}}}_{\mathfrak{H}}}}}{(\daleth\daleth-1)^{\sum_{\mathfrak{H}=1}^{\Finv}\frac{\overline{\overline{\mathrm{A}}}_{\mathfrak{H}}}{\sum_{\mathfrak{H}=1}^{\Finv}\overline{\overline{\mathrm{A}}}_{\mathfrak{H}}}-1}}\right) \leq 1-\log_{\daleth\daleth}\left(1+\frac{\prod_{\mathfrak{H}=1}^{\Finv}\left(\daleth\daleth^{1-u^{rp\,*}_{\Finv_{\mathfrak{H}}}}-1\right)^{\frac{\overset{\cdots}{\overline{\overline{\mathrm{A}}}}_{\mathfrak{H}}}{\sum_{\mathfrak{H}=1}^{\Finv}\overline{\overline{\mathrm{A}}}_{\mathfrak{H}}}}}{(\daleth\daleth-1)^{\sum_{\mathfrak{H}=1}^{\Finv}\frac{\overline{\overline{\mathrm{A}}}_{\mathfrak{H}}}{\sum_{\mathfrak{H}=1}^{\Finv}\overline{\overline{\mathrm{A}}}_{\mathfrak{H}}}-1}}\right).$$

In the same way, we find the unreal part, such as:

$$1-\log_{\daleth\daleth}\left(1+\frac{\prod_{\mathfrak{H}=1}^{\Finv}\left(\daleth\daleth^{1-u^{ip}_{\Finv_{\mathfrak{H}}}}-1\right)^{\frac{\overset{\cdots}{\overline{\overline{\mathrm{A}}}}_{\mathfrak{H}}}{\sum_{\mathfrak{H}=1}^{\Finv}\overline{\overline{\mathrm{A}}}_{\mathfrak{H}}}}}{(\daleth\daleth-1)^{\sum_{\mathfrak{H}=1}^{\Finv}\frac{\overline{\overline{\mathrm{A}}}_{\mathfrak{H}}}{\sum_{\mathfrak{H}=1}^{\Finv}\overline{\overline{\mathrm{A}}}_{\mathfrak{H}}}-1}}\right) \leq 1-\log_{\daleth\daleth}\left(1+\frac{\prod_{\mathfrak{H}=1}^{\Finv}\left(\daleth\daleth^{1-u^{ip\,*}_{\Finv_{\mathfrak{H}}}}-1\right)^{\frac{\overset{\cdots}{\overline{\overline{\mathrm{A}}}}_{\mathfrak{H}}}{\sum_{\mathfrak{H}=1}^{\Finv}\overline{\overline{\mathrm{A}}}_{\mathfrak{H}}}}}{(\daleth\daleth-1)^{\sum_{\mathfrak{H}=1}^{\Finv}\frac{\overline{\overline{\mathrm{A}}}_{\mathfrak{H}}}{\sum_{\mathfrak{H}=1}^{\Finv}\overline{\overline{\mathrm{A}}}_{\mathfrak{H}}}-1}}\right)$$

Furthermore, we have $v^{rp}_{\Finv_{\mathfrak{H}}} \geq v^{rp\,*}_{\Finv_{\mathfrak{H}}} \Rightarrow \daleth\daleth^{v^{rp}_{\Finv_{\mathfrak{H}}}} \geq \daleth\daleth^{v^{rp\,*}_{\Finv_{\mathfrak{H}}}} \Rightarrow \daleth\daleth^{v^{rp}_{\Finv_{\mathfrak{H}}}}-1 \geq \daleth\daleth^{v^{rp\,*}_{\Finv_{\mathfrak{H}}}}-1$

$$\Rightarrow \left(\daleth\daleth^{v^{rp}_{\Finv_{\mathfrak{H}}}}-1\right)^{\frac{\overline{\overline{\mathrm{A}}}_{\mathfrak{H}}}{\sum_{\mathfrak{H}=1}^{\Finv}\overline{\overline{\mathrm{A}}}_{\mathfrak{H}}}} \geq \left(\daleth\daleth^{v^{rp\,*}_{\Finv_{\mathfrak{H}}}}-1\right)^{\frac{\overline{\overline{\mathrm{A}}}_{\mathfrak{H}}}{\sum_{\mathfrak{H}=1}^{\Finv}\overline{\overline{\mathrm{A}}}_{\mathfrak{H}}}} \Rightarrow \prod_{\mathfrak{H}=1}^{\Finv}\left(\daleth\daleth^{v^{rp}_{\Finv_{\mathfrak{H}}}}-1\right)^{\frac{\overline{\overline{\mathrm{A}}}_{\mathfrak{H}}}{\sum_{\mathfrak{H}=1}^{\Finv}\overline{\overline{\mathrm{A}}}_{\mathfrak{H}}}}$$

$$\geq \prod_{\mathfrak{H}=1}^{\Finv}\left(\daleth\daleth^{v^{rp\,*}_{\Finv_{\mathfrak{H}}}}-1\right)^{\frac{\overline{\overline{\mathrm{A}}}_{\mathfrak{H}}}{\sum_{\mathfrak{H}=1}^{\Finv}\overline{\overline{\mathrm{A}}}_{\mathfrak{H}}}} \Rightarrow \frac{\prod_{\mathfrak{H}=1}^{\Finv}\left(\daleth\daleth^{v^{rp}_{\Finv_{\mathfrak{H}}}}-1\right)^{\frac{\overline{\overline{\mathrm{A}}}_{\mathfrak{H}}}{\sum_{\mathfrak{H}=1}^{\Finv}\overline{\overline{\mathrm{A}}}_{\mathfrak{H}}}}}{(\daleth\daleth-1)^{\sum_{\mathfrak{H}=1}^{\Finv}\frac{\overline{\overline{\mathrm{A}}}_{\mathfrak{H}}}{\sum_{\mathfrak{H}=1}^{\Finv}\overline{\overline{\mathrm{A}}}_{\mathfrak{H}}}-1}} \geq \frac{\prod_{\mathfrak{H}=1}^{\Finv}\left(\daleth\daleth^{v^{rp\,*}_{\Finv_{\mathfrak{H}}}}-1\right)^{\frac{\overline{\overline{\mathrm{A}}}_{\mathfrak{H}}}{\sum_{\mathfrak{H}=1}^{\Finv}\overline{\overline{\mathrm{A}}}_{\mathfrak{H}}}}}{(\daleth\daleth-1)^{\sum_{\mathfrak{H}=1}^{\Finv}\frac{\overline{\overline{\mathrm{A}}}_{\mathfrak{H}}}{\sum_{\mathfrak{H}=1}^{\Finv}\overline{\overline{\mathrm{A}}}_{\mathfrak{H}}}-1}}$$

$$\Rightarrow 1+\frac{\prod_{\mathfrak{H}=1}^{\Finv}\left(\daleth\daleth^{v^{rp}_{\Finv_{\mathfrak{H}}}}-1\right)^{\frac{\overline{\overline{\mathrm{A}}}_{\mathfrak{H}}}{\sum_{\mathfrak{H}=1}^{\Finv}\overline{\overline{\mathrm{A}}}_{\mathfrak{H}}}}}{(\daleth\daleth-1)^{\sum_{\mathfrak{H}=1}^{\Finv}\frac{\overline{\overline{\mathrm{A}}}_{\mathfrak{H}}}{\sum_{\mathfrak{H}=1}^{\Finv}\overline{\overline{\mathrm{A}}}_{\mathfrak{H}}}-1}} \geq 1+\frac{\prod_{\mathfrak{H}=1}^{\Finv}\left(\daleth\daleth^{v^{rp\,*}_{\Finv_{\mathfrak{H}}}}-1\right)^{\frac{\overline{\overline{\mathrm{A}}}_{\mathfrak{H}}}{\sum_{\mathfrak{H}=1}^{\Finv}\overline{\overline{\mathrm{A}}}_{\mathfrak{H}}}}}{(\daleth\daleth-1)^{\sum_{\mathfrak{H}=1}^{\Finv}\frac{\overline{\overline{\mathrm{A}}}_{\mathfrak{H}}}{\sum_{\mathfrak{H}=1}^{\Finv}\overline{\overline{\mathrm{A}}}_{\mathfrak{H}}}-1}}$$

$$\Rightarrow \log_{\daleth\daleth}\left(1+\frac{\prod_{\mathfrak{H}=1}^{\Finv}\left(\daleth\daleth^{v^{rp}_{\Finv_{\mathfrak{H}}}}-1\right)^{\frac{\overline{\overline{\mathrm{A}}}_{\mathfrak{H}}}{\sum_{\mathfrak{H}=1}^{\Finv}\overline{\overline{\mathrm{A}}}_{\mathfrak{H}}}}}{(\daleth\daleth-1)^{\sum_{\mathfrak{H}=1}^{\Finv}\frac{\overline{\overline{\mathrm{A}}}_{\mathfrak{H}}}{\sum_{\mathfrak{H}=1}^{\Finv}\overline{\overline{\mathrm{A}}}_{\mathfrak{H}}}-1}}\right) \geq \log_{\daleth\daleth}\left(1+\frac{\prod_{\mathfrak{H}=1}^{\Finv}\left(\daleth\daleth^{v^{rp\,*}_{\Finv_{\mathfrak{H}}}}-1\right)^{\frac{\overline{\overline{\mathrm{A}}}_{\mathfrak{H}}}{\sum_{\mathfrak{H}=1}^{\Finv}\overline{\overline{\mathrm{A}}}_{\mathfrak{H}}}}}{(\daleth\daleth-1)^{\sum_{\mathfrak{H}=1}^{\Finv}\frac{\overline{\overline{\mathrm{A}}}_{\mathfrak{H}}}{\sum_{\mathfrak{H}=1}^{\Finv}\overline{\overline{\mathrm{A}}}_{\mathfrak{H}}}-1}}\right).$$

In the same way, we find the unreal part, such as: $\log_{\beth}\left(1+\frac{\prod_{\mathfrak{H}=1}^{\daleth}\left(\beth^{v^{ip}_{\daleth_{\mathfrak{H}}}}-1\right)^{\frac{\overline{\overline{\mathring{A}}}_{\mathfrak{H}}}{\sum_{\mathfrak{H}=1}^{\daleth}\overline{\overline{\mathring{A}}}_{\mathfrak{H}}}}}{(\beth-1)^{\sum_{\mathfrak{H}=1}^{\daleth}\frac{\overline{\overline{\mathring{A}}}_{\mathfrak{H}}}{\sum_{\mathfrak{H}=1}^{\daleth}\overline{\overline{\mathring{A}}}_{\mathfrak{H}}}-1}}\right)$ $\geq \log_{\beth}\left(1+\frac{\prod_{\mathfrak{H}=1}^{\daleth}\left(\beth^{v^{ip^{*}}_{\daleth_{\mathfrak{H}}}}-1\right)^{\frac{\overline{\overline{\mathring{A}}}_{\mathfrak{H}}}{\sum_{\mathfrak{H}=1}^{\daleth}\overline{\overline{\mathring{A}}}_{\mathfrak{H}}}}}{(\beth-1)^{\sum_{\mathfrak{H}=1}^{\daleth}\frac{\overline{\overline{\mathring{A}}}_{\mathfrak{H}}}{\sum_{\mathfrak{H}=1}^{\daleth}\overline{\overline{\mathring{A}}}_{\mathfrak{H}}}-1}}\right)$. Then, with the presence of the score function and accuracy function, we can easily obtain our required result with $CIFFPA\left(\overline{\overline{I\mathfrak{T}}}_1,\overline{\overline{I\mathfrak{T}}}_2,\ldots,\overline{\overline{I\mathfrak{T}}}_{\daleth}\right)$ $\leq CIFFPA\left(\overline{\overline{I\mathfrak{T}}}^{*}_1,\overline{\overline{I\mathfrak{T}}}^{*}_2,\ldots,\overline{\overline{I\mathfrak{T}}}^{*}_{\daleth}\right)$. This proves the proposition. □

Proposition 3 (Boundedness). *If* $\overline{\overline{I\mathfrak{T}}}^{-}_{\mathfrak{H}}=\left(\left(\min_{\mathfrak{H}} u^{rp}_{\daleth_{\mathfrak{H}}},\min_{\mathfrak{H}} u^{ip}_{\daleth_{\mathfrak{H}}}\right),\left(\max_{\mathfrak{H}} v^{rp}_{\daleth_{\mathfrak{H}}},\max_{\mathfrak{H}} v^{ip}_{\daleth_{\mathfrak{H}}}\right)\right)$ *and* $\overline{\overline{I\mathfrak{T}}}^{+}_{\mathfrak{H}}=\left(\left(\max_{\mathfrak{H}} u^{rp}_{\daleth_{\mathfrak{H}}},\max_{\mathfrak{H}} u^{ip}_{\daleth_{\mathfrak{H}}}\right),\left(\min_{\mathfrak{H}} v^{rp}_{\daleth_{\mathfrak{H}}},\min_{\mathfrak{H}} v^{ip}_{\daleth_{\mathfrak{H}}}\right)\right)$, *then we have*

$$\overline{\overline{I\mathfrak{T}}}^{-}_{\mathfrak{H}}\leq CIFFPA\left(\overline{\overline{I\mathfrak{T}}}_1,\overline{\overline{I\mathfrak{T}}}_2,\ldots,\overline{\overline{I\mathfrak{T}}}_{\daleth}\right)\leq\overline{\overline{I\mathfrak{T}}}^{+}_{\mathfrak{H}}. \tag{22}$$

Proof. Using Propositions 1 and 2, we have $CIFFPA\left(\overline{\overline{I\mathfrak{T}}}_1,\overline{\overline{I\mathfrak{T}}}_2,\ldots,\overline{\overline{I\mathfrak{T}}}_{\daleth}\right)\leq CIFFPA\left(\overline{\overline{I\mathfrak{T}}}^{+}_1,\overline{\overline{I\mathfrak{T}}}^{+}_2,\ldots,\overline{\overline{I\mathfrak{T}}}^{+}_{\daleth}\right)=\overline{\overline{I\mathfrak{T}}}^{+}_{\mathfrak{H}}$ and $CIFFPA\left(\overline{\overline{I\mathfrak{T}}}_1,\overline{\overline{I\mathfrak{T}}}_2,\ldots,\overline{\overline{I\mathfrak{T}}}_{\daleth}\right)\geq CIFFPA\left(\overline{\overline{I\mathfrak{T}}}^{-}_1,\overline{\overline{I\mathfrak{T}}}^{-}_2,\ldots,\overline{\overline{I\mathfrak{T}}}^{-}_{\daleth}\right)=\overline{\overline{I\mathfrak{T}}}^{-}_{\mathfrak{H}}$. Then, $\overline{\overline{I\mathfrak{T}}}^{-}_{\mathfrak{H}}\leq CIFFPA\left(\overline{\overline{I\mathfrak{T}}}_1,\overline{\overline{I\mathfrak{T}}}_2,\ldots,\overline{\overline{I\mathfrak{T}}}_{\daleth}\right)\leq\overline{\overline{I\mathfrak{T}}}^{+}_{\mathfrak{H}}$. □

Definition 7. *The mathematical form of the CIFFPOA operator is shown below:*

$$\begin{aligned}CIFFPOA\left(\overline{\overline{I\mathfrak{T}}}_1,\overline{\overline{I\mathfrak{T}}}_2,\ldots,\overline{\overline{I\mathfrak{T}}}_{\daleth}\right)&=\left(\frac{\overline{\overline{\mathring{A}}}_1}{\sum_{\mathfrak{H}=1}^{\daleth}\overline{\overline{\mathring{A}}}_{\mathfrak{H}}}\right)\overline{\overline{I\mathfrak{T}}}_{o(1)}\oplus\left(\frac{\overline{\overline{\mathring{A}}}_2}{\sum_{\mathfrak{H}=1}^{\daleth}\overline{\overline{\mathring{A}}}_{\mathfrak{H}}}\right)\overline{\overline{I\mathfrak{T}}}_{o(2)}\oplus\ldots\oplus\left(\frac{\overline{\overline{\mathring{A}}}_{\daleth}}{\sum_{\mathfrak{H}=1}^{\daleth}\overline{\overline{\mathring{A}}}_{\mathfrak{H}}}\right)\overline{\overline{I\mathfrak{T}}}_{o(\daleth)}\\&=\oplus_{\mathfrak{H}=1}^{\daleth}\left(\frac{\overline{\overline{\mathring{A}}}_{\mathfrak{H}}}{\sum_{\mathfrak{H}=1}^{\daleth}\overline{\overline{\mathring{A}}}_{\mathfrak{H}}}\right)\overline{\overline{I\mathfrak{T}}}_{o(\mathfrak{H})}.\end{aligned}\tag{23}$$

With the values of $\overline{\overline{\mathring{A}}}_1=1$ and $\overline{\overline{\mathring{A}}}_{\mathfrak{H}}=\prod_{B=1}^{\mathfrak{H}-1}V_s\left(\overline{\overline{I\mathfrak{T}}}_B\right)$ and $o(\mathfrak{H})\leq o(\mathfrak{H}-1)$.

Theorem 2. *With the help of the data in Equation (23), we expose that the aggregated value of Equation (23) will again be in the form of CIFV, such as:*

$$CIFFPOA\left(\overline{\overline{I\mathfrak{T}}}_1,\overline{\overline{I\mathfrak{T}}}_2,\ldots,\overline{\overline{I\mathfrak{T}}}_{\daleth}\right)=\left(\begin{array}{c}\left(1-\log_{\beth}\left(1+\frac{\prod_{\mathfrak{H}=1}^{\daleth}\left(\beth^{1-u^{rp}_{\daleth_{o(\mathfrak{H})}}}-1\right)^{\frac{\overline{\overline{\mathring{A}}}_{\mathfrak{H}}}{\sum_{\mathfrak{H}=1}^{\daleth}\overline{\overline{\mathring{A}}}_{\mathfrak{H}}}}}{(\beth-1)^{\sum_{\mathfrak{H}=1}^{\daleth}\frac{\overline{\overline{\mathring{A}}}_{\mathfrak{H}}}{\sum_{\mathfrak{H}=1}^{\daleth}\overline{\overline{\mathring{A}}}_{\mathfrak{H}}}-1}}\right),1-\log_{\beth}\left(1+\frac{\prod_{\mathfrak{H}=1}^{\daleth}\left(\beth^{1-u^{ip}_{\daleth_{o(\mathfrak{H})}}}-1\right)^{\frac{\overline{\overline{\mathring{A}}}_{\mathfrak{H}}}{\sum_{\mathfrak{H}=1}^{\daleth}\overline{\overline{\mathring{A}}}_{\mathfrak{H}}}}}{(\beth-1)^{\sum_{\mathfrak{H}=1}^{\daleth}\frac{\overline{\overline{\mathring{A}}}_{\mathfrak{H}}}{\sum_{\mathfrak{H}=1}^{\daleth}\overline{\overline{\mathring{A}}}_{\mathfrak{H}}}-1}}\right)\right),\\\left(\log_{\beth}\left(1+\frac{\prod_{\mathfrak{H}=1}^{\daleth}\left(\beth^{v^{rp}_{\daleth_{o(\mathfrak{H})}}}-1\right)^{\frac{\overline{\overline{\mathring{A}}}_{\mathfrak{H}}}{\sum_{\mathfrak{H}=1}^{\daleth}\overline{\overline{\mathring{A}}}_{\mathfrak{H}}}}}{(\beth-1)^{\sum_{\mathfrak{H}=1}^{\daleth}\frac{\overline{\overline{\mathring{A}}}_{\mathfrak{H}}}{\sum_{\mathfrak{H}=1}^{\daleth}\overline{\overline{\mathring{A}}}_{\mathfrak{H}}}-1}}\right),\log_{\beth}\left(1+\frac{\prod_{\mathfrak{H}=1}^{\daleth}\left(\beth^{v^{ip}_{\daleth_{o(\mathfrak{H})}}}-1\right)^{\frac{\overline{\overline{\mathring{A}}}_{\mathfrak{H}}}{\sum_{\mathfrak{H}=1}^{\daleth}\overline{\overline{\mathring{A}}}_{\mathfrak{H}}}}}{(\beth-1)^{\sum_{\mathfrak{H}=1}^{\daleth}\frac{\overline{\overline{\mathring{A}}}_{\mathfrak{H}}}{\sum_{\mathfrak{H}=1}^{\daleth}\overline{\overline{\mathring{A}}}_{\mathfrak{H}}}-1}}\right)\right)\end{array}\right).\tag{24}$$

Proposition 4 (Idempotency) *If we use* $\overline{\overline{I\mathfrak{T}}}_{\mathfrak{H}}=\overline{\overline{I\mathfrak{T}}}=\left(\left(u^{rp}_{\daleth},u^{ip}_{\daleth}\right),\left(v^{rp}_{\daleth},v^{ip}_{\daleth}\right)\right)$, *then*

$$CIFFPOA\left(\overline{\overline{I\mathfrak{T}}}_1,\overline{\overline{I\mathfrak{T}}}_2,\ldots,\overline{\overline{I\mathfrak{T}}}_{\daleth}\right)=\overline{\overline{I\mathfrak{T}}}. \tag{25}$$

Proposition 5 (Monotonicity). *If* $\overline{\overline{I\mathfrak{T}}}_{\mathfrak{H}}=\left(\left(u^{rp}_{\daleth_{\mathfrak{H}}},u^{ip}_{\daleth_{\mathfrak{H}}}\right),\left(v^{rp}_{\daleth_{\mathfrak{H}}},v^{ip}_{\daleth_{\mathfrak{H}}}\right)\right)\leq\overline{\overline{I\mathfrak{T}}}^{*}_{\mathfrak{H}}=\left(\left(u^{rp\,*}_{\daleth_{\mathfrak{H}}},u^{ip\,*}_{\daleth_{\mathfrak{H}}}\right),\left(v^{rp\,*}_{\daleth_{\mathfrak{H}}},v^{ip\,*}_{\daleth_{\mathfrak{H}}}\right)\right)$, *then*

$$CIFFPOA\left(\overline{\overline{I\mathfrak{T}}}_1,\overline{\overline{I\mathfrak{T}}}_2,\ldots,\overline{\overline{I\mathfrak{T}}}_{\daleth}\right)\leq CIFFPOA\left(\overline{\overline{I\mathfrak{T}}}^{*}_1,\overline{\overline{I\mathfrak{T}}}^{*}_2,\ldots,\overline{\overline{I\mathfrak{T}}}^{*}_{\daleth}\right). \tag{26}$$

Proposition 6 (Boundedness). *If* $\overline{\overline{I\mathfrak{T}}}^{-}_{\mathfrak{H}}=\left(\left(\underset{\mathfrak{H}}{min}\,u^{rp}_{\daleth_{\mathfrak{H}}},\underset{\mathfrak{H}}{min}\,u^{ip}_{\daleth_{\mathfrak{H}}}\right),\left(\underset{\mathfrak{H}}{max}\,v^{rp}_{\daleth_{\mathfrak{H}}},\underset{\mathfrak{H}}{max}\,v^{ip}_{\daleth_{\mathfrak{H}}}\right)\right)$ *and* $\overline{\overline{I\mathfrak{T}}}^{+}_{\mathfrak{H}}=\left(\left(\underset{\mathfrak{H}}{max}\,u^{rp}_{\daleth_{\mathfrak{H}}},\underset{\mathfrak{H}}{max}\,u^{ip}_{\daleth_{\mathfrak{H}}}\right),\left(\underset{\mathfrak{H}}{min}\,v^{rp}_{\daleth_{\mathfrak{H}}},\underset{\mathfrak{H}}{min}\,v^{ip}_{\daleth_{\mathfrak{H}}}\right)\right)$, *then we have*

$$C\overline{\overline{I\mathfrak{T}}}^{-}_{\mathfrak{H}}\leq CIFFPOA\left(\overline{\overline{I\mathfrak{T}}}_1,\overline{\overline{I\mathfrak{T}}}_2,\ldots,\overline{\overline{I\mathfrak{T}}}_{\daleth}\right)\leq\overline{\overline{I\mathfrak{T}}}^{+}_{\mathfrak{H}} \tag{27}$$

Definition 8. *The mathematical form of the CIFFPG operator is shown below:*

$$\begin{aligned}CIFFPG\left(\overline{\overline{I\mathfrak{T}}}_1,\overline{\overline{I\mathfrak{T}}}_2,\ldots,\overline{\overline{I\mathfrak{T}}}_{\daleth}\right)&=\overline{\overline{I\mathfrak{T}}}_1^{\left(\frac{\overline{\overline{\dot{A}}}_1}{\sum_{\mathfrak{H}=1}^{\daleth}\overline{\overline{\dot{A}}}_{\mathfrak{H}}}\right)}\otimes\overline{\overline{I\mathfrak{T}}}_2^{\left(\frac{\overline{\overline{\dot{A}}}_2}{\sum_{\mathfrak{H}=1}^{\daleth}\overline{\overline{\dot{A}}}_{\mathfrak{H}}}\right)}\otimes\ldots\otimes\overline{\overline{I\mathfrak{T}}}_{\daleth}^{\left(\frac{\overline{\overline{\dot{A}}}_{\daleth}}{\sum_{\mathfrak{H}=1}^{\daleth}\overline{\overline{\dot{A}}}_{\mathfrak{H}}}\right)}\\&=\otimes_{\mathfrak{H}=1}^{\daleth}\overline{\overline{I\mathfrak{T}}}_{\mathfrak{H}}^{\left(\frac{\overline{\overline{\dot{A}}}_{\mathfrak{H}}}{\sum_{\mathfrak{H}=1}^{\daleth}\overline{\overline{\dot{A}}}_{\mathfrak{H}}}\right)}\end{aligned} \tag{28}$$

With the values of $\overline{\overline{\dot{A}}}_1=1$ and $\overline{\overline{\dot{A}}}_{\mathfrak{H}}=\prod_{B=1}^{\mathfrak{H}-1}V_s\left(\overline{\overline{I\mathfrak{T}}}_B\right)$.

Theorem 3. *With the help of the data in Equation (28), we expose that the aggregated value of Equation (28) will again be in the form of CIFV, such as:*

$$CIFFPG\left(\overline{\overline{I\mathfrak{T}}}_1,\overline{\overline{I\mathfrak{T}}}_2,\ldots,\overline{\overline{I\mathfrak{T}}}_{\daleth}\right)=\left(\begin{array}{c}\left(\log_{\beth}\left(1+\frac{\prod_{\mathfrak{H}=1}^{\daleth}\left(\beth^{u^{rp}_{\daleth_{\mathfrak{H}}}}-1\right)^{\frac{\overline{\overline{\dot{A}}}_{\mathfrak{H}}}{\sum_{\mathfrak{H}=1}^{\daleth}\overline{\overline{\dot{A}}}_{\mathfrak{H}}}}}{(\beth-1)^{\sum_{\mathfrak{H}=1}^{\daleth}\frac{\overline{\overline{\dot{A}}}_{\mathfrak{H}}}{\sum_{\mathfrak{H}=1}^{\daleth}\overline{\overline{\dot{A}}}_{\mathfrak{H}}}-1}}\right),\log_{\beth}\left(1+\frac{\prod_{\mathfrak{H}=1}^{\daleth}\left(\beth^{u^{ip}_{\daleth_{\mathfrak{H}}}}-1\right)^{\frac{\overline{\overline{\dot{A}}}_{\mathfrak{H}}}{\sum_{\mathfrak{H}=1}^{\daleth}\overline{\overline{\dot{A}}}_{\mathfrak{H}}}}}{(\beth-1)^{\sum_{\mathfrak{H}=1}^{\daleth}\frac{\overline{\overline{\dot{A}}}_{\mathfrak{H}}}{\sum_{\mathfrak{H}=1}^{\daleth}\overline{\overline{\dot{A}}}_{\mathfrak{H}}}-1}}\right)\right),\\\left(1-\log_{\beth}\left(1+\frac{\prod_{\mathfrak{H}=1}^{\daleth}\left(\beth^{1-v^{rp}_{\daleth_{\mathfrak{H}}}}-1\right)^{\frac{\overline{\overline{\dot{A}}}_{\mathfrak{H}}}{\sum_{\mathfrak{H}=1}^{\daleth}\overline{\overline{\dot{A}}}_{\mathfrak{H}}}}}{(\beth-1)^{\sum_{\mathfrak{H}=1}^{\daleth}\frac{\overline{\overline{\dot{A}}}_{\mathfrak{H}}}{\sum_{\mathfrak{H}=1}^{\daleth}\overline{\overline{\dot{A}}}_{\mathfrak{H}}}-1}}\right),1-\log_{\beth}\left(1+\frac{\prod_{\mathfrak{H}=1}^{\daleth}\left(\beth^{1-v^{ip}_{\daleth_{\mathfrak{H}}}}-1\right)^{\frac{\overline{\overline{\dot{A}}}_{\mathfrak{H}}}{\sum_{\mathfrak{H}=1}^{\daleth}\overline{\overline{\dot{A}}}_{\mathfrak{H}}}}}{(\beth-1)^{\sum_{\mathfrak{H}=1}^{\daleth}\frac{\overline{\overline{\dot{A}}}_{\mathfrak{H}}}{\sum_{\mathfrak{H}=1}^{\daleth}\overline{\overline{\dot{A}}}_{\mathfrak{H}}}-1}}\right)\right)\end{array}\right) \tag{29}$$

Proposition 7 (Idempotency). *If we use* $\overline{\overline{I\mathfrak{T}}}_{\mathfrak{H}}=\overline{\overline{I\mathfrak{T}}}=\left(\left(u^{rp}_{\daleth},u^{ip}_{\daleth}\right),\left(v^{rp}_{\daleth},v^{ip}_{\daleth}\right)\right)$, *then*

$$CIFFPG\left(\overline{\overline{I\mathfrak{T}}}_1,\overline{\overline{I\mathfrak{T}}}_2,\ldots,\overline{\overline{I\mathfrak{T}}}_{\daleth}\right)=\overline{\overline{I\mathfrak{T}}}. \tag{30}$$

Proposition 8 (Monotonicity). *If* $\overline{\overline{I\mathfrak{T}}}_{\mathfrak{H}}=\left(\left(u^{rp}_{\daleth_{\mathfrak{H}}},u^{ip}_{\daleth_{\mathfrak{H}}}\right),\left(v^{rp}_{\daleth_{\mathfrak{H}}},v^{ip}_{\daleth_{\mathfrak{H}}}\right)\right)\leq\overline{\overline{I\mathfrak{T}}}^{*}_{\mathfrak{H}}=\left(\left(u^{rp\,*}_{\daleth_{\mathfrak{H}}},u^{ip\,*}_{\daleth_{\mathfrak{H}}}\right),\left(v^{rp\,*}_{\daleth_{\mathfrak{H}}},v^{ip\,*}_{\daleth_{\mathfrak{H}}}\right)\right)$, *then*

$$CIFFPG\left(\overline{\overline{I\mathfrak{T}}}_1,\overline{\overline{I\mathfrak{T}}}_2,\ldots,\overline{\overline{I\mathfrak{T}}}_{\daleth}\right)\leq CIFFPG\left(\overline{\overline{I\mathfrak{T}}}^{*}_1,\overline{\overline{I\mathfrak{T}}}^{*}_2,\ldots,\overline{\overline{I\mathfrak{T}}}^{*}_{\daleth}\right). \tag{31}$$

Proposition 9 (Boundedness). *If* $\overline{\overline{I\mathfrak{T}}}_{\mathfrak{H}}^{-} = \left(\left(\underset{\mathfrak{H}}{min}u_{\daleth_{\mathfrak{H}}}^{rp}, \underset{\mathfrak{H}}{min}u_{\daleth_{\mathfrak{H}}}^{ip}\right), \left(\underset{\mathfrak{H}}{max}v_{\daleth_{\mathfrak{H}}}^{rp}, \underset{\mathfrak{H}}{max}v_{\daleth_{\mathfrak{H}}}^{ip}\right)\right)$ *and* $\overline{\overline{I\mathfrak{T}}}_{\mathfrak{H}}^{+} = \left(\left(\underset{\mathfrak{H}}{max}u_{\daleth_{\mathfrak{H}}}^{rp}, \underset{\mathfrak{H}}{max}u_{\daleth_{\mathfrak{H}}}^{ip}\right), \left(\underset{\mathfrak{H}}{min}v_{\daleth_{\mathfrak{H}}}^{rp}, \underset{\mathfrak{H}}{min}v_{\daleth_{\mathfrak{H}}}^{ip}\right)\right)$*, then we have*

$$\overline{\overline{I\mathfrak{T}}}_{\mathfrak{H}}^{-} \leq CIFFPG\left(\overline{\overline{I\mathfrak{T}}}_1, \overline{\overline{I\mathfrak{T}}}_2, \ldots, \overline{\overline{I\mathfrak{T}}}_{\daleth}\right) \leq \overline{\overline{I\mathfrak{T}}}_{\mathfrak{H}}^{+}. \tag{32}$$

Definition 9. *The mathematical form of the CIFFPOG operator is shown below:*

$$\begin{aligned} CIFFPOG\left(\overline{\overline{I\mathfrak{T}}}_1, \overline{\overline{I\mathfrak{T}}}_2, \ldots, \overline{\overline{I\mathfrak{T}}}_{\daleth}\right) &= \overline{\overline{I\mathfrak{T}}}_{o(1)}^{\left(\frac{\overline{\overline{\mathring{A}}}_1}{\sum_{\mathfrak{H}=1}^{\daleth}\overline{\overline{\mathring{A}}}_{\mathfrak{H}}}\right)} \otimes \overline{\overline{I\mathfrak{T}}}_{o(2)}^{\left(\frac{\overline{\overline{\mathring{A}}}_2}{\sum_{\mathfrak{H}=1}^{\daleth}\overline{\overline{\mathring{A}}}_{\mathfrak{H}}}\right)} \otimes \ldots \otimes \overline{\overline{I\mathfrak{T}}}_{o(\daleth)}^{\left(\frac{\overline{\overline{\mathring{A}}}_{\daleth}}{\sum_{\mathfrak{H}=1}^{\daleth}\overline{\overline{\mathring{A}}}_{\mathfrak{H}}}\right)} \\ &= \otimes_{\mathfrak{H}=1}^{\daleth}\overline{\overline{I\mathfrak{T}}}_{o(\mathfrak{H})}^{\left(\frac{\overline{\overline{\mathring{A}}}_{\mathfrak{H}}}{\sum_{\mathfrak{H}=1}^{\daleth}\overline{\overline{\mathring{A}}}_{\mathfrak{H}}}\right)} \end{aligned} \tag{33}$$

With the values of $\overline{\overline{\mathring{A}}}_1 = 1$ and $\overline{\overline{\mathring{A}}}_{\mathfrak{H}} = \prod_{\mathcal{B}=1}^{\mathfrak{H}-1} V_s\left(\overline{\overline{I\mathfrak{T}}}_{\mathcal{B}}\right)$ and $o(\mathfrak{H}) \leq o(\mathfrak{H}-1)$.

Theorem 4. *With the help of the data in Equation (33), we expose that the aggregated value of Equation (33) will again be in the form of CIFV, such as:*

$$CIFFPOG\left(\overline{\overline{I\mathfrak{T}}}_1, \overline{\overline{I\mathfrak{T}}}_2, \ldots, \overline{\overline{I\mathfrak{T}}}_{\daleth}\right) = \left(\begin{array}{c} \left(\log_{\daleth\daleth}\left(1 + \frac{\prod_{\mathfrak{H}=1}^{\daleth}\left(\daleth\daleth^{u_{\daleth_{o(\mathfrak{H})}}^{rp}} - 1\right)^{\frac{\overline{\overline{\mathring{A}}}_{\mathfrak{H}}}{\sum_{\mathfrak{H}=1}^{\daleth}\overline{\overline{\mathring{A}}}_{\mathfrak{H}}}}}{(\daleth\daleth - 1)^{\sum_{\mathfrak{H}=1}^{\daleth}\frac{\overline{\overline{\mathring{A}}}_{\mathfrak{H}}}{\sum_{\mathfrak{H}=1}^{\daleth}\overline{\overline{\mathring{A}}}_{\mathfrak{H}}} - 1}}\right), \log_{\daleth\daleth}\left(1 + \frac{\prod_{\mathfrak{H}=1}^{\daleth}\left(\daleth\daleth^{u_{\daleth_{o(\mathfrak{H})}}^{ip}} - 1\right)^{\frac{\overline{\overline{\mathring{A}}}_{\mathfrak{H}}}{\sum_{\mathfrak{H}=1}^{\daleth}\overline{\overline{\mathring{A}}}_{\mathfrak{H}}}}}{(\daleth\daleth - 1)^{\sum_{\mathfrak{H}=1}^{\daleth}\frac{\overline{\overline{\mathring{A}}}_{\mathfrak{H}}}{\sum_{\mathfrak{H}=1}^{\daleth}\overline{\overline{\mathring{A}}}_{\mathfrak{H}}} - 1}}\right)\right), \\ \left(1 - \log_{\daleth\daleth}\left(1 + \frac{\prod_{\mathfrak{H}=1}^{\daleth}\left(\daleth\daleth^{1 - v_{\daleth_{o(\mathfrak{H})}}^{rp}} - 1\right)^{\frac{\overline{\overline{\mathring{A}}}_{\mathfrak{H}}}{\sum_{\mathfrak{H}=1}^{\daleth}\overline{\overline{\mathring{A}}}_{\mathfrak{H}}}}}{(\daleth\daleth - 1)^{\sum_{\mathfrak{H}=1}^{\daleth}\frac{\overline{\overline{\mathring{A}}}_{\mathfrak{H}}}{\sum_{\mathfrak{H}=1}^{\daleth}\overline{\overline{\mathring{A}}}_{\mathfrak{H}}} - 1}}\right), 1 - \log_{\daleth\daleth}\left(1 + \frac{\prod_{\mathfrak{H}=1}^{\daleth}\left(\daleth\daleth^{1 - v_{\daleth_{o(\mathfrak{H})}}^{ip}} - 1\right)^{\frac{\overline{\overline{\mathring{A}}}_{\mathfrak{H}}}{\sum_{\mathfrak{H}=1}^{\daleth}\overline{\overline{\mathring{A}}}_{\mathfrak{H}}}}}{(\daleth\daleth - 1)^{\sum_{\mathfrak{H}=1}^{\daleth}\frac{\overline{\overline{\mathring{A}}}_{\mathfrak{H}}}{\sum_{\mathfrak{H}=1}^{\daleth}\overline{\overline{\mathring{A}}}_{\mathfrak{H}}} - 1}}\right)\right) \end{array}\right) \tag{34}$$

Proposition 10 (Idempotency). *If we use* $\overline{\overline{I\mathfrak{T}}}_{\mathfrak{H}} = \overline{\overline{I\mathfrak{T}}} = \left(\left(u_{\daleth}^{rp}, u_{\daleth}^{ip}\right), \left(v_{\daleth}^{rp}, v_{\daleth}^{ip}\right)\right)$*, then*

$$CIFFPOG\left(\overline{\overline{I\mathfrak{T}}}_1, \overline{\overline{I\mathfrak{T}}}_2, \ldots, \overline{\overline{I\mathfrak{T}}}_{\daleth}\right) = \overline{\overline{I\mathfrak{T}}}. \tag{35}$$

Proposition 11 (Monotonicity). *If* $\overline{\overline{I\mathfrak{T}}}_{\mathfrak{H}} = \left(\left(u_{\daleth_{\mathfrak{H}}}^{rp}, u_{\daleth_{\mathfrak{H}}}^{ip}\right), \left(v_{\daleth_{\mathfrak{H}}}^{rp}, v_{\daleth_{\mathfrak{H}}}^{ip}\right)\right) \leq \overline{\overline{I\mathfrak{T}}}_{\mathfrak{H}}^{*} = \left(\left(u_{\daleth_{\mathfrak{H}}}^{rp\,*}, u_{\daleth_{\mathfrak{H}}}^{ip\,*}\right), \left(v_{\daleth_{\mathfrak{H}}}^{rp\,*}, v_{\daleth_{\mathfrak{H}}}^{ip\,*}\right)\right)$*, then*

$$CIFFPOG\left(\overline{\overline{I\mathfrak{T}}}_1, \overline{\overline{I\mathfrak{T}}}_2, \ldots, \overline{\overline{I\mathfrak{T}}}_{\daleth}\right) \leq CIFFPOG\left(\overline{\overline{I\mathfrak{T}}}_1^{*}, \overline{\overline{I\mathfrak{T}}}_2^{*}, \ldots, \overline{\overline{I\mathfrak{T}}}_{\daleth}^{*}\right). \tag{36}$$

Proposition 12 (Boundedness). *If* $\overline{\overline{I\mathfrak{T}}}_{\mathfrak{H}}^{-} = \left(\left(\underset{\mathfrak{H}}{min}u_{\daleth_{\mathfrak{H}}}^{rp}, \underset{\mathfrak{H}}{min}u_{\daleth_{\mathfrak{H}}}^{ip}\right), \left(\underset{\mathfrak{H}}{max}v_{\daleth_{\mathfrak{H}}}^{rp}, \underset{\mathfrak{H}}{max}v_{\daleth_{\mathfrak{H}}}^{ip}\right)\right)$ *and* $\overline{\overline{I\mathfrak{T}}}_{\mathfrak{H}}^{+} = \left(\left(\underset{\mathfrak{H}}{max}u_{\daleth_{\mathfrak{H}}}^{rp}, \underset{\mathfrak{H}}{max}u_{\daleth_{\mathfrak{H}}}^{ip}\right), \left(\underset{\mathfrak{H}}{min}v_{\daleth_{\mathfrak{H}}}^{rp}, \underset{\mathfrak{H}}{min}v_{\daleth_{\mathfrak{H}}}^{ip}\right)\right)$*, then we have*

$$\overline{\overline{I\mathfrak{T}}}_{\mathfrak{H}}^{-} \leq CIFFPOG\left(\overline{\overline{I\mathfrak{T}}}_1, \overline{\overline{I\mathfrak{T}}}_2, \ldots, \overline{\overline{I\mathfrak{T}}}_{\daleth}\right) \leq \overline{\overline{I\mathfrak{T}}}_{\mathfrak{H}}^{+}. \tag{37}$$

4. CIF WASPAS Procedures

The main point of this section is to extend the theory of the WASPAS procedure to CIF information. The procedures of the WASPAS method contain various valuable and dominant steps. Before evaluating the normalization, we arrange a collection of CIF data which may be of a benefit type or cost type. If the data are of a benefit type, then good, otherwise, using the below theory, we normalize the information, such as:

$$\overline{\overline{I\mathfrak{T}}}_{0,\mathfrak{H}} = \left(\left(\max_{\mathfrak{H}} u^{rp}_{\daleth i\mathfrak{H}}, \max_{\mathfrak{H}} u^{ip}_{\daleth i\mathfrak{H}}\right), \left(\min_{\mathfrak{H}} v^{rp}_{\daleth i\mathfrak{H}}, \min_{\mathfrak{H}} v^{ip}_{\daleth i\mathfrak{H}}\right)\right), i, \mathfrak{H} = 1, 2, \ldots, \daleth, \daleth \quad (38)$$

$$\overline{\overline{I\mathfrak{T}}}'_{\mathfrak{H}} = \left(\left(u^{rp}_{\daleth i\mathfrak{H}}, u^{ip}_{\daleth i\mathfrak{H}}\right), \left(v^{rp}_{\daleth i\mathfrak{H}}, v^{ip}_{\daleth i\mathfrak{H}}\right)\right) = \begin{cases} 0 & otherwise \\ \frac{u^{rp}_{\daleth i\mathfrak{H}}}{1+u^{rp}_{\daleth 0,\mathfrak{H}}}, \frac{u^{ip}_{\daleth i\mathfrak{H}}}{1+u^{ip}_{\daleth 0,\mathfrak{H}}} & if\ u^{rp}_{\daleth i\mathfrak{H}} \le u^{rp}_{\daleth 0,\mathfrak{H}}, u^{ip}_{\daleth i\mathfrak{H}} \le u^{ip}_{\daleth 0,\mathfrak{H}} \\ \frac{v^{rp}_{\daleth i\mathfrak{H}}}{1+v^{rp}_{\daleth 0,\mathfrak{H}}}, \frac{v^{ip}_{\daleth i\mathfrak{H}}}{1+v^{ip}_{\daleth 0,\mathfrak{H}}} & for\ falsity\ grade\ (real\ a\daleth d\ u\daleth real\ parts) \end{cases} \quad (39)$$

After performing the above evaluation, we calculate the WSA and WPA with the help of derived theory, such as:

$$T^{WSA}_{\mathfrak{H}} = CIFFPA\left(\overline{\overline{I\mathfrak{T}}}_1, \overline{\overline{I\mathfrak{T}}}_2, \ldots, \overline{\overline{I\mathfrak{T}}}_{\daleth}\right) = \left(\begin{array}{c} \left(1-\log_{\gimel\gimel}\left(1+\frac{\prod_{\mathfrak{H}=1}^{\daleth}\left(\gimel\gimel^{1-u^{rp}_{\daleth\mathfrak{H}}}-1\right)^{\frac{\overline{\overline{\mathring{A}}}_{\mathfrak{H}}}{\sum_{\mathfrak{H}=1}^{\daleth}\overline{\overline{\mathring{A}}}_{\mathfrak{H}}}}}{(\gimel\gimel-1)^{\sum_{\mathfrak{H}=1}^{\daleth}\frac{\overline{\overline{\mathring{A}}}_{\mathfrak{H}}}{\sum_{\mathfrak{H}=1}^{\daleth}\overline{\overline{\mathring{A}}}_{\mathfrak{H}}}-1}}\right), 1-\log_{\gimel\gimel}\left(1+\frac{\prod_{\mathfrak{H}=1}^{\daleth}\left(\gimel\gimel^{1-u^{ip}_{\daleth\mathfrak{H}}}-1\right)^{\frac{\overline{\overline{\mathring{A}}}_{\mathfrak{H}}}{\sum_{\mathfrak{H}=1}^{\daleth}\overline{\overline{\mathring{A}}}_{\mathfrak{H}}}}}{(\gimel\gimel-1)^{\sum_{\mathfrak{H}=1}^{\daleth}\frac{\overline{\overline{\mathring{A}}}_{\mathfrak{H}}}{\sum_{\mathfrak{H}=1}^{\daleth}\overline{\overline{\mathring{A}}}_{\mathfrak{H}}}-1}}\right)\right), \\ \left(\log_{\gimel\gimel}\left(1+\frac{\prod_{\mathfrak{H}=1}^{\daleth}\left(\gimel\gimel^{v^{rp}_{\daleth\mathfrak{H}}}-1\right)^{\frac{\overline{\overline{\mathring{A}}}_{\mathfrak{H}}}{\sum_{\mathfrak{H}=1}^{\daleth}\overline{\overline{\mathring{A}}}_{\mathfrak{H}}}}}{(\gimel\gimel-1)^{\sum_{\mathfrak{H}=1}^{\daleth}\frac{\overline{\overline{\mathring{A}}}_{\mathfrak{H}}}{\sum_{\mathfrak{H}=1}^{\daleth}\overline{\overline{\mathring{A}}}_{\mathfrak{H}}}-1}}\right), \log_{\gimel\gimel}\left(1+\frac{\prod_{\mathfrak{H}=1}^{\daleth}\left(\gimel\gimel^{v^{ip}_{\daleth\mathfrak{H}}}-1\right)^{\frac{\overline{\overline{\mathring{A}}}_{\mathfrak{H}}}{\sum_{\mathfrak{H}=1}^{\daleth}\overline{\overline{\mathring{A}}}_{\mathfrak{H}}}}}{(\gimel\gimel-1)^{\sum_{\mathfrak{H}=1}^{\daleth}\frac{\overline{\overline{\mathring{A}}}_{\mathfrak{H}}}{\sum_{\mathfrak{H}=1}^{\daleth}\overline{\overline{\mathring{A}}}_{\mathfrak{H}}}-1}}\right)\right) \end{array}\right) \quad (40)$$

$$T^{WPA}_{\mathfrak{H}} = CIFFPG\left(\overline{\overline{I\mathfrak{T}}}_1, \overline{\overline{I\mathfrak{T}}}_2, \ldots, \overline{\overline{I\mathfrak{T}}}_{\daleth}\right) = \left(\begin{array}{c} \left(\log_{\gimel\gimel}\left(1+\frac{\prod_{\mathfrak{H}=1}^{\daleth}\left(\gimel\gimel^{u^{rp}_{\daleth\mathfrak{H}}}-1\right)^{\frac{\overline{\overline{\mathring{A}}}_{\mathfrak{H}}}{\sum_{\mathfrak{H}=1}^{\daleth}\overline{\overline{\mathring{A}}}_{\mathfrak{H}}}}}{(\gimel\gimel-1)^{\sum_{\mathfrak{H}=1}^{\daleth}\frac{\overline{\overline{\mathring{A}}}_{\mathfrak{H}}}{\sum_{\mathfrak{H}=1}^{\daleth}\overline{\overline{\mathring{A}}}_{\mathfrak{H}}}-1}}\right), \log_{\gimel\gimel}\left(1+\frac{\prod_{\mathfrak{H}=1}^{\daleth}\left(\gimel\gimel^{u^{ip}_{\daleth\mathfrak{H}}}-1\right)^{\frac{\overline{\overline{\mathring{A}}}_{\mathfrak{H}}}{\sum_{\mathfrak{H}=1}^{\daleth}\overline{\overline{\mathring{A}}}_{\mathfrak{H}}}}}{(\gimel\gimel-1)^{\sum_{\mathfrak{H}=1}^{\daleth}\frac{\overline{\overline{\mathring{A}}}_{\mathfrak{H}}}{\sum_{\mathfrak{H}=1}^{\daleth}\overline{\overline{\mathring{A}}}_{\mathfrak{H}}}-1}}\right)\right), \\ \left(1-\log_{\gimel\gimel}\left(1+\frac{\prod_{\mathfrak{H}=1}^{\daleth}\left(\gimel\gimel^{1-v^{rp}_{\daleth\mathfrak{H}}}-1\right)^{\frac{\overline{\overline{\mathring{A}}}_{\mathfrak{H}}}{\sum_{\mathfrak{H}=1}^{\daleth}\overline{\overline{\mathring{A}}}_{\mathfrak{H}}}}}{(\gimel\gimel-1)^{\sum_{\mathfrak{H}=1}^{\daleth}\frac{\overline{\overline{\mathring{A}}}_{\mathfrak{H}}}{\sum_{\mathfrak{H}=1}^{\daleth}\overline{\overline{\mathring{A}}}_{\mathfrak{H}}}-1}}\right), 1-\log_{\gimel\gimel}\left(1+\frac{\prod_{\mathfrak{H}=1}^{\daleth}\left(\gimel\gimel^{1-v^{ip}_{\daleth\mathfrak{H}}}-1\right)^{\frac{\overline{\overline{\mathring{A}}}_{\mathfrak{H}}}{\sum_{\mathfrak{H}=1}^{\daleth}\overline{\overline{\mathring{A}}}_{\mathfrak{H}}}}}{(\gimel\gimel-1)^{\sum_{\mathfrak{H}=1}^{\daleth}\frac{\overline{\overline{\mathring{A}}}_{\mathfrak{H}}}{\sum_{\mathfrak{H}=1}^{\daleth}\overline{\overline{\mathring{A}}}_{\mathfrak{H}}}-1}}\right)\right) \end{array}\right) \quad (41)$$

where, $\overline{\overline{\mathring{A}}}_1 = 1$ and $\overline{\overline{\mathring{A}}}_{\mathfrak{H}} = \prod_{B=1}^{\mathfrak{H}-1} V_s\left(\overline{\overline{I\mathfrak{T}}}_B\right)$.

Using the data in Equations (40) and (41), we calculate the aggregated measure based on convex theory, such as:

$$T_{\mathfrak{H}} = {}^{\circ}\mathrm{F} V_s T^{WSA}_{\mathfrak{H}} + (1 - {}^{\circ}\mathrm{F}) V_s T^{WPA}_{\mathfrak{H}}, {}^{\circ}\mathrm{F} \in [0, 1] \quad (42)$$

Before ranking the alternatives, we discuss the special cases of the WASPAS technique based on CIF information such as:

1. When $^{\circ}\mathrm{F} = 1$, we obtain the data in Equation (40);
2. When $^{\circ}\mathrm{F} = 0$, we obtain the data in Equation (41).

In last, we derive the ranking result for examining the best one from the family of finite preferences. Furthermore, we justify the supremacy and worth of the derived theory with the help of some suitable examples, such as:

$$\overline{\overline{I\mathfrak{T}}}_{\mathfrak{H}} = \begin{bmatrix} ((0.4,0.2),(0.1,0.2)) & ((0.4,0.2),(0.1,0.2)) & ((0.4,0.2),(0.1,0.2)) & ((0.1,0.2),(0.1,0.2)) \\ ((0.5,0.4),(0.2,0.3)) & ((0.1,0.4),(0.2,0.3)) & ((0.5,0.4),(0.2,0.3)) & ((0.5,0.4),(0.2,0.3)) \\ ((0.6,0.5),(0.2,0.3)) & ((0.6,0.5),(0.2,0.3)) & ((0.2,0.5),(0.2,0.3)) & ((0.6,0.5),(0.2,0.3)) \\ ((0.7,0.8),(0.1,0.1)) & ((0.7,0.8),(0.1,0.1)) & ((0.7,0.8),(0.1,0.1)) & ((0.7,0.8),(0.1,0.1)) \end{bmatrix}$$

Then, we find the positive ideal, such as:

$$\overline{\overline{I\mathfrak{T}}}_{0,\mathfrak{H}} = \{((0.7,0.8),(0.1,0.1)),((0.7,0.8),(0.1,0.1)),((0.7,0.8),(0.1,0.1)),((0.7,0.8),(0.1,0.1))\}$$

With the help of the $\overline{\overline{I\mathfrak{T}}}_{0,\mathfrak{H}}$ and the information in $\overline{\overline{I\mathfrak{T}}}_{\mathfrak{H}}$, we obtain the below theory, such as:

$$\overline{\overline{I\mathfrak{T}}}'_{\mathfrak{H}} = \begin{bmatrix} \begin{pmatrix} (0.2352,0.1111), \\ (0.0909,0.1818) \end{pmatrix} & \begin{pmatrix} (0.2353,0.1111), \\ (0.0909,0.1818) \end{pmatrix} & \begin{pmatrix} (0.2353,0.1111), \\ (0.0909,0.1818) \end{pmatrix} & \begin{pmatrix} (0.0588,0.1111), \\ (0.0909,0.2727) \end{pmatrix} \\ \begin{pmatrix} (0.2941,0.2222), \\ (0.1818,0.2727) \end{pmatrix} & \begin{pmatrix} (0.0588,0.2222), \\ (0.1818,0.2727) \end{pmatrix} & \begin{pmatrix} (0.2941,0.2222), \\ (0.1818,0.2727) \end{pmatrix} & \begin{pmatrix} (0.2941,0.2222), \\ (0.1818,0.2727) \end{pmatrix} \\ \begin{pmatrix} (0.3529,0.2777), \\ (0.1818,0.2727) \end{pmatrix} & \begin{pmatrix} (0.3529,0.2778), \\ (0.1818,0.2727) \end{pmatrix} & \begin{pmatrix} (0.1176,0.2778), \\ (0.1818,0.2727) \end{pmatrix} & \begin{pmatrix} (0.3529,0.2778), \\ (0.1818,0.2727) \end{pmatrix} \\ \begin{pmatrix} (0.4117,0.4444), \\ (0.0909,0.0909) \end{pmatrix} & \begin{pmatrix} (0.4118,0.4444), \\ (0.0909,0.0909) \end{pmatrix} & \begin{pmatrix} (0.4118,0.4444), \\ (0.0909,0.0909) \end{pmatrix} & \begin{pmatrix} (0.4118,0.4444), \\ (0.0909,0.0909) \end{pmatrix} \end{bmatrix}$$

After performing the above evaluation, we calculate the WSA and WPA with the help of derived theory, such as: $T_1^{WSA} = ((0.2375,0.1154),(0.0933,0.1846))$, $T_2^{WSA} = ((0.229,0.1146),(0.0933,0.1846))$, $T_3^{WSA} = ((0.2373,0.1154),(0.0933,0.1846))$, $T_4^{WSA} = ((0.0456,0.1055),(0.0866,0.1764))$, and $T_1^{WPA} = ((0.2373,0.1140),(0.0944,0.1853))$, $T_2^{WPA} = ((0.2241,0.1136),(0.0944,0.1849))$, $T_3^{WPA} = ((0.2373,0.1141),(0.0944,0.1853))$, $T_4^{WPA} = ((0.0532,0.1058),(0.0865,0.1778))$. Using the data in Equations (40) and (41) with $^\circ\mathrm{F} = 0.2$, we calculate the aggregated measure based on convex theory, such as: $T_1 = 0.03621, T_2 = 0.0302$, $T_3 = 0.036, T_4 = -0.053$. According to the score values of the four alternatives, the ranking results is with $T_1 > T_3 > T_2 > T_4$. Thus, the best optimal is T_1 according to the score values of alternatives.

5. Application in MADM Method

The MADM technique is the valuable and dominant part of the decision-making procedure. The main theme of this section is to utilize the theory of the MADM technique based on the presented information for CIF set theory. To examine the above problem, we collect the finite family of alternatives $\overline{\overline{I\mathfrak{T}}}_1, \overline{\overline{I\mathfrak{T}}}_2, \overline{\overline{I\mathfrak{T}}}_3, \overline{\overline{I\mathfrak{T}}}_4, \ldots, \overline{\overline{I\mathfrak{T}}}_{\daleth}$ and their attributes $\overline{\overline{I\mathfrak{T}}}_{a-1}, \overline{\overline{I\mathfrak{T}}}_{a-2}, \overline{\overline{I\mathfrak{T}}}_{a-3}, \overline{\overline{I\mathfrak{T}}}_{a-4}, \overline{\overline{I\mathfrak{T}}}_{a-\daleth}$. Based on the above alternatives and their attributes, we compute the matrix of information whose term is computed in the form of CIF values such that the CIFS with a truth grade $\left(u_{\daleth}^{rp}(x), u_{\daleth}^{ip}(x)\right)$ and falsity grade $\left(v_{\daleth}^{rp}(x), v_{\daleth}^{ip}(x)\right)$ must be implementing the following rule: $0 \leq u_{\daleth}^{rp}(x) + v_{\daleth}^{rp}(x) \leq 1$ and $0 \leq u_{\daleth}^{ip}(x) + v_{\daleth}^{ip}(x) \leq 1$. The notion of neutral grade is stated by $r(x) = \left(r^{rp}(x), r^{ip}(x)\right) = \left(1 - \left(u_{\daleth}^{rp}(x) + v_{\daleth}^{rp}(x)\right), 1 - \left(u_{\daleth}^{ip}(x) + v_{\daleth}^{ip}(x)\right)\right)$ and the representation of the CIFVs is with $\overline{\overline{I\mathfrak{T}}}_{\mathfrak{H}} = \left(\left(u_{\daleth_{\mathfrak{H}}}^{rp}, u_{\daleth_{\mathfrak{H}}}^{ip}\right), \left(v_{\daleth_{\mathfrak{H}}}^{rp}, v_{\daleth_{\mathfrak{H}}}^{ip}\right)\right), \mathfrak{H} = 1, 2, \ldots, \daleth$. Furthermore, to proceed with the above information, we compute a technique of decision-making, whose major steps are shown below:

Step 1: Before evaluating the normalization, we arrange a collection of CIF data which may be of a benefit type or cost type. If the data are of a benefit type, then good, otherwise, using the below theory, we normalize the information, such as:

$$C = \begin{cases} \left(\left(u^{rp}_{\daleth_{\mathfrak{H}}}, u^{ip}_{\daleth_{\mathfrak{H}}}\right), \left(v^{rp}_{\daleth_{\mathfrak{H}}}, v^{ip}_{\daleth_{\mathfrak{H}}}\right)\right) & \textit{for benefit} \\ \left(\left(v^{rp}_{\daleth_{\mathfrak{H}}}, v^{ip}_{\daleth_{\mathfrak{H}}}\right), \left(u^{rp}_{\daleth_{\mathfrak{H}}}, u^{ip}_{\daleth_{\mathfrak{H}}}\right)\right) & \textit{for cost.} \end{cases}$$

Step 2: After performing the above evaluation, we calculate the CIFFPA operator and CIFFPG operator with the help of the derived theory.

Step 3: Evaluate the score or accuracy values of the aggregated information.

Step 4: Examine the ranking values in the presence of the score information.

In the last, we aim to show the supremacy and worth of the above procedure with the help of illustrating some numerical examples.

Illustrative Example: An investment enterprise wants to invest in an enterprise to increase or grow its income. There are five potential enterprises as alternatives, which are $\overline{\overline{I\mathfrak{T}}}_1, \overline{\overline{I\mathfrak{T}}}_2, \overline{\overline{I\mathfrak{T}}}_3, \overline{\overline{I\mathfrak{T}}}_4$ and $\overline{\overline{I\mathfrak{T}}}_5$. Four attributes are employed to resolve the problem in order to find the best preference from our five alternatives, including $\overline{\overline{I\mathfrak{T}}}_{a-1}$: growth analysis, $\overline{\overline{I\mathfrak{T}}}_{a-2}$: social-political impact, $\overline{\overline{I\mathfrak{T}}}_{a-3}$: environmental impact, and $\overline{\overline{I\mathfrak{T}}}_{a-4}$: development of society. Furthermore, to proceed with the above information, we compute a technique of decision-making, whose major steps are shown below:

Step 1: Before evaluating the normalization, we arrange a collection of CIF data in the form of Table 1, which may be of a benefit type or cost type. If the data are of a benefit type, then good, otherwise, using the below theory, we normalize the information, such as:

$$C = \begin{cases} \left(\left(u^{rp}_{\daleth_{\mathfrak{H}}}, u^{ip}_{\daleth_{\mathfrak{H}}}\right), \left(v^{rp}_{\daleth_{\mathfrak{H}}}, v^{ip}_{\daleth_{\mathfrak{H}}}\right)\right) & \textit{for benefit} \\ \left(\left(v^{rp}_{\daleth_{\mathfrak{H}}}, v^{ip}_{\daleth_{\mathfrak{H}}}\right), \left(u^{rp}_{\daleth_{\mathfrak{H}}}, u^{ip}_{\daleth_{\mathfrak{H}}}\right)\right) & \textit{for cost.} \end{cases}$$

However, the data in Table 2 is not required to be normalized.

Table 2. Original CIF information matrix.

	$\overline{\overline{I\mathfrak{T}}}_{a-1}$	$\overline{\overline{I\mathfrak{T}}}_{a-2}$	$\overline{\overline{I\mathfrak{T}}}_{a-3}$	$\overline{\overline{I\mathfrak{T}}}_{a-4}$
$\overline{\overline{I\mathfrak{T}}}_1$	$((0.4, 0.3), (0.1, 0.3))$	$((0.41, 0.31), (0.11, 0.31))$	$((0.42, 0.32), (0.12, 0.32))$	$((0.43, 0.33), (0.13, 0.33))$
$\overline{\overline{I\mathfrak{T}}}_2$	$((0.6, 0.7), (0.2, 0.1))$	$((0.61, 0.71), (0.21, 0.11))$	$((0.62, 0.72), (0.22, 0.12))$	$((0.63, 0.73), (0.23, 0.13))$
$\overline{\overline{I\mathfrak{T}}}_3$	$((0.3, 0.2), (0.3, 0.4))$	$((0.31, 0.21), (0.31, 0.41))$	$((0.32, 0.22), (0.32, 0.42))$	$((0.33, 0.23), (0.33, 0.43))$
$\overline{\overline{I\mathfrak{T}}}_4$	$((0.7, 0.4), (0.2, 0.3))$	$((0.71, 0.41), (0.21, 0.31))$	$((0.72, 0.42), (0.22, 0.32))$	$((0.73, 0.43), (0.23, 0.33))$
$\overline{\overline{I\mathfrak{T}}}_5$	$((0.7, 0.7), (0.1, 0.1))$	$((0.71, 0.71), (0.11, 0.11))$	$((0.72, 0.72), (0.12, 0.12))$	$((0.73, 0.73), (0.13, 0.13))$

Step 2: After performing the above evaluation, we calculate the CIFFPA operator and CIFFPG operator with the help of the derived theory, and see Table 3.

Table 3. Aggregated values.

	CIFFPA	*CIFFPG*
$\overline{\overline{I\mathfrak{T}}}_1$	$((0.4084, 0.3071), (0.0969, 0.2941))$	$((0.3941, 0.2941), (0.1039, 0.3071))$
$\overline{\overline{I\mathfrak{T}}}_2$	$((0.7369, 0.8279), (0.108, 0.043))$	$((0.4986, 0.6188), (0.2794, 0.1486))$
$\overline{\overline{I\mathfrak{T}}}_3$	$((0.3007, 0.2007), (0.3002, 0.1003))$	$((0.3002, 0.2003), (0.3007, 0.1006))$
$\overline{\overline{I\mathfrak{T}}}_4$	$((0.7448, 0.4393), (0.1694, 0.2647))$	$((0.6776, 0.364), (0.2249, 0.3329))$
$\overline{\overline{I\mathfrak{T}}}_5$	$((0.8753, 0.8753), (0.0258, 0.0258))$	$((0.5741, 0.5741), (0.1738, 0.1738))$

Step 3: Evaluate the score or accuracy values of aggregated information, and see Table 4.

Table 4. Score values.

	CIFFPA	*CIFFPG*
$\overline{\overline{\overline{I\mathfrak{T}}}}_1$	0.1623	0.1386
$\overline{\overline{\overline{I\mathfrak{T}}}}_2$	0.7069	0.3447
$\overline{\overline{\overline{I\mathfrak{T}}}}_3$	0.0504	0.0496
$\overline{\overline{\overline{I\mathfrak{T}}}}_4$	0.375	0.2419
$\overline{\overline{\overline{I\mathfrak{T}}}}_5$	0.8495	0.4003

Step 4: Examine the ranking values of the score information, and see Table 5.

Table 5. Ranking information.

Methods	*Ranking Results*
CIFFPA	$\overline{\overline{\overline{I\mathfrak{T}}}}_5 > \overline{\overline{\overline{I\mathfrak{T}}}}_2 > \overline{\overline{\overline{I\mathfrak{T}}}}_4 > \overline{\overline{\overline{I\mathfrak{T}}}}_1 > \overline{\overline{\overline{I\mathfrak{T}}}}_3$
CIFFPG	$\overline{\overline{\overline{I\mathfrak{T}}}}_5 > \overline{\overline{\overline{I\mathfrak{T}}}}_2 > \overline{\overline{\overline{I\mathfrak{T}}}}_4 > \overline{\overline{\overline{I\mathfrak{T}}}}_1 > \overline{\overline{\overline{I\mathfrak{T}}}}_3$

The valuable and best preference is $\overline{\overline{\overline{I\mathfrak{T}}}}_5$, according to the theory of CIFFPA and CIFFPG operators. Furthermore, by excluding the phase term, we have checked the stability and supremacy of the derived information. Thus, we remove the phase information from the data in Table 2 in which their score values are shown in Table 6.

Table 6. Score values (without phase term).

	CIFFPA	*CIFFPG*
$\overline{\overline{\overline{I\mathfrak{T}}}}_1$	0.1558	0.1451
$\overline{\overline{\overline{I\mathfrak{T}}}}_2$	0.3144	0.1096
$\overline{\overline{\overline{I\mathfrak{T}}}}_3$	0.0002	0.00002
$\overline{\overline{\overline{I\mathfrak{T}}}}_4$	0.2877	0.2263
$\overline{\overline{\overline{I\mathfrak{T}}}}_5$	0.4247	0.2001

Furthermore, we examine the ranking values of the score information, and see Table 7.

Table 7. Ranking information.

Methods	*Ranking Results*
CIFFPA	$\overline{\overline{\overline{I\mathfrak{T}}}}_5 > \overline{\overline{\overline{I\mathfrak{T}}}}_2 > \overline{\overline{\overline{I\mathfrak{T}}}}_4 > \overline{\overline{\overline{I\mathfrak{T}}}}_1 > \overline{\overline{\overline{I\mathfrak{T}}}}_3$
CIFFPG	$\overline{\overline{\overline{I\mathfrak{T}}}}_4 > \overline{\overline{\overline{I\mathfrak{T}}}}_5 > \overline{\overline{\overline{I\mathfrak{T}}}}_1 > \overline{\overline{\overline{I\mathfrak{T}}}}_2 > \overline{\overline{\overline{I\mathfrak{T}}}}_3$

The valuable and best preference is $\overline{\overline{\overline{I\mathfrak{T}}}}_5$ according to the theory of the CIFFPA operator. Furthermore, the best preference is $\overline{\overline{\overline{I\mathfrak{T}}}}_4$ according to the theory of the CIFFPG operator. Additionally, we find the comparisons between the proposed and existing data with the help of data in Table 2.

6. Comparative Analysis

In this section, we select some existing operators based on various prevailing ideas. We then try to compare their obtained results with the obtained results of our proposed works. The comparative analysis is one of the most effective and dominant techniques because without comparison we fail to show the supremacy and validity of the derived theory. For this, we consider different types of information, such as aggregation operators (AOs) for IFSs [23], geometric AOs for IFSs [24], the complex fuzzy credibility Frank AOs [26]. Additionally, Yu [31] examined the theory of generalized prioritized AOs for intuitionistic fuzzy environments, and Lin, et al. [32] derived the fuzzy number intuitionistic fuzzy prioritized AOs and their application in decision-making procedures. Furthermore, Garg and Rani [33] exposed the averaging operators for CIFSs. Garg and Rani [34] evaluated the geometric operators for CIFSs, and Mahmood, et al. [35] examined the Aczel–Alsina aggregation operators for CIFSs. Using data in Table 2, the comparison information is listed in Table 8.

Table 8. Comparative analysis.

Methods	Score Information	Ranking Information
Xu [23]	****************************	****************************
Xu and Yager [24]	****************************	****************************
Yahya, et al. [26]	****************************	****************************
Yu [31]	****************************	****************************
Lin, et al. [32]	****************************	****************************
Garg and Rani [33]	0.1506, 0.5008, 0.0506, 0.3005, 0.6010	$\overline{\overline{I\mathfrak{T}}}_5 > \overline{\overline{I\mathfrak{T}}}_2 > \overline{\overline{I\mathfrak{T}}}_4 > \overline{\overline{I\mathfrak{T}}}_1 > \overline{\overline{I\mathfrak{T}}}_3$
Garg and Rani [34]	0.1497, 0.4998, 0.0496, 0.2997, 0.5998	$\overline{\overline{I\mathfrak{T}}}_5 > \overline{\overline{I\mathfrak{T}}}_2 > \overline{\overline{I\mathfrak{T}}}_4 > \overline{\overline{I\mathfrak{T}}}_1 > \overline{\overline{I\mathfrak{T}}}_3$
Mahmood, et al. [35]	0.1506, 0.5007, 0.0505, 0.3005, 0.6009	$\overline{\overline{I\mathfrak{T}}}_5 > \overline{\overline{I\mathfrak{T}}}_2 > \overline{\overline{I\mathfrak{T}}}_4 > \overline{\overline{I\mathfrak{T}}}_1 > \overline{\overline{I\mathfrak{T}}}_3$
CIFFPA	$0.1623, 0.7069, 0.0504, 0.375, 0.8495$	$\overline{\overline{I\mathfrak{T}}}_5 > \overline{\overline{I\mathfrak{T}}}_2 > \overline{\overline{I\mathfrak{T}}}_4 > \overline{\overline{I\mathfrak{T}}}_1 > \overline{\overline{I\mathfrak{T}}}_3$
CIFFPG	$0.1386, 0.3447, 0.0496, 0.2419, 0.4003$	$\overline{\overline{I\mathfrak{T}}}_5 > \overline{\overline{I\mathfrak{T}}}_2 > \overline{\overline{I\mathfrak{T}}}_4 > \overline{\overline{I\mathfrak{T}}}_1 > \overline{\overline{I\mathfrak{T}}}_3$

The valuable and best preference is $\overline{\overline{I\mathfrak{T}}}_5$, according to the theory of CIFFPA, CIFFPG operators, Garg and Rani [33,34], and Mahmood, et al. [35]. However, the theory of AOs for IFSs [23], geometric AOs for IFSs [24], the complex fuzzy credibility of Frank AOs [26], and Yu [31] examined the theory of generalized prioritized AOs for intuitionistic fuzzy environments with the limitation that fails to evaluate it. Similarly, Lin, et al. [35] derive the fuzzy number intuitionistic fuzzy prioritized AOs and their application in decision-making procedures also with the limitation and restriction, because they fail to evaluate it. It is possible if we use the data in Table 2, however, without phase information, then the comparison information is listed in Table 9.

Table 9. Comparative analysis (without phase terms).

Methods	Score Information	Ranking Information
Xu [23]	$0.1504, 0.5005, 0.0504, 0.3003, 0.6006$	$\overline{\overline{I\mathfrak{T}}}_5 > \overline{\overline{I\mathfrak{T}}}_2 > \overline{\overline{I\mathfrak{T}}}_4 > \overline{\overline{I\mathfrak{T}}}_1 > \overline{\overline{I\mathfrak{T}}}_3$
Xu and Yager [24]	$0.1498, 0.4999, 0.0497, 0.2998, 0.5999$	$\overline{\overline{I\mathfrak{T}}}_5 > \overline{\overline{I\mathfrak{T}}}_2 > \overline{\overline{I\mathfrak{T}}}_4 > \overline{\overline{I\mathfrak{T}}}_1 > \overline{\overline{I\mathfrak{T}}}_3$
Yahya, et al. [26]	****************************	****************************
Yu [31]	$0.1614, 0.6901, 0.0504, 0.3687, 0.8319$	$\overline{\overline{I\mathfrak{T}}}_5 > \overline{\overline{I\mathfrak{T}}}_2 > \overline{\overline{I\mathfrak{T}}}_4 > \overline{\overline{I\mathfrak{T}}}_1 > \overline{\overline{I\mathfrak{T}}}_3$
Lin, et al. [32]	$0.1394, 0.3526, 0.0496, 0.2452, 0.4097$	$\overline{\overline{I\mathfrak{T}}}_5 > \overline{\overline{I\mathfrak{T}}}_2 > \overline{\overline{I\mathfrak{T}}}_4 > \overline{\overline{I\mathfrak{T}}}_1 > \overline{\overline{I\mathfrak{T}}}_3$
Garg and Rani [33]	0.1502, 0.2002, 0.0001, 0.2502, 0.3003	$\overline{\overline{I\mathfrak{T}}}_5 > \overline{\overline{I\mathfrak{T}}}_4 > \overline{\overline{I\mathfrak{T}}}_2 > \overline{\overline{I\mathfrak{T}}}_1 > \overline{\overline{I\mathfrak{T}}}_3$
Garg and Rani [34]	0.1499, 0.1999, 0.0001, 0.2499, 0.2999	$\overline{\overline{I\mathfrak{T}}}_5 > \overline{\overline{I\mathfrak{T}}}_4 > \overline{\overline{I\mathfrak{T}}}_2 > \overline{\overline{I\mathfrak{T}}}_1 > \overline{\overline{I\mathfrak{T}}}_3$
Mahmood, et al. [35]	0.1502, 0.2001, 0.00009, 0.2502, 0.3003	$\overline{\overline{I\mathfrak{T}}}_5 > \overline{\overline{I\mathfrak{T}}}_4 > \overline{\overline{I\mathfrak{T}}}_2 > \overline{\overline{I\mathfrak{T}}}_1 > \overline{\overline{I\mathfrak{T}}}_3$
CIFFPA	$0.1558, 0.3144, 0.0002, 0.2877, 0.4247$	$\overline{\overline{I\mathfrak{T}}}_5 > \overline{\overline{I\mathfrak{T}}}_2 > \overline{\overline{I\mathfrak{T}}}_4 > \overline{\overline{I\mathfrak{T}}}_1 > \overline{\overline{I\mathfrak{T}}}_3$
CIFFPG	$0.1451, 0.1096, 0.00002, 0.2263, 0.2001$	$\overline{\overline{I\mathfrak{T}}}_4 > \overline{\overline{I\mathfrak{T}}}_5 > \overline{\overline{I\mathfrak{T}}}_1 > \overline{\overline{I\mathfrak{T}}}_2 > \overline{\overline{I\mathfrak{T}}}_3$

The valuable and best preference is $\overline{\overline{I\mathfrak{T}}}_5$ according to the theory of the CIFFPA operator, Xu [23], Xu and Yager [24], Yu [31], Lin, et al. [32], Garg and Rani [33,34], and Mahmood, et al. [35]. However, the most valuable and best preference is $\overline{\overline{I\mathfrak{T}}}_4$ according to the theory of the CIFFPG operator. However, the complex fuzzy credibility Frank AOs [26] have limitations and restrictions, and because it failed to evaluate it. Therefore, the proposed work is effective and valid for evaluating most of the CIFS information.

7. Conclusions

The idea of CIFS is the modified version of the complex fuzzy set theory, which covered the grade of truth and falsity in the form of polar coordinates. Furthermore, the theory of Frank and prioritized aggregation operators is also very famous and valuable because they are the modified version of the simple averaging and geometric aggregation operators. Motivated by the above information, in this manuscript, we examined the following ideas:

1. We evaluated the Frank operational laws for the theory of CIF information;
2. We examined the theory of the CIFFPA, CIFFPOA, CIFFPG, and CIFFPOG operators, and their properties of idempotency, monotonicity, and boundedness;
3. We derived the WASPAS under the presence of the CIFFPA and CIFFPG operators;
4. We demonstrated the MADM procedures based on the invented theory for CIF information;
5. We compared the derived theory with various existing information to show the validity and worth of the discovered approaches.

In the future, we aim to develop new aggregation operators based on Frank operational laws and then we aim to employ them in the field of game theory, neural networks, clustering, pattern recognition, and decision-making to enhance the worth of the derived information.

Author Contributions: Conceptualization, Z.A. and T.M.; methodology, Z.A. and M.-S.Y.; validation, Z.A. and T.M.; formal analysis, Z.A., T.M. and M.-S.Y.; investigation, Z.A., T.M. and M.-S.Y.; writing—original draft preparation, Z.A. and T.M.; writing—review and editing, M.-S.Y.; visualization, Z.A. and T.M.; supervision, T.M. and M.-S.Y.; funding acquisition, M.-S.Y. All authors have read and agreed to the published version of the manuscript.

Funding: This work was supported in part by the National Science and Technology Council, Taiwan, under Grant MOST-111-2118-M-033-001-.

Data Availability Statement: Not applicable.

Conflicts of Interest: The authors declare no conflict of interest.

References

1. Zadeh, L.A. Fuzzy sets. *Inf. Control* **1965**, *8*, 338–353. [CrossRef]
2. Chang-Chien, S.J.; Hung, W.L.; Yang, M.S. On mean shift-based clustering for circular data. *Soft Comput.* **2012**, *16*, 1043–1060. [CrossRef]
3. Chaomurilige, C.; Yu, J.; Yang, M.S. Deterministic annealing Gustafson-Kessel fuzzy clustering algorithm. *Inf. Sci.* **2017**, *417*, 435–453. [CrossRef]
4. Chen, J.; Lin, Y.; Mi, J.; Li, S.; Ding, W. A spectral feature selection approach with kernelized fuzzy rough sets. *IEEE Trans. Fuzzy Syst.* **2022**, *30*, 2886–2901. [CrossRef]
5. Xie, T.; Gong, Z. A hesitant soft fuzzy rough set and its applications. *IEEE Access* **2019**, *7*, 167766–167783. [CrossRef]
6. Mahmood, T.; Ali, Z. Fuzzy superior mandelbrot sets. *Soft Comput.* **2022**, *26*, 9011–9020. [CrossRef]
7. Atanassov, K. Intuitionistic fuzzy sets. *Fuzzy Sets Syst.* **1986**, *20*, 87–96. [CrossRef]
8. Hwang, C.M.; Yang, M.S. New construction for similarity measures between intuitionistic fuzzy sets based on lower, upper and middle fuzzy sets. *Int. J. Fuzzy Syst.* **2013**, *15*, 359–366.
9. Yang, M.S.; Hussian, Z.; Ali, M. Belief and plausibility measures on intuitionistic fuzzy sets with construction of belief-plausibility TOPSIS. *Complexity* **2020**, *2020*, 7849686. [CrossRef]
10. Kaushal, M.; Lohani, Q.M.D. Intuitionistic fuzzy c-ordered means clustering algorithm. *IEEE Access* **2022**, *10*, 26271–26281. [CrossRef]
11. Ramot, D.; Milo, R.; Friedman, M.; Kandel, A. Complex fuzzy sets. *IEEE Trans. Fuzzy Syst.* **2002**, *10*, 171–186. [CrossRef]
12. Selvachandran, G.; Quek, S.G.; Lan, L.T.H.; Son, L.H.; Giang, N.L.; Ding, W.; Abdel-Basset, M.; de Albuquerque, V.H.C. A new design of Mamdani complex fuzzy inference system for multiattribute decision making problems. *IEEE Trans. Fuzzy Syst.* **2021**, *29*, 716–730. [CrossRef]
13. Alkouri, A.M.D.J.S.; Salleh, A.R. Complex intuitionistic fuzzy sets. *AIP Conf. Proc. Am. Inst. Phys.* **2012**, *1482*, 464–470.
14. Gulzar, M.; Mateen, M.H.; Alghazzawi, D.; Kausar, N. A novel applications of complex intuitionistic fuzzy sets in group theory. *IEEE Access* **2020**, *8*, 196075–196085. [CrossRef]
15. Frank, M.J. On the simultaneous associativity of F(x,y) and x + y − F(x, y). *Aequ. Math.* **1979**, *19*, 194–226. [CrossRef]

16. Łukasiewicz, J. O logice trójwartościowej. In *Ruch Filozoficzny 5*; English translation: On three-valued, logic; Borkowski, L., Ed.; Selected works by Jan Łukasiewicz: North-Holland, The Amsterdam, 1970; pp. 170–171, 87–88.
17. Yager, R.R. Prioritized aggregation operators. *Int. J. Approx. Reason.* **2008**, *48*, 263–274. [CrossRef]
18. Yu, X.; Xu, Z. Prioritized intuitionistic fuzzy aggregation operators. *Inf. Fusion* **2013**, *14*, 108–116. [CrossRef]
19. Zavadskas, E.K.; Turskis, Z.; Antucheviciene, J.; Zakarevicius, A. Optimization of weighted aggregated sum product assessment. *Elektron. Ir Elektrotechnika* **2012**, *122*, 3–6. [CrossRef]
20. Zavadskas, E.K.; Antucheviciene, J.; Šaparauskas, J.; Turskis, Z. Multi-criteria assessment of facades' alternatives: Peculiarities of ranking methodology. *Procedia Eng.* **2013**, *57*, 107–112. [CrossRef]
21. Mishra, A.R.; Rani, P.; Prajapati, R.S. Multi-criteria weighted aggregated sum product assessment method for sustainable biomass crop selection problem using single-valued neutrosophic sets. *Appl. Soft Comput.* **2021**, *113*, 108038. [CrossRef]
22. Debnath, B.; Bari, A.B.M.M.; Haq, M.M.; Pacheco, D.A.D.J.; Khan, M.A. An integrated stepwise weight assessment ratio analysis and weighted aggregated sum product assessment framework for sustainable supplier selection in the healthcare supply chains. *Supply Chain. Anal.* **2023**, *1*, 100001. [CrossRef]
23. Xu, Z. Intuitionistic fuzzy aggregation operators. *IEEE Trans. Fuzzy Syst.* **2007**, *15*, 1179–1187.
24. Xu, Z.; Yager, R.R. Some geometric aggregation operators based on intuitionistic fuzzy sets. *Int. J. Gen. Syst.* **2006**, *35*, 417–433. [CrossRef]
25. Zhang, X.; Liu, P.; Wang, Y. Multiple attribute group decision making methods based on intuitionistic fuzzy Frank power aggregation operators. *J. Intell. Fuzzy Syst.* **2015**, *29*, 2235–2246. [CrossRef]
26. Yahya, M.; Abdullah, S.; Almagrabi, A.O.; Botmart, T. Analysis of S-box based on image encryption application using complex fuzzy credibility Frank aggregation operators. *IEEE Access* **2022**, *10*, 88858–88871. [CrossRef]
27. Qin, J.; Liu, X.; Pedrycz, W. Frank aggregation operators and their application to hesitant fuzzy multiple attribute decision making. *Appl. Soft Comput.* **2016**, *41*, 428–452.
28. Tang, X.; Yang, S.; Pedrycz, W. Multiple attribute decision-making approach based on dual hesitant fuzzy Frank aggregation operators. *Appl. Soft Comput.* **2018**, *68*, 525–547.
29. Ye, J. Prioritized aggregation operators of trapezoidal intuitionistic fuzzy sets and their application to multicriteria decision-making. *Neural Comput. Appl.* **2014**, *25*, 1447–1454. [CrossRef]
30. Ali, Z.; Mahmood, T.; Aslam, M.; Chinram, R. Another view of complex intuitionistic fuzzy soft sets based on prioritized aggregation operators and their applications to multiattribute decision making. *Mathematics* **2021**, *9*, 1922.
31. Yu, D. Multi-Criteria Decision Making Based on Generalized Prioritized Aggregation Operators under Intuitionistic Fuzzy Environment. *Int. J. Fuzzy Syst.* **2013**, *15*, 47–54.
32. Lin, R.; Zhao, X.; Wei, G. Fuzzy number intuitionistic fuzzy prioritized operators and their application to multiple attribute decision making. *J. Intell. Fuzzy Syst.* **2013**, *24*, 879–888. [CrossRef]
33. Garg, H.; Rani, D. Some generalized complex intuitionistic fuzzy aggregation operators and their application to multicriteria decision-making process. *Arab. J. Sci. Eng.* **2019**, *44*, 2679–2698. [CrossRef]
34. Garg, H.; Rani, D. Generalized geometric aggregation operators based on t-norm operations for complex intuitionistic fuzzy sets and their application to decision-making. *Cogn. Comput.* **2020**, *12*, 679–698. [CrossRef]
35. Mahmood, T.; Ali, Z.; Baupradist, S.; Chinram, R. Complex intuitionistic fuzzy Aczel-Alsina aggregation operators and their application in multi-attribute decision-making. *Symmetry* **2022**, *14*, 2255. [CrossRef]
36. Sarfraz, M.; Ullah, K.; Akram, M.; Pamucar, D.; Božanić, D. Prioritized aggregation operators for intuitionistic fuzzy information based on Aczel–Alsina t-norm and t-conorm and their applications in group decision-making. *Symmetry* **2022**, *14*, 2655. [CrossRef]
37. Poryazov, S.; Andonov, V.; Saranova, E.; Atanassov, K. Two approaches to the traffic quality intuitionistic fuzzy estimation of service compositions. *Mathematics* **2022**, *10*, 4439. [CrossRef]
38. Dai, S. Linguistic complex fuzzy sets. *Axioms* **2023**, *12*, 328. [CrossRef]

mathematics

MDPI

Article

Building Up of Fuzzy Evaluation Model of Life Performance Based on Type-II Censored Data

Kuo-Ching Chiou

Department of Finance, Chaoyang University of Technology, Taichung 413310, Taiwan; kcchiou@cyut.edu.tw

Abstract: The semiconductor industry is a rapidly growing sector. As collection technologies for production data continue to improve and the Internet of Things matures, production data analysis improves, thus accelerating progress towards smart manufacturing. This not only enhances the process quality, but also increases product lifetime and reliability. Under the assumption of exponential distribution, the ratio of lifetime and warranty has been proposed as a lifetime performance index for electronic products. As unknown parameters of the index, to use point estimates to assess lifetime performance may cause misjudgment due to sampling errors. In addition, cost and time limitations often lead to small sample sizes that can affect the results of the analysis. Type-II censored data are widely applied in production and manufacturing engineering. Thus, this paper proposes an unbiased and consistent estimator of lifetime performance based on type-II censored data. The $100(1-\alpha)\%$ confidence interval of the proposed index is derived based on its probability density function. Overly small sample sizes not only make the length estimates of lifetime performance index intervals for electronic products too long, but they also increase sampling errors, which distort the estimation and test results. We therefore used the aforementioned interval to construct a fuzzy test model for the assessment of product lifetime and further help manufacturers to be more prudent and precise to evaluate the performance of product life cycles. A numerical example illustrates the applicability of the proposed model.

Keywords: relative lifetime performance index; type II censoring data; unbiased estimator; consistent estimator; confidence-interval-based fuzzy testing method

MSC: 62A86

Citation: Chiou, K.-C. Building Up of Fuzzy Evaluation Model of Life Performance Based on Type-II Censored Data. *Mathematics* **2023**, *11*, 3686. https://doi.org/10.3390/math11173686

Academic Editor: Aleksandar Aleksić

Received: 11 August 2023
Revised: 22 August 2023
Accepted: 24 August 2023
Published: 27 August 2023

1. Introduction

The semiconductor industry is involved in the wafer manufacturing, integrated circuit (IC) design, packaging, and peripheral components necessary for end products such as smartphones, tablet computers, and smart internet end devices [1,2]. Industry clusters in Taiwan represent a crucial industry chain for consumer electronics worldwide [3–6]. Offering good product quality not only enhances its product lifespan and reliability, but also bolsters user satisfaction and willingness to use it [2,7]. As the collection technologies for production data continue to improve and the Internet of Things matures, production data analysis improves, thus accelerating progress toward smart manufacturing. This not only enhances the process quality, but also increases the product lifetime and reliability [8]. Furthermore, owing to the limitation of the cost and time, the estimation accuracy of the samples in the study leads to not being significant. Thus, in order to increase its estimation accuracy and eliminate the uncertain measurement data, confidence interval-based fuzzy evaluation models were built up via the confidence interval of indices in the study [9,10]. In order to prevent the risk of misjudgment caused not only by sampling errors but also by factoring in expert experiences and past data into consideration, it becomes necessary to increase the accuracy of each case with smaller sample sizes and analyze data with confidence intervals. Constructing a fuzzy test model to evaluate the product lifetime is also a way to compensate for sampling errors in small sample sizes [7,9].

As for product marketing, the term of warranty is shown to be a crucial index. Chen and Yu [11] indicated that whether customers feel satisfied with the products and be willing to use them lies in the good quality of the product with longer product lifetime and its reliability. Many researchers have confirmed the convenience and efficacy of process capability indices (PCIs) for the assessment of process quality in practice [12]. PCIs have also been applied to the lifetime and reliability of electronic products [13]. On the basis of some studies shown, it has been proved that ameliorating the process of quality check is able to shun off some unnecessary cost caused by the rework and defective products. Furthermore, it is also able to decrease energy consumption and carbon emissions [14,15]. It is of importance to manufacture all parts of the product with high quality. In order to make all the final products meet the quality standard, forming stringent requirements becomes necessary [16].

Additionally, in the industrial field, on account of the limitation of cost and time, noticing small-size samples implemented in the survey is not uncommon [17]. According to some previous studies conducted by the experts, it has been argued that utilizing the analyzing tool, fuzzy evaluation model, to analyze the sample with small data is able to make the result of the survey reach its reliability and validity [9,10]. Additionally, in order to lower the risk of misjudgment caused by sampling errors, putting interval estimates into practice has been proved to be much more accurate compared to the point estimates [7].

Product lifetime is exponentially distributed with mean λ. Tong et al. [18] proposed the following lifetime performance index C_L:

$$C_L = \frac{\mu_T - L}{\mu_T} = 1 - \frac{L}{\lambda} \tag{1}$$

where L denotes the minimum number of time units that the lifetime of each electronic component is required to reach, and parameter λ is the expected value μ_T of the electronic component lifetime. We assume that the lifetime of the electronic component (T) follows an exponential distribution with the mean λ; thus, the probability density function of T is as follows:

$$f_T(t) = \frac{1}{\lambda} e^{-\frac{t}{\lambda}},\ t > 0 \tag{2}$$

As noted by Chen and Yu [19], when the mean lifetime of the electronic component $\lambda \geq L$, then the lifetime performance index $C_L \geq 0$. Clearly, the greater the lifetime performance index C_L is, the better its lifetime performance is. However, the warranty period of a product is generally only three years (L = 3), yet only when the mean lifetime λ approaches infinity does the lifetime performance index C_L of the electronic component approach 1. This does not fit the conventions of the industry. Chen et al. [20] therefore proposed a relative lifetime performance index. This index is defined as follows:

$$\beta_L = \frac{\mu_T}{L} = \frac{\lambda}{L} \tag{3}$$

As noted by Chen and Yu [19], the lifetime performance index is the ratio of λ and L. The one-to-one relationship between both index β_L and C_L is $\beta_L = (1 - C_L)^{-1}$. If the relative lifetime is $X = T/L$, then (1) when random variable $X < 1$, the lifetime of electronic component is denoted as equal to the warranty ($T < L$), (2) when random variable $X = 1$, the lifetime of the electronic component is denoted as equal to the warranty ($T = L$), (3) when random variable $X > 1$, the lifetime of the electronic component is denoted as longer than the warranty ($T > L$). Thus, X is the only value required for managers to assess if product lifetime is sufficient.

The probability density function of relative lifetime X is as follows:

$$f_X(x) = \frac{1}{\beta_L} exp\left(-\frac{x}{\beta_L}\right),\ x > 0 \tag{4}$$

Relative lifetime X is an exponential distribution with mean β_L. Therefore, the failure rate is $r_X(x) = \beta_L^{-1}$ and product reliability $p_r = S_X(1) = \exp\left(\beta_L^{-1}\right)$ where $S_X(x)$ is the survival function of relative lifetime X as follows:

$$S_X(x) = p(X > x) = exp\left(-\frac{x}{\beta_L}\right),\ x > 0 \tag{5}$$

As pointed out by Chen et al. [21], the unknown parameters in the index decrease its accuracy if the point estimates are simply utilized to evaluate the data with small-size samples [7,19,21–23]. As the results of statistical tests tend to vary with sample size, censoring can be applied to achieve consistent results in a short time [22–26]. Type-II censoring is widely applied in production and manufacturing data. Thus, this paper proposes an unbiased and consistent estimator for the lifetime performance index β_L based on type-II censored data. The $100(1-\alpha)\%$ confidence interval of the index β_L is derived based on its probability density function. Using this interval and the method proposed by Chen and Yu [19], a fuzzy test model is constructed to assess whether product lifetime performance reaches the required level. The application of the model proposed in the study is demonstrated through a numerical example. The final section presents our conclusions.

The rest of the present paper would be arranged as follows. In Section 2, we derive the estimator and find the confidence interval of the lifetime performance index. Section 3 presents the fuzzy test method for lifetime performance index. We employ an application to demonstrate the efficacy of the proposed approach in Section 4. Conclusions are given in Section 5.

2. Estimation of Ratio for Lifetime Performance Index

Incomplete data collection due to external or human factors during product development can reduce the reliability of analysis results. Censoring type is a form of data collection that is accurate as well as cost-effective and quick [14]. Censoring type can be divided into three types: type-I censoring, type-II censoring, and random censoring [27]. Type-II censoring is the most widely applied in production and manufacturing engineering [14,27]. Furthermore, type-II progressive censoring has become a common approach to the analysis of lifetime data for highly reliable products [14,28–31].

The proposed index must be estimated based on sample data. The lifetime T follows an exponential distribution with mean λ, denoted as $T \sim exp(\lambda)$. The relative lifetime $X = T/L$ is an exponential distribution with mean β_L, denoted as $X \sim exp(\beta_L)$. $(T_1, T_2, \ldots, T_n)$ and $(X_1, X_2, \ldots, X_n)$ are random samples of T and X, respectively. $(Y_1, Y_2, \ldots, Y_n)$ is a sample set of the type-II censored data, $Y_j = min\left(X_j,\ X_{(r)}\right) = min\left(T_j/L,\ T_{(r)}/L\right), j = 1, 2, \ldots, n$, where the number of uncensored data is denoted by r and the order statistics are denoted by $X_{(r)}$ and $T_{(r)}$. The estimator $\hat{\beta}_L$ of β_L is as follows:

$$\hat{\beta}_L = \frac{\hat{\lambda}}{L} = \frac{1}{r}\sum_{i=1}^{n} Y_i \tag{6}$$

where

$$\hat{\lambda} = \frac{L}{r}\sum_{i=1}^{n} Y_i \tag{7}$$

If random variable $W = 2r\hat{\beta}_L/\beta_L$, according to Chiou and Chen [14], W follows a chi-square distribution with $2r$ degrees of freedom, denoted by $W \sim \chi^2_{(2r)}$. Therefore, the expected value of the estimator $\hat{\beta}_L$ is as follows:

$$E\left[\hat{\beta}_L\right] = E[W] \times \left(\frac{\beta_L}{2r}\right) = (2r) \times \left(\frac{\beta_L}{2r}\right) = \beta_L \tag{8}$$

$\hat{\beta}_L$ is an unbiased estimator of the lifetime performance index β_L. Its variance is calculated as follows:

$$Var[\hat{\beta}_L] = Var[W] \times \left(\frac{\beta_L}{2r}\right)^2 = (4r) \times \left(\frac{\beta_L^2}{4r^2}\right) = \frac{\beta_L^2}{r} \quad (9)$$

For large samples,

$$\lim_{n\to\infty} E(\hat{\beta}_L - \beta_L)^2 = \lim_{n\to\infty} Var(\hat{\beta}_L) = \lim_{r\to\infty} \frac{\beta_L^2}{r} = 0 \quad (10)$$

Based on Equations (8) and (10), $\hat{\beta}_L$ is an unbiased and consistent estimator of the lifetime performance index β_L. The 100(1 − α)% confidence interval of the lifetime performance index β_L is derived as follows:

$$\begin{aligned} 1-\alpha = p\left\{\chi^2_{(2r),\,\alpha/2} \le W \le \chi^2_{(2r),\,1-\alpha/2}\right\} &= p\left\{\chi^2_{(2r),\,\alpha/2} \le \frac{2r\hat{\beta}_L}{\beta_L} \le \chi^2_{(2r),\,1-\alpha/2}\right\} \\ &= p\left\{\left(\frac{2r}{\chi^2_{(2r),\,1-\alpha/2}}\right)\hat{\beta}_L \le \beta_L \le \left(\frac{2r}{\chi^2_{(2r),\,\alpha/2}}\right)\hat{\beta}_L\right\} \end{aligned} \quad (11)$$

where $\chi^2_{(2r),\alpha/2}$ is the lower $\alpha/2$ quantiles of $\chi^2_{(2r)}$ and $\chi^2_{(2r),1-\alpha/2}$ is the lower $1 - \alpha/2$ quantiles of $\chi^2_{(2r)}$. Therefore, the lower confidence of the lifetime performance index β_L is

$$L\beta_L = \left(\frac{2r}{\chi^2_{(2r),\,1-\alpha/2}}\right)\hat{\beta}_L \quad (12)$$

Similarly, the upper confidence of the lifetime performance index β_L is

$$U\beta_L = \left(\frac{2r}{\chi^2_{(2r),\,\alpha/2}}\right)\hat{\beta}_L \quad (13)$$

The length of the 100(1 − α)% confidence interval of the lifetime performance index β_L is

$$l\beta_L = U\beta_L - L\beta_L = \left(\frac{2r}{\chi^2_{(2r),\,\alpha/2}} - \frac{2r}{\chi^2_{(2r),\,1-\alpha/2}}\right)\hat{\beta}_L \quad (14)$$

Since $\hat{\beta}_L$ is an unbiased estimator of the lifetime performance index β_L, the following defines the expected length of the 100(1 − α)% confidence interval $l\beta_L$:

$$E(l\beta_L) = \left(\frac{2r}{\chi^2_{(2r),\,\alpha/2}} - \frac{2r}{\chi^2_{(2r),\,1-\alpha/2}}\right)\beta_L \quad (15)$$

For fixed (1 − α) × 100% = 95%, sample n = 100, r = 10 (10) 100, and β_L = 1 (1) 5, the expected value $E(l\beta_L)$ is shown in Figure 1, where r = 10 (10) 100 indicates that the value of r begins at 10 and increases by 10 each time until its value equals 100. Similarly, index β_L = 1 (1) 5 means that the value of the index β_L begins at 1 and increases by 1 each time until its value equals 5.

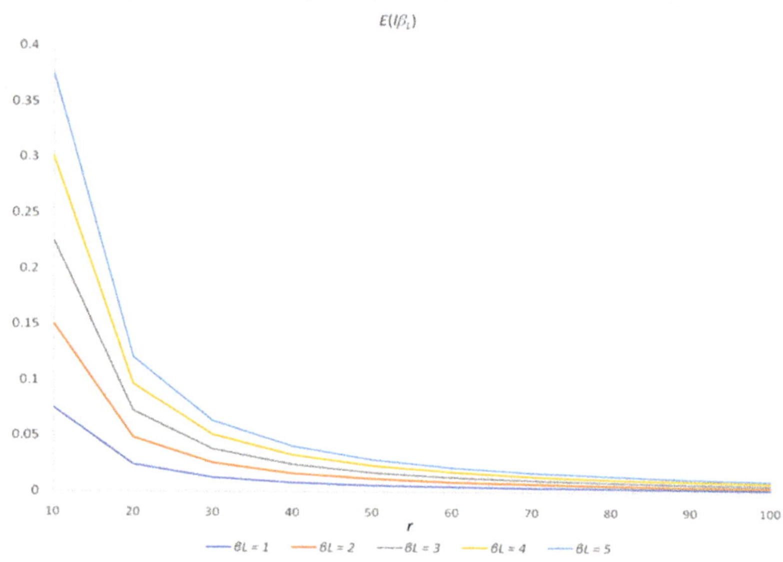

Figure 1. $E(l\beta_L)$ curves for β_L = 1 (1) 5, r = 10 (10) 100, and α = 0.05.

Given confidence level $((1 - \alpha) \times 100\%)$ and sample size n, the smaller the mean length of confidence interval $E(l\beta_L)$ is, the better estimation of the index β_L under different numbers of uncensored data r is. As noted in Figure 1, when index β_L is fixed, the mean length of the confidence interval $E(l\beta_L)$ is inversely proportional to the number of uncensored data r. This means that the better the estimate of the index β_L is, the more uncensored data have been collected.

3. Fuzzy Test Method for Lifetime Performance Index

In this section, for the purpose of determining whether lifetime performance meets its requirement, a fuzzy test method is utilized. The hypothesis is $H_0{:}\beta_L \geq c$ vs. $H_1{:}\beta_L < c$ [19], where c is the minimal value of relative lifetime performance index β_L required by customers. The following statistical testing rules are taken into consideration:

(1) If $\hat{\beta}_L < C_R$, then $\beta_L < c$ (i.e., the null hypothesis is rejected).
(2) If $\hat{\beta}_L \geq C_R$, then $\beta_L \geq c$ (i.e., the null hypothesis is not rejected).

C_R is the critical value determined by

$$p\{\hat{\beta}_L < C_R \,|\beta_L = c \in H_0\} = p\left\{W < \frac{2r\,C_R}{c}\right\} = \alpha \tag{16}$$

Hence, C_R can be calculated as follows:

$$C_R = \frac{c\,\chi^2_{(2r),\,\alpha}}{2r} \tag{17}$$

If we let $y_1, y_2, \ldots, y_n$ be the observed value of $Y_1, Y_2, \ldots, Y_n$, then the observed value of the estimator is

$$\hat{\beta}_{L0} = \frac{\hat{\lambda}_0}{L} = \frac{1}{r}\sum_{i=1}^{n} y_i \tag{18}$$

where

$$\hat{\lambda}_0 = \frac{L}{r}\sum_{i=1}^{n} y_i \tag{19}$$

As noted by Buckley [32], the α-cuts of triangular fuzzy numbers $\widetilde{\beta}_{L0}$ are [19,22]

$$\widetilde{\beta}_{L0}[\alpha] = \begin{cases} \left[\hat{\beta}_{L01}(\alpha), \hat{\beta}_{L02}(\alpha)\right], \text{ for } 0.01 \le \alpha \le 1 \\ \left[\hat{\beta}_{L01}(0.01), \hat{\beta}_{L02}(0.01)\right], \text{ for } 0 \le \alpha \le 0.01 \end{cases} \tag{20}$$

where

$$\hat{\beta}_{L01}(\alpha) = \frac{2r}{\chi^2_{(2r),1-\alpha/2}}\hat{\beta}_{L0} \tag{21}$$

and

$$\hat{\beta}_{L02}(\alpha) = \frac{2r}{\chi^2_{(2r),\,\alpha/2}}\hat{\beta}_{L0} \tag{22}$$

Obviously, the value of $\hat{\beta}_{L01}(\alpha)$ is not equal to the value of $\hat{\beta}_{L02}(\alpha)$ with $\alpha < 1$. As $\alpha = 1$, $\hat{\beta}_{L01}(1) = \hat{\beta}_{L02}(1) = \left(\frac{2r}{\chi^2_{(2r),\,0.5}}\right)\hat{\beta}_{L0} \neq \hat{\beta}_{L0}$.

Therefore, this paper let

$$\beta^*_{L0} = \frac{\chi^2_{(2r),\,0.5}}{2r}\hat{\beta}_{L0} \tag{23}$$

Then, the α-cuts of new triangular fuzzy numbers $\widetilde{\beta}^*_{L0}$ are

$$\widetilde{\beta}^*_{L0}[\alpha] = \begin{cases} \left[\beta^*_{L01}(\alpha), \beta^*_{L02}(\alpha)\right], \text{ for } 0.01 \le \alpha \le 1 \\ \left[\beta^*_{L01}(0.01), \beta^*_{L02}(0.01)\right], \text{ for } 0 \le \alpha \le 0.01 \end{cases} \tag{24}$$

where

$$\beta^*_{L01}(\alpha) = \frac{\chi^2_{(2r),0.5}}{\chi^2_{(2r),1-\alpha/2}}\hat{\beta}_{L0} \tag{25}$$

and

$$\beta^*_{L02}(\alpha) = \frac{\chi^2_{(2r),0.5}}{\chi^2_{(2r),\,\alpha/2}}\hat{\beta}_{L0} \tag{26}$$

Obviously, the value of $\beta^*_{L01}(\alpha)$ is equal to the value of $\beta^*_{L02}(\alpha)$ with $\alpha = 1$ ($\beta^*_{L01}(1) = \beta^*_{L02}(1) = \hat{\beta}_{L0}$) and there is a new triangular fuzzy number, denoted as $\widetilde{\beta}^{**}_{L0} = \Delta(\beta_{L0}, \beta_{M0}, \beta_{R0})$, where $\beta_{M0} = \hat{\beta}_{L0}$,

$$\beta_{L0} = \frac{\chi^2_{(2r),0.5}}{\chi^2_{(2r),0.995}}\hat{\beta}_{L0} \tag{27}$$

and

$$\beta_{R0} = \frac{\chi^2_{(2r),0.5}}{\chi^2_{(2r),0.005}}\hat{\beta}_{L0} \tag{28}$$

The following defines the membership function of fuzzy number $\widetilde{\beta}^{**}_{L0}$:

$$h(x) = \begin{cases} 0 & , if\ x < \beta_{L0} \\ 2\left(1 - F_W\left(\frac{\hat{\beta}_{L0}}{x}\chi^2_{(2r),0.5}\right)\right) & , if\ \beta_{L0} \le x < \hat{\beta}_{L0} \\ 1 & , if\ x = \hat{\beta}_{L0} \\ 2F_W\left(\frac{\hat{\beta}_{L0}}{x}\chi^2_{(2r),0.5}\right) & , if\ \hat{\beta}_{L0} < x \le \beta_{R0} \\ 0 & , if\ \theta_{R0} < x \end{cases} \tag{29}$$

where the cumulative distribution function of random variable W is denoted by F_W. Similarly to fuzzy numbers $\widetilde{\beta}^*_{L0}$, the α-cuts of triangular fuzzy critical values $\widetilde{C}_R$ are

$$\widetilde{C}_R[\alpha] = \begin{cases} [C_{R1}(\alpha), C_{R2}(\alpha)], \text{ for } 0.01 \le \alpha \le 1 \\ [C_{R1}(0.01), C_{R2}(0.01)], \text{ for } 0 \le \alpha \le 0.01 \end{cases} \tag{30}$$

where

$$C_{R1}(\alpha) = \frac{\chi^2_{(2r),0.5}}{\chi^2_{(2r),1-\alpha/2}} C_R \tag{31}$$

and

$$C_{R2}(\alpha) = \frac{\chi^2_{(2r),0.5}}{\chi^2_{(2r),\,\alpha/2}} C_R \tag{32}$$

Obviously, the value of $C_{R1}(\alpha)$ is equal to the value of $C_{R2}(\alpha)$ with $\alpha = 1$ ($C_{R1}(1) = C_{R2}(1) = C_R$) and the new triangular fuzzy number is $\widetilde{C}_0 = \Delta(C_{LR}, C_{MR}, C_{RR})$, where $C_{MR} = C_R$,

$$C_{LR} = \frac{\chi^2_{(2r),0.5}}{\chi^2_{(2r),0.995}} C_R \tag{33}$$

and

$$C_{RR} = \frac{\chi^2_{(2r),0.5}}{\chi^2_{(2r),0.005}} C_R \tag{34}$$

The following defines the membership function of fuzzy $\widetilde{C}_R$:

$$g(x) = \begin{cases} 0 & , if\ x < C_{LR} \\ 2\left(1 - F_W\left(\frac{C_R}{x}\chi^2_{(2r),0.5}\right)\right) & , if\ C_{LR} \le x < C_R \\ 1 & , if\ x = C_R \\ 2F_W\left(\frac{C_R}{x}\chi^2_{(2r),0.5}\right) & , if\ C_R < x \le C_{RR} \\ 0 & , if\ C_{RR} < x \end{cases} \tag{35}$$

As noted, the cumulative distribution function of random variable W is denoted by F_W. Membership functions $h(x)$ and $g(x)$ are presented in Figure 2:

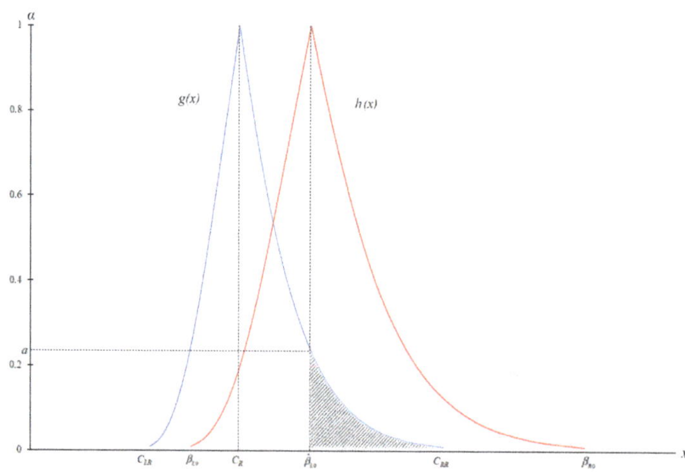

Figure 2. Membership functions $h(x)$ and $g(x)$.

Based on Chen and Yu [19], this paper let set B_T be the area under the graph of $g(x)$. That is,

$$B_T = \{ (x, \alpha) | C_{R1}(\alpha) \leq x \leq C_{R2}(\alpha), 0 \leq \alpha \leq 1 \} \tag{36}$$

As noted by Chen and Chang [13] and Chen and Yu [19], it is difficult to use integration to calculate the area of a set B_T. The approach, trapezoidal rule, is implemented in the study in order to build up the area of the block B_T. The procedures are following: (1) we classify the block B_T, n = 100, into several equal horizontal blocks. (2) Each section of the blocks would be calculated through the approximate trapezoid area. Then, (3) the sum of the areas for these 100 horizontal blocks is calculated. For this reason, $i = [100 \times \alpha]$ is considered. Then, $i = 0, 1, 2, \ldots, 100$ for $0 \leq \alpha \leq 1$, where $[100 \times \alpha]$ represents the largest integer less than or equal to $100 \times \alpha$. Similarly, $\alpha = i \times 0.01$, $i = 0, 1, 2, \ldots, 100$. These 101 horizontal lines are cut B_T into 100 trapezoidal blocks. Then, the following denotes the ith block:

$$B_{Ti} = \{ (x, \alpha) | C_{R1}(0.01 \times i) \leq x \leq C_{R2}(0.01 \times i), 0.01 \times (i-1) \leq \alpha \leq 0.01 \times i \}, i = 1, \ldots, 100 \tag{37}$$

The following definition for the length of ith horizontal line d_i as follows:

$$d_i = \left(\frac{\chi^2_{(2r),0.5}}{\chi^2_{(2r),\ 0.005\times i}} - \frac{\chi^2_{(2r),0.5}}{\chi^2_{(2r),\ 1-0.005\times i}} \right) C_R = 1, 2, \ldots, 100 \tag{38}$$

Obviously, $d_0 = d_1$ and $d_{100} = 0$, so the area B_T is

$$B_T = \sum_{i=1}^{100} (0.01) \times \left(\frac{d_{i-1} + d_i}{2} \right) = 0.01 \left(\frac{d_1}{2} + \sum_{i=1}^{99} d_i \right) \tag{39}$$

If B_R denotes the area under graph $g(x)$ to the right of $x = \hat{\beta}_{L0}$, then

$$B_R = \{ (x, \alpha) | \hat{\beta}_{L0} \leq x \leq C_{R2}(\alpha), 0 \leq \alpha \leq a \} \tag{40}$$

where $\alpha = a$ such that $C_{R2}(a) = \hat{\beta}_{L0}$. Similarly B_T, $k = [100 \times a]$. Then, for $0 \leq \alpha \leq a$, where $[100 \times a]$ represents the largest integer less than or equal to $100 \times a$. Obviously, $a = 0.01 \times k$ and $\alpha = i \times 0.01$, ($i = 0, 1, 2, \ldots, k$) horizontal lines cut B_R into k trapezoidal blocks. Then, the ith block can be expressed as follows:

$$B_{Ri} = \{ (x, \alpha) | \hat{\beta}_{L0} \leq x \leq C_{R2}(0.01 \times i), 0.01 \times (i-1) \leq \alpha \leq 0.01 \times i \} = 1, 2, \ldots, k \tag{41}$$

The following defines the length of ith horizontal line r_i:

$$r_i = \frac{\chi^2_{(2r),0.5}}{\chi^2_{(2r),\ 0.005\times i}} C_R - \hat{\beta}_{L0} = 1, 2, \ldots, k \tag{42}$$

This indicates that $r_0 = r_1$ and $r_k = 0$, so the area of B_R is

$$B_R = \sum_{i=1}^{k} (0.01) \times \left(\frac{r_{i-1} + r_i}{2} \right) = 0.01 \left(\frac{r_1}{2} + \sum_{i=1}^{k-1} r_i \right) \tag{43}$$

The ratio of B_R/B_T can be usefully applied to fuzzy decision-making:

$$B_R/B_T = \frac{0.01 \left(\frac{r_1}{2} + \sum_{i=1}^{k-1} r_i \right)}{0.01 \left(\frac{d_1}{2} + \sum_{i=1}^{99} d_i \right)} \tag{44}$$

However, calculation of B_R/B_T is complicated.

According to Equations (39) and (43), these have calculated that, respectively, obtaining the block areas of B_T and B_R is extremely complicated. Therefore, for the purpose of simplifying the complicated calculating process of ratio B_R/B_T, the technique, membership functions $g(x)$ and $h(x)$ with asymmetry (in Figure 2), proposed by Chen and Chang [13] is utilized in the present study. The method suggested by Chen and Chang [13], to replace d_R (the length of the base of the set B_R) with the area of B_R, facilitates industrial applications. Similarly, d_T (the length of the base of the set B_T) is replaced with the area of B_T. As the membership functions are asymmetric, $d_T = 2(C_{RR} - C_R)$ on the basis of Chen and Chang [13] and Chen et al. [20]. In Figure 2, by using the principle of similar shapes, the square of the side length ratio is equal to the area ratio. Next, B_R/B_T was replaced with d_R/d_T as the fuzzy evaluation tool, where d_R and d_T are calculated as follows [13,20,33]:

$$d_R = C_{RR} - \hat{\beta}_{L0} = \frac{\chi^2_{(2r),0.5}}{\chi^2_{(2r),\ 0.995}} C_R - \hat{\beta}_{L0} \tag{45}$$

and

$$d_T = \ 2(C_{RR} - C_R = 2\left(\frac{\chi^2_{(2r),0.5}}{\chi^2_{(2r),0.005}} C_R - C_R\right) \tag{46}$$

Based on their past experiences originating from other experts and the past data over the certain products [34], manufacturing engineers are allowed to define the values of δ_1 and δ_2. The following two numbers $0 < \delta_1 < \delta_2 < 0.5$ and $\delta = d_R/d_T$, the fuzzy test rules are as follows [13,20,35]:

(1) If $\delta < \delta_1$, then conclude that $\beta_L \geq c$ (i.e., do not reject H_0).
(2) If $\delta_1 \leq \delta \leq \delta_2$, then make no decision; more information is needed.
(3) If $\delta_2 < \delta < 0.5$, then conclude that $\beta_L < c$ (i.e., reject H_0).

4. Practical Example

This section presents a numerical example to demonstrate the proposed fuzzy test method. The required value of the lifetime performance index is at least 3; thus, the null hypothesis is H_0:$\beta_L \geq 3$ vs. the alternative hypothesis H_1:$\beta_L < 3$ [19]. If $y_1, y_2, \ldots, y_{30}$ is the observed value of $Y_1, Y_2, \ldots, Y_{30}$ with number of the uncensored data $r = 18$ ($r/n = 60\%$), then the observed value of the estimator is

$$\hat{\beta}_{L0} = \frac{1}{r}\sum_{i=1}^{n} y_i = \frac{41.6894}{18} = 2.316 \tag{47}$$

The values of β_{L0} and β_{R0} are then calculated as follows:

$$\beta_{L0} = \frac{\chi^2_{(18),0.5}}{\chi^2_{(18),0.995}} \times \ 2.316 \ = \ 1.329 \tag{48}$$

and

$$\beta_{R0} = \frac{\chi^2_{(18),0.5}}{\chi^2_{(18),0.005}} \times \ 2.316 \ = \ 4.576 \tag{49}$$

Furthermore, the membership function of fuzzy numbers $\tilde{\beta}^{**}_{L0}$ is

$$h(x) = \begin{cases} 0 & , if\ x < 1.329 \\ 2\left(1 - F_W\left(\frac{1.3290}{x} \times \chi^2_{(36),0.5}\right)\right) & , if\ 1.329 \leq x < \ 2.316 \\ 1 & , if\ x = 2.316 \\ 2F_W\left(\frac{1.3290}{x} \times \chi^2_{(36),0.5}\right) & , if\ 2.316 < x \leq 4.576 \\ 0 & , if\ 4.576 < x \end{cases} \tag{50}$$

where $\chi^2_{(36),0.005}$ = 17.887, $\chi^2_{(36),0.995}$ = 61.581, and $\chi^2_{(36),0.5}$ = 35.336. As the significance level is $\alpha = 0.05$, then

$$C_R = \frac{c\chi^2_{(2r),\,\alpha}}{2r} = \frac{3\chi^2_{(36),0.05}}{36} = 1.939 \tag{51}$$

The values of C_{LR} and C_{RR} are calculated as follows:

$$C_{LR} = \frac{\chi^2_{(36),0.5}}{\chi^2_{(36),0.995}} \times 1.9391 = 1.113 \tag{52}$$

and

$$C_{RR} = \frac{\chi^2_{(36),0.5}}{\chi^2_{(36),0.005}} \times 1.9391 = 3.831 \tag{53}$$

Furthermore, the membership function of fuzzy number $\widetilde{C}_R$ is

$$g(x) = \begin{cases} 0 & ,if\ x \le 1.113 \\ 2\left(1 - F_W\left(\frac{1.9391}{x}\chi^2_{(36),0.5}\right)\right) & ,if\ 1.113 < x < 1.939 \\ 1 & ,if\ x = 1.939 \\ 2F_W\left(\frac{1.9391}{x}\chi^2_{(36),0.5}\right) & ,if\ 1.939 < x \le 3.831 \\ 0 & ,if\ 3.831 < x \end{cases} \tag{54}$$

By Equations (50) and (54), we have the graphs of $h(x)$ and $g(x)$ in Figure 3. From Equation (54), we obtain α = $g(x)$. When x = $\hat{\beta}_{L0}$ = 2.316, $\alpha \in$ (0.46, 0.47), a = 0.468 could be obtained by interpolation method.

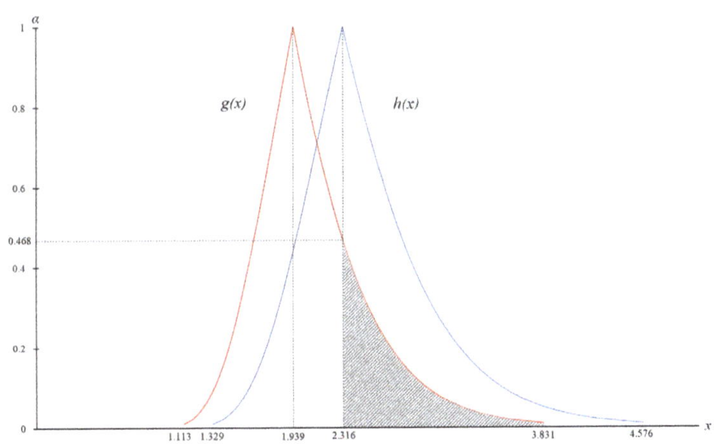

Figure 3. Membership functions $g(x)$ and $h(x)$ for numerical example.

The values of d_R and d_T are calculated as follows [13,20,33]:

$$d_R = C_{RR} - \hat{\beta}_{L0} = 3.831 - 2.316 = 1.515 \tag{55}$$

and

$$d_T = 2(C_{RR} - C_R) = 2(3.831 - 1.939) = 3.784 \tag{56}$$

Therefore,

$$\delta = d_R/d_T = 1.515/3.784 = 0.4004 \tag{57}$$

This leads to the conclusion that for $\hat{\beta}_{L0} = 2.316 > C_R = 1.939$, $\beta_L \geq 3$ (i.e., the null hypothesis should not be rejected). However, $\hat{\beta}_{L0} = 2.316$ is far less than $\beta_L = 3$. Thus, for $\delta_1 = 0.2$ and $\delta_2 = 0.4$ [13], $\beta_L < 3$ (i.e., the null hypothesis should be rejected). This is the risk of misjudgment caused by sampling errors in small sample sizes [7,9]. The proposed fuzzy method therefore provides a more reasonable conclusion.

5. Conclusions

This paper proposes an evaluation approach for product lifetime performance under type-II censoring. This evaluation enables the improvement of lifetime performance, which enhances the value of products as well as attains green goals such as energy efficiency and waste reduction. The proposed index is easy to use as its value increases with performance. Examination of the probability density function, cumulative distribution function, and reliability function of relative lifetime X indicated that reliability increased with the value of the index, as did the probability of the product lifetime surpassing the minimum with value L. An unbiased consistent estimator of the proposed index is also presented alongside a fuzzy test model based on the derived confidence interval. This model reduces the probability of misjudgment caused by sampling errors [7,9]. Additionally, many benefits will be gained by seizing the chance to improve, such as decreasing the testing cost and meeting the certain requirements in a short time. Furthermore, doing so is said to expand the possibility of using less paper, saving social resources, decreasing the carbon footprint and so forth [36]. In the electronics industry, passive components have long been indispensable parts that stimulate peripheral equipment industries. The proposed model thus focuses on passive components, with applicability demonstrated through a numerical example.

Funding: This research received no external funding.

Institutional Review Board Statement: Not applicable.

Informed Consent Statement: Not applicable.

Data Availability Statement: Not applicable.

Conflicts of Interest: The author declares no conflict of interest.

References

1. Huang, C.F.; Huang, D.H.; Lin, Y.K. System reliability analysis for a cloud-based network under edge server capacity and budget constraints. *Ann. Oper. Res.* **2022**, *312*, 217–234. [CrossRef]
2. Lin, K.P.; Yu, C.M.; Chen, K.S. Production data analysis system using novel process capability indices-based circular economy. *Ind. Manag. Data Syst.* **2019**, *119*, 1655–1668.
3. Suppipat, S.; Hu, A.H. Achieving sustainable industrial ecosystems by design: A study of the ICT and electronics industry in Taiwan. *J. Clean. Prod.* **2022**, *3691*, 133393. [CrossRef]
4. Borazon, E.Q.; Huang, Y.C.; Liu, J.M. Green market orientation and organizational performance in Taiwan's electric and electronic industry: The mediating role of green supply chain management capability. *J. Bus. Ind. Mark.* **2022**, *37*, 1475–1496.
5. Chen, K.S.; Yu, C.M.; Huang, M.L. Fuzzy selection model for quality-based IC packaging process outsourcers. *IEEE Trans. Semicond. Manuf.* **2022**, *35*, 102–109. [CrossRef]
6. Wang, C.T.; Chiu, C.S. Competitive strategies for Taiwan's semiconductor industry in a new world economy. *Technol. Soc.* **2014**, *36*, 60–73.
7. Chen, K.S.; Yu, C.M. Fuzzy test model for performance evaluation matrix of service operating systems. *Comput. Ind. Eng.* **2020**, *140*, 106240.
8. Wang, G.; Shao, M.; Lv, S.; Kong, X.; He, Z.; Vining, G. Process parameter optimization for lifetime improvement experiments considering warranty and customer satisfaction. *Reliab. Eng. Syst. Saf.* **2022**, *221*, 108369. [CrossRef]
9. Yu, C.M.; Chen, K.S.; Lai, K.K.; Hsu, C.H. Fuzzy supplier selection method based on smaller-the-better quality characteristic. *Appl. Sci.* **2020**, *10*, 3635. [CrossRef]
10. Chen, K.S. Fuzzy testing of operating performance index based on confidence intervals. *Ann. Oper. Res.* **2022**, *311*, 19–33.
11. Chen, K.S.; Yu, C.M. Special Issue: Smart service technology for industrial applications. *Appl. Sci.* **2022**, *12*, 10259. [CrossRef]
12. Wu, J.W.; Hong, C.W. The assessment of quality performance of lifetime index of exponential products with fuzzy data under progressively type ii right censored sample. *ICIC Express Lett. Part B Appl.* **2018**, *9*, 1101–1107.
13. Chen, K.S.; Chang, T.C. Fuzzy testing model for the lifetime performance of products under consideration with exponential distribution. *Ann. Oper. Res.* **2022**, *312*, 87–98. [CrossRef]

14. Chiou, K.C.; Chen, K.S. Lifetime performance evaluation model based on quick response thinking. *Eksploat. I Niezawodn.* **2022**, *24*, 1–6. [CrossRef]
15. Yang, C.M. An improved multiple quality characteristic analysis chart for simultaneous monitoring of process mean and variance of steering knuckle pin for green manufacturing. *Qual. Eng.* **2021**, *33*, 383–394. [CrossRef]
16. Wu, M.F.; Chen, H.Y.; Chang, T.C.; Wu, C.F. Quality evaluation of internal cylindrical grinding process with multiple quality characteristics for gear products. *Int. J. Prod. Res.* **2019**, *57*, 6687–6701.
17. Chen, K.S.; Wang, C.H.; Tan, K.H.; Chiu, S.F. Developing one-sided specification six-sigma fuzzy quality index and testing model to measure the process performance of fuzzy information. *Int. J. Prod. Econ.* **2019**, *208*, 560–565.
18. Tong, L.I.; Chen, K.S.; Chen, H.T. Statistical testing for assessing the performance of lifetime index of electronic components with exponential distribution. *Int. J. Qual. Reliab. Manag.* **2002**, *19*, 812–824. [CrossRef]
19. Chen, K.S.; Yu, C.M. Lifetime performance evaluation and analysis model of passive component capacitor products. *Ann. Oper. Res.* **2022**, *311*, 51–64. [CrossRef]
20. Chen, K.S.; Wu, C.F.; Tsaur, R.C.; Huang, T.H. Fuzzy evaluation and improvement decision-making model for machining operation Performance. *Appl. Sci.* **2023**, *13*, 1430. [CrossRef]
21. Chen, K.S.; Yu, C.M.; Hsu, T.H.; Cai, S.R.; Chiou, K.C. A model for evaluating the performance of the bearing manufacturing process. *Appl. Sci.* **2019**, *9*, 3105. [CrossRef]
22. Chen, K.S. Two-tailed Buckley fuzzy testing for operating performance index. *J. Comput. Appl. Math.* **2019**, *361*, 55–63. [CrossRef]
23. Chen, K.S.; Yu, C.M.; Chang, T.C.; Chen, H.Y. Fuzzy evaluation of product reliability based on ratio-based lifetime performance index. *Ann. Oper. Res.* **2022**. [CrossRef]
24. Wu, S.F.; Wu, Y.C.; Wu, C.H.; Chang, W.T. Experimental design for the lifetime performance index of weibull products based on the progressive type i interval censored sample. *Symmetry* **2021**, *13*, 1691. [CrossRef]
25. Krishna, H.; Kumar, K. Reliability estimation in generalized inverted exponential distribution with progressively type II censored sample. *J. Stat. Comput. Simul.* **2013**, *83*, 1007–1019. [CrossRef]
26. Tong, L.I.; Chiou, K.C. Estimating the censored observations under random censoring model for the Exponential distribution. *J. Chin. Inst. Ind. Eng.* **1999**, *16*, 85–92.
27. Miller, R.G. *Survival Analysis*; John Wiley and Sons: New York, NY, USA, 1981.
28. Shafay, A.R. Step-stress partially accelerated life tests with progressive type-II censored sample from two-parameter inverted exponential distribution. *J. Stat. Appl. Probab.* **2022**, *11*, 981–990.
29. Salem, M.; Amin, Z.; Ismail, M. Progressively censored reliability sampling plans based on mean product lifetime. *Sankhya B* **2020**, *82*, 1–33. [CrossRef]
30. Mondal, S.; Kundu, D. On the joint type-II progressive censoring scheme. *Commun. Stat.—Theory Methods* **2020**, *49*, 958–976. [CrossRef]
31. Aslam, M.; Yen, C.H.; Chang, C.H.; Al-Marshadi, A.H.; Jun, C.H. A multiple dependent state repetitive sampling plan based on performance index for lifetime data with type II censoring. *IEEE Access* **2019**, *7*, 49377–49391. [CrossRef]
32. Buckley, J.J. Fuzzy statistics: Hypothesis testing. *Soft Comput.* **2005**, *9*, 512–518. [CrossRef]
33. Chen, K.S.; Liu, C.C.; Chen, C.H. Fuzzy evaluation of process quality with process yield index. *Mathematics* **2022**, *10*, 2514. [CrossRef]
34. Chen, K.S.; Lai, Y.L.; Huang, M.C.; Chang, T.C. Fuzzy judgement model for assessment of improvement effectiveness to performance of processing characteristics. *Int. J. Prod. Res.* **2023**, *61*, 1591–1605. [CrossRef]
35. Chen, K.S.; Huang, T.H.; Tsaur, R.C.; Kao, W.Y. Fuzzy evaluation models for accuracy and precision indices. *Mathematics* **2022**, *10*, 3961. [CrossRef]
36. Chen, K.S.; Wang, C.H.; Tan, K.H. Developing a fuzzy green supplier selection model using six sigma quality indices. *Int. J. Prod. Econ.* **2019**, *212*, 1–7.

mathematics

Article

Assessing Knowledge Quality Using Fuzzy MCDM Model

Chiu-Chi Wei *, Chih-Chien Tai, Shun-Chin Lee and Meng-Ling Chang

Ph.D. Program of Management, Chung Hua University, Hsinchu City 30012, Taiwan;
ccd12261973@gmail.com (C.-C.T.); sglee@tpcu.edu.tw (S.-C.L.); b1777@tpech.gov.tw (M.-L.C.)
* Correspondence: a0824809@gmail.com

Abstract: The purpose of knowledge management is to excavate the tacit knowledge accumulated by each enterprise member through the knowledge proposal system. Each knowledge proposal must be assessed, and after passing the quality assessment, the knowledge proposal will be stored in the knowledge repository and shared with other employees who need the knowledge at work. In the long run, the capabilities of all employees will gradually enhance and the competitiveness of enterprises will naturally increase. The correct assessment of knowledge quality is the key to the success of knowledge management. Some scholars propose to use the AHP (analytical hierarchical process) to determine the quality of knowledge. The problem with this approach is that the AHP can only obtain the relative quality of all knowledge, not the actual quality of knowledge. Therefore, this study proposes a fuzzy assessment model to measure knowledge quality, which includes a knowledge quality fuzziness index (KQFI) and a checking gate. First, experts conduct linguistic evaluation on the weight of criteria and knowledge quality. All linguistic evaluations are then integrated into a knowledge quality fuzziness index (KQFI), which is compared with a fuzzy threshold (FT); then, the level of goodness of KQFI to FT is obtained. If it is greater than 0.5, it means that the quality of the knowledge proposal is qualified; otherwise, it means that the quality of the knowledge proposal is unqualified. This study uses a case including five experts and nine knowledge proposals to demonstrate the applicability of the method. The results show that the method finally judges six knowledge instances as qualified and three as unqualified. The results show that the proposed method can indeed assist enterprises to effectively screen knowledge proposals.

Keywords: knowledge management; fuzzy theory; multi-criteria decision making

MSC: 03B52; 03E72; 90B50

Citation: Wei, C.-C.; Tai, C.-C.; Lee, S.-C.; Chang, M.-L. Assessing Knowledge Quality Using Fuzzy MCDM Model. *Mathematics* **2023**, *11*, 3673. https://doi.org/10.3390/math11173673

Academic Editor: Aleksandar Aleksić

Received: 20 July 2023
Revised: 21 August 2023
Accepted: 23 August 2023
Published: 25 August 2023

1. Introduction

Enterprises are facing stiff global competition, and the best way to effectively lead competitors is to develop new products, new services, and new business models [1]. One of the ways to achieve this goal is to promote knowledge management. First, establish a culture of knowledge sharing; then, collect, review, store, and reuse existing knowledge within the enterprise. Employees can socialize, externalize, combine, and internalizing existing knowledge to create new knowledge [2,3]. Enterprise knowledge includes tacit knowledge and explicit knowledge; each can be subdivided into personal knowledge and organizational knowledge [4,5]. Explicit organizational knowledge includes operating standards, procedures, and manuals. Tacit organizational knowledge refers to the knowledge that a group of people can effectively complete a project. Explicit personal knowledge includes notes and computer files. Tacit personal knowledge includes work experience, work skills, techniques, etc. Tacit organizational knowledge and tacit personal knowledge generally remain in the brains of individuals. Therefore, when personnel retire or leave the enterprise, this knowledge is permanently lost. Most importantly, tacit knowledge is the most critical knowledge that enterprises can win the competition. Therefore, enterprises must be equipped with knowledge management systems to preserve these tacit knowledge [6,7]. The method most often used to preserve tacit knowledge is to set up a knowledge community (community of practice) with a reward system to establish a collaborative work culture so that knowledge can be transferred from one person's brain to another [8,9]. In this way, problems that only one person could solve in the past can now be solved by many people and everyone's ability will gradually improve. When the company needs to cope with competition, the company's personnel can quickly and effectively develop new products and new services to overcome challenges.

The community of practice must correspond to the strategy of the enterprise. For example, the enterprise hopes to establish independent product design and development capabilities; so, several related communities of practice can be established and each with a clear objective to achieve. Each community of practice has a management team and a process for receiving and reviewing knowledge proposals. Finally, people are encouraged to participate in communities of practice and rewards are offered for knowledge proposals. The knowledge proposals put forward by community members will be reviewed by senior personnel in related fields, and the knowledge that passes the review will be shared, published, and stored in the knowledge base. Obviously, if the judgment on the quality of knowledge is wrong, poorer knowledge will also pass the review; so, the overall ability of the enterprise will not be improved. On the contrary, if quality knowledge fails to pass the review all the time, after a period, the ability of the enterprise will become worse and worse.

The criteria of measuring knowledge quality in different fields may be different but most of them must be evaluated from multiple aspects at the same time. In the past, scholars mentioned that measuring knowledge quality should include correctness, completeness, consistency, relevance, etc. [10]. Some studies mentioned that measuring knowledge quality should include certainty, accuracy, and operability [11], while Arora et al. (2013) believed that measuring knowledge quality should include completeness, timeliness, accuracy, transparency, and relevancy [12]. The assessment of knowledge quality is a multi-attribute decision-making problem. In practice, when experts assess the quality of knowledge, they do not only make qualitative and subjective judgments; more importantly, it is difficult to make a completely correct quality judgment while considering multiple aspects at the same time. Further, academically, there have been very few papers on knowledge quality assessment in the past. The methods used mainly included the AHP and statistical analysis of questionnaires [10,13,14]. The problem with the AHP is that it is necessary to compare the relative importance of all knowledge proposals at the same time. When there are many knowledge proposals, it is difficult to know what is wrong with the large matrix obtained by pairwise comparison. If the consistency checks fail, the adjustment process will be very complicated, not to mention that the quality of knowledge should be an absolute judgment of good or bad rather than a comparison of relative good or bad. Moreover, most papers on statistical analysis of questionnaires discuss the impact of knowledge quality on corporate innovation and operational performance rather than measuring the quality of knowledge.

Obviously, a method that can correctly assess knowledge quality is nonexistent. In addition, it is difficult for humans to give specific numbers to measure the quality of knowledge. It is relatively easy to use fuzzy linguistic assessment of very good, good, fair, poor, and very poor to evaluate the quality of knowledge. Therefore, the objective of this study is to develop a multi-criteria knowledge quality assessment model, including a knowledge quality fuzziness index and a checking gate, to effectively determine the quality of knowledge proposals. This study can improve the shortcomings of existing methods, assist enterprises to screen out high-quality knowledge, and improve the performance of enterprise knowledge management.

2. Literature Review

This section reviews literature related to this study, including knowledge quality, and fuzzy set and group decision making.

2.1. Knowledge Quality

Quality is defined by the Oxford Dictionary as the degree of excellence of a thing; relative nature or kind or character of a thing; or class or grade of something determined by this [1]. Additionally, qualities need to be described using some attributes. For example, the quality of products is expressed in terms of functionality and reliability while the quality of services is measured in terms of responsiveness and empathy [15]. Enterprises determine the specifications of products and services, while quality is judged by customers. Enterprises usually use many methods to try to tap customers' needs for products and services, and hope that products and services can meet customers' requirements 100%. However, it has been proved in practice that this is almost an impossible task. For example, even for a company as large as Microsoft, the development of Windows Vista still cannot meet customer needs and is called the biggest failure of Microsoft ever [16]. Product development not only involves product-related knowledge but also involves marketing-related knowledge. Both must conform to the overall strategy of the enterprise, and the quality of knowledge must be accurately evaluated to ensure that the retained and stored knowledge can enhance the competitiveness of the enterprise [13,14].

Chakrabarti et al. (2018) proposed an approach to relate knowledge quality with elements that consist of attributes; thus, an enterprise can discover which element provides the most effective direction to improve knowledge quality [10]. Lim et al. (2013) examined the relationship between sentiment and quality of knowledge shared among knowledge workers and job performance [17]. It was indicated that data quality and information quality are often used to assess knowledge quality, and a reliable knowledge quality measure is nonexistent [11]. Zhou et al (2022) explored the impact of knowledge transfer among supply chain members on firm innovation and operational performance, and how knowledge quality affects the relationship between them [18]. Abdollahbeigi and Salehi (2021) found that the quality of knowledge significantly affects the innovation of enterprises, and the ability of innovation will have a significant impact on non-financial performance [19]. Ganguly et al. (2019) concluded that the transfer of tacit knowledge and the quality of knowledge are positively related to the innovation ability of enterprises [20]. Zhou et al. (2022) discovered that the knowledge quality of relational capital and cognitive capital positively affects product innovation performance but structural social capital does not affect the quality of knowledge [18,21]. From these past literatures, it can be found that up to now, no effective assessment method of knowledge quality has been proposed that can assist enterprises to correctly measure the quality of knowledge.

2.2. Fuzzy Set and Group Decision Making

Deterministic and quantifiable information with values between 0 and 1 are usually handled using classical set theory; however, when the information contains uncertainties that cannot be quantified, classical sets cannot be used. The evaluation of knowledge proposals is uncertain, and quantitative methods cannot be used to judge whether they meet the quality requirements. Therefore, classical set theory is not applicable. This situation where the value can vary continuously between zero and one is where fuzzy theory comes in handy. Fuzzy theory uses membership functions to express the degree of membership between components and sets. A fuzzy set A can be expressed as Equation (1).

$$A = \{(x, \mu_A(x))|x \in X\} \tag{1}$$

where X is a universe of discourse and $\mu_A(x)$ indicates the degree of membership between component x and fuzzy set A.

Because the triangular fuzzy function has been proven to be very suitable for dealing with the imprecision and uncertainty of the multi-criteria decision-making process [12], this study uses the triangular fuzzy function to evaluate the attribute weight and knowledge quality. Let triangular fuzzy number A $= (a, b, c)$; then, the membership function of A can be expressed as Equation (2) [22–24].

$$\mu_A = \begin{cases} \frac{(x-a)}{(b-a)} & a \le x \le b \\ \frac{(c-x)}{(c-b)} & b \le x \le c \\ 0 & otherwise \end{cases} \tag{2}$$

The α-cut is applied to convert a fuzzy number into a crisp set, and the α-cut of a triangular fuzzy number A can be described as Equation (3), where A_α is a crisp set [25,26].

$$A_\alpha = \{x | \mu_A(x) \ge \alpha\} \quad \alpha \in \{0, 1\} \tag{3}$$

The confidence of interval α-level of A_α can be described as Equation (4), and A_α implies the confidence level of a decision maker in the evaluation outcome.

$$A = [(b-a)\alpha + a, c - (c-b)\alpha] \tag{4}$$

Chen (2000) indicated that precise quantified information and solving real problems are not necessarily relevant [22]. Li et al. (2022) pointed out that linguistic variables can be used in fuzzy theory to manipulate the operations [27]. When assessors judge the importance of the criteria using fuzzy method, the linguistic variables can be converted into a fuzzy number and results can be obtained using fuzzy algebra [28]. For instance, the linguistic variables of very good, good, fair, poor, and very poor in Table 1 can be used to assess the quality of the knowledge proposal, and the assessment of a knowledge proposal can be obtained using the triangular fuzzy number (TFN).

Table 1. Weight, quality, and TFN of knowledge proposal.

Weight	Quality	TFN
VL (Very low)	VP (Very poor)	(0, 0, 0.1)
L (Low)	P (Poor)	(0, 0.1, 0.3)
ML (Medium low)	MP (Medium poor)	(0.1, 0.3, 0.5)
M (Medium)	M (Medium)	(0.3, 0.5, 0.7)
MH (Medium high)	MG (Medium good)	(0.5, 0.7, 0.9)
H (High)	G (Good)	(0.7, 0.9, 1.0)
VH (Very high)	VG (Very good)	(0.9, 1.0, 1.0)

Because the measurement of knowledge quality involves imprecise information that cannot be quantified, and the quality of knowledge must be evaluated by a group of experts and several attributes at the same time, it is suitable to use the fuzzy multi-criteria group decision making method [29]. Past studies mostly used the average value as the final group decision [26,30–33]; however, average of opinion cannot accurately reflect the overall judgment. Therefore, Hsu and Chen proposed a similarity aggregation method (SAM) [34,35], and Lee developed an optimal aggregation method (OAM), to help obtain the consistence of fuzzy opinion [1,36]. Because OAM is an effective method for integrating expert opinions, this study adopts OAM to consolidate the opinions of experts. The steps of OAM can be described as below:

(1) Let the fuzzy number of the expert's opinion of A and B be $\widetilde{A} = (a_1, a_2, a_3)$ and $\widetilde{B} = (b_1, b_2, b_3)$; then, the distance between $\widetilde{A}$ and $\widetilde{B}$ can be computed using Equation (5), and the similarity between $\widetilde{A}$ and $\widetilde{B}$ can be obtained using Equation (6).

$$d_2(\widetilde{A}, \widetilde{B}) = \sqrt{\sum_{i=1}^{3} (|a_i - b_i|)^2} \quad (5)$$

$$S_2(\widetilde{A}, \widetilde{B}) = 1 - \frac{\left(d_2(\widetilde{A}, \widetilde{B})\right)^2}{4u^2} \quad (6)$$

where $u = \max(U) - \min(U)$, U is the universe of discourse, and $0 \le S_2(\widetilde{A}, \widetilde{B}) \le 1$.

(2) Set the initial aggregated weight as the weight of the first expert. $0 < w_i^{(0)} < 1$ and $\sum_{i=1}^{n} w_i^{(0)} = 1, i = 1, 2, \ldots, n$, n is the number of criteria, and iteration $l = 0,\ 1,\ 2,\ \cdots$.

$$\sum_{i=1}^{n} w_i^l = 1 \quad (7)$$

(3) Compute aggregated opinion using Equation (8); $\widetilde{R}_i$ is the ith expert's individual opinion.

$$\widetilde{R}^{(l+1)} = \frac{\sum_{i=1}^{n} \left(w_i^{(l)}\right)^m \widetilde{R}_i}{\sum_{i=1}^{n} \left(w_i^{(l)}\right)^m}, \text{ where } m \text{ is an exponential weight.} \quad (8)$$

(4) Let aggregated weight $W^{(l)} = (w_1^{(l)}, w_2^{(l)}, \ldots, w_n^{(l)})$, and compute $W^{(l+1)}$ using Equation (9).

$$w_i^{(l+1)} = \frac{\left(\frac{1}{(c - S_2(\widetilde{R}_i, \widetilde{R}^{(l+1)}))}\right)^{\frac{1}{m-1}}}{\sum_{j=1}^{n} \left(\frac{1}{(c - S_2(\widetilde{R}_j, \widetilde{R}^{(l+1)}))}\right)^{\frac{1}{m-1}}}, \text{ where } c \text{ is an integer constant.} \quad (9)$$

(5) If $\left\| W^{(l+1)} - W^{(l)} \right\| \le \varepsilon$, stop; otherwise, $l = l + 1$, go to Step (3).

In terms of literature on multi-criteria decision-making (MCDM), Stojčić et al. (2019) reviewed the application of MCDM methods in sustainable engineering. From a review of 108 related literatures from 2008 to 2018, they found that MCDM methods are very suitable for sustainable engineering [37]. Zavadskas et al. (2014) also reviewed the relevant literature on MCDM and believed that there is a need for research to compare the strengths and weaknesses of different decision-making methods [38]. Jamwal et al. (2021) explored how MCDM methods were applied in sustainable manufacturing and found that most of the methods used are based on fuzzy theory [39]. Kumar et al. (2017) developed an insight into various MCDM methods, and the application progress in renewable energy and prospects [40].

From past fuzzy theory literature and knowledge management literature, no papers have been found that use fuzzy MCDM to measure the quality of knowledge; thus, this study should be the latest attempt.

3. Model Formulation

This section describes the proposed method, including knowledge quality fuzziness index, fuzzy gate selection, and implementation procedures.

3.1. Knowledge Quality Fuzziness Index

Based on the fuzzy weight average (FWA) method [30–32,41–45], this study proposes a knowledge quality fuzziness index (*KQFI*) to help make the Go/No go decision of the knowledge proposal. *KQFI* can be described as Equation (10), where r_i and w_i are the fuzzy assessment and fuzzy weight of the knowledge proposal, respectively, and i denotes the criteria for evaluating the knowledge proposal.

$$KQFI = \frac{\sum_{i=1}^{n} (r_i \otimes w_i)}{\sum_{i=1}^{n} w_i} \tag{10}$$

Because the computation of FWA can reach $O(n \ \log \ n)$ [35,46], Kao and Liu proposed a fractional programming approach (FPA) to solve the above problem [47]. Conducting α-cut to r_i and w_i of *KQFI* produces $(r_i)_\alpha = [(r_i)_\alpha^L, (r_i)_\alpha^U]$ and $(w_i)_\alpha = [(w_i)_\alpha^L, (w_i)_\alpha^U]$, and let $t = \frac{1}{\sum_{i=1}^{n} w_i}$ and $v_i = tw_i$; then, the membership function of *KOFI* can be obtained using Equations (11) and (12) by employing different values of α-cut.

$$\begin{aligned} KQFI_\alpha^L = Min \quad & \sum_{i=1}^{n} v_i (r_i)_\alpha^L \\ s.t. \quad & t(w_i)_\alpha^L \le v_i \le t(w_i)_\alpha^U \\ & \sum_{i=1}^{n} v_i = 1 \\ & t \ge 0 \end{aligned} \tag{11}$$

$$\begin{aligned} KQFI_\alpha^U = Max \quad & \sum_{i=1}^{n} v_i (r_i)_\alpha^U \\ s.t. \quad & t(w_i)_\alpha^L \le v_i \le t(w_i)_\alpha^U \\ & \sum_{i=1}^{n} v_i = 1 \\ & t \ge 0 \end{aligned} \tag{12}$$

For example, a triangular membership function (0.42, 0.57, 0.69) of KQFI is obtained by solving Equations (11) and (12) at α-cut = 0 and α-cut = 1. When α-cut = 0, one can obtain $KQFI_\alpha^L$ = 0.42 and $KQFI_\alpha^U$ = 0.69; when α-cut = 1, one can obtain $KQFI_\alpha^L$ = 0.57 and $KQFI_\alpha^U$ = 0.57.

3.2. *Fuzzy Gate Selection*

Based on previous researches [48–52], this study proposes a checking gate to screen the knowledge proposals (see Figure 1). Enterprises can choose a fuzzy threshold (*FT*) according to strategic objectives, and *FT* is used to decide if a knowledge proposal meets the minimum quality level. Figure 1 shows that the farther the *FT* is to the right, the higher the standard for reviewing knowledge quality. The dotted line in the figure represents FT.

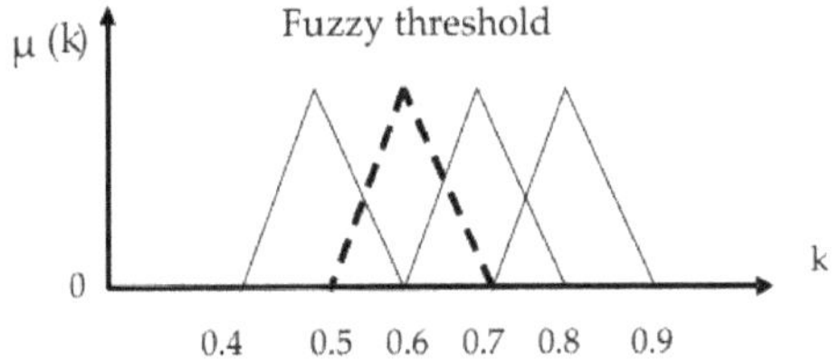

Figure 1. Checking gate.

The procedure for making the Go/No go decision of a knowledge proposal is listed as below:

(a) Compute fuzzy preference z based on Equation (13).
(b) Conduct α-cut to z and obtain $z_\alpha^l < 0$ and $z_\alpha^u > 0$ as in Figure 2 and Equation (14).

(c) Compute the level of goodness e_{KFT} of *KQFI* to *FT* using Equations (15) and (16); if $e_{KFT} > 0.5$, then the knowledge proposal is qualified and accepted.

$$z = KQFI - FT \tag{13}$$

$$z_a = \left[z_\alpha^l, z_\alpha^u\right] \tag{14}$$

$$e_{KFT} = \frac{S_1}{S_1 + S_2} \tag{15}$$

$$\text{where } S_1 = \int_{x>0} \mu_{\tilde{z}}(x)dx,\ S_2 = \int_{x<0} \mu_{\tilde{z}}(x)dx \tag{16}$$

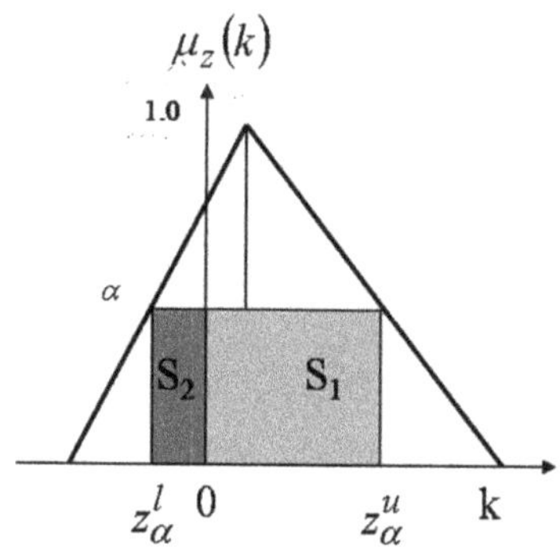

Figure 2. e_{KFT}.

3.3. Implementation Procedures

The flowchart of the proposed model is depicted in Figure 3, and the implementation procedures are described as below:

a. The expert penal receives the knowledge proposal and decides the assessment criteria, the linguistic variables, and the fuzzy number.
b. Experts assess the criteria weight and quality performance of knowledge proposals, and obtain the consensus of the expert decision.
c. Obtain the membership function of the *KQFI* for each knowledge proposal.
d. Specify the *FT* value according to enterprise strategic objectives.
e. Compute fuzzy preference z and level of goodness e_{KFT}, and make a Go/No go decision for each knowledge proposal.

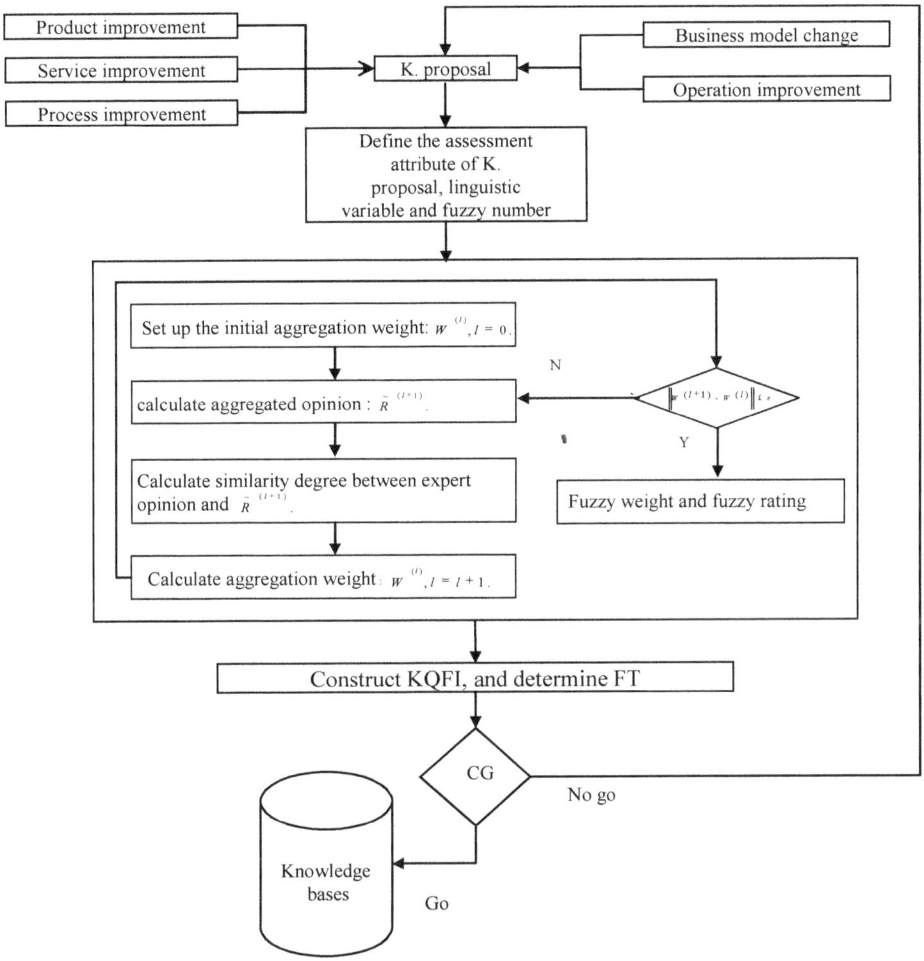

Figure 3. Flowchart of the proposed model.

3.4. Size of Expert Panel

Past studies did not agree on the number of group decision-making experts. Lynn (1986) believed that 5 experts was enough to produce good judgments, and even as few as 3 experts was possible [4]. Hashmi et al. (2023) mentioned that the optimal number is 5 to 7 people [53]. Emmerling and Rooders (2022) believed that the number of experts should be controlled at 3 to 5 [54]. Axtell (2018) researched that the optimal number of participants in a meeting is 8 people [55,56]. From the above research, it can be found that the number of people making group decisions is between 3 and 8 people. Therefore, this study uses five experts to measure the quality of knowledge proposals.

4. Case Implementation

This section uses an example to demonstrate the applicability of the proposed model. The knowledge assessment expert panel of a company wants to select qualified knowledge from nine knowledge proposals, and the detailed steps are as follows:

a. The expert panel composed of five experts receives nine knowledge proposals; decides the assessment criteria of knowledge quality as (1) originality (A_1), (2) applicability (A_2), (3) practicality (A_3), (4) value (A_4), and (5) uniqueness (A_5); and uses linguistic variables of very good, good, fair, poor, and very poor, and the triangular fuzzy number listed in Table 1, to assess the criteria weight and knowledge quality.

b. Experts assess the criteria weight and quality performance of knowledge proposals (Table 2) and obtain the consensus of the expert opinion. D_1 to D_5 indicate experts and K_1 to K_9 represent knowledge proposals in Table 2. The consensus of the expert decision is described below.

   A_1, K_8 (Table 2), and Table 1 are used to obtain the consensus of the expert opinion.

   (i) Let c = 1.5, m = 2, and $u = \max(U) - \min(U) = 0.7$, where the initial aggregated weight is set as the weight of the first expert, i.e., $W^{(0)} = (1,0,0,0,0)$. Then, the aggregated opinion $\widetilde{R}^{(1)}$ can be obtained using Equation (8), and $\widetilde{R}_i$ is the opinion of an individual expert.

$$\widetilde{R}^{(1)} = \frac{\sum_{i=1}^{5} (w_i)^2 \widetilde{R}_i}{\sum_{i=1}^{5} (w_i)^2} = W(1)^{(0)} \otimes \widetilde{R}_1 + W(2)^{(0)} \otimes \widetilde{R}_2 + W(3)^{(0)} \otimes \widetilde{R}_3 + W(4)^{(0)} \otimes \widetilde{R}_4 + W(5)^{(0)} \otimes \widetilde{R}_5$$
$$= 1 \otimes \mathrm{M} + 0 \otimes \mathrm{M} + 0 \otimes \mathrm{MG} + 0 \otimes \mathrm{M} + 0 \otimes \mathrm{MG}$$
$$= 1 \otimes (0.3, 0.5, 0.7) + 0 \otimes (0.3, 0.5, 0.7) + 0 \otimes (0.5, 0.7, 0.9) + 0 \otimes (0.3, 0.5, 0.7) + 0 \otimes (0.5, 0.7, 0.9)$$
$$= (0.3, 0.5, 0.7)$$

   (ii) The similarity between the individual $\widetilde{R}_i$ and the aggregated $\widetilde{R}^{(1)}$ are computed using Equation (6), where u = 0.7.

$$S_2(\widetilde{R}_1, \widetilde{R}^{(1)}) = 1 - \frac{(d_2(\widetilde{R}_1, \widetilde{R}^{(1)}))^2}{4u^2} \quad S_2(\widetilde{R}_2, \widetilde{R}^{(1)}) = 1 - \frac{(d_2(\widetilde{R}_2, \widetilde{R}^{(1)}))^2}{4u^2}$$
$$S_2(\widetilde{R}_3, \widetilde{R}^{(1)}) = 1 - \frac{(d_2(\widetilde{R}_3, \widetilde{R}^{(1)}))^2}{4u^2} \quad S_2(\widetilde{R}_4, \widetilde{R}^{(1)}) = 1 - \frac{(d_2(\widetilde{R}_4, \widetilde{R}^{(1)}))^2}{4u^2}$$
$$S_2(\widetilde{R}_5, \widetilde{R}^{(1)}) = 1 - \frac{(d_2(\widetilde{R}_5, \widetilde{R}^{(1)}))^2}{4u^2}$$

   (iii) The new aggregated weight can be computed using Equation (9) and obtained as $W^{(1)} = (0.2864, 0.2864, 0.1811, 0.2462, 0.2864)$. $W^{(l)} = (w_1^{(l)}, w_2^{(l)}, \ldots, w_n^{(l)})$ and $w_1^{(1)}$ can be computed as below:

$$w_1^{(1)} = \frac{\left(\frac{1}{(1.5 - S_2(\widetilde{R}_1, \widetilde{R}^{(1)}))}\right)^{\frac{1}{2-1}}}{\sum_{j=1}^{n}\left(\frac{1}{(1.5 - S_2(\widetilde{R}_j, \widetilde{R}^{(1)}))}\right)^{\frac{1}{2-1}}}, \mathrm{c} = 1.5, \mathrm{m} = 2$$

   (iv) The new aggregated opinion can be computed using Equation (8) as below:

$$\widetilde{R}^{(2)} = \frac{\sum_{i=1}^{n} \left(w_i^{(1)}\right)^m \widetilde{R}_i}{\sum_{i=1}^{n} \left(w_i^{(1)}\right)^m} = \frac{\sum_{i=1}^{n} \left(w_i^{(1)}\right)^2 \widetilde{R}_i}{\sum_{i=1}^{n} \left(w_i^{(1)}\right)^2}, \mathrm{m} = 2,$$

   $\widetilde{R}^{(2)}$ can be obtained as $(0.4583, 0.6583, 0.6583, 0.8444)$.

   (v) Repeat the above steps, the consensus of the fuzzy weight of A_1 can be obtained, and it will converge to (0.4334, 0.6334, 0.6334, 0.8144) (Table 3) at $l = 8$, meaning that the consensus of the experts toward the criteria has been reached. Table 4 lists the results.

c. Obtain the membership function of $KQFI_1$ to $KQFI_9$ using Equations (11) and (12) at different values of α-cut. The results are listed in Table 5.

d. Specify *FT* value as (0.5, 0.6, 0.7) according to the enterprise strategic objectives.

e. Compute fuzzy preference z and level of goodness e_{KFT} using Equations (13) and (15); $KQFI_8$ is demonstrated at α-cut = 0.5 in Figure 4, and the values of z, $z_{0.5}$, and

Go/No go decisions are listed in Table 5. The level of goodness e_{KFT} of knowledge K_6, K_7, and K_8 are less than 0.5; therefore, they are rejected.

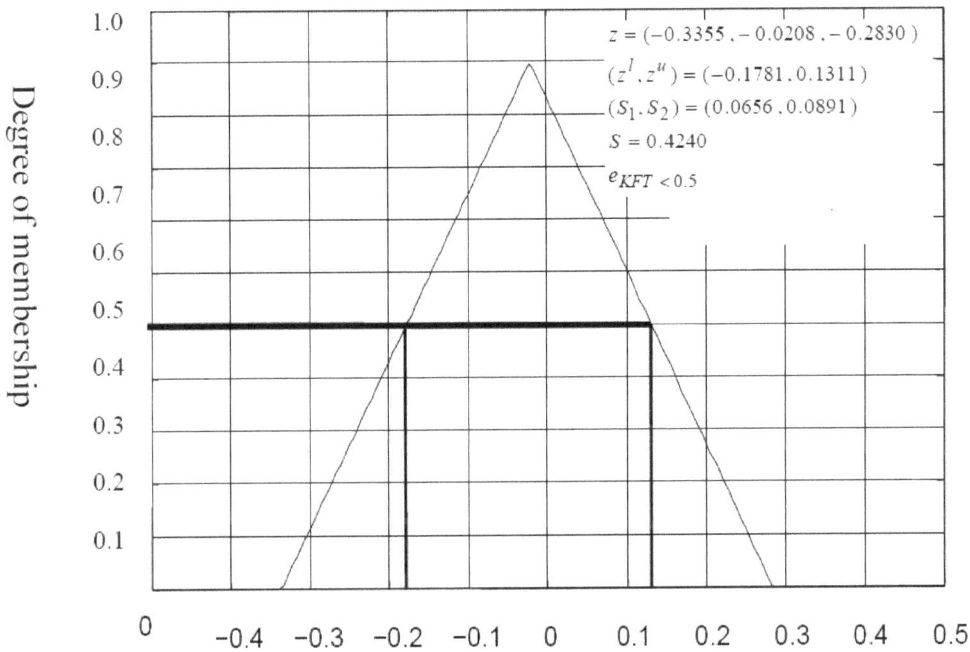

Figure 4. The e_{KFT} of knowledge proposal K_8 (α-cut = 0.5).

4.1. Summary

This section summarizes the method proposed in this study as follows: (1) Experts assign weights to the criteria and evaluate the quality of knowledge proposal based on the fuzzy linguistic scale. (2) Aggregate the weights and evaluations, including (a) setting the initial integrated weight, (b) computing the aggregated evaluation, (c) computing the similarity between the individual evaluation and the integrated evaluation, (d) computing the new integrated weight according to the similarity, (e) computing the new integrated evaluation based on the integrated weight, and repeating these steps until the integrated weight and the integrated evaluation converge to one unchanged state. (3) Use the integrated weight and integrated evaluation to find the fuzzy membership function of *KQFI*. (4) Compute z = *KQFI—FT* (fuzzy threshold). (5) Obtain S_1 and S_2 according to the z membership function; then, calculate e_{KFT}. If $e_{KFT} > 0.5$, it indicates that the knowledge proposal is qualified; otherwise, it is not qualified.

The whole process described above can be computerized and combined with the existing knowledge management system of the enterprise; every time the system receives a new knowledge proposal, it will notify the experts to evaluate the knowledge proposal. Experts only need to conduct linguistic fuzzy evaluation on the weight and knowledge quality. The rest can be completed by the computer, because the fuzzy threshold can be set in advance; the system will finally make a qualified or unqualified judgment on the knowledge proposal, and the qualified knowledge proposal will be automatically stored in the knowledge repository so that the burden on personnel can be greatly reduced and the efficiency of the knowledge management system can be significantly improved.

Table 2. Criteria weight and fuzzy assessment of knowledge proposal.

Expert / Criteria			W					K_1					K_2					K_3					K_4		
	D_1	D_2	D_3	D_4	D_5	D_1	D_2	D_3	D_4	D_5	D_1	D_2	D_3	D_4	D_5	D_1	D_2	D_3	D_4	D_5	D_1	D_2	D_3	D_4	D_5
A_1	M	M	H	MH	M	G	VG	G	MG	MG	G	M	MG	M	MG	M	MP	M	MG	M	G	VG	MG	MG	MG
A_2	MH	M	H	VH	MH	G	G	M	M	G	G	G	MG	G	G	M	M	VG	VG	G	MG	MG	M	M	G
A_3	H	VH	MH	H	H	G	G	MG	MG	G	G	M	M	M	G	G	MG	G	MG	G	MG	G	M	MG	G
A_4	MH	M	H	H	M	G	G	G	MG	G	G	G	G	MG	MG	M	MG	MG	MG	M	G	G	G	MG	G
A_5	MH	M	H	H	M	MP	MG	M	MG	M	M	MG	MG	MG	M	M	M	G	MG	MG	MG	MG	MG	G	MG
Expert / Criteria			K_5					K_6					K_7					K_8					K_9		
	D_1	D_2	D_3	D_4	D_5	D_1	D_2	D_3	D_4	D_5	D_1	D_2	D_3	D_4	D_5	D_1	D_2	D_3	D_4	D_5	D_1	D_2	D_3	D_4	D_5
A_1	M	M	MG	G	MG	G	G	M	G	MG	MP	M	M	M	MG	M	M	MG	M	MG	G	M	MG	M	MG
A_2	M	M	MG	M	G	M	M	M	G	G	MP	M	G	M	G	MG	G	MG	M	G	MG	G	G	MG	M
A_3	G	G	MG	M	MG	M	MP	M	M	G	G	MG	MG	MG	M	MG	M	M	MG	M	M	G	M	M	M
A_4	MP	MG	MG	M	M	MP	M	MG	M	G	M	MG	MG	MG	G	MP	M	M	M	G	MP	M	M	MG	M
A_5	MG	MG	MG	G	MG	M	MG	MG	MG	M	M	M	M	MG	G	MG	MP	G	M	M	M	MG	MG	MG	M

Table 3. Consensus of criteria A_1.

l	$W^{(l)}$					$R^{(l+1)}$			
0	1	0	0	0	0	0.3000	0.5000	0.5000	0.7000
1	0.2865	0.2865	0.1812	0.2463	0.2865	0.3981	0.5981	0.5981	0.7854
2	0.2651	0.2651	0.2073	0.2630	0.2650	0.4228	0.6228	0.6228	0.8058
3	0.2592	0.2592	0.2152	0.2668	0.2591	0.4302	0.6302	0.6302	0.8118
4	0.2574	0.2574	0.2176	0.2679	0.2573	0.4324	0.6324	0.6324	0.8134
5	0.2569	0.2569	0.2184	0.2682	0.2568	0.4331	0.6331	0.6331	0.8141
6	0.2567	0.2567	0.2186	0.2683	0.2566	0.4333	0.6333	0.6333	0.8143
7	0.2567	0.2567	0.2187	0.2684	0.2567	0.4334	0.6334	0.6334	0.8144
8	0.2567	0.2567	0.2187	0.2684	0.2567	0.4334	0.6334	0.6334	0.8144

Table 4. Consensus of fuzzy weight and fuzzy assessment.

Knowledge \ Criteria	A_1	A_2	A_3	A_4	A_5
Weight	(0.4332, 0.6332, 0.8142)	(0.6131, 0.7931, 0.9155)	(0.6002, 0.8002, 0.9503)	(0.6552, 0.8552, 0.9776)	(0.5686, 0.7686, 0.9158)
K1	(0.7065, 0.8837, 0.9516)	(0.5002, 0.7002, 0.8502)	(0.6001, 0.8001, 0.9501)	(0.6601, 0.8601, 0.9801)	(0.3572, 0.5476, 0.7287)
K2	(0.4332, 0.6334, 0.8142)	(0.6601, 0.8601, 0.9801)	(0.3576, 0.5574 0.7432)	(0.6601, 0.8601, 0.9801)	(0.3652, 0.5652, 0.7652)
K3	(0.3001, 0.5001, 0.7001)	(0.5883, 0.7402, 0.8443)	(0.6001, 0.8001, 0.9501)	(0.4586, 0.6584, 0.8586)	(0.4586, 0.6584, 0.8586)
K4	(0.6328, 0.8141, 0.9475)	(0.4001, 0.6001, 0.8001)	(0.5016, 0.7018, 0.8797)	(0.6601, 0.8601, 0.9801)	(0.4332, 0.6334, 0.8144)
K5	(0.3186, 0.5186, 0.7141)	(0.3436, 0.5436, 0.7436)	(0.5688, 0.7686, 0.9158)	(0.3652, 0.5652, 0.7652)	(0.5018, 0.7016, 0.8795)
K6	(0.6424, 0.8424, 0.9568)	(0.3574, 0.5576, 0.7432)	(0.2537, 0.4539, 0.6537)	(0.3001, 0.5001, 0.7001)	(0.4586, 0.6584, 0.8586)
K7	(0.2547, 0.4547, 0.6547)	(0.3204, 0.5203, 0.7032)	(0.5447, 0.7448, 0.9226)	(0.4561, 0.6564, 0.8562)	(0.346, 0.5436, 0.7436)
K8	(0.3416, 0.5414, 0.7416)	(0.5016, 0.7016, 0.8797)	(0.4001, 0.6001, 0.8001)	(0.2537, 0.4537, 0.6537)	(0.4023, 0.6025, 0.7811)
K9	(0.4332, 0.6332, 0.8142)	(0.6001, 0.8001, 0.9501)	(0.3574, 0.5574, 0.7432)	(0.3001, 0.5001, 0.7001)	(0.4586, 0.6584, 0.8586)

Table 5. Decision outcome of knowledge proposal.

$\alpha = 0.5$	$KQFI$	z	(z^l, z^u)	(S_1, S_2)	e_{KFT}	Result	Decision
K1	(0.5352, 0.7561, 0.9101)	(−0.1648, 0.1561, 0.4102)	(−0.0042, 0.2833)	(0.1415, 0.0023)	0.9851	>0.5	Go
K2	(0.4721, 0.7007, 0.8838)	(−0.1648, 0.1563, 0.4102)	(−0.0635, 0.2422)	(0.1213, 0.0317)	0.7923	>0.5	Go
K3	(0.4675, 0.6786, 0.8614)	(−0.2323, 0.0786, 0.3616)	(−0.0768, 0.2202)	(0.1101, 0.0385)	0.7412	>0.5	Go
K4	(0.5028, 0.7207, 0.8986)	(−0.1972, 0.1207, 0.3984)	(−0.0382, 0.2598)	(0.1297, 0.0192)	0.8718	>0.5	Go
K5	(0.4019, 0.6226, 0.8247)	(−0.2982, 0.0226, 0.3247)	(−0.1378, 0.1737)	(0.0868, 0.0688)	0.5578	>0.5	Go
K6	(0.3642, 0.5901, 0.7982)	(−0.3358, −0.0097, 0.2982)	(−0.1728, 0.1443)	(0.0722, 0.0865)	0.4548	<0.5	No go
K7	(0.3694, 0.5912,0.8047)	(−0.3306, −0.0088, 0.3047)	(−0.1698, 0.1481)	(0.0741, 0.0847)	0.4657	<0.5	No go
K8	(0.3646, 0.5791, 0.7831)	(−0.3354, −0.0207, 0.2831)	(−0.1782, 0.1312)	(0.0657, 0.0892)	0.4241	<0.5	No go
K9	(0.4092, 0.6272, 0.8276)	(−0.2909, 0.0273, 0.3276)	(−0.1318, 0.1775)	(0.0886, 0.0658)	0.5736	>0.5	Go

4.2. Comparisons with Past Method

In the past, one scholar used the AHP to measure the quality of knowledge [10]. Therefore, this section compares the proposed fuzzy method with the AHP method. Table 6 shows the results of the comparison. It can be seen from the table that both methods need to form an expert panel and both need to decide the criteria of knowledge quality. To measure the weight of criteria, this research invites experts to directly conduct intuitive fuzzy linguistic assessments while AHP requires a pairwise comparison of criteria. Pairwise comparisons can easily lead to inconsistent importance. When there are many criteria, the adjustments can be challenging. Furthermore, for the evaluation of knowledge proposals, the proposed method asks experts to evaluate the quality of knowledge proposals one by one according to the criteria, while AHP needs to conduct pairwise evaluation of all knowledge proposals for each criterion. Relative comparisons, therefore, will produce inconsistencies. Most importantly, the proposed method directly evaluates the quality of each knowledge and, thus, obtains absolute knowledge quality while AHP makes an indirect relative comparison and obtains relative knowledge quality. The problem with the quality of relative knowledge is that knowledge may not be very useful; in fact, because other knowledge is worse, it is considered relatively good knowledge, which may not help the enterprise at all in the long run. Finally, this study has a very clear Go/No checking threshold. Knowledge above the threshold is qualified knowledge. On the contrary, AHP has no qualified judgment threshold. Therefore, how to screen out relatively high-quality knowledge from the comparison results in the end is a difficult task because the relative ranking of knowledge does not represent the quality of knowledge. Based on the above comparison, one can find that the proposed method can indeed measure the knowledge quality of enterprises more accurately than AHP.

Table 6. Comparisons between the proposed method and AHP.

Method	Expert Penal	Criteria	Weight	Knowledge Evaluation	Knowledge Quality	Go/No Threshold
Proposed	v	v	Linguistic assessment	direct	absolute	v
AHP	v	v	pairwise comparison	indirect	relative	x

5. Conclusions

Knowledge management has been proven to be an effective management method for enterprises to enhance core capabilities, and one of the important tasks is to screen quality knowledge into the knowledge repository. Current industrial and academic methods cannot ensure that quality knowledge is screened out. This study proposes a multi-attribute fuzzy group decision-making model, which can effectively integrate the opinions of experts into a knowledge quality fuzziness index and then compare it with the fuzzy threshold set by the enterprise; finally, it can discover whether the quality of knowledge meets the enterprise quality threshold. There have been very few studies on measuring knowledge quality in the literature. The contribution of this study is twofold. First, this study can fill the academic research gap. Secondly, the systematic method of this study can assist enterprises to screen out genuine quality knowledge, directly promote the success of knowledge management, and improve the innovation ability and business performance of enterprises. The implication of this study to management practice is that the quality of knowledge proposals is the key to whether an enterprise can promote its competitiveness, and the correct knowledge evaluation method must be adopted in order to allow the enterprise's knowledge repository to accumulate knowledge that can truly enhance core competencies.

Finally, this study uses the triangular fuzzy function to measure the criteria weight and knowledge quality; however, it cannot be determined whether it is the best choice. Subsequent research can use other fuzzy membership functions to verify and compare the results.

Author Contributions: Conceptualization, C.-C.W. Methodology, S.-C.L.; Validation, M.-L.C.; Supervision, C.-C.T. All authors have read and agreed to the published version of the manuscript.

Funding: This research received no external funding.

Data Availability Statement: All data used for this research are available in the paper.

Conflicts of Interest: The authors declare no conflict of interest.

References

1. Liu, W.; Wang, Y.; Li, L. Research on the optimal aggregation method of judgment matrices based on spatial steiner-weber point. *J. Syst. Sci. Complex.* **2023**, *36*, 1228–1249. [CrossRef]
2. Ko, A.; Vas, R.; Kovacs, T.; Szabo, I. Knowledge Creation from the perspective of the supply chain, the role of ICT. *Soc. Econ.* **2019**, *41*, 311–329.
3. Nonaka, I. *A Dynamic Theory of Organizational Knowledge Creation*; Institute of Business Research, Hitotsubashi University: Kunitachi, Japan, 1994; pp. 14–37.
4. Lynn, M.R. Determination and quantification of content validity. *Nurs. Res.* **1986**, *35*, 382–386. [CrossRef]
5. Mohajan, H.K. The impact of knowledge management models for the development of organizations. *J. Environ. Treat. Tech.* **2017**, *5*, 12–33.
6. Chen, W.; Tan, J.S.H.; Pi, Z. The spiral model of collaborative knowledge improvement: An exploratory study of a networked collaborative classroom. *Int. J. Comput. Support. Collab. Learn.* **2021**, *16*, 7–35. [CrossRef]
7. Spangler, S.C.; Skovira, R.J.; Kohun, F.G. Key factors in a successful knowledge management model. *Online J. Appl. Knowl. Manag.* **2015**, *3*, 51–60.
8. Aljuwaiber, A. Communities of practice as an initiative for knowledge sharing in business organizations: A literature review. *J. Knowl. Manag.* **2016**, *20*, 731–748. [CrossRef]

9. Choi, H.J.; Ahn, J.C.; Jung, S.H.; Kim, J.H. Communities of practice and knowledge management systems: Effects on knowledge management activities and innovation performance. *Knowl. Manag. Res. Pract.* **2020**, *18*, 53–68. [CrossRef]
10. Chakrabarti, D.; Arora, M.; Sharma, P. Evaluating knowledge quality in knowledge management systems. *J. Stat. Appl. Probab.* **2018**, *7*, 75–84. [CrossRef]
11. Waheed, M.; Kaur, K. Knowledge quality: A review and a revised conceptual model. *Inf. Dev.* **2016**, *32*, 271–284. [CrossRef]
12. Arora, M.; Chakrabarti, D. Knowledge Quality Assessment in Knowledge Management Systems. *Int. J. Knowl. Manag. Pract.* **2014**, *2*, 1.
13. Xiang, Y.D.; Zhang, P.Z.; Wu, S. Content-based knowledge quality assessment and its application in health management system. *J. Shanghai Jiaotong Univ.* **2021**, *26*, 116–128. [CrossRef]
14. Yoo, D.K.; Vonderembse, M.A.; Ragu-Nathan, T.S. Knowledge quality: Antecedents and consequence in project teams. *J. Knowl. Manag.* **2011**, *15*, 329–343.
15. Leonardo, M.A.; Szczerbicki, E.; Sanín, C. A proposal for a knowledge market based on quantity and quality of knowledge. *Cybern. Syst.* **2013**, *44*, 118–132.
16. Krishnaraj, N. Windows Vista: Why did it fail? *TechTalkers*, 22 June 2020.
17. Lim, R.A.; Siew Lee, H.A.; Lim, T.M. A Study on Knowledge Quality and Job Performance of Knowledge Workers by Analyzing Content of Social Network Sites Using Sentiment Network Analysis. *Inf. Manag. Bus. Rev.* **2013**, *5*, 525–530.
18. Zhou, X.; Min, M.; Zhang, Z. Research on the social capital, knowledge quality and product innovation performance of knowledge-intensive firms in China. *Front. Psychol.* **2022**, *13*, 946062. [CrossRef] [PubMed]
19. Abdollahbeigi, B.; Salehi, F. Knowledge quality and non-financial performance-A Malaysian experience. *Knowl. Process Manag.* **2021**, *29*, 12–22. [CrossRef]
20. Ganguly, A.; Talukdar, A.; Chatterjee, D. Evaluating the role of social capital, tacit knowledge sharing, knowledge quality and reciprocity in determining innovation capability of an organization. *J. Knowl. Manag.* **2019**, *23*, 1105–1135. [CrossRef]
21. Zubielqui, G.; Lindsay, N.; Lindsay, W. Knowledge quality, innovation and firm performance: A study of knowledge transfer in SMEs. *Small Bus. Econ.* **2019**, *53*, 145–164. [CrossRef]
22. Chen, C.T. Extensions of the TOPIS for group decision-making under fuzzy environment. *Fuzzy Sets Syst.* **2000**, *114*, 1–9. [CrossRef]
23. Clement, A.M.; Bharatraj, J. Theory of triangular fuzzy number. In Proceedings of the National Conference on Advanced Trends in Mathematics, Tamil Nadu, India, 27 March 2017.
24. Sudha, T.; Jayalalitha, G. Fuzzy triangular numbers in—Sierpinski triangle and right-angle triangle. *J. Phys. Conf. Ser.* **2022**, *1597*. [CrossRef]
25. Tamilmani, C. Arithmetic operation of fuzzy numbers using A-cut method. *Int. J. Innov. Sci. Eng. Technol.* **2015**, *2*, 299–315.
26. Zhang, G.; Lu, J. An integrated group decision-making method dealing with fuzzy preference for alternatives and individual judgments for selection criteria. *Group Decis. Negot.* **2003**, *12*, 501–515. [CrossRef]
27. Li, H.L.; Yang, J.Q.; Xiang, Z.Q. A fuzzy linguistic multi-criteria decision-making approach to assess emergency suppliers. *Sustainability* **2022**, *14*, 13114. [CrossRef]
28. Kukkurainen, P. Fuzzy logic and Zadeh algebra. *Adv. Pure Math.* **2017**, *7*, 350–365. [CrossRef]
29. Efe, B. An integrated fuzzy multi criteria group decision making approach for ERP system selection. *Appl. Soft Comput.* **2016**, *38*, 106–117. [CrossRef]
30. Lin, C.; Tan, B.; Hsieh, P.J. Application of the fuzzy weighted average in strategic portfolio management. *Decis. Sci.* **2005**, *36*, 489–510. [CrossRef]
31. Pavlacka, O.; Pavlackova, M.; Hetflei, V. Fuzzy weighted average as a fuzzified aggregation operator and its properties. *Kybernetika* **2017**, *53*, 137–160. [CrossRef]
32. Pavlacka, O.; Pavlackova, M. On the properties of the fuzzy weighted average of fuzzy numbers with normalized fuzzy weights. *Iran. J. Fuzzy Syst.* **2021**, *18*, 1–17.
33. Wang, R.C.; Chu, S.J. Group decision-making using a fuzzy linguistic approach for evaluating the flexibility in a manufacturing system. *Eur. J. Oper. Res.* **2004**, *154*, 563–572. [CrossRef]
34. Kaushik, M.; Kumar, M. Distance and similarity-based aggregation method in intuitionistic fuzzy fault tree analysis. *SSRN* **2023**. [CrossRef]
35. Lee, D.H.; Park, D. An efficient algorithm for fuzzy weight average. *Fuzzy Sets Syst.* **1997**, *87*, 39–45. [CrossRef]
36. Lee, H.S. Optimal consensus of fuzzy opinions under group decision making environment. *Fuzzy Sets Syst.* **2002**, *132*, 303–315. [CrossRef]
37. Stojčić, M.; Zavadskas, E.K.; Pamučar, D.; Stević, Ž.; Mardani, A. Application of MCDM Methods in Sustainability Engineering: A Literature Review 2008–2018. *Symmetry* **2019**, *11*, 350. [CrossRef]
38. Zavadskas, E.K.; Turskis, Z.; Kildienė, S. State of art surveys of overviews on MCDM/MADM methods. *Technol. Econ. Dev. Econ.* **2014**, *20*, 165–179. [CrossRef]
39. Jamwal, A.; Agrawal, R.; Sharma, M.; Kumar, V. Review on multi-criteria decision analysis in sustainable manufacturing decision making. *Int. J. Sustain. Eng.* **2021**, *14*, 202–225. [CrossRef]
40. Kumar, A.; Sah, B.; Singh, A.R.; Deng, Y.; He, X.N.; Kumar, P.; Bansal, R.C. A review of multi criteria decision making (MCDM) towards sustainable renewable energy development. *Renew. Sustain. Energy Rev.* **2017**, *69*, 596–609. [CrossRef]

41. Chang, P.T.; Hung, K.C. Applying the fuzzy-weighted-average approach to evaluate network security systems. *Comput. Math. Appl.* **2005**, *49*, 1797–1814. [CrossRef]
42. Leonardo, M.A.; Sanín, C.; Szczerbicki, E. Quality assessment of experiential knowledge. *Cybern. Syst.* **2012**, *43*, 96–113.
43. Lin, C.; Hsieh, P.J. A fuzzy decision support system for strategic portfolio management. *Decis. Support Syst.* **2004**, *38*, 383–398. [CrossRef]
44. Lin, C.T.; Chen, C.T. Bid/no-bid decision-making—A fuzzy linguistic approach. *Int. J. Proj. Manag.* **2004**, *22*, 585–593. [CrossRef]
45. Verma, R.; Sharma, B. Intuitionistic fuzzy Einstein prioritized weighted average operators and their application to multiple attribute group decision making. *Appl. Math. Inf. Sci.* **2015**, *9*, 3095–3107.
46. Dong, W.M.; Wong, F.S. Fuzzy weighted averages and implementation of the extension principle. *Fuzzy Sets Syst.* **1987**, *21*, 183–199. [CrossRef]
47. Kao, C.; Liu, S.T. Fractional programming approach to fuzzy weighted average. *Fuzzy Sets Syst.* **2001**, *120*, 435–444. [CrossRef]
48. Cheng, S.L.; Kuan, Y.W. A fuzzy logic-based knowledge management performance measurement system for SMEs. *Cybern. Syst.* **2017**, *48*, 277–302.
49. Cheng, C.H.; Lin, Y. Evaluating the main battle tank using fuzzy decision theory with linguistic criteria evaluation. *Eur. J. Oper. Res.* **2002**, *142*, 174–186. [CrossRef]
50. Chu, P.V. A decision support for project portfolio selection. *Comput. Ind.* **1996**, *32*, 141–149. [CrossRef]
51. Fan, Z.P. An approach to multiple attribute decision making based on fuzzy preference information on alternative. *Fuzzy Sets Syst.* **2002**, *131*, 101–106. [CrossRef]
52. Hsu, H.M.; Chen, C.T. Aggregation of fuzzy opinions under group decision making. *Fuzzy Sets Syst.* **1996**, *79*, 279–285.
53. Hashmi, N.; Shankaranarayanan, G.; Malone, T.W. Is bigger better? A study of the effect of group size on collective intelligence in online groups. *Decis. Support Syst.* **2023**, *167*, 113914. [CrossRef]
54. Emmerling, T.; Rooders, D. 7 strategies for better group decision-making. *Harvard Business Review*, 22 September 2022.
55. Axtell, P. The most productive meetings have fewer than 8 people. *Harvard Business Review*, 22 June 2018.
56. Taylor, E. We Agree, Don't We? The Delphi method for health environments research. *Health Environ. Res. Des. J.* **2020**, *13*, 11–23. [CrossRef] [PubMed]

mathematics

Article

Integrity on m-Polar Fuzzy Graphs and Its Application

Ghulam Muhiuddin [1,*,†], Tanmoy Mahapatra [2,†], Madhumangal Pal [2,†], Ohoud Alshahrani [1,†] and Ahsan Mahboob [3,†]

1 Department of Mathematics, Faculty of Science, University of Tabuk, P.O. Box 741, Tabuk 71491, Saudi Arabia; 411008512@stu.ut.edu.sa
2 Department of Applied Mathematics with Oceanology and Computer Programming, Vidyasagar University, Midnapore 721102, India; tmahapatrapmath@gmail.com (T.M.); mmpalvu@gmail.com (M.P.)
3 Department of Mathematics, Madanapalle Institute of Technology & Science, Madanapalle 517325, India; khanahsan56@gmail.com
* Correspondence: chishtygm@gmail.com
† These authors contributed equally to this work.

Abstract: Integrity for crisp graph theory is a well-defined topic. However, the integrity concept for fuzzy graphs has only recently been defined and extensively researched. However, in m-polar fuzzy graphs (mPFG), each node as well as edges has m components. So, defining integrity in the mPF environment needs a new concept. As in the m-polar fuzzy environment, each node and edge has m components, so we have more flexibility to address the uncertainty rather than fuzzy as well as other uncertain environments. In this article, we developed a brand-new idea known as node integrity on mPFG and went in-depth on a few of their related properties. We have thoroughly covered some of their related properties as well as a brand-new idea called dominating integrity on mPFG. Different types of integrity on mPFG such as node integrity, dominating integrity, and edge integrity are discussed thoroughly along with some of its interesting facts have been introduced. Under isomorphism, their properties have also been studied. We also discussed the interrelation between them. A new type of mPFG called efficient mPFG which is directly related to dominating integrity concept has also been introduced. Several facts about efficient mPFG have also been studied here along with details descriptions. Finally, a real-world mobile network application that is directly related to the integrity of the mPFG concept has been discussed.

Keywords: m-polar fuzzy graph; node integrity; edge integrity; dominating set; dominating integrity

MSC: 05C72; 03E72

Citation: Muhiuddin, G.; Mahapatra, T.; Pal, M.; Alshahrani, O.; Mahboob, A. Integrity on m-Polar Fuzzy Graphs and Its Application. *Mathematics* **2023**, *11*, 1398. https://doi.org/10.3390/math11061398

Academic Editor: Aleksandar Aleksic

Received: 3 February 2023
Revised: 5 March 2023
Accepted: 7 March 2023
Published: 14 March 2023

1. Introduction

1.1. Research Background and Related Works

In technical development, fuzzy graph (FG) theory has an important role. The way of many rule-based expert systems for engineers have been made from FG theory. It is seen in the maximum time that graph theory is found as an essential part of connectivity in some fields of geometry, algebra, topology, number theory, computer science, operations research and optimization. In real life, many problems have been solved using data that come from different origins or sources. This type of data collection represents multi-polarity. In this type of polarity, we can not be structured well by the conception of fuzzy models or bipolar fuzzy models. For example, if we consider a mobile networking model which assures minimum installations of towers to cover the whole area so that no loss of signal throughout the area. For this, we assign the node membership value (MV) based on the situation of the mobile towers as (maximum capacity of a tower unit, distance covered for signaling systems, and material used for installation of a mobile tower). In nature, these terms are uncertain. To represent this situation, we need to use the 3-polar fuzzy model. Using a fuzzy model or an intuitionistic fuzzy model, or a bipolar fuzzy model, this situation can not be handled.

To address uncertainty and fuzziness, Zadeh [1] introduced fuzzy sets. From that point forward, the hypothesis of fuzzy sets has become a space of exploration in different orders. In 1994, Zhang [2,3] expanded the thought of fuzzy sets to bipolar fuzzy sets, which have numerous applications in mathematical speculations just as real-life problems. The set concept in a m-dimension fuzzy environment was then introduced by Chen et al. [4] as an extension of bipolar fuzzy sets. Utilizing Zadeh's procedure, Kauffman [5] fostered an incredible strategy of fuzzy graph theory in 1973. After that, Rosenfeld [6] gave another explained idea of a fuzzy graph. Several concepts on fuzzy graphs have been investigated on [7–11]. In 2014, the influential thought of the m-PFS was proposed by Chen et al. [4], which is an augmentation of bipolar fuzzy sets to manage these sorts of properties of algebraic and graphical models. Some basic properties on mPFG has been inspected by Ghorai and Pal [12,13]. Different properties of graphs under m-polar fuzzy environment have been studied by Akram et al. [14–16]. The conception of domination integrity and efficient FG was initiated by Mariappan et al. [17]. The conception of Concept of integrity as well as its value of FGs was introduced by Saravanana et al. [18]. They also developed a study on regular FGs and the integrity of FG [19]. Next, Mahapatra et al. [20,21] presented coloring on different uncertainty environments.

In real life, many problems have been solved using data that come from different origins or sources. This type of data collection represents multi-polarity. In this type of polarity, we can not be structured well by the conception of fuzzy models or bipolar fuzzy models. For example, we consider a mobile networking model which assures minimum installations of the tower to cover the whole area so that no loss of signal throughout the area. For this, we assign the node membership value (MV) based on the situation of the mobile towers as (maximum capacity of a tower unit, distance covered for signaling systems, and material used for installation of the mobile tower). In nature, these terms are uncertain. To represent this situation, we need to use the 3-polar fuzzy model. Using a fuzzy model or an intuitionistic fuzzy model, or a bipolar fuzzy model, this situation can not be handled. Thus, the integrity concept on FGs is not appropriate for operation in these kinds of situations. To overcome this situation, we are interested in working using the concepts of integrity concept in a m-polar fuzzy environment. Anyone can analyze the MVs in a multi-polar fuzzy environment in a certain way. Since in our consideration, we consider three components for each node as well as edges, therefore we can not handle this type of situation using a fuzzy model as there is a single component for this concept. Again, we can not apply a bipolar or intuitionistic fuzzy graph model as each edge or node have just two components. Thus, these mPFG models give more efficient fuzziness results than another fuzzy model. Furthermore, it is very interesting to develop and analyze such types of mPFGs with examples and related theorems. These definitions and theorems are definitely improving the existing concepts of mPFGs and are more reliable for solving any complicated real-life problem.

1.2. Contribution of This Study

The formation of this article is as follows: Section 2 mentions some useful concepts which are essential for this article. In Section 3, we have defined the different types of integrity on mPFG and provided some theorems on their aspect. In Section 3.1, Node integrity on mPFG is a brand-new idea that we have introduced along with some of its related properties. In Section 3.2, We have thoroughly covered some of their related properties as well as a brand-new idea called dominating integrity on mPFG. Here, we also investigated some relations between dominating integrity and node integrity. In Section 3.3, we have introduced a new concept called node integrity on mPFG and also discussed some of their related properties thoroughly. We investigated some features based on the above conception. In Section 4, we have discussed new types of mPFG called efficient mPFG along with its details description. In Section 5, we have discussed an application based on a mobile networking problem. In Section 7, we have discussed the advantages and limitations of the proposed model. Finally, the conclusion of the study has been presented in Section 8.

2. Preliminaries

Here, we briefly call again some definitions connected to mPFG, such as complete mPFG, strong mPFG, and path in mPFG.

Throughout this article $p_i : [0,1]^m \to [0,1]$ indicated ith material of projection mapping. Furthermore, $i = 1(1)m$ indicates that $i = 1, 2, \ldots, m$.

Definition 1 ([13]). *An mPFG* $\Gamma = (\tilde{A}, \sigma, \mu)$ *having underlying crisp graph (UCG)* $\Gamma^* = (\tilde{A}, \tilde{B})$, *where* $\sigma : \tilde{A} \to [0,1]^m$ *and* $\mu : \tilde{A} \times \tilde{A} \to [0,1]^m$ *indicate an mPFS of* $\tilde{A}$ *and* $\tilde{A} \times \tilde{A}$, *respectively, and which follows the relation such that for all* $i = 1(1)m$, $p_i \circ \mu(b,c) \leq \{p_i \circ \sigma(b) \wedge p_i \circ \sigma(c)\}$ *for all* $(b,c) \in \tilde{A} \times \tilde{A}$ *as well as* $\mu(b,c) = 0$ *for all* $(b,c) \in (\tilde{A} \times \tilde{A} - \tilde{B})$.

Definition 2 ([22]). $\Gamma = (V, \sigma, \mu)$ *is conferred as complete mPFG provided* $p_i \circ \mu(b,d) = \{p_i \circ \sigma(b) \wedge p_i \circ \sigma(d)\}, i = 1(1)m, \forall b, d \in V$.

Definition 3 ([13]). $\Gamma = (V, \sigma, \mu)$ *is conferred as a mPF strong graph if*

$$p_i \circ \mu(b,d) = \{p_i \circ \sigma(b) \wedge p_i \circ \sigma(d)\},$$

$i = 1(1)m, \forall (b,d) \in E$.

Definition 4 ([15]). *Let* $\Gamma = (V, \sigma, \mu)$ *is an mPFG as well as* $P : b_1, b_2, \ldots, b_k$ *be a path in* G. $S(P)$ *indicates the strength of P, defined as* $S(P) = (\min_{1 \leq i < j \leq k} p_1 \circ \mu(b_i, b_j), \min_{1 \leq i < j \leq k} p_2 \circ \mu(b_i, b_j), \ldots, \min_{1 \leq i < j \leq k} p_m \circ \mu(b_i, b_j)) = (\mu_1^n(b_i, b_j), \mu_2^n(b_i, b_j), \ldots, \mu_m^n(b_i, b_j))$.

The strength of connectedness (SC) of the path in between b_1 *and* b_k *is given in the following way:* $CONN_G(b_1, b_k) = (p_1 \circ \mu(b_i, b_j)^\infty, p_2 \circ \mu(b_i, b_j)^\infty, \ldots, p_m \circ \mu(b_i, b_j)^\infty)$, *where* $(p_i \circ \mu(b_i, b_j)^\infty) = \max_{n \in N}(\mu_i^n(b_i, b_j))$.

Definition 5 ([20]). *An edge* $(b,d), b, d \in V$ *is regarded as independently strong for an mPFG* $\Gamma = (V, \sigma, \mu)$ *if* $\frac{1}{2}\{p_i \circ \sigma(b) \wedge p_i \circ \sigma(d)\} \leq p_i \circ \mu(b,d), i = 1(1)m$. *It is considered weak independently if not. In order to determine how strong the edge* (b,d) *is,*

$$p_i \circ I(b,d) = \frac{p_i \circ \mu(b,d)}{p_i \circ \sigma(b) \wedge p_i \circ \sigma(d)}, i = 1(1)m.$$

Definition 6 ([22]). *Consider* Γ *and* Γ' *be two mPFGs of the UCG* $\Gamma^* = (\tilde{A}, \tilde{B})$ *and* $\Gamma'^* = (\tilde{A}', \tilde{B}')$, *respectively. A bijective mapping is an isomorphism between* Γ *and* Γ *is* $f : \tilde{A} \to \tilde{A}'$ *which satisfies*

$$p_i \circ \sigma(t) = p_i \circ \sigma'(f(t)) \text{ and } p_i \circ \mu(t,u) = p_i \circ \mu'(f(t), f(u))$$

$\forall t, u \in \tilde{A}$ *and every* $i = 1(1)m$. *Then* Γ *is called to be isomorphic with* Γ'.

3. Differents Types of Integrity on m-Polar Fuzzy Graph

Here, we will discuss a new concept of mPFGs such as node integrity, edge integrity, and dominating integrity such that they fulfill a specific criterion on the node set or edge set or dominating set (DS).

3.1. Node Integrity on m-Polar Fuzzy Graph

In this subsection, we will discuss a new idea of mPFGs node integrity such that they fulfill a specific criterion on the node set.

Definition 7. *Let Γ be an mPFG having UGC $\Gamma^* = (V, E)$. Suppose $A \subset V$ is the set of node bumps whose omission disconnects Γ^*. Now, $p_i \circ m(\Gamma - A)$ indicates the set of the order of the topmost element in the graph $V - A$. Then the node integrity of Γ is indicated by $\tilde{I}(\Gamma)$ and is conferred by $p_i \circ \tilde{I}(\Gamma) = \min_{A \subset V}\{p_i \circ |A| + p_i \circ m(\Gamma - A)\}, \forall i = 1(1)m$.*

Example 1. *Let Γ be a 3PFG displayed in Figure 1. The set of nodes $V = \{a, b, c, d\}$ such that $\sigma(a) = (0.5, 0.4, 0.3), \sigma(b) = (0.3, 0.3, 0.4), \sigma(c) = (0.3, 0.5, 0.2), \sigma(d) = (0.6, 0.5, 0.2)$ with $\mu(a, b) = (0.1, 0.1, 0.05), \mu(b, c) = (0.3, 0.3, 0.2), \mu(c, d) = (0.3, 0.5, 0.2), \mu(d, a) = (0.5, 0.4, 0.2), \mu(b, d) = (0.3, 0.3, 0.2)$.*

The node integrity value in 3PFG Γ, shown in Figure 1 is $\tilde{I}_V(\Gamma) = (1.4, 1.3, 0.9)$ and integrity set $\tilde{I}_V$ of Γ is $\{b, d\}$ which disconnects the graph Γ.

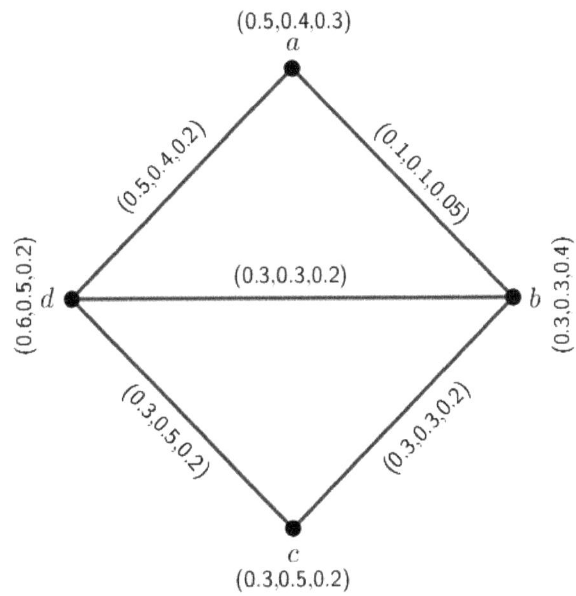

Figure 1. Node integrity of 3PFG Γ.

Theorem 1. *Let P be an mPF subgraph of Γ. Then $\tilde{I}(P) \leq \tilde{I}(\Gamma)$.*

Proof. Assume P is a full mPF subgraph of Γ. Then there is a node, let us say a, of P that has a lower MV than the MV of Γ; otherwise, P has fewer nodes than Γ. Consequently $|P| < |\Gamma|$. However, for any mPFG P, $\tilde{I}(P) \leq |P| < |\Gamma|$. Let an integrity set A of P be such that $\tilde{I}(\Gamma) \leq \tilde{I}(P)$. As a result, $p_i \circ \tilde{I}(P) = \{p_i \circ |A| + p_i \circ m(P - A)\}$, for all $i = 1(1)m$. As a result, $p_i \circ \tilde{I}(P) - \{p_i \circ |A| \geq p_i \circ m(P - A)\}$, for all $i = 1(1)m$. If A is also an integrity set of Γ, then $p_i \circ m(P - A) \leq p_i \circ m(\Gamma - A)$, for all $i = 1(1)m$, which is not feasible, as P is subset of Γ. If A is without integrity set of Γ then $p_i \circ \tilde{I}(\Gamma) - p_i \circ |A| \leq p_i \circ m(\Gamma - A)$, for all $i = 1(1)m$, which is a contradictory. Hence, the theorem. □

Theorem 2. *Node integrity is invariant under isomorphism between two mPFG.*

Proof. Let Γ_1 and Γ_2 be two mPFG and $\phi : \Gamma_1 \to \Gamma_2$ be the isomorphism, that is, $\forall i = 1(1)m$.

1. $p_i \circ \sigma_1(b) = p_i \circ \sigma_2(\phi(b)), \forall a \in V_1$.
2. $p_i \circ \mu_1(b,d) = p_i \circ \mu_2(\phi(b), \phi(d)), \forall (b,d) \in \widetilde{V_1 \times V_1}$.

Suppose, A_1 be set in Γ_1 such that $p_i \circ \tilde{I}(\Gamma_1) = \min_{A_1 \subset V}\{p_i \circ |A_1| + p_i \circ m(G - A_1)\}$. Since Γ_1 and Γ_2 are isomorphic therefore from Equations (i) and (ii) its preserves the node and edge MVs. Hence, clearly $p_i \circ \tilde{I}(\Gamma_2) = \min_{\phi(A_1) \subset V}\{p_i \circ |\phi(A_1)| + p_i \circ m(\Gamma_2 - \phi(A_1))\}$ holds for $\phi(A_1)$. Therefore, Γ_2 is also a node integrity. □

Theorem 3. *Node integrity of a complete mPFG is its order.*

Proof. Let Γ be the full mPFG. Every node is then adjacent to the rest of the remaining nodes. Removing any number of p nodes from Γ yields a single component graph with many $n - p$ nodes. Therefore, for each subset A, the value of $\{p_i \circ |A| + p_i \circ m(\Gamma - A)\}$, $\forall i = 1(1)m$ remains fixed as well as equals to the total MV of the node. Thus, the node integrity of Γ is the same as the order of Γ. □

Theorem 4. *An integrity of a star mPFG $K_{1,n}$ is equal to the $max p_i \circ \sigma(a) + p_i \circ \sigma(c)$, for each $i = 1(1)m$ and $(a,c) \in E$.*

Proof. Each edge of Γ is connected to an intermediate node on the star diagram. So removing this intermediate node leaves all other nodes isolated. So, according to Theorem 3, the maximum value of the remaining nodes MV defines the maximum degree of the remaining graph. Summing this with the MVs of the intermediate nodes gives the completeness of the star graph $K_{1,n}$. Since this central node is linked to this node, it is defined as the integrity of the star graph $K_{1,n}$ is equal to the $max p_i \circ \sigma(a) + p_i \circ \sigma(d), \forall i = 1(1)m$ and $(a,d) \in E$. □

3.2. Dominating Integrity on m-Polar Fuzzy Graph

In this section, we'll talk about a novel notion of mPFGs dominating integrity by meeting a particular DS criterion.

Definition 8. *Let us assume that Γ be an mPFG having UGC $\Gamma^* = (V, E)$. Assume that $A \subset V$ is a DS. The set of the largest component's order in the graph $\Gamma - A$ is now indicated by the expression $m(\Gamma - A)$. Then the dominating integrity (DI) of Γ is indicated by $\tilde{DI}(\Gamma)$ and is conferred by $p_i \circ \tilde{DI}(\Gamma) = \min_{A \subset V}\{p_i \circ |A| + p_i \circ m(\Gamma - A)\}, \forall i = 1(1)m$.*

Definition 9. *If every node in $V - A$ has at least one strong nbd in A, then the set A is said to be the DS of an mPFG Γ. The minimal cardinality of such a DS is the dominance number of Γ denoted by $\gamma(\Gamma)$, and such a set is called minimal dominance.*

Definition 10. *Any subset of A of $V(\Gamma)$ that is an integrity set $\tilde{I}$-set of Γ if $p_i \circ \tilde{DI}(\Gamma) = \{p_i \circ |A| + p_i \circ m(\Gamma - A)\}, \forall i = 1(1)m$.*

Example 2. *Let Γ be a 3PFG depcited in Figure 2. The set of nodes $V = \{a,b,c,d\}$ such that $\sigma(a) = (0.5, 0.4, 0.3), \sigma(b) = (0.3, 0.3, 0.4), \sigma(c) = (0.3, 0.5, 0.2), \sigma(d) = (0.6, 0.5, 0.2)$ with $\mu(a,b) = (0.1, 0.1, 0.05), \mu(b,c) = (0.3, 0.3, 0.2), \mu(c,d) = (0.3, 0.5, 0.2)$, $\mu(d,a) = (0.5, 0.4, 0.2), \mu(b,d) = (0.3, 0.3, 0.2)$.*

Clearly, the arcs $(a,d), (b,d), (b,c)$ and (c,d) are strong. The DS $A, \Gamma - A, m(\Gamma - A), |A|$ and $|A| + m(\Gamma - A)$ all are given in Table 1.

From Table 1, we see that $min(|A| + m(\Gamma - A)) = (1.4, 1.3, 0.9)$, which corresponds to the domination set $\{b,d\}$. As a result, $\{b,d\}$ is the minimal DI set of 3PFG Γ.

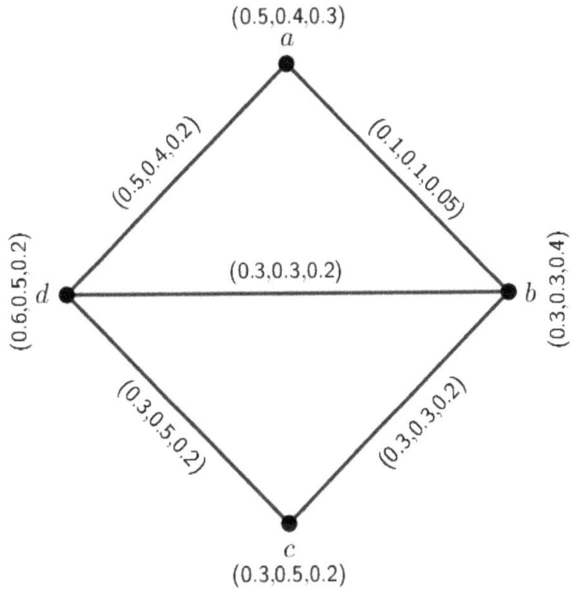

Figure 2. Domination integrity of 3PFG Γ.

Table 1. Domination integrity of 3PFG Γ.

A	$\Gamma - A$	$m(\Gamma - A)$	$\|A\|$	$\|A\| + m(\Gamma - A)$
$\{d\}$	$a - b - c$	$(1.1, 1.2, 0.9)$	$(0.6, 0.5, 0.2)$	$(1.7, 1.7, 1.1)$
$\{a, b\}$	$d - c$	$(0.9, 1, 0.4)$	$(0.8, 0.7, 0.7)$	$(1.7, 1.7, 1.1)$
$\{a, c\}$	$d - b$	$(0.9, 0.8, 0.6)$	$(0.8, 0.9, 0.5)$	$(1.7, 1.7, 1.1)$
$\{a, d\}$	$b - c$	$(0.6, 0.8, 0.6)$	$(1.1, 0.9, 0.5)$	$(1.7, 1.7, 1.1)$
$\{c, d\}$	$b - a$	$(0.8, 0.7, 0.7)$	$(0.9, 1, 0.4)$	$(1.7, 1.7, 1.1)$
$\{b, d\}$	$\{a\}, \{c\}$	$(0.5, 0.5, 0.3)$	$(0.9, 0.8, 0.6)$	$(1.4, 1.3, 0.9)$
$\{a, b, c\}$	$\{d\}$	$(0.6, 0.5, 0.2)$	$(1.1, 1.2, 0.9)$	$(1.7, 1.7, 1.1)$
$\{a, c, d\}$	$\{b\}$	$(0.3, 0.3, 0.4)$	$(1.4, 1.4, 0.7)$	$(1.7, 1.7, 1.1)$
$\{b, c, d\}$	$\{a\}$	$(0.5, 0.4, 0.3)$	$(1.2, 1.3, 0.8)$	$(1.7, 1.7, 1.1)$
$\{a, b, d\}$	$\{c\}$	$(0.3, 0.5, 0.2)$	$(1.4, 1.2, 0.9)$	$(1.7, 1.7, 1.1)$
$\{a, b, c, d\}$	ϕ	$(0, 0, 0)$	$(1.7, 1.7, 1.1)$	$(1.7, 1.7, 1.1)$

Theorem 5. *For any mPFG G, $p_i \circ \gamma(G) \leq p_i \circ \tilde{DI}(G)$, for $i = 1(1)m$.*

Proof. Only the cardinality of DS affects the dominance of mPFG. The dominance number, on the other hand, is influenced by DS A and the greatest order of the corresponding components, $G - S$. From this $p_i \circ \gamma(G) \leq p_i \circ \tilde{DI}(G)$, for $i = 1(1)m$. Equality holds for the mPFG with isolated nodes only. Equality applies only to mPFG with isolated nodes. The entire set of nodes is the only DS in graphs with isolated nodes. For this set $p_i \circ m(G^* - A) = 0$, for $i = 1(1)m$. □

Remark 1. *The integrity of G is equal to the maximum MV of G nodes if an mPFG G is null.*

Theorem 6. *DI is invariant under isomorphism between two mPFG.*

Proof. Let Γ_1 and Γ_2 be two mPFG and $\phi : \Gamma_1 \to \Gamma_2$ be the isomorphism, that is, $\forall i = 1(1)m$

1. $p_i \circ \sigma_1(b) = p_i \circ \sigma_2(\phi(b)), \forall a \in V_1$.
2. $p_i \circ \mu_1(b,d) = p_i \circ \mu_2(\phi(b), \phi(d)), \forall (b,d) \in \widetilde{V_1 \times V_1}$.

Suppose, A_1 be a DS in Γ_1 such that $p_i \circ \tilde{DI}(\Gamma_1) = \min_{A_1 \subset V}\{p_i \circ |A_1| + p_i \circ m(\Gamma_1 - A_1)\}$. Since Γ_1 and Γ_2 are isomorphic therefore from Equations (i) and (ii) its preserves the node and edge MVs. Hence, clearly $p_i \circ \tilde{DI}(\Gamma_2) = \min_{\phi(A_1) \subset V}\{p_i \circ |\phi(A_1)| + p_i \circ m(\Gamma_1 - \phi(A_1))\}$ holds for $\phi(A_1)$. Therefore, Γ_2 is also a DI. □

Theorem 7. *Let Γ be an mPFG. $p_i \circ \tilde{DI}(\Gamma) = t$, for $i = 1(1)m$ iff Γ is either complete mPFG or $\bar{\Gamma}$, where $(t, t, \ldots, t)$ is the order of the mPFG.*

Proof. Let Γ be the complete mPFG. Any dominating subset V of Γ, and all that is left of the graph is a single component made up of all the remaining nodes. As a result, the DI of Γ is only the order of Γ. Assume that the entire mPFG is complemented by Γ. Then, the graph with a set of isolated nodes is indicated as G. Thus, $N(a) = \phi$; for all $a \in V$. The total node set is the only DS of Γ. For this DS A, is $p_i \circ m(\Gamma - A) = 0$, for $i = 1(1)m$. DI of Γ is therefore nothing more than Γ order. Thus, the only DS is V. As a result, $p_i \circ \tilde{DI}(\Gamma) = t$, for $i = 1(1)m$.

Consider, however, that A is the DS of Γ and that the order of Γ is the DI number. Since the sum of the MV of A as well as $m(\Gamma - A)$ is not greater than the order of Γ, if $\Gamma - A$ has more than one connected component, then $\Gamma - A$ must only have one connected component. This holds for all DS, but especially for any set of singletons. Given that this singleton node set controls all of Γ's other nodes, it is possible that Γ is a full mPFG. □

Theorem 8. *Let Γ be a complete bi-partiated mPFG $K_{\{\sigma_1, \sigma_2\}}$. Then*

$$p_i \circ \tilde{DI}(\Gamma) = min\{p_i \circ |V_1| + p_i \circ max|\sigma_2|, p_i \circ |V_2| + p_i \circ max|\sigma_1|\},$$

where $V = V_1 \cup V_2$.

Proof. Since Γ is a bi-partiated mPFG, therefore its node set can be partitioned into V_1, V_2. Suppose Γ_1 and Γ_2. Let A be a DS in Γ. Then three cases may arise which are discussed as follows:

Case 1: Suppose $A \in V_1$. The collection of all isolated nodes in V_2 is then called $(\Gamma - A)$. Hence, $p_i \circ m(\Gamma - A) = p_i \circ max|\sigma_2(V_2)|$.

Case 2: Suppose $A \in V_2$. The collection of all isolated nodes in V_1 is then called $(\Gamma - A)$. Hence, $p_i \circ m(\Gamma - A) = p_i \circ max|\sigma_1(V_1)|$.

Case 3: If A is a DS with a node from V_1 and the remainder from V_2, $\Gamma - A$ will still be the only connected component. With one node from V_1 and the remaining nodes from V_2, we can disregard the DS A by taking into account the minimum value for the three DSs mentioned above.

Combining all these cases we get

$$p_i \circ \tilde{DI}(\Gamma) = min\{p_i \circ |V_1| + p_i \circ max|\sigma_2|, p_i \circ |V_2| + p_i \circ max|\sigma_1|\},$$

where $V = V_1 \cup V_2$. □

Theorem 9. *Assume that Γ is a strong mPFG and that $\bar{\Gamma}$ is Γs complement. When Γ and $\bar{\Gamma}$ are united, the node integrity of the union equals the order of the complete mPFG on Γ.*

Proof. As Γ is a strong mPFG, according to the Definition of complement of mPFG, $\bar{\Gamma}$ also qualifies as a strong mPFG. Each node in the resulting graph is next to every other node and has an edge MV that is lower than the MVs of its neighbors. As a result, we know that the order of the mPFG determines the integrity of the mPFG produced by the union of Γ and its complement. As a result, the order of the entire mPFG form in Γ is the integrity of the union of Γ and its complement. □

3.3. Edge Integrity on m-Polar Fuzzy Graph

In this subsection, we will discuss a new idea of mPFGs edge integrity such that they fulfill a specific criterion on the edge set.

Definition 11. *Let us assume that Γ is a mPFG with UGC $\Gamma^* = (V, E)$. Assume that $A \subset E$ is an edge set that, if removed, would render Γ^* as disconnected. The set of the order of the largest component in the graph $\Gamma - A$ is now indicated by $p_i \circ m(\Gamma - A)$. Then the edge integrity of Γ is indicated by $\tilde{I}_E(\Gamma)$ and granted by $p_i \circ \tilde{I}_E(\Gamma) = \min_{A \subset E}\{p_i \circ |A| + p_i \circ m(\Gamma - A)\}, \forall i = 1(1)m$.*

Example 3. *Let Γ be a 3PFG depicted in Figure 3. The nodes sets $V = \{a, b, c, d\}$ such that $\sigma(a) = (0.2, 0.5, 0.3), \sigma(b) = (0.6, 0.4, 0.2), \sigma(c) = (0.5, 0.3, 0.2), \sigma(d) = (0.2, 0.3, 0.2)$ with $\mu(a, b) = (0.2, 0.4, 0.1), \mu(b, c) = (0.5, 0.3, 0.2), \mu(c, d) = (0.2, 0.3, 0.2), \mu(d, a) = (0.1, 0.2, 0.2)$, $\mu(a, c) = (0.2, 0.3, 0.2)$.*

Let $B_1 = \{(a, d), (a, c), (b, c)\}$ and $B_2 = \{(d, c), (a, c), (a, b)\}$ be two subsets of edge set of G which disconnect the graph G. Now, $|B_1| = (0.8, 0.8, 0.6)$, $|B_2| = (0.6, 1, 0.5)$, $m(G - B_1) = (0.2, 0.4, 0.2)$ and $m(G - B_2) = (0.5, 0.3, 0.2)$. For the set B_1, the edge integrity of G is $\tilde{I}_{B_1}(G) = (1, 1.2, 0.8)$ and for B_2 is $\tilde{I}_{B_2}(G) = (1.1, 1.3, 0.7)$. Thus, the edge integrity value of the 3PFG G is $\tilde{I}_E(G) = (1, 1.2, 0.7)$ and the integrity sets are $B_1 = \{(a, d), (a, c), (b, c)\}$ and $B_2 = \{(d, c), (a, c), (a, b)\}$.

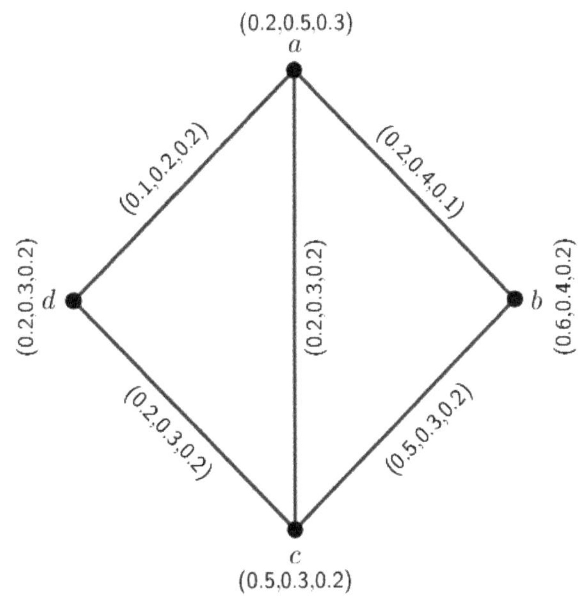

Figure 3. Edge integrity of 3PFG Γ.

Theorem 10. *Let Γ be a complete mPFG. $max\{p_i \circ \sigma(v)\}$ $\forall vs. \in V$ is the edge integrity of complement of the complete mPFG, $\forall vs. \in V$.*

Proof. Complement of complete mPFGs is, obviously, a mPFG without edges. As a result, $\bar{\Gamma} = (V, \bar{\sigma}, \bar{\mu})$ is a set of nodes without edges. Maximum node MV in $\bar{\Gamma} = (V, \bar{\sigma}, \bar{\mu})$

determines the edge integrity of this graph. Thus, for all $I_E(\tilde{\Gamma}) = max\{p_i \circ \sigma(v)\}, v \in V$. □

Theorem 11. *The edge integrity of a complete mPFG is its order.*

Proof. Let Γ be the full mPFG. Each node is then connected to the rest of the other nodes. Removing any number of p nodes from Γ yields a single component graph with many $n-p$ nodes. Therefore, for each subset A, the value of $\{p_i \circ |A| + p_i \circ m(\Gamma - A)\}, \forall i = 1(1)m$ remains fixed as well as equals to the total value of node MV. As a result, the order of Γ is equal to the edge integrity of Γ. □

Theorem 12. *Let us have a strong mPFG* Γ. *Order of the entire mPFG formed in* Γ *is the edge integrity of the union of* Γ *as well as* $\bar{\Gamma}$.

Proof. The union of Γ and $\bar{\Gamma}$ is understood to be a complete mPFG. Theorem 11 establishes the order of the edge integrity of the entire mPFG. As a result, the edge integrity of the union of Γ as well as $\bar{\Gamma}$ depends on the order of the entire mPFG created in Γ. □

4. Efficient m-Polar Fuzzy Graph

Each UCG is undeniably a member of a unique class of mPFG. A mPFG becomes a crisp graph if the σ and μ values are the same for each node and edge. The possibility of the DSs of the mPFG and the UCG being identical is an intriguing discovery. We create the effective mPFG class of mPFGs as a result.

Definition 12. *An efficient mPFG is an mPFG that shares equal DS as its crisp graph, other than V.*

Example 4. *Assume that* Γ *is the 3PFG depicted in Figure 4. The collection of nodes is* $V = \{a, b, c\}$ *where* $\sigma(a) = (0.3, 0.5, 0.4), \sigma(b) = (0.2, 0.3, 0.5), \sigma(c) = (0.6, 0.5, 0.3)$ *with* $\mu(a, b) = (0.2, 0.2, 0.3), \mu(b, c) = (0.2, 0.3, 0.3), \mu(c, a) = (0.3, 0.5, 0.2)$.

Each arc of Γ *is strong. There is no doubt that the sets* $\{a\}$, $\{b\}$, $\{c\}$, $\{a, b\}$, $\{b, c\}$, $\{c, a\}$ *are the domination sets of* Γ^*. *Now,* $N_A(a) = \{b, c\}$, $N_A(b) = \{c, a\}$, $N_A(c) = \{a, b\}$, $N_A(\{a, b\}) = \{c\}$, $N_A(\{b, c\}) = \{a\}$, $N_A(\{c, a\}) = \{b\}$, *which are also the DSs of* Γ. *Therefore the DSs of* Γ^* *and* Γ *are the same. Hence* Γ *is an efficient 3PFG.*

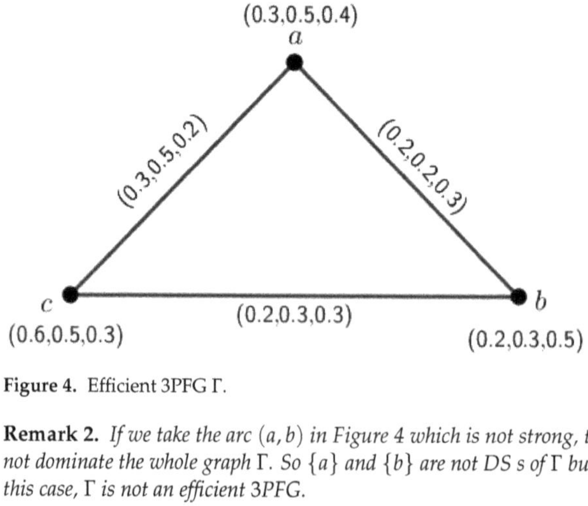

Figure 4. Efficient 3PFG Γ.

Remark 2. *If we take the arc* (a, b) *in Figure 4 which is not strong, then the set* $\{a\}$ *and* $\{b\}$ *can not dominate the whole graph* Γ. *So* $\{a\}$ *and* $\{b\}$ *are not DS s of* Γ *but they are also DS s of* Γ^*. *In this case,* Γ *is not an efficient 3PFG.*

Theorem 13. *Every complete mPFG is an efficient mPFG.*

Proof. Every node set in a complete graph is a DS of Γ. Every edge in the given mPFG Γ is a strong arc since it is a complete mPFG. The edge and node sets of Γ and Γ^* are identical. As a result, they share a DS. Therefore, each complete mPFG is an efficient mPFG. □

Theorem 14. *Every mPFG with a constant μ value is a mPFG that works effectively.*

Proof. According to the idea of a strong arc, arcs are β strong if all of their MV same. All of the arcs are strengthened as a result. Therefore, both the crisp graph and the mPFG graph have the same closed nbd set for each node. As a result of Γ and Γ^* sharing the same DS, Γ is an efficient mPFG. □

Theorem 15. *Let Γ be a mPFG having UCG as a path P_n. Then it is efficient.*

Proof. Since Γ is an mPFG having UCG a path. Therefore we get two cases:-
Case 1: Every arc is an effective arc Hence, Γ and UCG Γ^* share equal node sets and edge sets. Therefore, they contain equal DS. Hence, Γ is an efficient mPFG.
Case 2: Let us say all the arcs are ineffective. Here, we need to demonstrate that all non-efficient arcs must also be strong arcs. In P_n, there is unquestionably one and only one path connecting any two nodes. The graph becomes disconnected when an edge is removed. Therefore, between adjacent nodes, the SC will decrease. Therefore, the edge needs to be strong. Therefore, the strong arcs in P_n are the ineffective arcs. As a result, Γ is an effective graph. □

5. Application

The mPFG is a key mathematical framework for visualizing interconnected real-world phenomena, where nodes and edges are represented by mPF information. In this section, we have illustrated a mobile network installation issue using the idea of integrity on mPFG.

5.1. Model Construction

Suppose you need to cover a group of villages on your mobile phone Telephone network tower. Let us say every village has at least one tower. The graph theory problem looks like this: Every village is a landmark. Villages are connected by edges if the tower installed in the village is close by. Now the issue is identifying the dominant group. However, we cannot guarantee that the best models are produced by a dominant set. Everyone is at risk of network failure during a natural disaster. You will need to cover the biggest area, even if some of the towers fail. As a result, network providers need to think about their networks. A team that covers the most ground. As a graph consistency, this idea came into existence. Both honesty and dominance are taken into consideration by the new parameters. The advantage of the integrity set (set of towers) is that it ensures complete coverage and a more stable network with the widest possible coverage area. The situation with multiple towers in real time

Here, we use the seven villages a, b, c, d, e, f, g as the installation nodes for mobile towers. If the mobile network covers the signal from the closest mobile towers, there will be an edge between two nodes. To solve the allocation issue in this case, dominating integrity in 3PFG $G = (V, \sigma, \mu)$ is used. Three factors are taken into account when calculating the node MVs. Those criteria are as follows: {maximum capacity of a tower unit, distance covered for signaling systems, material used for installation of mobile tower}. Three factors are taken into account when calculating the edge MV. These are listed below: {signal strength between two towers, internet speed, hazards occurrence due to unavoidable circumstances}. The 3PFG model is displayed in Figure 5.

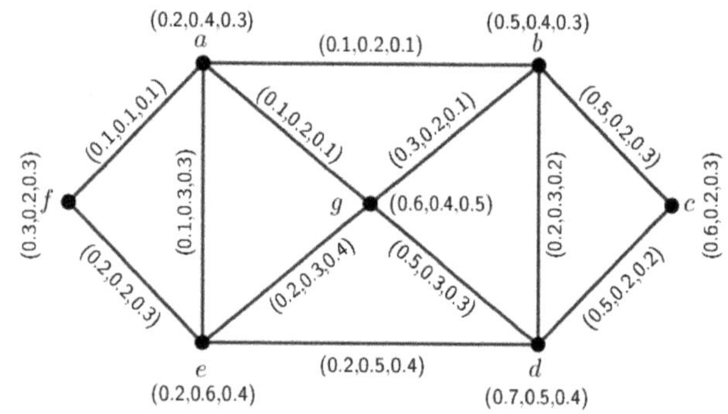

Figure 5. Domination integrity of 3PFG G.

5.2. Illustration of Membership Value

Clearly, the arcs $(e,g),(g,d),(e,d),(b,c),(d,c),(e,f)$ and (a,e) are strong arcs. The DS $A, G-A, m(G-A), |A|$ and $|A|+m(G-A)$ all are given in Table 2.

Table 2. Model 3PF domination integrity graph G.

A	$G-A$	$m(G-A)$	$\lvert A\rvert$	$\lvert A\rvert+m(G-A)$
$\{e,b,d\}$	$a-f-g,\{c\}$	$(1.1,1,1.1)$	$(1.4,1.5,1.1)$	$(2.5,2.5,2.2)$
$\{e,d,c\}$	$f-a-g-b$	$(1.6,1.4,1.4)$	$(1.5,1.3,1.1)$	$(3.1,2.7,2.5)$
$\{e,g,b\}$	$a-f,d-c$	$(1.3,0.7,0.7)$	$(0.7,1,0.7)$	$(2,1.7,1.4)$
$\{e,g,c\}$	$f-a-b-d$	$(1.7,1.5,1.3)$	$(1.4,1.2,1.2)$	$(3.1,2.7,2.3)$
$\{a,b,g,f\}$	$e-d-c$	$(1.5,1.3,1.1)$	$(1.6,1.4,1.4)$	$(3.1,2.7,2.5)$
$\{f,e,g,b\}$	$\{a\},d-c$	$(1.3,0.7,0.7)$	$(1.6,1.6,1.5)$	$(2.9,2.3,2.2)$
$\{a,g,e,c\}$	$b-d,\{f\}$	$(1.2,0.9,0.7)$	$(1.6,1.6,1.5)$	$(2.8,2.5,2.2)$
$\{f,e,d,c\}$	$a-g-b$	$(1.3,1,0.7)$	$(1.8,1.5,1.4)$	$(3.1,2.5,2.1)$
$\{f,a,b,c\}$	$e-g-d$	$(1.5,1.5,1.3)$	$(1.6,1.2,1.2)$	$(3.1,2.7,2.5)$
$\{a,c,d,e\}$	$b-g,\{f\}$	$(1.1,0.8,0.8)$	$(1.7,1.7,1.4)$	$(2.8,2.5,2.2)$
$\{b,c,d,e\}$	$f-a-g$	$(1.1,1,1.1)$	$(2,1.7,1.4)$	$(3.1,2.7,2.5)$
$\{b,d,e,f\}$	$a-g,\{c\}$	$(0.8,0.8,0.8)$	$(1.7,1.7,1.4)$	$(2.5,2.5,2.2)$
$\{a,b,c,d,e\}$	$\{f\},\{g\}$	$(0.6,0.4,0.5)$	$(2.2,2.1,1.7)$	$(2.8,2.5,2.3)$
$\{a,c,d,e,f\}$	$b-g$	$(1.1,0.8,0.8)$	$(2,1.9,1.7)$	$(3.1,2.7,2.5)$
$\{a,d,e,f,g\}$	$b-c$	$(1.1,0.6,0.6)$	$(2,2.1,1.9)$	$(3.1,2.7,2.5)$
$\{b,c,d,e,f\}$	$a-g$	$(0.8,0.8,0.8)$	$(2.3,1.9,1.7)$	$(3.1,2.7,2.5)$
$\{b,d,e,f,g\}$	$\{a\},\{c\}$	$(0.6,0.4,0.3)$	$(2.3,2.1,1.9)$	$(2.9,2.5,2.2)$
$\{c,d,e,f,g\}$	$a-b$	$(0.7,0.8,0.6)$	$(2.4,1.9,1.9)$	$(3.1,2.7,2.5)$
$\{a,b,c,d,e,f\}$	$\{g\}$	$(0.6,0.4,0.5)$	$(2.5,2.3,2)$	$(3.1,2.7,2.5)$
$\{a,c,d,e,f,g\}$	$\{b\}$	$(0.5,0.4,0.3)$	$(2.4,2.3,2.2)$	$(2.9,2.7,2.5)$
$\{b,c,d,e,f,g\}$	$\{a\}$	$(0.2,0.4,0.3)$	$(2.7,2.3,2.2)$	$(2.9,2.7,2.5)$
V	ϕ	$(0,0,0)$	$(2.9,2.7,2.5)$	$(2.9,2.7,2.5)$

5.3. Decision Making

From Table 2, we see that $min(|A|+m(G-A)) = (2,1.7,1.4)$, which corresponds to the domination set $\{e,g,b\}$. Hence $\{e,g,b\}$ is the minimal dominating integrity set of 3PFG G. As a result, these sets are trustworthy and dominant in the network. The way the tower was constructed. Compared to other nodes, these nodes offer better protection even during natural disasters.

Through the discussion above, we can conclude that dominating integrity on mPFG actually plays a significant role in this kind of allocation problem. Furthermore, we acknowledge that in the allocation problem, dominating integrity on mPFG is more applicable than dominating integrity on FG.

6. Comparative Study

At first Saravanan et al. [19] introduced an integrity idea for fuzzy graph theory. Next, Mariappan et al. [17] studied the integrity concept and its value in fuzzy graphs. Mariappan et al. [18] then also discussed dominating integrity and efficient fuzzy graph concept. So, all the results discussed earlier are not applicable when the model is considered in another environment like in m-polar fuzzy sets. If we consider a mobile networking model which assures minimum installations of towers to cover the whole area so that no loss of signal throughout the area. For this, we assign the node membership value (MV) based on the situation of the mobile towers as (maximum capacity of a tower unit, distance covered for signaling systems, and material used for installation of the mobile tower). In nature, these terms are uncertain. To represent this situation, we need to use the 3-polar fuzzy model. Using a fuzzy model or an intuitionistic fuzzy model, or a bipolar fuzzy model, this situation can not be handled. Thus, the integrity concept of FGs is not appropriate for operating in these kinds of situations. To overcome this situation, we are interested in working using the concepts of integrity concept in a m-polar fuzzy environment. This is why the proposed model in this paper plays a significant role in such situations to give better results.

7. Advantages and Limitations of the Proposed Work

Some of the advantages of the proposed are as follows:

1. This work mainly depends on m-polar fuzzy logic network system.
2. Many important definitions and theorems are presented in this study which is very useful.
3. A real application of m-polar fuzzy integrity is presented on a resource power controlling system.

Some of the limitations of this study are given as follows:

1. This work mainly focuses on m-polar fuzzy graph.
2. If the membership value of the characters is given in different interval-valued m-polar fuzzy environment, then m-polar fuzzy integrity graph cannot be used.
3. This type of proposed work mains used in control systems.

8. Conclusions

In this article, we developed a brand-new idea known as node integrity on mPFG and went in-depth on a few of their related properties. We have thoroughly covered some of their related properties as well as a brand-new idea called dominating integrity on mPFG. Here, we also investigated some relations between dominating integrity and node integrity. We investigated some features based on the above conception. We studied a new type of mPFG called efficient mPFG along with its details description. A new type of mPFG called efficient mPFG which is directly related to dominating integrity concept has also been introduced. Several facts about efficient mPFG have also been studied here. Finally, a real-world mobile network application that is directly related to the integrity of the mPFG concept has been discussed. We will extend our research work on a more generalized mPFG concept.

Author Contributions: Conceptualization, G.M. and M.P.; Methodology, T.M. and A.M.; Software, O.A.; Validation, T.M.; Investigation, A.M.; Data curation, T.M., A.M. and O.A.; Writing—original draft, G.M.; Visualization, M.P. All authors have read and agreed to the published version of the manuscript.

Funding: The authors extend their appreciation to the Deanship of Scientific Research at University of Tabuk for funding this work through Research no. S-0136-1443.

Data Availability Statement: Not applicable.

Acknowledgments: The authors would like to express their sincere thanks to the anonymous reviewers for their valuable comments and helpful suggestions which greatly improved the quality of this paper. The authors extend their appreciation to the Deanship of Scientific Research at University of Tabuk for funding this work through Research no. S-0136-1443.

Conflicts of Interest: The authors declare no conflict of interest.

References

1. Zadeh, L.A. Fuzzy sets. *Inf. Control* **1965**, *8*, 338–353. [CrossRef]
2. Zhang, R.W. Bipolar fuzzy sets and relations:a computational fremework for cognitive modeling and multiagent decision analysis. In Proceedings of the First International Joint Conference of the North American Fuzzy Information Processing Society Biannual Conference—The Industrial Fuzzy Control and Intellige, San Antonio, TX, USA, 18–21 December 1994; pp. 305–309.
3. Zhang, R.W. Bipolar fuzzy sets. In Proceedings of the 1998 IEEE International Conference on Robotics and Automation, Washington, DC, USA, 20–20 May 1998; pp. 835–840.
4. Chen, J.; Li, S.; Ma, S.; Wang, X. *m*-Polar fuzzy sets: An extension of bipolar fuzzy sets. *Hindwai Publ. Corp. Sci. World J.* **2014**, *2014*, 416530. [CrossRef] [PubMed]
5. Kauffman, A. *Introduction a la Theorie des Sous-Emsembles Flous*; Mansson: Paris, France, 1973.
6. Rosenfeld, A. *Fuzzy Graphs, Fuzzy Sets and Their Application*; Academic Press: New York, NY, USA, 1975; pp. 77–95.
7. Mathew, S.; Sunitha, M.S. *Fuzzy Graphs: Basics, Concepts and Applications*; Lap Lambert Academic Publishing: Saarbruecken, Germany, 2012.
8. Mordeson, J.N.; Nair, P.S. *Fuzzy Graph and Fuzzy Hypergraphs*; Physica-Verlag: Heidelberg, Germany, 2000.
9. Nair, P.S.; Cheng, S.C. Cliques and fuzzy cliques in fuzzy graphs. In Proceedings of the IFSA World Congress and 20th NAFIPS International Conference, Vancouver, BC, Canada, 25–28 July 2001; Volume 4, pp. 2277–2280.
10. Muhiuddin, G.; Hameed, S.; Rasheed, A.; Ahmad, U. Cubic Planar Graphs with Application to Road Network. *Math. Probl. Eng.* **2022**, *2022*, 5251627. [CrossRef]
11. Muhiuddin, G.; Takallo, M.M.; Jun, Y.B.; Borzooei, R.A. Cubic graphs and their application to a traffic flow problem. *Int. J. Comput. Intell. Syst.* **2020**, *13*, 1265–1280. [CrossRef]
12. Pal, M.; Samanta, S.; Ghorai, G. *Modern Trends in Fuzzy Graph Theory*; Springer: Berlin/Heidelberg, Germany, 2020.
13. Ghorai, G.; Pal, M. Some properties of *m*-polar fuzzy graphs. *Pac. Sci. Rev. Nat. Sci. Eng.* **2016**, *18*, 38–46. [CrossRef]
14. Akram, M.; Adeel, A. *m*-Polar fuzzy graphs and *m*-polar fuzzy line graphs. *J. Discret. Math. Sci. Cryptogr.* **2017**, *20*, 1597–1617. [CrossRef]
15. Akram, M.; Wassem, N.; Dudek, W.A. Certain types of edge *m*-polar fuzzy graph. *Iran. J. Fuzzy Syst.* **2016**, *14*, 27–50.
16. Akram, M. *m-Polar Fuzzy Graphs, Theory, Methods, Application*; Springer International Publishing: Berlin/Heidelberg, Germany, 2019. [CrossRef]
17. Mariappan, S.; Sujatha, R.; Sundareswaran, R.; Sahoo, S.; Pal, M. Concept of integrity and its value of fuzzy graphs. *J. Intell. Fuzzy Syst.* **2018**, *34*, 2429–2439.
18. Mariappan, S.; Ramalingam, S.; Raman, S.; Bacak-Turan, G. Domination integrity and efficient fuzzy graphs. *Neural Comput. Appl.* **2020**, *32*, 10263–10273. [CrossRef]
19. Saravanan, M.; Sujatha, R.; Sundareswaran, R. A study on regular FGs and integrity of FG. *Int. J. Appl. Eng. Res.* **2015**, *10*, 160–164.
20. Mahapatra, T.; Pal, M. Fuzzy colouring of *m*-polar fuzzy graph and its application. *J. Intell. Fuzzy Syst.* **2018**, *35*, 6379–6391. [CrossRef]
21. Mahapatra, T.; Ghorai, G.; Pal, M. Competition graphs under interval-valued *m*-polar fuzzy environment and its application. *Comp. Appl. Math* **2022**, *41*, 285. [CrossRef]
22. Ghorai, G.; Pal, M. On some operations and density of *m*-polar fuzzy graphs. *Pac. Sci. Rev. Nat. Sci. Eng.* **2015**, *17*, 14–22. [CrossRef]

mathematics

Article

Industrial Application of the ANFIS Algorithm—Customer Satisfaction Assessment in the Dairy Industry

Nikolina Ljepava [1], Aleksandar Jovanović [2] and Aleksandar Aleksić [2,*]

[1] College of Business Administration, American University in the Emirates, Dubai P.O. Box 503000, United Arab Emirates; nikolina.ljepava@aue.ae
[2] Faculty of Engineering, University of Kragujevac, 34000 Kragujevac, Serbia; a.jovanovic@kg.ac.rs
* Correspondence: aaleksic@kg.ac.rs

Abstract: As a part of the food industry, the dairy industry is one of the most important sectors of the process industry, keeping in mind the number of employees in that sector, the share in the total industrial production, and the overall value added. Many strategies have been developed over time to satisfy customer needs and assess customer satisfaction. This paper proposes an innovative model based on adaptive neuro-fuzzy inference system (ANFIS) and elements of the ACSI (American customer satisfaction index) for assessing and monitoring the level of customer satisfaction in a dairy manufacturing company where there are no large seasonal variations. In terms of an innovative approach, the base of fuzzy logic rules is determined by applying the fuzzy Delphi technique for the application of the ANFIS algorithm and assessment of customer satisfaction. The verification of the model is delivered by testing a real sample from a company of the dairy industry. As decisions on the strategic company level may be impacted by customer satisfaction, the company management should choose the most precise methodology for customer satisfaction assessment. The results are compared with other methods in terms of mean absolute deviation (MAD), mean squared error (MSE), and mean absolute percentage error (MAPE). Results show that ANFIS outperformed other methods used for assessing the level of customer satisfaction, such as case-based reasoning and multiple linear regression.

Keywords: customer satisfaction; fuzzy logic system; neural networks; multiple linear regression; ANFIS

MSC: 03E72

Citation: Ljepava, N.; Jovanović, A.; Aleksić, A. Industrial Application of the ANFIS Algorithm—Customer Satisfaction Assessment in the Dairy Industry. *Mathematics* **2023**, *11*, 4221. https://doi.org/10.3390/math11194221

Academic Editor: Antonin Dvorak

Received: 25 August 2023
Revised: 4 October 2023
Accepted: 6 October 2023
Published: 9 October 2023

1. Introduction

Measuring customer satisfaction is vital for companies in today's competitive business landscape. Companies' operational functions and top-level management activities can be significantly strengthened by understanding and analyzing customer satisfaction. It is crucial that organizations collect customer feedback on their performance across various business activities, analyze this feedback, and make decisions based on the customer insights [1]. Satisfied customers are more likely to become loyal customers, and high customer satisfaction can increase customer retention and brand loyalty [2]. The competitive market can put a lot of pressure on the company since numerous activities, such as testing and defining the quality of raw materials and finished products, packaging, distribution, and sales, are carried out in food processing. In most cases, food processing involves physical, chemical, and microbiological treatment and transformation of the raw materials [3]. Most processed products consist of raw materials and ingredients that are produced separately; some of them are pre-disintegrated (by crushing into smaller particles) and used as ingredients for making finished products. At the same time, dairy products can be easily transformed into a condition that is not suitable for eating [4] and can cause potential health hazards for consumers, costly product recalls, and negatively impact brand reputation [5]. Therefore, strict control of the hygienic correctness of the process equipment and work surfaces, the shelf life of products at different storage temperatures, as well as trends in monitoring specific quality parameters are conducted during the production process.

To improve customer satisfaction and identify any factors that may impact it, the marketing department needs to understand the customer demands and expectations and subsequently translate these insights into product features [1]. The motivation for conducting the present research stands for cultivating strong customer relationships, and maintaining loyal returning customers can strengthen an organization's resilience and overall performance [6]. The customers expect quality products in the foreground and demand that such quality be maintained at a high level throughout the whole product's lifecycle. To meet those demands, different models based on fuzzy sets are developed for managing quality KPIs [7]. At the same time, customer expectations come because of customers' primary requirement that the food is supposed to be healthy and that the food is not subject to unwanted changes during the time of consumption.

At first, the company's marketing department needs to understand the customer demands and expectations and then understand how to transfer these demands to the products' features [1]. In the current business conditions, a good relationship with the customer may even help increase the resilience of the company. To accurately predict customer purchasing behavior, company managers should first explore customer needs and wants and provide them with adequate stimuli. Additionally, customers should be provided with enough information about a set of products and how the product will meet their needs. In the beginning, customers will form an opinion (perception) about the products and how the product can meet their needs. After that, specific priorities are formed, and only then will the purchase decision follow. The manager should not neglect the customer's feelings after the purchase because it affects customer retention and loyalty along with the impressions that the customer will convey to other potential customers through recommendations and word-of-mouth marketing [8]. However, in the contemporary marketing funnel, the final phase marketers want to reach is customer advocacy [9]. The feedback conveyed by satisfied customers to potential ones through recommendations and word-of-mouth marketing plays a pivotal role in attracting new customers and shaping a company's reputation [8,9].

The important question is, what are the gaps or missing links that need to be addressed in the field of customer satisfaction assessment? The existing literature suggests various theoretical models for evaluating customer satisfaction [10,11]. An important feature of economic theories is that they are based on many assumptions about customer intentions [12], and the main assumptions might include rational behavior, preferences, information, budget constraints, unmet needs, risk, and perceived quality [13,14]. Modern economists rightly point out that customers have the final say in the market, and the practice indicates that most customers consider the amount of money they have at their disposal, the time they should spend on the purchase, and the effort they would have to invest during the purchase. However, economic and sociological explanations of customer behavior do not consider the personal, individual-human factor [15]; individual characteristics greatly influence consumer purchasing behavior.

Extensive analysis of the literature indicates that customer satisfaction can be analyzed through a variety of approaches, such as the MCDM approach [16], customer satisfaction index calculation [11], and different models based on statistics, for example, SERVQUAL [17], artificial neural networks [18], different metaheuristics, for example, the particle swarm optimization (PSO) and rough sets-based ANFIS approach [19]. The main research gap can be identified through the shortcomings of the mentioned research. When speaking about MCDM methods, the main shortcoming is the bias of the decision-makers while choosing appropriate criteria and proposing criteria weights. The main shortcoming of statistical analysis, while assessing the customer satisfaction index, is the need for a large amount of input data originating from evidence that might be inaccurate and outdated. The main shortcoming of neural network application in the treated area seems to be the inaccuracy of the output due to network training issues.

The mentioned research gaps can be overcome through the application of a reliable model for the decision-makers to make assessments of the objective level of customer satisfaction.

To address these research gaps, a reliable model that can provide more accurate output data compared to the existing models and enable decision-makers to objectively evaluate customer satisfaction levels is needed. Following the stated, the goal of this research is to provide a model based on the ANFIS approach since it is based on the fuzzy logic system that needs fewer input data points compared to the statistical methods and neural networks, and at the same time, it provides more accurate output data. The model proposed in the scope of this research is based on neural networks. Many authors [20] suggest that the problem of customer satisfaction assessment could be solved by applying a nonlinear fuzzy neural network model (NN) since it: (i) enables a better understanding of the interaction between variables in a more precise manner compared to multivariate statistic analysis, and (ii) minimizes the number of variables.

The use of this approach offers the following advantages [20]: (1) it is capable of approximating various nonlinear functions, as demonstrated in this study dealing with a strongly nonlinear function; (2) all quantitative and qualitative information is uniformly distributed and saved in the neural network, which ensures a high degree of robustness; (3) the approach employs parallel distributed processing, which enables fast data processing. This makes it particularly suitable for performing complicated behavioral science analysis in the area of business management; (4) this model requires minimal setup. It only necessitates data for the input and output layers to analyze the dependencies between variables; (5) the number of learning epochs adapts to the relationships between variables without requiring prior understanding or a predefined sample. The research methodology used in this paper is based on the implementation of the adaptive neuro-fuzzy inference system (ANFIS) with MATLAB.

The paper is organized in the following manner. After Section 1, which is dedicated to the introduction, Section 2 presents the literature review from the analyzed field. The research methodology is presented in Section 3, while Section 4 presents the numerical results of the research. Section 5 sets the discussion and conclusion.

2. Literature Review

Product features and customer preferences significantly impact customer satisfaction. Various mathematical models are used to determine and manage customer satisfaction, such as (i) statistical methods [10–13,21–23], (ii) multi-criteria decision-making methods [24,25], fuzzy logic [26,27], and (iii) neural network and artificial neural network [18,19,28–30]. Authors typically determine the number of factors (independent variables) according to which they evaluate customer satisfaction (dependent variable) based on the literature and/or best practices. In most cases, the values of the variables were obtained based on questionnaires defined in each were analyzed, where customers expressed their estimates using pre-defined measurement scales, and survey data collection was conducted. To analyze consumers' behavioral intentions, a model based on service quality and satisfaction can be employed [12]. It is important to note that service quality and satisfaction are dependent on a variety of factors that are not mutually independent. The properties of all factors of the proposed model were further evaluated by confirmatory factor analysis. The proposed model is then tested on a representative sample, evenly distributed by gender and mirroring the population, with a slight under-representation of respondents aged 56 years and older. The obtained results are analyzed by applying statistical analysis, and chi squared test is used to perform the comparison of the results obtained by the proposed model and other models.

The analysis of consumers' behavioral intentions may also include the risk and the perceived quality [13]. The mentioned model is adopted to be tested in the web environment, so 211 respondents articulated their assessment by using the seven-point measurement scale. In the first step, the authors confirmed a relationship between the tested factors and customer satisfaction by applying an exploratory factor analysis. In the second step, the covariance analysis of linear structural equations procedure defined in SPSS was employed to determine the correlation coefficients between several variables that can be either directly observed variables or unobserved hypothetical variables. Finally, the acceptability of the proposed model was tested according to different indices.

Another research proposed the robust customer satisfaction index (RCSI), a modified adaptation of the American customer satisfaction index (ACSI) to measure air transportation traveler satisfaction [31]. The study included 503 respondents, traveling on domestic Colombian flights. Following the reliability and validity analysis, the structural equation model (SEM) was employed to evaluate the adequacy of the proposed RCSI. The results of the study indicated that the perceived quality and perceived value are strong predictors of overall passenger satisfaction.

In addition to examining consumers' behavioral intentions, a significant number of research studies pointed to the assessment and ranking of the factors that impact customer satisfaction. In most cases, these problems have been stated as multi-criteria optimization tasks (MCDM) in an uncertain environment. In compliance with the mentioned, there is an assumption that enhancement of customer satisfaction might be achieved by taking defined actions that address the factors ranked in the first place. For example, the problem of customer assessment in the Kuwait banking sector [16] has been analyzed through a sample of 863 participants. To ensure sample representativeness, it was decided to distribute as many questionnaires as available resources would permit and reach customers in different locations. The accuracy of the input data was tested by applying statistical tests, and the priority of factors that impact customer satisfaction was determined using a multiple-attribute approach.

Similarly, by applying MCDM, the customer satisfaction index might be calculated. In the case of the banking sector in Serbia, the factors with corresponding subfactors that impact customer satisfaction have been analyzed [26]. In this case, all uncertainties in the relative importance of factors, subfactors, and values of subfactors were described by triangular fuzzy numbers, and the weights vector of subfactors and factors was calculated by using extent analysis [32]. The aggregated values of factors were calculated by using a fuzzy order-weighted operator, and the value of the customer satisfaction index was based on fuzzy logic IF-THAN rules.

In the model for determining the European customer satisfaction index [11], the assessment of the input variable values is based on the questionnaire containing 23 questions. Besides the model questions, demographic questions (e.g., age, gender, education level, etc.) are integral to the questionnaire. In the first step of the research, it was assumed that customer loyalty may be described through the linear multi-regression model. In the second step, all the treated factors were described by polynomial regression curves. The statistical analysis of the results obtained by the proposed nonlinear regression model by [11] indicates that the provided information is essential for managers in developing marketing strategies as well as is a tool for measuring performance and benchmarking.

The literature indicates that the customer satisfaction index [10] for smart services was proposed on a survey questionnaire developed through several rounds utilizing the Delphi method [33]. The expert assessment was conducted using a ten-point measurement scale. This research is conducted in two steps. In the first step, the convergent validity and reliability were assessed by using composite reliability, factor loadings, and average variance; additionally, the correlation coefficient between dependent variables was determined. In the second step, the success of the prediction is measured by applying the chi-square test.

Different models based on statistics, such as SERVQUAL [17], are also helpful in determining customer satisfaction. The mentioned model was used for determining customer satisfaction in the domain of public transportation [23]. Besides SERVQUAL, the service quality and customer satisfaction might be analyzed by applying other models, such as the ACSI model. The ACSI model was used for the research in the mobile telecommunication industry [22], where the input data were obtained through the questionnaire and denoted as primary data. A structured questionnaire was conducted with closed-ended questions to customers of the mobile operators. The secondary data were collected from the directories of the companies that were used in this study, online articles, and journals. The data analysis was based on descriptive statistics and regression analysis.

Different mathematical bases might be employed for the customer satisfaction assessment using artificial neural networks (ANNs). In the research related to customer satisfaction and loyalty in the pharmaceutical industry [18], 19 specific dimensions of products and services have been analyzed. The analysis of the obtained data is performed by using linear regression analysis and ANNs. In the presented research, the advantages of using ANNs are highlighted, and the managerial implications of using ANN modeling to identify the key drivers and set priorities for improvements are demonstrated. Similarly, the determination of customer satisfaction from mobile phone users is analyzed [28]. In the first step of that research, the multiple linear regression (MLR) model with 12 variables was conducted. The measure of the adequacy of the stated model is 0.41, which might be seen as in compliance with other cross-sectional studies. Based on the correlation coefficients obtained, the set of input variables was determined, so 46 participants were included. The input data were divided into two groups: (1) factors with a positive impact on customer satisfaction and (2) factors with a negative impact on customer satisfaction. In compliance with the stated, it has been suggested that multiple linear regression is not adequately precise for customer satisfaction assessment, and a better result might be obtained by applying ANNs.

The adaptive neuro-fuzzy model (ANFIS) has been widely used in the literature to solve various engineering problems [34]. Further, ANFIS was used to stabilize the operation of aerobic granular sludge [35]. The mentioned research indicated ANFIS to be superior to other methods in solving this problem. In other research areas, such as dealing with the problem of optimal extraction of groundwater resources, ANFIS was used to predict optimal groundwater exploitation [36]. It has also been applied for very short-term and accurate energy consumption forecasts for educational buildings [37]. Recently, ANFIS was found as a suitable method for solving the problem of diagnosis of fault nodes in wireless sensor networks [38].

The customer satisfaction assessment might be analyzed by applying different metaheuristics such as particle swarm optimization (PSO) and rough sets-based ANFIS approach [19]. The proposed research involved the following: rough sets theory is introduced to determine indispensable design attributes for generating customer satisfaction models. The PSO-based ANFIS approach was introduced to develop nonlinear customer satisfaction models. The proposed model has been tested on a large set of data. By comparing the training error, mean validation errors, and mean absolute deviation obtained by applying the fuzzy least-squares regression, fuzzy regression, and genetic programming-based fuzzy regression, it has been shown that the proposed model has a certain advantage over other treated models proposed. The mentioned research is extended with a more comprehensive model [39].

The main difference between the analyzed model [39] and our model is the larger number of hierarchical data levels, which is propagated to the more complex structure of the ANFIS model. There, the learning algorithm of the mentioned ANFIS is determined by using PSO [40], rough sets theory, PSO, and the least square estimation [39]. In our research, it is accomplished by a Delphi technique with TFNs and a fuzzy logic system (FLS).

In the ANFIS application for determining customer satisfaction, the calculated customer satisfaction is compared with other methods by using different errors. PSO-based ANFIS [40] was compared with fuzzy regression, ANFIS, and GA-based ANFIS by using mean absolute error, a variance of errors, and testing error. The rough sets and PSO-based ANFIS approaches [39] were compared with fuzzy least-squares regression, fuzzy regression, and genetic programming-based fuzzy regression by using mean absolute error and variance of errors. The results of our research are compared with case base reasoning (CBR), MLR, ANN, and FLS by using MAD, MSE, and MAPE.

3. The Research Methodology

In this paper, five methods for the estimation of the customer satisfaction index (CSI) are applied. The emphasis is on the hybrid algorithm ANFIS. The model presented in this research incorporates ANFIS, while other methods are used to verify and compare the obtained results. It is worth mentioning that ANFIS represents the hybrid algorithm of ANN and FLS. The type of the proposed research is characterized as descriptive quantitative research employing survey techniques for obtaining the input data. The graphical representation of the proposed research is presented in Figure 1.

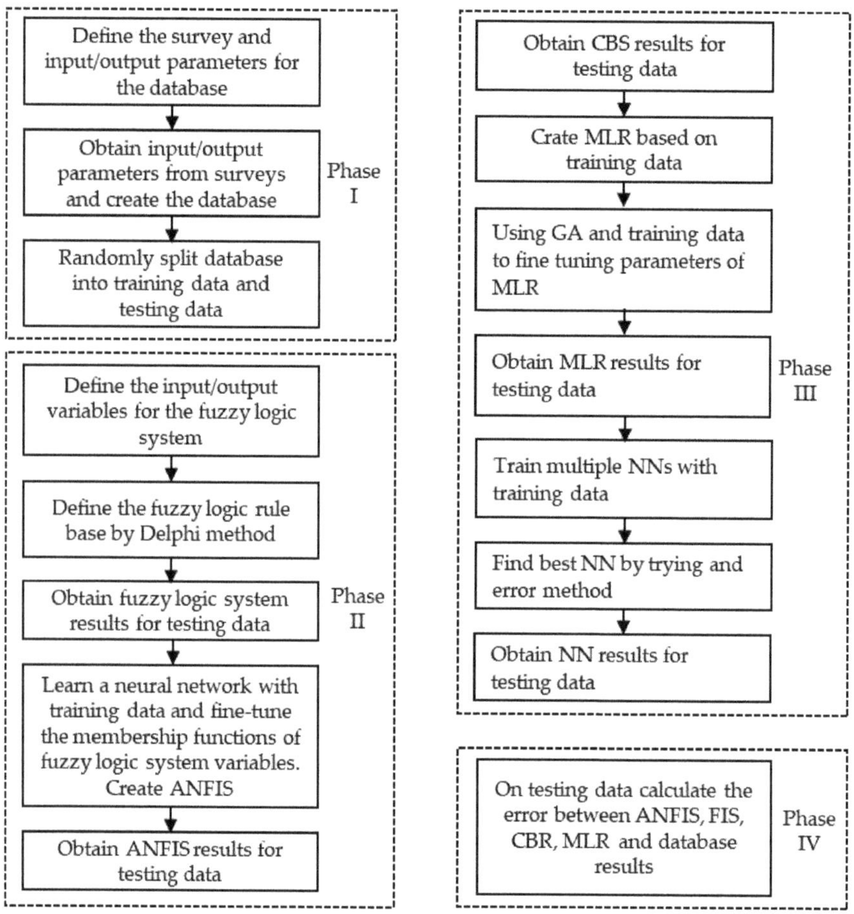

Figure 1. The model for determining the objective CSI.

The determination of the CSI is based on the delivery of ANFIS calculation. The calculation is dependent on fuzzy logic base rules and input data for determining CSI, which are defined in the form of a survey. When the value of objective CSI is obtained, it should be compared with other methods' results for verification and comparison of the obtained results. The least deviation of the obtained results compared to the calculation on the real data indicates which method provides the most precise results.

The proposed methodology consists of four phases which are further explained.

Phase 1: The survey for the determination of customer satisfaction is defined in compliance with the ACSI methodology. A sample of 109 surveys is collected. The data filtering and cleaning procedures are performed, so eight surveys have been removed from further analysis due to suspicious answers. With 101 regular surveys, the input database is formed as well as the testing database. The explanation in detail is provided in Section 3.1.

Phase 2: At the beginning of this phase, the definition of linguistic expressions based on fuzzy sets is performed. The group of decision-makers used those linguistic expressions while delivering the fuzzy Delphi algorithm to define the rule base for FLS. Further, we use ANFIS for fine-tuning the FLS membership functions.

Phase 3: To justify ANFIS employment, other quantitative methods are used for the assessment of CSI. For this purpose, MLR, CBR, and ANNs are used. This is explained in more detail in Section 3.4.

Phase 4: The obtained results from ANFIS and other solution methods are compared by using well-known calculated errors: MAD, MSE, and MAPE.

3.1. *The Definition of Survey for Customer Satisfaction Assessment in Dairy Production Enterprise and Data Collection*

In compliance with the ACSI model [22], three variables have been analyzed with the purpose of determining CSI: customer expectations (CEs), perceived quality (PQ), and perceived value (PV). In this case, the CSI variable is dependent. The values for the denoted variables are determined according to the survey results considering customers' answers. A production company from the dairy industry that operates in Serbia with retail stores has asked their customers to fill out the surveys and provide the input data for testing the proposed methodology.

As usual in this kind of research, cross-sectional surveys are used since the intention is to collect data from different types of groups of customers that may be denoted as adults at a single time. The sampling procedure to be applied is simple random sampling where individuals are chosen randomly. The products that are used by customers belong to the dairy industry, so practically there are no specific target groups that consume these products more than others. The customers are randomly selected at different store locations, and they are informed that their opinions will be used for scientific research. As written surveys require minimum usage of resources in terms of time and costs and at the same time, they are very convenient for obtaining the needed information, an approach is employed. Written surveys enable the absence of direct contact between the person who is collecting data and respondents, so errors induced in their interaction are avoided.

The survey is defined according to the results of the best practice. The survey is designed in a way to include only benefit and cost-type questions rated on a scale of 1–7. Value 1 stands for the least degree of belief and value 7 stands for the greatest degree of belief that the statement is true. The objective is to collect more than 100 surveys that may be used for the testing and training data, as well as for the control input data. The surveys are examined carefully so all samples with identified errors and suspicious content are removed. The process of obtaining the input data is presented in Figure 2.

The value of each variable in the model is obtained as a mean value of all answers corresponding to that variable at the customer expectations level of each customer that filled out the survey. The group of 90 randomly selected surveys was used for the calculation and training of the ANN, FLS, and ANFIS. In this way, the objective CSI is obtained, so any new customer that fills the survey might be checked if his satisfaction significantly deviates from the large group of his predecessors. This can be very useful when modification of the product is introduced to the market. The group of 11 randomly selected surveys was used to compare the proposed methods' results with real data.

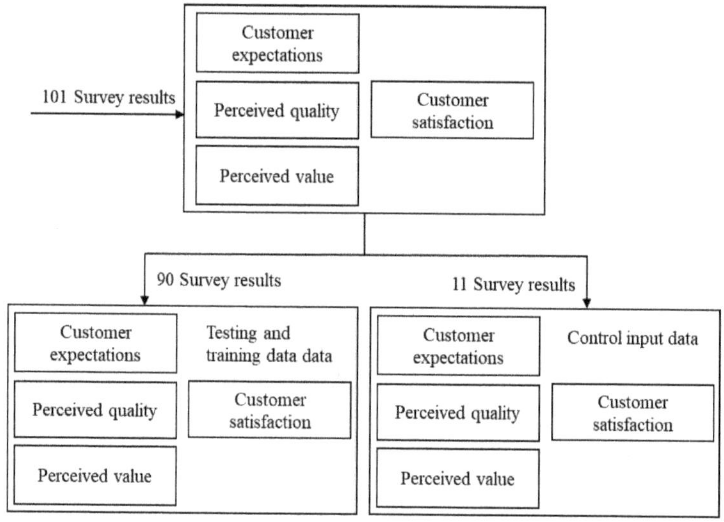

Figure 2. The definition of survey for customer satisfaction assessment in dairy production enterprise and data collection.

3.2. *The Definition of Fuzzy Logic Rules—Fuzzy Delphi*

The fuzzy logic rules that are integrated into the ANFIS algorithm are defined by applying the fuzzy Delphi method proposed by this research. The used fuzzy linguistic expressions (Figure 3) for the fuzzy Delphi method are distinguished from FLS, which is used for calculating CSI. These expressions are used to obtain crisp values representing the input in the rule base (Table 1) for FLS and ANFIS.

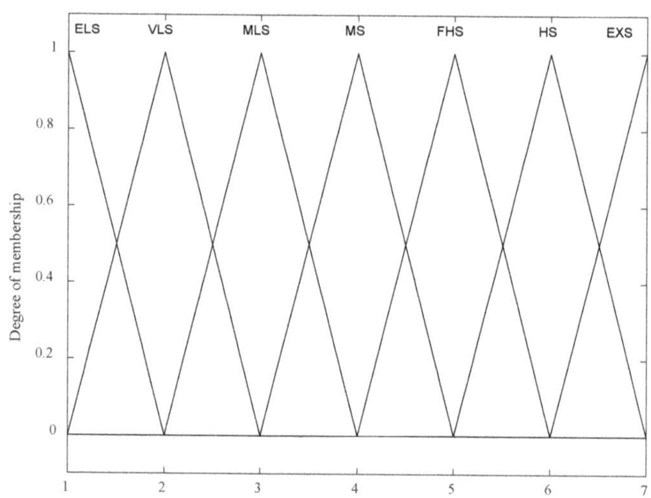

Figure 3. The proposed linguistic expressions.

Table 1. Fuzzy logic rule base.

Rule #	If *Quality* Is	And *Expectation* Is	And *Value* Is	Then *CSI* Is
1	Low	Low/Medium/High	Low	1
2	Low	Low/Medium	Medium	1.5
3	Low	High	High	2
4	Low	Low/Medium	High	2.5
5	Medium	High	Low	3
6	Medium	Medium	Low	3
7	Medium	Low	Low	3.5
8	Medium	Low	Medium	4
9	Medium	Low	Medium	4.5
10	High	High	Low	5
11	High	Medium	Low	5.5
12	High	Low	Low	6
13	High	Medium/High	Medium	6.5
14	High	Low/Medium/High	High	7

It is assumed that the level of customer satisfaction depends on variables that can be described using three linguistic expressions: low, medium, and high. The fuzzy rule base is based on the fuzzy Delphi technique, where E decision-makers (DMs) participate. They may be formally presented by the set of indices $\{1, \ldots, e, \ldots, E\}$, and the index of DM is denoted as e, $e = 1, \ldots, E$. It is worth mentioning that there is anonymity of DMs who participate in the assessment of CSI since the survey does not reflect who has filled it. They submit their assessment in the written form after each iteration to the rest of the DMs. By respecting the values of input variables, DMs assess the level of customer satisfaction by using one of seven linguistic expressions (Figure 3), which are modeled by triangular fuzzy numbers (TFNs):

Extremely low satisfaction (ELS)—$(1, 1, 2)$;
Very low satisfaction (VLS)—$(1, 2, 3)$;
Medium low satisfaction (MLS)—$(2, 3, 4)$;
Medium satisfaction (MS)—$(3, 4, 5)$;
Firmly high satisfaction (FHS)—$(4, 5, 6)$;
High satisfaction (HS)—$(5, 6, 7)$;
Extremely high satisfaction (EHS)—$(6, 7, 7)$.

The values in the bracket represent the membership function parameters of the fuzzy set.

The domains of these TFNs are defined on the measurement scale $(1–7)$. Value 1 and value 7 denote that the customer is completely unsatisfied or completely satisfied, respectively.

The proposed fuzzy Delphi algorithm is presented in Figure 4.

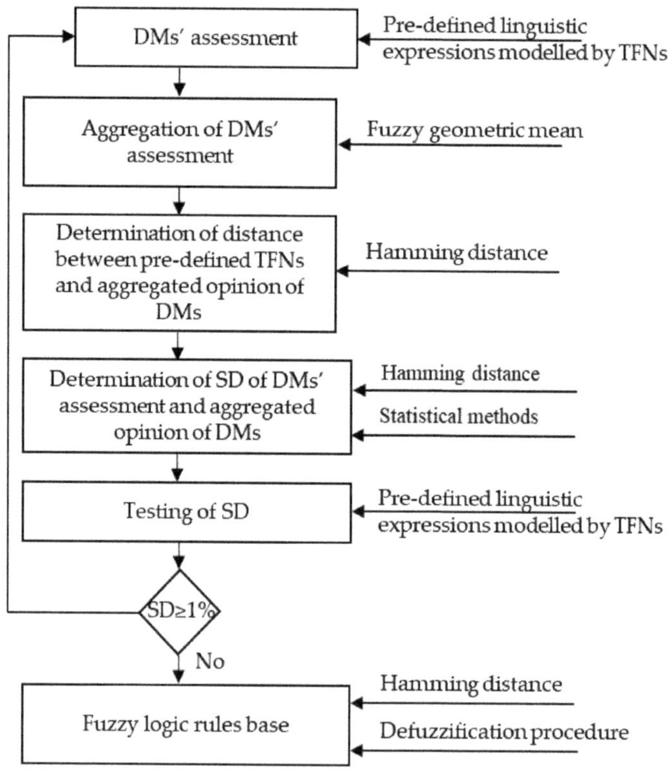

Figure 4. The proposed fuzzy Delphi algorithm.

The proposed fuzzy Delphi algorithm can be realized through the following steps.

Step 1. DM e, $e = 1, \ldots, E$ based on their knowledge and experience, respecting the values of input variables evaluates CSI, $\tilde{x}_e = (l_e, m_e, u_e)$. In the presented research, there were $E = 6$ DMs: marketing manager, sales manager, chief executive officer, chief financial officer, quality manager, and production manager.

Step 2. The aggregated value of DMs' assessment, $\tilde{x}^*$, is obtained by applying the method of fuzzy geometric mean:

$$\tilde{x}^* = \left(\sqrt[E]{\prod_{e=1,\ldots,E} l_e}, \sqrt[E]{\prod_{e=1,\ldots,E} m_e}, \sqrt[E]{\prod_{e=1,\ldots,E} u_e} \right) = (l^*, m^*, u^*), \tag{1}$$

where the lowest and upper bounds of fuzzy numbers are denoted as l_e and u_e, respectively. The modal value is denoted as m_e.

Step 3. The distance TFN $\tilde{x}^*$ should be determined for each of the seven defined TFN $\tilde{x}$, which are modeled by pre-defined linguistic expressions, $d(\tilde{x}, \tilde{x}^*)$, so that: $d(\tilde{x}, \tilde{x}^*)$ is the normalized Hamming distance [41].

Step 4. The standard deviation (SD) should be determined:

$$SD = \frac{1}{E-1} \cdot \sum_{e=1,\ldots,E} \left(d(\tilde{x}_e, \tilde{x}^*) \right)^2, \tag{2}$$

where $d(\tilde{x}_e, \tilde{x}^*)$ represents the normalized Hamming distance.

Step 5. It may be assumed that the SD limit value, where the consensus is reached, is 0.01, or the total deviation of the DMs' assessment compared to the aggregated value is 1% at most. Let us test the hypothesis at the 5% risk level that the obtained SD value is less than the assumed limit value. If the hypothesis is not fulfilled, then the process of assessment is brought back to step 1. Otherwise, step 6 is applied.

Step 6. Let us find the smallest deviation of the aggregated value of CSI from pre-defined linguistic expressions:

$$\min_{1,\ldots,7} d(\tilde{x}, \tilde{x}^*) \tag{3}$$

The aggregated CSI value can be at least away from the pre-defined linguistic expression $\tilde{x}'$.

Step 7. The defuzzified value of the linguistic expression $\tilde{x}'$ should be found by using the method of maximum possibility [42]. It represents the crisp value of the CSI index (Table 1).

As mentioned, Table 1 represents the input for FLS and ANFIS.

3.3. The ANFIS

The calculation of CSI is based on "Sugeno" FLS with three input fuzzy sets and one output crisp set. All input fuzzy sets—Quality, Expectations, and Value—are described by three Gaussian membership functions (gmf): low gmf (1.247, 1), medium gmf (1.247, 4) and high gmf (1.247, 7), as shown in Figure 5. The gmf is the abbreviation for the Gaussian membership function, which is characterized by two parameters: function center (c) and function width (σ).

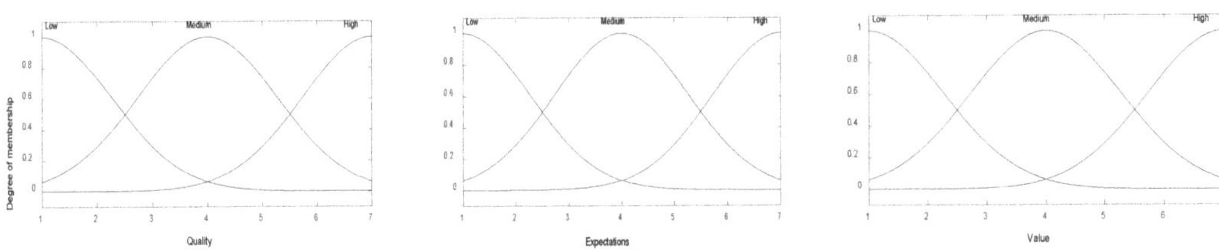

Figure 5. Input fuzzy sets.

FLS was created based on "IF-THEN" rules. The "IF" part of the rule is a premise, while "THEN" represents a consequence. In our problem constellation, "THEN" represents a crisp value of CSI. The rule base is explained in Table 1.

Due to the nonlinear dependence of the output variable on the input variables and the pronounced sensitivity of this dependence, it is necessary to fine tune the membership functions of the fuzzy logic system.

A hybrid algorithm that uses neural networks to optimize the shape of the fuzzy logic system membership functions was proposed by [43]. Figure 6 reveals the layers of the ANFIS, with the 14 rules which are shown in the following form:

$$\begin{array}{c}\text{IF } x_1 \text{ is } A_{11} \text{ and } x_2 \text{ is } B_{11} \text{ and } x_3 \text{ is } C_{11} \\ \text{THEN y is } f_1(x_1, x_2, x_3) = p_1x_1 + q_1x_2 + r_1x_3 + s_1,\end{array} \tag{4}$$

where A_{11}, B_{11}, and C_{11} are one of three membership functions of the input variables X_1 (quality), X_2 (expectation), and X_3 (value), respectively.

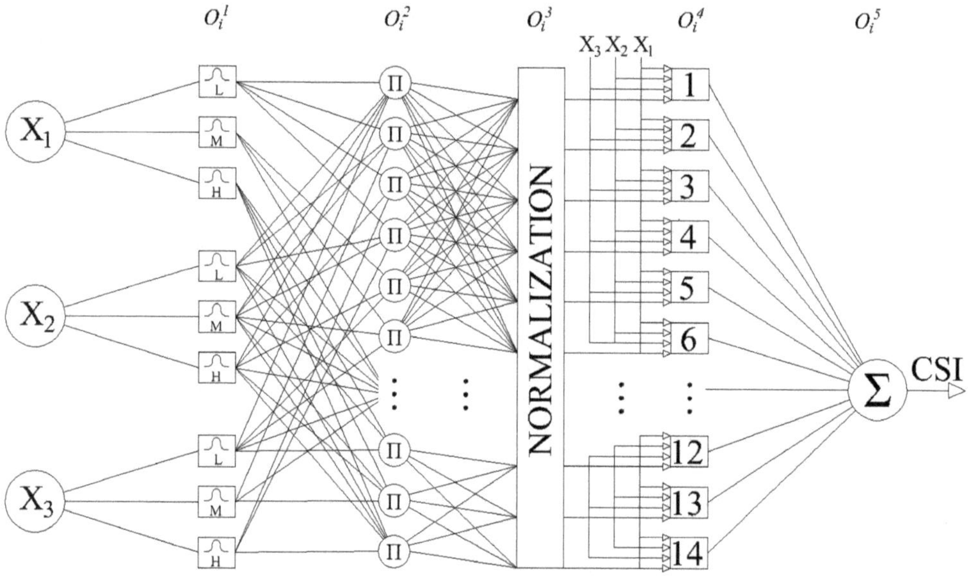

Figure 6. ANFIS architecture.

Let O_i^j denote node i in layer j. The functions of each ANFIS layer from the network presented in Figure 4 are the following.

Layer 1. The nodes of the first layer represent quantified values of input data sets. Each node is described by one of the Gaussian membership functions $\mu_{xi}(x_i)$, $i = 1,2,3$, which are characterized by the two parameters—c (the function centre) and σ (the function width):

$$Gaussian\ (x, c, \sigma) = e^{-\frac{1}{2}\cdot\left(\frac{x-c}{\sigma}\right)^2} \tag{5}$$

Layer 2. Each node of this layer calculates the minimum value of three input values of the fuzzy neural network. The output values of the layer 2 nodes are the rule signification:

$$O_i^2 = \omega_i = \mu_{Ai}(x_1) \times \mu_{Bi}(x_2) \times \mu_{Ci}(x_3), \tag{6}$$

Layer 3. Each i-node in this layer calculates the total weight of i-rule from the rule base by the following equation:

$$O_i^3 = \omega_i = \frac{\omega_i}{\sum_i \omega_i}, \tag{7}$$

Layer 4. This layer has 14 nodes that represent the output value CSI. Each node of this layer is connected with the normalized neuron from the previous layer. A defuzzification neuron computed the weighted consequent value of a given rule as:

$$O_i^4 = \omega_i \cdot f_i = \omega_i \cdot (p_1x_1 + q_1x_2 + r_1x_3 + s_1), i = 1, 2 \ldots n, \tag{8}$$

where n is the total number of rules in the fuzzy rules' base, while p, q, and r are the consequence parameters of the i-th rule.

Layer 5. The only node of the fifth layer is the fixed node in which the output result of the ANFIS is calculated. It is a fuzzy set with determining membership degrees of the possible values of CSI for the determined customer. Defuzzification is performed in the fifth level node. The output value is a real number which is in the interval (1–7):

$$O_i^5 = \sum_{i=1}^{14} \omega_i \cdot f_i = \frac{\sum_{i=1}^{14} \omega_i \cdot (p_1 x_1 + q_1 x_2 + r_1 x_3 + s_1)_i}{\sum_{i=1}^{14} \omega_{ii}} \tag{9}$$

ANFIS Learning

The output ANFIS function is described in linear form from the consequences part of fuzzy rules.

$$f = \omega_1 \cdot f_1 + \omega_2 \cdot f_2 + \omega_3 \cdot f_3 \tag{10}$$

The error E is calculated in the same manner as in the case of ANN application to the subject problem. The number of epochs is set to 1000 (Figure 7).

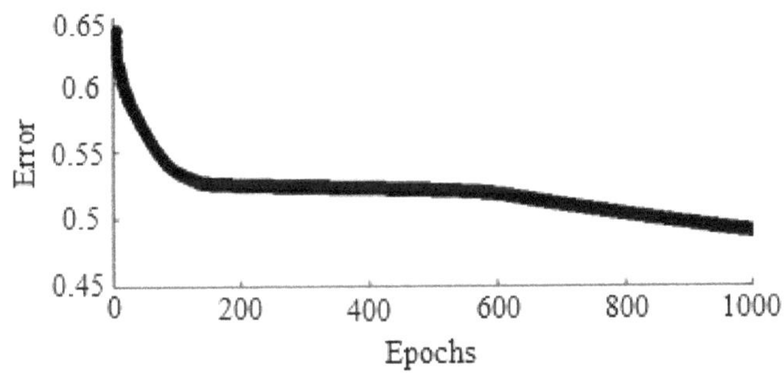

Figure 7. The error evaluation in the ANFIS learning process.

During the learning process, the membership functions have changed their shape. The parameters of the input value X_1 are: low gmf (1.098, 1.591); medium gmf (0.847, 3.83); and high gmf (0.239, 6.426). The parameters of the input value X_2 are: low gmf (1.641, 1.934); medium gmf (1.631, 4.232); and high gmf (0.166, 6.32). The parameters of the input value X_3 are: low gmf (0.941, 2.544); medium gmf (0.995, 3.117); and high gmf (0.453, 6.271). The new look of the functions is presented in Figure 8.

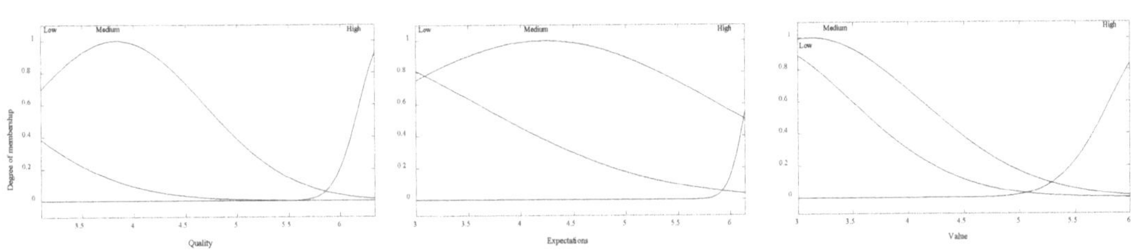

Figure 8. Membership functions after the ANFIS process.

By application of the proposed ANFIS algorithm, the CSI values are obtained. Those values are represented in Table 2.

Table 2. The results on data used for models' testing.

#	CSI^{DB}	CSI^{CBR}	CSI^{MLR}	CSI^{ANN}	CSI^{FLS}	CSI^{ANFIS}
1	5.00	4.99	4.90	4.91	4.70	4.92
2	4.33	4.84	4.60	4.57	4.75	4.64
3	4.90	4.51	4.70	4.73	4.89	4.77
4	4.67	4.96	4.68	4.68	4.61	4.79
5	4.67	4.77	4.77	4.74	4.79	4.84
6	3.85	4.32	4.17	4.17	4.39	3.77
7	4.67	4.95	4.49	4.49	4.25	4.57
8	5.00	4.8	4.85	4.86	4.63	4.89
9	4.33	4.3	4.79	4.82	4.78	4.84
10	5.67	5.3	5.49	5.39	5.80	5.74
11	5.33	5.8	5.45	5.34	5.43	5.32
MAD		0.28	0.19	0.18	0.27	**0.15**
MSE		0.12	0.05	0.06	0.11	**0.05**
MAPE		6.65	4.59	4.39	6.47	**3.72**

3.4. Other Solution Methods

Customers have a wide choice of products at their disposal. The customers demand their product expectations to be met, so company management must continually seek new, innovative ways to meet the demands and expectations of its customers, all to survive in the marketplace.

Customer satisfaction is the opinion they have about a company or product, and it is based on comparing the perceived characteristics of the product with their expectations. Therefore, satisfaction needs to be measured to understand the experience that customers have while using the product. Satisfaction measurement involves data collection and analysis and is usually measured through a questionnaire. There are several methods of collecting data on customer satisfaction using a survey questionnaire, and the goal is to maximize the number of responses and have a representative sample.

3.4.1. Case-Based Reasoning (CBR)

CBR is a well-known method that provides a relatively quick and straightforward solution compared to other methods, making it very popular and applicable in solving a wide range of problems [44].

The learning database consists of 90 input/output data pars. As there are 90 (out of learning database) input variables values, marked as a column vector $[X_1,\ X_2,\ X_3]$, the CBR algorithm is suited for finding the most similar input variables from the learning database compared to $[X_1,\ X_2,\ X_3]$ vector. The input part of the learning database can be presented in the matrix form $(x_{ij})_{90x3}$ and the output part as the column vector $[Y_1, Y_2, \ldots\ Y_{90}]T$.

The measure of comparing mathematically can be formulated in the following way:

$$\min_i \sum_{j=1}^{n} \left| (X_j)_{1xn} - (X_{ij})_{mxn} \right|, \tag{11}$$

where $(X_j)_{1xn}$ is associated with a new entry, and $(X_{ij})_{mxn}$ is associated with the training base.

In each following step, it is necessary to generate s values that correspond to the criteria in Formula (1). Let $s = 3$. Each of the obtained s couples of the input variables is associated with one value of the output variable Y. In this way, three output values were obtained, arranged from the lowest to the greatest: Y_{s1}, Y_{s2}, and Y_{s3}. Each of these three values need to be assigned weight coefficients w_1, w_2, and w_3. The final value CSI^{CBR} is obtained as the weighted average:

$$CSI^{CBR} = \frac{Y_{s1} \cdot w_1 + Y_{s2} \cdot w_2 + Y_{s3} \cdot w_3}{w_1 + w_2 + w_3} \quad (12)$$

Adopted values for coefficients w_1, w_2, and w_3 are 0.4, 0.35, and 0.25, respectively.

3.4.2. Multiple Linear Regression (MLR)

The method of machine learning that could be applied before more sophisticated tools of artificial intelligence such as FLS or ANN is MLR. According to this, the following linear equation is introduced:

$$CSI = a + b_1 \cdot X_1 + b_2 \cdot X_2 + b_3 \cdot X_3 \quad (13)$$

where b_i are the coefficients and a is an intercept.

For the optimization of the regression coefficients and intercept, the following model of combinatory optimization is set:

Minimize

$$FF = \sqrt{\frac{\sum_{i=1}^{m} \left(CSI_i^{DB} - CSI_i^{MLR}\right)^2}{m}} \quad (14)$$

subject to:

$$amin \leq a \leq amax \quad (15)$$

$$bjmin \leq bj \leq bjmax \quad (16)$$

The fitness function (FF) (14) minimizes the difference between the CSI from the learning database (CSI^{DB}) and the CSI obtained by MLR (CSI^{MLR}) for all the m pairs of the training data. In the presented case, m is equal to 90. The constraint (15) defines the interval for the feasible intercept values. The constraint (16) defines the interval for the feasible coefficients.

The following equation of multiple linear regression has been obtained:

$$CSI = 0.849 + 0.131 \cdot X_1 + 0.328 \cdot X_2 + 0.415 \cdot X_3 \quad (17)$$

The multiple coefficients of correlation R_{mul} are calculated thus:

$$R_{mul} = \frac{SD\left(s^{MLR}\right)}{SD\left(s^{SIM}\right)} = 0.585 \quad (18)$$

where SD is the marked standard deviation. Adjusted R (R_{adj}) is calculated as follows:

$$R_{adj} = 1 - \frac{1 - R_{mul}^2 \cdot (n-1)}{n-k-1} = 0.319 \quad (19)$$

where n is the number of training data and k is the number of variables.

3.4.3. Artificial Neural Network (ANN)

ANN represents the universal approximator that maps one set of independent variables to another set of dependent variables. Since traditional analytical methods, such as CBR and MLR, do not always provide satisfactory results, it is necessary to examine the possibility of applying more sophisticated methods of machine learning.

The method of trial and error led to the configuration of a neural network with two hidden layers and 20 neurons in each of the layers. The first layer of a neural network consists of input variables, X_1—quality; X_2—expectations; X_3—value, while the last, fourth layer, represents the output—CSI. The obtained configuration is presented in Figure 9.

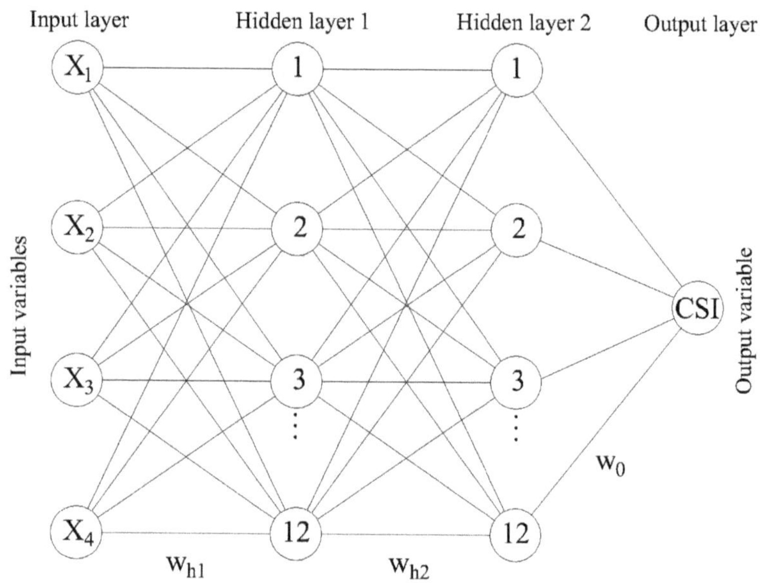

Figure 9. The adopted ANN architecture.

Each of the neural network layers consists of a number of neurons that are associated with the next layer by its weight. This connection (interdependence) extends in the direction from the input to the output layer. By fine-tuning weights w_{hi} and w_o, the neural network is learning by estimating an error in each epoch (learning iteration).

The most widely used algorithm for the neural network learning process is backpropagation, which estimates an error (E) during the training as follows:

$$E = \sum_{n=1}^{N} \frac{1}{2} \cdot \left\| d_p - o_p \right\|^2 \tag{20}$$

where: n—the neuron index, $n = 1, 2, \ldots, N$, d_p—desired output, and o_p—obtained output. The ANFIS training was realized in 1000 epochs (Figure 10).

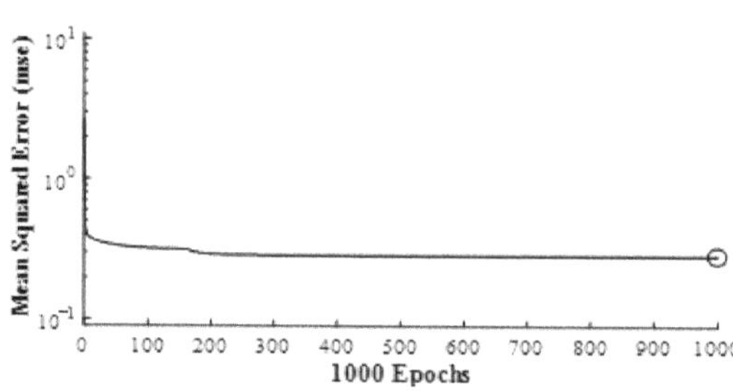

Figure 10. The error evaluation through the epochs.

More details about the backpropagation algorithm can be found in the work of the authors [45] who initially proposed it.

4. Model Testing

The time duration of the data collection was one month since the data practice indicated no seasonality in demand during the year. All surveys that showed flaws and suspicious answers were removed from the research during data cleaning. The selected 101 surveys were used for data testing. The surveys that were divided into two groups with a ratio of 90:11 was randomly selected.

The output of the conducted calculations is the value of CSI regarding the products belonging to dairy products that are produced by one company in Central Serbia. It can be concluded that the customers are quite satisfied with the dairy products as the assessed CSI is relatively high no matter which methodology is used. However, there are small differences between the obtained results of each applied methodology, so results need to be compared. The management of the production company should choose the most precise methodology since their strategy decision might rely on customer satisfaction.

This section is set to compare the performance of the five solution methods CBR, MLR, ANN, FLS, and ANFIS algorithms to determine CSI. A numerical example on which the suggested methods are tested represents m input data ($m = 11$), which have not been taken into consideration for the CBR, MLR, ANN, FLS, and ANFIS models. In other words, a comparison of the five models will be performed on the unknown input parameters.

In Table 2, the testing results are shown for each of the suggested methods. The errors for each of these suggested methods were marked respectively. The error represents the gap between the database results and results that come from the proposed methods.

$$MAD = \frac{|x_i - f_i|}{m} \tag{21}$$

$$MSE = \frac{\sum_{i=1,\ldots,m}(x_i - f_i)^2}{m-1} \tag{22}$$

$$MAPE = \frac{1}{m} \cdot \frac{\sum_{i=1,\ldots,m}|x_i - f_i|}{x_i} \tag{23}$$

where:

$x_i, i = 1, \ldots m$ is the value of CSI which is calculated based on the estimated values of customer satisfaction from the questionnaire,
$f_i, i = 1, \ldots m$ is the calculated value of CSI, which is calculated by applying all the analyzed methods.

Results in Table 2 reveal that ANFIS yielded a decrease of 44.44%, 54.54%, and 42.50% in MAD, MSE, and MAPE, respectively, compared to FLS.

Based on Table 2, the linear dependence between the obtained data from all methods and data from the database is given in Figure 11. The R^2 values are 0.5756, 0.8072, 0.8169, 0.5814, and 0.8365 for CBR, MLR, ANN, FLS, and ANFIS, respectively. In comparison with the other methods, it can be concluded that ANFIS fits the database in the best way.

Taking into consideration MAD, MSE, and MAPE, ANFIS yielded a decrease of 16.67%, 16.67%, and 15.26%, respectively, compared to ANN. Also, ANFIS yielded a decrease of 21.05%, 0%, and 18.95% in MAD, MSE, and MAPE, respectively, compared to MLR. Finally, considering MAD, MSE, and MAPE, ANFIS yielded a decrease of 46.43%, 58.33%, and 44.06%, respectively, compared to CBR.

In the end, taking into consideration the sum of all error measures (MAD, MSE, and MAPE), the closest method to ANFIS is ANN (with an increase of 15.33%), while the farthest solution is provided by the CBR method (with an increase of 44.40%).

The graphical representation of MAD, MSE, and MAPE is presented in Figure 12.

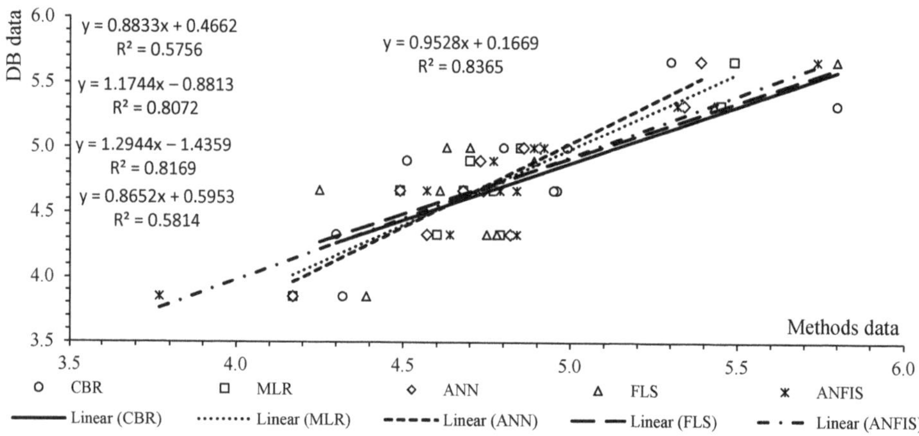

Figure 11. Linear dependence between the database and applied methods results.

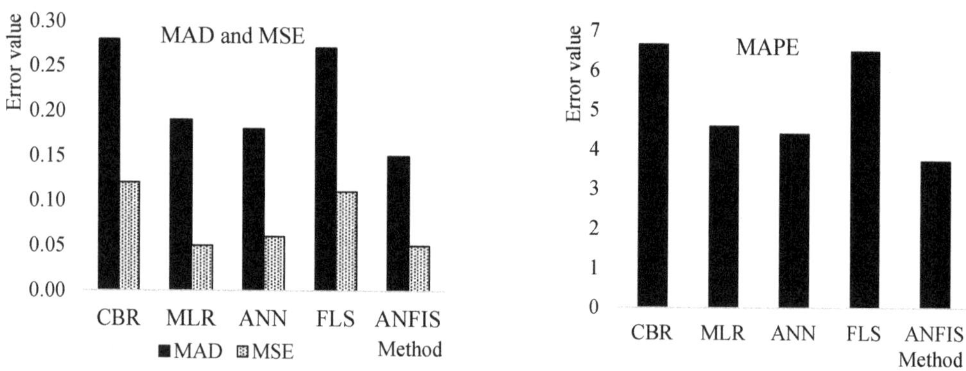

Figure 12. Graphical presentation of errors.

For the comparison of linear and nonlinear models, some authors consider the Akaike information criterion (AIC) to be the appropriate measure [46]. For small samples, AIC_c could be calculated as [47]:

$$AIC_c = m \cdot ln\left(\frac{\sum_{i=1,\ldots,m}(x_i - f_i)^2}{m}\right) + 2 \cdot k + \frac{2 \cdot k \cdot (k+1)}{m-k-1} + m \cdot \ln(2\pi) + m \quad (24)$$

where k is the number of parameters of the analyzed methods.

According to $AICc$ values of 16.3086, 7.6606, 8.1295, 15.4976, and 5.6321 for CBR, MLR, ANN, FLS, and ANFIS, respectively, it could be concluded that ANFIS is better than MLR, as the second-best method, for 26.48%.

To describe the relationship between the input and output variables before (FLS, upper part of Figure 13) and after (ANFIS, lower part of Figure 13) the learning process, we show the inner sensibility of the model.

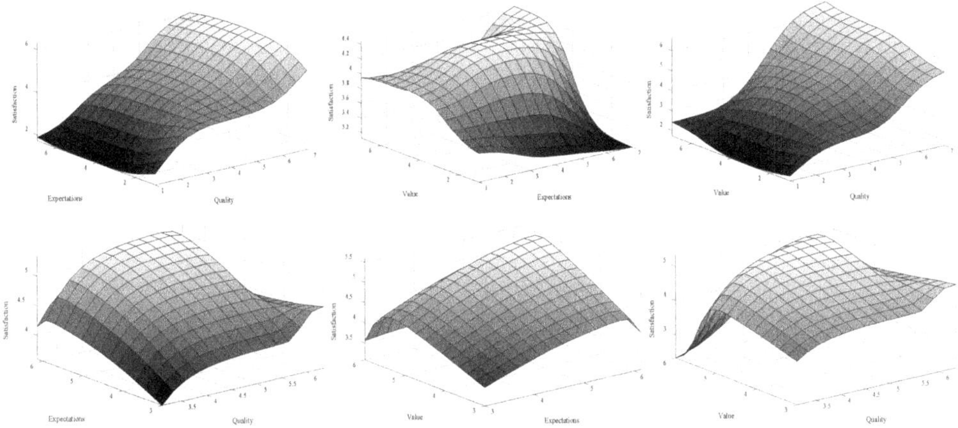

Figure 13. 3D dependence of FLS (**upper**) and ANFIS (**lower**).

In the end, ANFIS can obtain reasonable and better data-fitted results compared to other methods: CBR, MLR, ANN, and FLS. Based on the insights and conclusions, decision-makers can make informed choices and develop strategies. These decisions can pertain to a wide range of areas, such as product development and marketing.

5. Discussion and Conclusions

Creating stronger and long-term relationships with customers and attracting and attaining loyal customers is the goal of every company. Loyal customers contribute to increased sales and profits and can serve as the company's advocates, helping increase the market share. Satisfaction stems from expectations based on the degree of satisfaction of needs and desires, successful problem-solving, and satisfying consumer demands. A brand that customers recognize as the one that can meet their most important needs and wants will enjoy a high level of customer satisfaction and will build a base of loyal returning customers.

Positive customer feedback and testimonials can become a powerful marketing tool. This is of particular significance nowadays, with social networks as one of the main channels for customers to share their product feedback. Additionally, customer satisfaction insights can help identify areas for improvement and positively contribute to the company and brand. The proposed model bridges the identified gap in research, as explained in the introduction section. The shortcomings of the models based on pure MADM techniques, statistical analyses, and neural networks are overcome, which is shown in the previous section. The presented research deepens the understanding of the methods currently used for evaluating customer satisfaction and proposes a more accurate method for customer satisfaction assessment.

Research limitations and practical implications of the present study could be further presented. In the existing literature, only a few research studies assessed customer satisfaction without using self-reported measures collected through questionnaires. The determination of customer satisfaction through the questionnaires demands the engagement of resources in terms of personnel, money, and time with a noticeable constraint. The constraints are related to obtaining a representative sample in the survey study and the susceptibility to various types of biases due to the nature of survey studies, which requires a continuous validity check of the obtained data.

The FLS and ANFIS models are based on the fuzzy logic system that is obtained by employing the fuzzy Delphi technique. This fuzzy logic system is the base for future calculations, so those constraints are mastered. In the proposed model, the methods used for the determination of customer satisfaction are tested on the obtained input data whose validity is checked.

The practical implications of the research are summarized as follows. Five different methods are applied to calculate CSI. As fuzzy logic base is trained on the sample of 90 filled surveys, 11 are used for model testing. The gap between the actual value of customer CSI and the values obtained by applying the proposed methods is calculated by applying MAD, MSE, and MAPE. Considering the data provided in Table 2, it may be concluded that the ANFIS algorithm provides the lowest gap between the real and calculated values of CSI. Successful companies define their strategic actions by data-driven decision-making. This is not a one-time process since it requires continuous monitoring and adjustment based on new data and changing circumstances. This iterative approach is allowed by the application of the proposed ANFIS model. It is worth mentioning that the obtained results could even be improved by testing it on a larger sample of input data so FLS could be enhanced.

Social implications are an essential aspect of the research since the proposed model is not a substitute for the original methodology of the customer satisfaction assessment. In this way, the CSI can be tracked throughout the whole year, so significant disruptions may be identified in real time. This can be used as a base for an agile and resilient response of the company. Further, the model can be applied when new products are introduced to the market or when some products are considered for pulling off the market. As customer satisfaction may have a positive impact even on the resilience of the company, this research is delivered in the scope of the project "Coping with unpredictable disruptions in the domain of Engineering Management—Organizational resilience enhancement—CODEMO", supported by the University of Kragujevac.

The constraints of the research are mainly attributed to the need to conduct original research on customer satisfaction based on ACSI. Future research could analyze if all aspects of the customer satisfaction variables are embraced by the defined questions or those should be enhanced or expanded. Further, the consistency check of the input data might be delivered by some techniques. Future research could also investigate different industries, specifically the ones where there is a seasonality in customer demand, to compare the effectiveness of different methods for customer satisfaction assessment. Furthermore, other techniques, such as multi-gene genetic programming, which uses the genetic algorithm to find the best possible multiple regression model, will be used to solve the subject problem.

Author Contributions: Conceptualization, A.A. and N.L.; methodology, A.J.; software, A.J.; validation, A.A. and N.L.; formal analysis, N.L.; investigation, A.A.; resources, A.A.; data curation, A.J.; writing—original draft preparation, A.A. and N.L.; writing—review and editing, A.J.; visualization, A.J.; supervision, A.A.; project administration, A.A. All authors have read and agreed to the published version of the manuscript.

Funding: This research received no external funding.

Data Availability Statement: Not applicable.

Acknowledgments: This research is delivered in the scope of the project "Coping with unpredict-able disruptions in the domain of Engineering Management—Organizational resilience enhance-ment—CODEMO", supported by the University of Kragujevac.

Conflicts of Interest: The authors declare no conflict of interest.

References

1. Ali, M.M.; Doumbouya, M.B.; Louge, T.; Rai, R.; Karray, M.H. Ontology-Based Approach to Extract Product's Design Features from Online Customers' Reviews. *Comput. Ind.* **2020**, *116*, 103175. [CrossRef]
2. Javed, S.; Rashidin, S.; Xiao, Y. Investigating the Impact of Digital Influencers on Consumer Decision-Making and Content Outreach: Using Dual AISAS Model. *Econ. Res. Ekon. Istraž.* **2022**, *35*, 1183–1210. [CrossRef]
3. Chen, X.; Voigt, T. Implementation of the Manufacturing Execution System in the Food and Beverage Industry. *J. Food Eng.* **2020**, *278*, 109932. [CrossRef]

4. Harwood, W.S.; Drake, M.A. Manufacture of Milk and Whey Products: Impact of Processing on Sensory Characteristics of Milk and Dairy Products. In *Encyclopedia of Dairy Sciences*, 3rd ed.; McSweeney, P.L.H., McNamara, J.P., Eds.; Academic Press: Oxford, UK, 2022; pp. 103–117. ISBN 978-0-12-818767-8.
5. Wu, W.; Zhang, A.; van Klinken, R.D.; Schrobback, P.; Muller, J.M. Consumer Trust in Food and the Food System: A Critical Review. *Foods* **2021**, *10*, 2490. [CrossRef] [PubMed]
6. Otto, A.S.; Szymanski, D.M.; Varadarajan, R. Customer Satisfaction and Firm Performance: Insights from over a Quarter Century of Empirical Research. *J. Acad. Mark. Sci.* **2020**, *48*, 543–564. [CrossRef]
7. Nestić, S.; Gojković, R.; Petrović, T.; Tadić, D.; Mimović, P. Quality Performance Indicators Evaluation and Ranking by Using TOPSIS with the Interval-Intuitionistic Fuzzy Sets in Project-Oriented Manufacturing Companies. *Mathematics* **2022**, *10*, 4174. [CrossRef]
8. Kotler, P.T.; Armstrong, G. *Principles of Marketing*, 18th ed.; Pearson Higher Education: London, UK, 2021; ISBN 978-1-292-34113-2.
9. Kotler, P.; Kartajaya, H.; Setiawan, I. *Marketing 5.0: Technology for Humanity*; John Wiley & Sons: Hoboken, NJ, USA, 2021; ISBN 978-1-119-66851-0.
10. Shin, D.-H. Measuring the Quality of Smartphones: Development of a Customer Satisfaction Index for Smart Services. *Int. J. Mob. Commun.* **2014**, *12*, 311–327. [CrossRef]
11. Turkyilmaz, A.; Oztekin, A.; Zaim, S.; Fahrettin Demirel, O. Universal Structure Modeling Approach to Customer Satisfaction Index. *Ind. Manag. Data Syst.* **2013**, *113*, 932–949. [CrossRef]
12. Cronin, J.J.; Brady, M.K.; Hult, G.T.M. Assessing the Effects of Quality, Value, and Customer Satisfaction on Consumer Behavioral Intentions in Service Environments. *J. Retail.* **2000**, *76*, 193–218. [CrossRef]
13. Udo, G.J.; Bagchi, K.K.; Kirs, P.J. An Assessment of Customers' e-Service Quality Perception, Satisfaction and Intention. *Int. J. Inf. Manag.* **2010**, *30*, 481–492. [CrossRef]
14. Pakurár, M.; Haddad, H.; Nagy, J.; Popp, J.; Oláh, J. The Service Quality Dimensions That Affect Customer Satisfaction in the Jordanian Banking Sector. *Sustainability* **2019**, *11*, 1113. [CrossRef]
15. Sheth, J.N. Demographics in Consumer Behavior. *J. Bus. Res.* **1977**, *5*, 129–138. [CrossRef]
16. Al-Eisa, A.S.; Alhemoud, A.M. Using a Multiple-attribute Approach for Measuring Customer Satisfaction with Retail Banking Services in Kuwait. *Int. J. Bank Mark.* **2009**, *27*, 294–314. [CrossRef]
17. Parasuraman, A.P.; Zeithaml, V.; Berry, L. SERVQUAL A Multiple-Item Scale for Measuring Consumer Perceptions of Service Quality. *J. Retail.* **1988**, *64*, 12–40.
18. Grnholdt, L.; Martensen, A. Analysing Customer Satisfaction Data: A Comparison of Regression and Artificial Neural Networks. *Int. J. Mark. Res.* **2005**, *47*, 121–130. [CrossRef]
19. Jiang, H.; Kwong, C.K.; Law, M.C.; Ip, W.H. Development of Customer Satisfaction Models for Affective Design Using Rough Set and ANFIS Approaches. *Procedia Comput. Sci.* **2013**, *22*, 104–112. [CrossRef]
20. Lin, W.-B. The Exploration of Customer Satisfaction Model from a Comprehensive Perspective. *Expert Syst. Appl.* **2007**, *33*, 110–121. [CrossRef]
21. Hallencreutz, J.; Parmler, J. Important Drivers for Customer Satisfaction—From Product Focus to Image and Service Quality. *Total Qual. Manag. Bus. Excell.* **2021**, *32*, 501–510. [CrossRef]
22. Angelova, B.; Zeqiri, C. Measuring Customer Satisfaction with Service Quality Using American Customer Satisfaction Model (ACSI Model). *Int. J. Acad. Res. Bus. Soc. Sci.* **2011**, *1*, 232–258. [CrossRef]
23. Wisniewski, M. Using SERVQUAL to Assess Customer Satisfaction with Public Sector Services. *Manag. Serv. Qual. Int. J.* **2001**, *11*, 380–388. [CrossRef]
24. Celik, E.; Bilisik, O.N.; Erdogan, M.; Gumus, A.T.; Baracli, H. An Integrated Novel Interval Type-2 Fuzzy MCDM Method to Improve Customer Satisfaction in Public Transportation for Istanbul. *Transp. Res. Part E Logist. Transp. Rev.* **2013**, *58*, 28–51. [CrossRef]
25. Ahani, A.; Nilashi, M.; Yadegaridehkordi, E.; Sanzogni, L.; Tarik, A.R.; Knox, K.; Samad, S.; Ibrahim, O. Revealing Customers' Satisfaction and Preferences through Online Review Analysis: The Case of Canary Islands Hotels. *J. Retail. Consum. Serv.* **2019**, *51*, 331–343. [CrossRef]
26. Tadic, D.; Aleksic, A.; Mimovic, P.; Puskaric, H.; Misita, M. A Model for Evaluation of Customer Satisfaction with Banking Service Quality in an Uncertain Environment. *Total Qual. Manag. Bus. Excell.* **2018**, *29*, 1342–1361. [CrossRef]
27. Lamrani Alaoui, Y.; Tkiouat, M. Modeling Customer Satisfaction in Microfinance Sector: A Fuzzy Bayesian Networks Approach. *Int. J. Eng. Bus. Manag.* **2019**, *11*, 1847979019869533. [CrossRef]
28. Goode, M.M.H.; Davies, F.; Moutinho, L.; Jamal, A. Determining Customer Satisfaction from Mobile Phones: A Neural Network Approach. *J. Mark. Manag.* **2005**, *21*, 755–778. [CrossRef]
29. Kalinić, Z.; Marinković, V.; Kalinić, L.; Liébana-Cabanillas, F. Neural Network Modeling of Consumer Satisfaction in Mobile Commerce: An Empirical Analysis. *Expert Syst. Appl.* **2021**, *175*, 114803. [CrossRef]
30. Kumar, S.; Zymbler, M. A Machine Learning Approach to Analyze Customer Satisfaction from Airline Tweets. *J. Big Data* **2019**, *6*, 62. [CrossRef]
31. Munoz, C.; Laniado, H.; Córdoba, J. Development of a Robust Customer Satisfaction Index for Domestic Air Journeys. *Res. Transp. Bus. Manag.* **2020**, *37*, 100519. [CrossRef]

32. Chen, Y.H. Fuzzy Ratings in Mechanical Engineering Design—Application to Bearing Selection. *Proc. Inst. Mech. Eng. Part B J. Eng. Manuf.* **1996**, *210*, 49–53. [CrossRef]
33. Green, K.; Armstrong, J.S.; Graefe, A. Methods to Elicit Forecasts from Groups: Delphi and Prediction Markets Compared. *SSRN J.* **2008**, 1–6. [CrossRef]
34. Salleh, M.N.M.; Talpur, N.; Hussain, K. Adaptive Neuro-Fuzzy Inference System: Overview, Strengths, Limitations, and Solutions. In *Proceedings of the Data Mining and Big Data;* Tan, Y., Takagi, H., Shi, Y., Eds.; Springer International Publishing: Cham, Switzerland, 2017; pp. 527–535.
35. Zaghloul, M.S.; Hamza, R.A.; Iorhemen, O.T.; Tay, J.H. Comparison of Adaptive Neuro-Fuzzy Inference Systems (ANFIS) and Support Vector Regression (SVR) for Data-Driven Modelling of Aerobic Granular Sludge Reactors. *J. Environ. Chem. Eng.* **2020**, *8*, 103742. [CrossRef]
36. Milan, S.G.; Roozbahani, A.; Arya Azar, N.; Javadi, S. Development of Adaptive Neuro Fuzzy Inference System–Evolutionary Algorithms Hybrid Models (ANFIS-EA) for Prediction of Optimal Groundwater Exploitation. *J. Hydrol.* **2021**, *598*, 126258. [CrossRef]
37. Ghenai, C.; Al-Mufti, O.A.A.; Al-Isawi, O.A.M.; Amirah, L.H.L.; Merabet, A. Short-Term Building Electrical Load Forecasting Using Adaptive Neuro-Fuzzy Inference System (ANFIS). *J. Build. Eng.* **2022**, *52*, 104323. [CrossRef]
38. Rajan, M.S.; Dilip, G.; Kannan, N.; Namratha, M.; Majji, S.; Mohapatra, S.K.; Patnala, T.R.; Karanam, S.R. Diagnosis of Fault Node in Wireless Sensor Networks Using Adaptive Neuro-Fuzzy Inference System. *Appl. Nanosci.* **2023**, *13*, 1007–1015. [CrossRef]
39. Jiang, H.; Kwong, C.K.; Siu, K.W.M.; Liu, Y. Rough Set and PSO-Based ANFIS Approaches to Modeling Customer Satisfaction for Affective Product Design. *Adv. Eng. Inform.* **2015**, *29*, 727–738. [CrossRef]
40. Jiang, H.M.; Kwong, C.K.; Ip, W.H.; Wong, T.C. Modeling Customer Satisfaction for New Product Development Using a PSO-Based ANFIS Approach. *Appl. Soft Comput.* **2012**, *12*, 726–734. [CrossRef]
41. Grzegorzewski, P. Distances between Intuitionistic Fuzzy Sets and/or Interval-Valued Fuzzy Sets Based on the Hausdorff Metric. *Fuzzy Sets Syst.* **2004**, *148*, 319–328. [CrossRef]
42. Dubois, D.; Prade, H. Systems of Linear Fuzzy Constraints. *Fuzzy Sets Syst.* **1980**, *3*, 37–48. [CrossRef]
43. Jang, J.-S.R. ANFIS: Adaptive-Network-Based Fuzzy Inference System. *IEEE Trans. Syst. Man Cybern.* **1993**, *23*, 665–685. [CrossRef]
44. Aamodt, A.; Plaza, E. Case-Based Reasoning: Foundational Issues, Methodological Variations, and System Approaches. *AI Commun.* **1994**, *7*, 39–59. [CrossRef]
45. Rumelhart, D.E.; Hinton, G.E.; Williams, R.J. *Learning Internal Representations by Error Propagation;* Elsevier: Amsterdam, The Netherlands, 1988; pp. 399–421.
46. Cavanaugh, J.E.; Neath, A.A. The Akaike Information Criterion: Background, Derivation, Properties, Application, Interpretation, and Refinements. *WIREs Comput. Stat.* **2019**, *11*, e1460. [CrossRef]
47. Portet, S. A Primer on Model Selection Using the Akaike Information Criterion. *Infect. Dis. Model.* **2020**, *5*, 111–128. [CrossRef] [PubMed]

mathematics

Article

A Hybrid Fuzzy MCDM Approach to Open Innovation Partner Evaluation

Sanja Puzović [1,*], Jasmina Vesić Vasović [1], Dragan D. Milanović [2] and Vladan Paunović [1]

1 Faculty of Technical Sciences Čačak, University of Kragujevac, 32000 Čačak, Serbia; jasmina.vesic@ftn.kg.ac.rs (J.V.V.); vladan.paunovic@ftn.kg.ac.rs (V.P.)
2 Faculty of Mechanical Engineering, University of Belgrade, 11000 Belgrade, Serbia; ddmilanovic@mas.bg.ac.rs
* Correspondence: sanja.puzovic@ftn.kg.ac.rs

Abstract: Even though interest in open innovation (OI) both as a research field and as an industrial practice for creating competitively advantageous innovation through collaboration has grown exponentially over the last decades, the issue of how to transform OI strategy into a sustainable competitive advantage is still an open research question. Selecting partners capable of operationally and strategically contributing to the OI project goals is a strategic decision for companies striving to effectively implement the OI concept. The study is aimed at defining a structured and methodology-supported decision-making process for OI partner selection based on a novel hybrid Multi-Criteria Decision-Making (MCDM) model which is enhanced by interval type-2 fuzzy sets (IT2F) to deal with the inherent uncertainty. The model combines IT2F Delphi (IT2FD), IT2F Analytical Hierarchy Process (IT2F AHP), and IT2F Preference Ranking Organization METHod for Enrichment of Evaluations (IT2F PROMETHEE). The study provides a comprehensive framework of the OI partner performance indicators; additionally, it provides a contingent approach to identifying evaluation criteria depending on the nature of the company's innovation processes, contextual conditions, and innovation strategy. The case study is used to verify the feasibility and effectiveness of the proposed process. The study's results highlight the significance of specific factors related to the partners' technological competencies.

Keywords: open innovation projects; MCDM; IT2FD; IT2F AHP; IT2F PROMETHEE

MSC: 90B50

Citation: Puzović, S.; Vesić Vasović, J.; Milanović, D.D.; Paunović, V. A Hybrid Fuzzy MCDM Approach to Open Innovation Partner Evaluation. *Mathematics* **2023**, *11*, 3168. https://doi.org/10.3390/math11143168

Academic Editor: Ignacio Javier Perez Galvez

Received: 25 June 2023
Revised: 14 July 2023
Accepted: 17 July 2023
Published: 19 July 2023

1. Introduction

Increased technological complexity, a fusion of different technologies, resource scarcity, and market unpredictability have resulted in companies adopting a more open, cooperative approach to generating competitive advantage [1,2], which has led to more flexible business models based on more open interactions with the external environment. This prompted companies to transform their centralized research and development (R&D) systems by spreading their innovation processes across a global network of external partners and locations [3,4].

Dominantly closed innovation systems, exclusively concentrated inside organizational boundaries, have proven to be unsustainable, unreliable, rigid, costly, too sluggish, and incapable of generating technologically superior and market-sustainable innovations. Instead, companies are encouraged to use input from outsiders and find external opportunities for the commercialization of products and technologies to strengthen internal innovation processes [5].

This gave birth to the concept of open innovation (OI), introducing a radical transition in the way companies manage their innovation processes. OI is defined as a distributed innovation process that relies on purposefully managed knowledge flows across organizational boundaries, using pecuniary and nonpecuniary mechanisms in line with the business model to guide and motivate knowledge sharing [6].

Such relations tend to combine complementary resources and build synergies while simultaneously being characterized by an intensive exchange of knowledge and learning processes [7]. The OI concept enables companies to be more efficient in terms of creating and capturing value, collective intelligence acquisition, saving costs and time, and accelerating new revenue opportunities. It also enables overcoming geographical, institutional, and disciplinary barriers thanks to the openness of research and development, the diffusion of technology, and the open exchange of knowledge [6,8].

The OI concept has evolved into one of today's most important business paradigms. According to Bogers et al. [9], OI will play a critical role in the world's developed economies in the coming decades. They cite new technological trends such as blockchain and digitalization, genome editing, and sustainable development goals promoted at the international level as key incentives for OI. OI is also often mentioned as an accelerant of the new industrial initiative Industry 4.0. In fact, according to Hizam-Hanafiah and Soomro [10], the OI and initiatives referring to the external exploitation of knowledge are fully in line with the needs of an integrated digital business model.

These trends encourage companies to implement the OI paradigm through partnerships and collaborations among firms in R&D projects. Relying on key OI principles such as the use of external ideas and technologies that reduce R&D costs and time while improving the overall efficiency of the company and the ability to acquire, perceive, and use new knowledge faster than competitors [11].

Despite the fact that interest in the OI concept in the creation of technological innovations as a field of research has grown exponentially during the last decades, the questions of how to promote the adaptation of companies to OI practices and how to transform an OI strategy into a sustainable competitive advantage remain open research questions. Namely, Carmona-Lavadoa et al. [12] argue that openness in itself cannot be a determinant of performance unless it is supported by complementary means, such as coordination ability and complementarity of innovation partners, which, it turns out, are essential in ensuring the successful transfer and integration of critical know-how and in creating value through collaboration [13].

Although methods and strategies for establishing cooperation between innovation partners have been the subject matter of interest for many research studies [14,15], the studies pertaining to the development of methodologically supported approaches to OI optimal partner selection have not led to the development of a dominant reference approach; consequently, this process, in practice, is still mostly carried out on an ad hoc basis.

Current methodological approaches in this area have several deficiencies to some extent, including: a lack of the necessary holistic approach; a lack of operational indications; a lack of flexibility in the system of evaluation criteria to take into account the company's business circumstances, innovation policies, or strategies; and a lack of systematized and structured decision-making process.

The OI partner selection problem can be analyzed in a multi-dimensional space of different parameters and objectives in order to cope with complexity; as such, it could be considered a multi-criteria decision-making (MCDM) problem. The use of MCDM methods for these problems could provide a reliable compromising solution regarding various objectives, aspects, and criteria. It provides some advantages such as integrating a large number of different and often conflicting criteria and making alternatives evaluation much more flexible, objective, and acceptable.

In addition, when a detailed literature review is made it is seen that there are many different application areas of MCDM techniques for OI management-related issues. MCDM applications in the OI environment are comprehensively reviewed; Table 1 denotes the papers with a research focus on the application of MCDM methods to OI management-related issues.

Table 1. Application of MCDM Methods in the OI Environment.

MCDM Methods	Research Focus	Studies
Fuzzy AHP	Collaboration network partner selection with integration business, social, and environmental goals	[16]
Fuzzy AHP	Evaluation of process of innovation-oriented knowledge under the open innovation paradigm	[17]
Fuzzy Delphi, Fuzzy DEMATEL, DANP	Prioritizing and analyzing interrelationships among factors affecting Foreign Direct Investment attractiveness and open innovation	[18]
Fuzzy TOPSIS	Ranking the indicators of open innovation adoption based on new product development factors	[19]
Fuzzy AHP, Fuzzy VIKOR	Determination of an appropriate open innovation model for logistics firms	[20]
Delphi, Fuzzy ANP	End-to-end analysis of an open innovation setup for determining a suitable innovation structure	[21]
ANP, PROMETHEE	Ranking the moderating factors that have contributed to the degree of small and mid-size enterprises' participation in open innovation activities	[22]
DEMATEL	Determination the best ranking of effective factors in open innovation success in manufacturing enterprises	[23]
AHP ISM	Investigation mechanisms for improving supply chain open innovation networks	[24]
IT2F AHP	Supporting the effective selection of partners for collaborative technological R&D projects	[25]

Analytic Network Process: ANP; Analytical Hierarchy Process: AHP; DEMATEL-based Analytic Network Process: DANP; Decision-Making Trial and Evaluation Laboratory: DEMATEL; Interpretative Structural Modeling: ISM; Interval Type-2 Fuzzy: IT2F; Preference Ranking Organization METHod for Enrichment of Evaluations: PROMETHEE; Technique for the Order Preference by Similarity to Ideal Solution: TOPSIS; VIšeKriterijumska Optimizacija i kompromisno Rešenje: VIKOR.

The paper introduces a hybrid MCDM model for OI partner selection that provides a systematic and structured approach that may facilitate the generation of relevant decision-making factors and an assessment of the relative importance of various decision-making elements.

The study provides a certain methodological advancement in OI partner selection. First, a comprehensive framework of OI partner performance indicators is provided, encompassing the essential technological, operational, and strategic evaluation aspects. In addition, it provides a contingent approach to identifying OI partner evaluation criteria, considering the nature of the company's innovation processes, contextual conditions, and innovation strategy.

The increasing complexity of the social and economic environment, along with the vagueness of the inherently subjective nature of human thinking, leads to the impossibility of describing the input data of the decision-making process with crisp values [26]. Given the capabilities of interval type-2 fuzzy (IT2F) sets in representing vague preferences and dealing with the hesitation of human perception, the proposed MCDM model is extended in the context of the IT2F set theory to make the proposed approach more convenient for modeling different sources of vagueness and uncertainty in real-life decision-making problems.

The rest of the paper is organized as follows: The problem of OI partner selection is analyzed in Section 2, with a literature review. Section 3 presents the theoretical basis of the methods involved in this paper and describes the proposed decision-making process. In Section 4, a case study is presented to help understand the methodology proposed for OI partner selection and demonstrate its practicality and feasibility. Finally, discussion and directions for future research are presented in Section 5.

2. The Problem of OI Partner Selection

The various types of methodological support developed thus far lack the necessary holistic approach and thus only reflect the relationships between OI partners superficially.

Otherwise, it is difficult to follow them from the perspective of an enterprise due to a lack of operational indications, for which they require great expertise and experience on the analyst's part. Additionally, they often provide generic systems of criteria, whereby their adaptability to the company's business circumstances, innovation policies, and strategies has not been given much attention.

A significant number of the studies dealing with this issue are directed towards the development of methodologies intended to discover data about potential partners, whose contribution is limited to providing support for their identification and the analysis of their competencies and capacities but does not provide systematized and structured approaches for making the final decision on partner selection.

Yoon and Song [27] identify three methodological approaches in OI partner selection, namely: (1) the mathematical programming methods that deal with the theoretical process for formulating variables and equations; (2) the approaches to the evaluation and ranking based upon the analyst's evaluations; and (3) the approaches based upon the application of the artificial intelligence techniques that served to process a large amount of data to find a well-matched partner.

The approaches to evaluation and ranking based on the analyst's evaluations are the most suitable to apply in the practical operations of an enterprise. The academic and professional literature, however, lacks consensus on which aspects of potential partner evaluation should be included in deciding on the selection. In that context, several characteristic approaches are possible to identify.

The first group consists of approaches exclusively limited to the technological aspects of partner competency and complementarity while simultaneously most frequently relying on the analysis of information about registered patents, which in this case are considered to be the most relevant data for researching the innovation activities of potential partners. For example, Manotungvorapun and Gerdsri [28] propose an approach that utilizes patent information by applying morphology analysis and generative topology maps. Park and Yoon [29] used a multistage patent citation analysis method that included bibliographic coupling and the keyword vector mapping information visualization method. Wang [30] and Jeon et al. [31] take a similar approach.

The approaches based on the consideration of partners' technological compatibility illustrate the affinity for their technological knowledge [32] and may be suitable if technologies are becoming increasingly complex or distributed over various sources, or when technology fusion is recognized as an important part of collaborative innovation [31]. They are, however, criticized for excluding taking the strategic perspective into account.

Many authors point out the need for a multidimensional assessment, namely, beside technical compatibility as an indicator of the compatibility of the relevant knowledge and technology capabilities that determine the absorptive potential, an important emphasis is placed on nontechnical compatibility as well as the indicators of the congruence of the goals and the cooperation process, which affect the stability and organizational harmonization of OI partner relations and generate trust and commitment [33–35]. Namely, OI success does not only depend on the efficient integration of internal and external technologies and types of knowledge, but it also depends on factors such as strategic goals, organizational culture, the R&D strategy, the top management style, an attitude towards cooperation in R&D activities, the innovation partners' protocols, and the innovation environment [28,30]. Büyüközkan et al. [36] emphasize that two dimensions should necessarily be included in potential partner assessment: the first pertaining to their strategic excellence and the second pertaining to their business, i.e., their operational excellence. Holmberg and Cummings [37] suggest that, when selecting OI partners, potential partners' motivation for innovation cooperation must be considered apart from the criteria related to the partners' attributes [38].

Consistent evaluation outcomes necessitate the use of the contingent system of evaluation criteria, which could be adapted to the company's business circumstances, innovation policies, and strategies. According to Shah and Swaminathan [39], if the interpretability of the OI project outcome is low, i.e., if there are limited possibilities of interpreting and understanding them, priority should be given to the criteria pertaining to the compatibility of the identity, culture, goals, mutual trust, and so forth. The same is also applicable if it is difficult to establish control over the process. On the contrary, outcome-oriented criteria (such as financial cost-effectiveness) will dominate in OI partner selection. According to Sarkar et al. [40], if the OI aim is focused on achieving product novelty or technological superiority, the technological complementarity criteria should be prioritized. If the emphasis is on accelerating and smoothing the innovation process and establishing a supportive collaborative environment, nontechnical criteria are heavily weighted.

3. Materials and Methods

The study contributes to the development of an improved decision-making methodology framework for OI partner selection, and the following are some benefits that are represented in it:

(i) For the first time, a comprehensive framework of OI partner performance indicators is provided, consisting of five critical dimensions and twenty-seven indicators, encompassing all relevant technological, operational, and strategic evaluation aspects.
(ii) The novel hybrid IT2F MCDM model combining IT2FD, IT2F AHP, and IT2F PROMETHEE is established, which provides a contingent approach to identifying OI partner evaluation criteria considering the nature of the company's innovation processes, contextual conditions, and innovation strategy; and yields precise multi-criteria alternatives evaluation under a high uncertainty level.

This results in a structured and methodology-supported, five-stage decision-making process for OI partner selection. In addition, by identifying the key OI partners' performance indicators, the study offers valuable guidance for managers in generating management strategies and best practices to maximize synergy of collaboration in OI projects.

The study primarily provides a comprehensive framework that encompasses all relevant technological, operational, and strategic aspects of OI partner evaluation, ensuring a holistic approach to the problem. This leads to a more integrated and coherent list of potential OI partner evaluation criteria (Table 2). The research process included a review of the existing methodological frameworks for decision support in the selection of OI partners, studies on inter-organizational knowledge transfer and generation, as well as studies on inter-organizational relationships, from which the key indicators that model the quality of cooperation among OI partners have been identified. As a result, five critical dimensions were identified, including (i) technological competencies; (ii) resource complementarity; (iii) financial terms of collaboration; (iv) cooperative capability; and (v) strategic alignments. A comprehensive set of twenty-seven indicators of OI partner performance covering these five dimensions was established.

In parallel with the literature analysis, as a second part of the research process, interviews were conducted with 18 managers with at least two years of experience in managing OI projects and an engineering or business academic background. The managers were asked to state the factors that, in their opinion, had the most significant influence on the quality of partner cooperation during the OI projects they were involved in. After revising the collected statements and rationalizing the different nomenclature, it was concluded that the indicators recommended by the practitioners are already included in the literature review-defined list.

Table 2. The potential OI partner evaluation criteria list.

Criteria		Studies
Technological competencies	Technological innovation level	[41,42]
	Technological complementarity	[34]
	Product experience	[43]
	Number of patents held	[27]
	Expected capabilities of abstraction	[43]
	Technology transfer capability	[44]
Resource complementarity	Overlapping knowledge base	[35]
	Product-specific knowledge	[45]
	Market knowledge complementarity	[35,45]
	Expected knowledge maturity	[43]
	Past experiences	[34]
	Financial assets	[45]
Financial terms	Expected debt ratio and refund ability	[46]
	Financial resources demand of the project	[34]
	Return of investment	[46]
Cooperative capability	Collaborative behavior	[28]
	Mutual trust and commitment	[35,39,45,47,48]
	Management and organizational culture	[34,35,42,46,49,50]
	Previous relationship	[28]
	Propensity to change	[35]
	Geographical proximity	[28,43]
	Symmetry of scale and scope	[45,46]
Strategic alignments	Compatibility of corporation strategies	[46]
	Convergence of expectations between partners	[34]
	Motivation and goal correspondence	[35,50]
	Strategic objectives of intellectual property management	[28]
	Market complementarity	[34]

The proposed methodology framework is implemented in a five-phase hybrid process (Figure 1) combining the advantages of three methods, namely Delphi, the Analytical Hierarchy Process (AHP), and the Preference Ranking Organization METHod for Enrichment of Evaluations (PROMETHEE). As far as we know, this is the first time the proposed hybrid MCDM model has been applied.

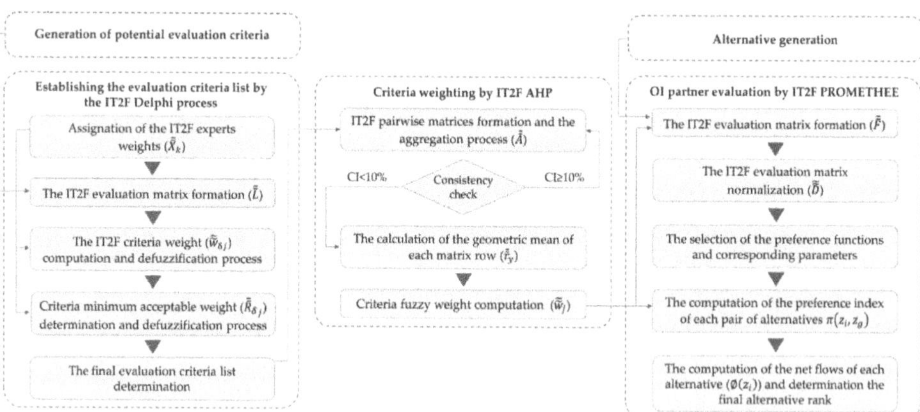

Figure 1. The proposed methodology framework.

The synthesis of the conventional Delphi approach and the IT2F set theory (IT2FD) enables the inclusion of expert subjective judgments, enabling a more comprehensive and inclusive approach to elicit the most essential evaluation criteria from the list of potential ones (Table 2), depending on the nature of the company's innovation processes as well as the contextual conditions and the corporate innovation strategy. The proposed IT2FD process results in a contingent and reliable set of OI partner performance indicators, enhancing the validity and practicality of the chosen evaluation criteria in OI partner selection. It provides companies with a way to optimize the evaluation criteria and make consistent decisions.

The framework also provides a methodology for rational and reliable criteria weighting based on the vague linguistic evaluations of multiple experts by applying the IT2F AHP approach. The strength of this method is based on the ability to express preferences using linguistic statements, which is more similar to the human way of thinking, the concept of systematic pairwise criteria comparison, and a mathematically simple synthesis of the obtained results for deriving criteria weights. Additionally, the framework incorporates the methodology for multi-criteria evaluation and ranking of alternative OI partners based on the IT2F PROMETHEE approach, which is suitable for solving complex decision-making problems that require a range of human perceptions and judgments, especially when there are significant differences in participants' perceptions of the decision-making process. Moreover, the criteria properties are treated in a proper way by involving different types of preference functions as well as the associated parameters for each criterion.

The conventional forms of the MCDM methods, modeled with crisp input data, are incompatible with the human thinking process, rendering them inapplicable under the highly vague and uncertain decision environment derived by the subjective nature of the preferences, the impossibility of expressing preference relations using crisp measures, the hesitation of human perception, and a lack of quantitative criteria.

A typical MCDM problem involves several qualitative and quantitative measurements. Given the fact that these measurements are frequently impossible to be precisely presented and precisely anticipated based upon objective pieces of information or directly and analytically explained, they will instead be based on the objective assessments made by the representatives of personalized types of knowledge and experiences specific to a particular criterion. Modeling the uncertainties and imprecisions that arise in that case in this study is performed by the application of the mathematical models developed in an IT2F environment.

To efficiently resolve the ambiguity frequently arising from the available pieces of information and to do more justice to the essential vagueness in human judgment and preference, fuzzy set theory is used to establish ill-defined MCDM problems [51]. Indubitably, the value of the MCDM methods will be improved if the properties of human adaptively, intransitivity, and dynamic adjustment of preferences can be considered in the decision process [52]. Fuzzy set theory is oriented towards the conversion of human perceptions given as linguistic statements into an arithmetical form by representing vague data using fuzzy numbers [51].

The literature notes a large number of different MCDM models integrated with fuzzy sets theory used for modeling uncertain and imprecise data in different decision-making problems [53–60]. Moreover, some studies [26,61] that provide insight into research in the field of MCDM encompassing the application of IT2F sets, confirm the high level of use of IT2F sets-based MCDM models in the domains of engineering and management, assuming that the trend in research in IT2F MCDM will remain stable in the future.

3.1. Multi-Criteria Decision-Making and Interval Type-2 Fuzzy Sets

Due to the power of fuzzy logic to overcome the problems of indeterminacy and inconsistency, fuzzy sets are used in decision-making processes in which their application allows decision-makers to convert the linguistic terms or responses to be evaluated with a degree of certainty [62]. To accommodate ambiguity, fuzzy sets allow for membership inside an interval between two real values. Accordingly, fuzzy sets allow decision-making problems to be resolved in a more flexible and precise manner.

The choice the shape of the membership function can be considered a problem in itself. A special case of generalized type-2 fuzzy IT2F sets has been seen as the most useful since they are more manageable in terms of calculations [63]. Moreover, IT2F sets are often chosen as a viable alternative due to their numerous superiorities over conventional type-1 fuzzy sets. Namely, in IT2F sets, linguistic statements are modeled more efficiently in comparison to type-1 fuzzy sets, which are defined with a two-dimensional membership function, while the IT2F set membership function is three-dimensional, providing additional degrees of freedom for better dealing with vague data. Therefore, type-2 fuzzy sets are proposed [64] as more applicable to real-life decision-making problems.

IT2F sets have been successfully implemented with MCDM methods that involve expressing decision-makers' preferences using a linguistic scale to fully describe the inherent uncertainties, and make it more convenient for applying in a highly vague and uncertain decision-making environment. In the past two decades, research in the field of IT2F MCDM has experienced intense growth. So far, a large number of studies have been focused on the development and implementation of IT2F MCDM models for real-world problems in an uncertain and ambiguous environment.

Mathew et al. [55] introduced the IT2F MCDM model based on the AHP and the Technique for Order of Preference by Similarity to Ideal Solution (TOPSIS), which can effectively handle the degree of uncertainty in group decision-making process of selecting the optimal industrial asset maintenance strategy. Ecer [56] utilized an extension to AHP under an IT2F environment to better cope with ambiguity for supplier selection, considering green concepts. The study [25] employed the same approach, aimed at supporting the effective selection of partners for collaborative technological R&D projects. Gölcük [65] introduced a novel risk assessment model by combining the IT2F Best–Worst Method (BWM) with perceptual reasoning for the evaluation of risk in digital transformation projects. Wu et al. [66] proposed an investment decision-making framework based on IT2F sets and the PROMETHEE-II model. In the study [67], the IT2F-based MCDM approach established by integrating TOPSIS and Decision-Making Trial and Evaluation Laboratory (DEMATEL) methods is utilized for the SWOT-based strategy selection problem by means of preparatory efforts to develop a renewed strategic plan for the industrial engineering department. Boral et al. [68] suggested a novel integrated framework comprising IT2F AHP, IT2F DEMATEL, and IT2F Measurement of Alternatives and Ranking according to COmpromise Solution (MARCOS) for prioritizing the risks associated with human error in the context of Failure Modes and Effects Analysis (FMEA)-based risk analysis approach. Bera et al. [69] used Multi-Objective Optimization on the basis of Ratio Analysis (MOORA) and TOPSIS methods in an IT2F environment for supplier selection, considering both subjective and objective factors. Karagöz et al. [70] utilized an extension of the Additive Ratio ASsessment (ARAS) method under the IT2F environment for solving the end-of-life vehicle recycling facility location problem. The study [71] constructs a barrier evaluation framework for forest carbon sink project implementation by introducing a hybrid MCDM model encompassing the BWM and PROMETHEE II in the IT2F environment.

3.2. Interval Type-2 Fuzzy Sets

The interval type-2 fuzzy (IT2F) sets, first introduced by Zadeh [64], are defined by the interval, a three-dimensional membership function that is fuzzy by itself, which makes it much more competent for modeling the ambiguities inherent to MCDM problems since it is described by both the primary and the secondary membership functions, which provides a higher degree of flexibility. According to Mendel and John [72], the main sources of these ambiguities include:

- The meanings of the used linguistic terms and the consequences of the rules can be uncertain;
- Consequents may have a histogram of the values associated with them, especially when knowledge is extracted from a group of experts who do not have a unified attitude;
- The measurements that activate type-1 fuzzy logic may be uncertain;
- The data used to tune the parameters of the type-1 fuzzy logic system may be noisy.

According to Aleksic and Tadic [26], handling uncertainties by using type-2 numbers implies making fewer assumptions, which results in more realistic solutions to real-life decision-making problems.

In the following, a brief review of some definitions of IT2F sets is presented [63].

Definition 1. *Let* $(\overset{\approx}{A})$ *be a type-2 fuzzy number characterized by the membership function* $\mu_{\overset{\approx}{A}}(x,u)$:

$$\overset{\approx}{A} = \left\{ \left((x,u), \mu_{\overset{\approx}{A}}(x,u) \right) \forall x \in X,\ \forall u \in J_x \subseteq [0,1], 0 \le \mu_{\overset{\approx}{A}}(x,u) \le 1 \right\} \tag{1}$$

which can also be interpreted as in Equation (2).

$$\begin{gathered} \overset{\approx}{A} = \int_{x\in X}\int_{u\in J_x} \frac{\mu_{\overset{\approx}{A}}(x,u)}{(x,u)}, J_x \subseteq [0,1] = \int_{x\in X}\left[\int_{u\in J_x} \frac{\mu_{\overset{\approx}{A}}(x,u)}{u}\right]/x \\ UMF\left(\overset{\approx}{A}\right) = sup\left\{u \middle| u \in [0,1], \mu_{\overset{\approx}{A}}(x,u) > 0\right\}, \forall x \in X \\ LMF\left(\overset{\approx}{A}\right) = inf\left\{u \middle| u \in [0,1], \mu_{\overset{\approx}{A}}(x,u) > 0\right\}, \forall x \in X \end{gathered} \tag{2}$$

Definition 2. *If it is further assumed that each* $\mu_{\overset{\approx}{A}}(x,u)$ *is equal to 1, then* $\overset{\approx}{A}$ *can be considered as an IT2F number which can be interpreted as in Equations (3) and (4).*

$$\overset{\approx}{A} = \int_{x\in X}\int_{u\in J_x} \frac{1}{(x,u)},\ J_x \subseteq [0,1] \tag{3}$$

$$\begin{gathered} \overset{\approx}{A} = \left(\overset{\sim U}{A}, \overset{\sim L}{A}\right) \\ = \left(a_1^U, a_2^U, a_3^U, a_4^U; H_1\left(\overset{\sim U}{A}\right), H_2\left(\overset{\sim U}{A}\right)\right)\left(a_1^L, a_2^L, a_3^L, a_4^L; H_1\left(\overset{\sim L}{A}\right), H_2\left(\overset{\sim L}{A}\right)\right) \end{gathered} \tag{4}$$

- $a_1^U, a_2^U, a_3^U, a_4^U, a_1^L, a_2^L, a_3^L, a_4^L$—*reference points of IT2F number* $\overset{\approx}{A}$,
- $H_j\left(\overset{\sim U}{A}\right) \in [0,1]\ 1 \le j \le 2$—*value of* $a_{(j+1)}^U$ *in upper trapezoidal membership function.*
- $H_j\left(\overset{\sim L}{A}\right) \in [0,1];\ 1 \le j \le 2$—*value of* $a_{(j+1)}^L$ *in lower trapezoidal membership function.*

which can graphically be interpreted as in Figure 2.

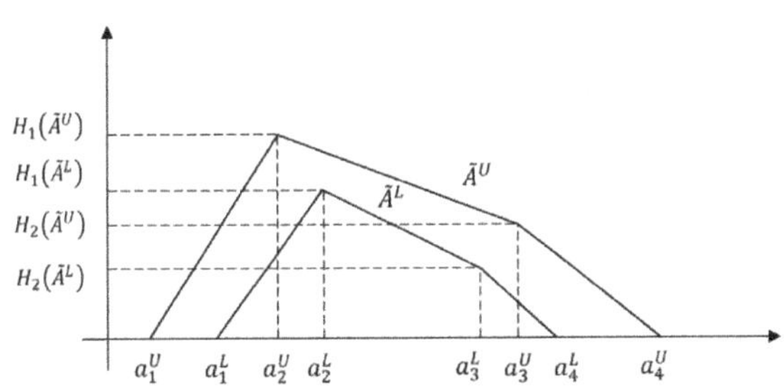

Figure 2. The IT2F number.

Let $\overset{\approx}{A_1} = \left(\overset{\sim}{A_1}^U, \overset{\sim}{A_1}^L\right)$ and $\overset{\approx}{A_2} = \left(\overset{\sim}{A_2}^U, \overset{\sim}{A_2}^L\right)$ be two IT2F numbers, whose form is interpreted in Equation (4), then the arithmetic operations between them can be defined as follows:

Definition 3. *The addition operation between two IT2F numbers $\overset{\approx}{A_1}$ and $\overset{\approx}{A_2}$ is defined as in Equation (5)*

$$\overset{\approx}{A_1} \oplus \overset{\approx}{A_2} = \left(\overset{\sim}{A_1}^U, \overset{\sim}{A_1}^L\right) \oplus \left(\overset{\sim}{A_2}^U, \overset{\sim}{A_2}^L\right)$$
$$= \left(\begin{matrix} a_{11}^U + a_{21}^U,\ a_{12}^U + a_{22}^U,\ a_{13}^U + a_{23}^U,\ a_{14}^U + a_{24}^U; \\ min\left(H_1\left(\overset{\sim}{A_1}^U\right), H_1\left(\overset{\sim}{A_2}^U\right)\right), min\left(H_2\left(\overset{\sim}{A_1}^U\right), H_2\left(\overset{\sim}{A_2}^U\right)\right)\end{matrix}\right), \left(\begin{matrix} a_{11}^L + a_{21}^L,\ a_{12}^L + a_{22}^L,\ a_{13}^L + a_{23}^L,\ a_{14}^L + a_{24}^L; \\ min\left(H_1\left(\overset{\sim}{A_1}^L\right), H_1\left(\overset{\sim}{A_2}^L\right)\right), min\left(H_2\left(\overset{\sim}{A_1}^L\right), H_2\left(\overset{\sim}{A_2}^L\right)\right)\end{matrix}\right) \quad (5)$$

Definition 4. *The Subtraction operation between two IT2F numbers $\overset{\approx}{A_1}$ and $\overset{\approx}{A_2}$ is defined as in Equation (6)*

$$\overset{\approx}{A_1} \ominus \overset{\approx}{A_2} = \left(\overset{\sim}{A_1}^U, \overset{\sim}{A_1}^L\right) \ominus \left(\overset{\sim}{A_2}^U, \overset{\sim}{A_2}^L\right)$$
$$= \left(\begin{matrix} a_{11}^U - a_{21}^U,\ a_{12}^U - a_{22}^U,\ a_{13}^U - a_{23}^U,\ a_{14}^U - a_{24}^U; \\ min\left(H_1\left(\overset{\sim}{A_1}^U\right), H_1\left(\overset{\sim}{A_2}^U\right)\right), min\left(H_2\left(\overset{\sim}{A_1}^U\right), H_2\left(\overset{\sim}{A_2}^U\right)\right)\end{matrix}\right), \left(\begin{matrix} a_{11}^L - a_{21}^L,\ a_{12}^L - a_{22}^L,\ a_{13}^L - a_{23}^L,\ a_{14}^L - a_{24}^L; \\ min\left(H_1\left(\overset{\sim}{A_1}^L\right), H_1\left(\overset{\sim}{A_2}^L\right)\right), min\left(H_2\left(\overset{\sim}{A_1}^L\right), H_2\left(\overset{\sim}{A_2}^L\right)\right)\end{matrix}\right) \quad (6)$$

Definition 5. *The multiplication operation between two IT2F numbers $\overset{\approx}{A_1}$ and $\overset{\approx}{A_2}$ is defined as in Equation (7)*

$$\overset{\approx}{A_1} \otimes \overset{\approx}{A_2} = \left(\overset{\sim}{A_1}^U, \overset{\sim}{A_1}^L\right) \otimes \left(\overset{\sim}{A_2}^U, \overset{\sim}{A_2}^L\right)$$
$$= \left(\begin{matrix} a_{11}^U \cdot a_{21}^U,\ a_{12}^U \cdot a_{22}^U,\ a_{13}^U \cdot a_{23}^U,\ a_{14}^U \cdot a_{24}^U; \\ min\left(H_1\left(\overset{\sim}{A_1}^U\right), H_1\left(\overset{\sim}{A_2}^U\right)\right), min\left(H_2\left(\overset{\sim}{A_1}^U\right), H_2\left(\overset{\sim}{A_2}^U\right)\right)\end{matrix}\right), \left(\begin{matrix} a_{11}^L \cdot a_{21}^L,\ a_{12}^L \cdot a_{22}^L,\ a_{13}^L \cdot a_{23}^L,\ a_{14}^L \cdot a_{24}^L; \\ min\left(H_1\left(\overset{\sim}{A_1}^L\right), H_1\left(\overset{\sim}{A_2}^L\right)\right), min\left(H_2\left(\overset{\sim}{A_1}^L\right), H_2\left(\overset{\sim}{A_2}^L\right)\right)\end{matrix}\right) \quad (7)$$

Definition 6. *The arithmetic operation between crisp value s and an IT2F number $\overset{\approx}{A_1}$ is defined as in Equation (8).*

$$s \otimes \overset{\approx}{A_1} = s \otimes \left(\overset{\sim}{A_1}^U, \overset{\sim}{A_1}^L\right) = \left(\begin{matrix} \left(s \cdot a_{11}^U, s \cdot a_{12}^U, s \cdot a_{13}^U, s \cdot a_{14}^U; H_1\left(\overset{\sim}{A_1}^U\right), H_2\left(\overset{\sim}{A_1}^U\right)\right), \\ \left(s \cdot a_{11}^L, s \cdot a_{12}^L, s \cdot a_{13}^L, s \cdot a_{14}^L; H_1\left(\overset{\sim}{A_1}^L\right), H_2\left(\overset{\sim}{A_1}^L\right)\right)\end{matrix}\right) \quad (8)$$

Definition 7. *The reciprocal operation of an IT2F number $\overset{\approx}{A_1}$ is defined as in Equation (9).*

$$\left(\overset{\approx}{A_1}\right)^{-1} = \left(\left(\frac{1}{a_{14}^U}, \frac{1}{a_{13}^U}, \frac{1}{a_{12}^U}, \frac{1}{a_{11}^U}; H_1\left(\overset{\sim}{A_1}^U\right), H_2\left(\overset{\sim}{A_1}^U\right)\right)\left(\frac{1}{a_{14}^L}, \frac{1}{a_{13}^L}, \frac{1}{a_{12}^L}, \frac{1}{a_{11}^L}; H_1\left(\overset{\sim}{A_1}^L\right), H_2\left(\overset{\sim}{A_1}^L\right)\right)\right) \quad (9)$$

3.3. Interval Type-2 Delphi Model

The Delphi method is a formal communication strategy or approach originally conceived as a systematic interactive predictive process based on an expert panel [73]. Today, it is extensively used as part of hybrid MCDM models, to identify critical decision factors. The study employs the IT2FD-based approach for eliciting the most important evaluation criteria.

The synthesis of type-1 fuzzy set theory and conventional Delphi is often used as a part of the hybrid MCDM frameworks to resolve the ambiguity and vagueness of an expert's judgments, which are issues that the conventional Delphi approach has always suffered from. However, it can be noticed that the extensions of the Delphi method in the context of IT2F sets have not been taken much into consideration and is still in their infancy. There are only a few studies suggesting the application of an IT2FD-based approach, for instance; Shringi et al. [62] developed a hybrid IT2F Delphi-AHP model for analyzing critical factors for effective knowledge acquisition in construction safety training. Deveci et al. [74] introduced the IT2FD-based approach to rank indicators affecting site selection of vehicle shredding facilities. In study [75], the critical competencies for lifelong learning were assessed using the fuzzy model for sustainable education, whereby the IT2FD approach was employed to aggregate students' opinions into unique marks, during the assessment process. While Ayyildiz et al. [76] utilized IT2FD to determine the most important criteria that affect the credit evaluation process.

In order to provide a more intuitive and convenient way to address uncertain and ambiguous information during the Delphi process, the paper suggests an extension of the Delphi model in the context of IT2F sets. In this regard, the modified fuzzy Delphi model proposed by Gupta et al. [77] is used, in which IT2F sets are used instead of triangular type-1 fuzzy sets.

Since the experts ($E_k, k = 1, 2, \ldots, K$), engaged in the Delphi process participate in different phases of the innovation process (IP) and have different experiences, qualifications, and designations, their judgments should be assigned different weights. For instance, the opinion of the expert with more experience, a higher designation, or more qualifications could be considered more trustworthy; therefore, the weight factors ($\overset{\approx}{X}_k$) will be assigned to the experts on this basis. The weight factor reflects the expert's competencies for dealing with the considered problem. The linguistic variables describing experts' experience, qualification, designation, and the phase of the innovation process (IP) they are involved in (which will be used as expert evaluation criteria in this study) can be quantified using IT2F numbers, according to Table 3. The weight factor for each expert is then formed as an aggregation of these variables and represents the arithmetic mean of the assigned IT2F numbers.

Table 3. IT2F scale for the expert evaluation criteria.

Experience	Qualification	Designation	IP Phase	Linguistic Variables	IT2F Numbers
≤5	Under graduate	Up to executive	Launch and market penetration	Low	(0, 0.1, 0.15, 0.3; 1, 1) (0.05, 0.1, 0.15, 0.2; 0.9, 0.9)
5–10	Graduate	Executive to Specialist	Idea generation	Medium	(0.3, 0.5, 0.55, 0.7; 1, 1) (0.4, 0.5, 0.55, 0.6; 0.9, 0.9)
10–15	Master graduation	Specialist to Manager	Concept development	High	(0.7, 0.85, 0.9, 1; 1, 1) (0.8, 0.85, 0.9, 0.95; 0.9, 0.9)
≥15	Post graduate	Manager to GM	Product development	Very high	(0.9, 1, 1, 1; 1, 1) (0.95, 1, 1, 1; 0.9, 0.9)

In the experts' opinions ($\overset{\approx}{I}_{jk}$) on the importance of considering each of the identified criteria (C_j, $j = 1, 2, \ldots, n$), an IT2F evaluation matrix is established:

$$\tilde{\tilde{L}} = \left(\tilde{\tilde{l}}_{jk}\right)_{nxK} = \begin{matrix} & E_1 & E_2 & \cdots & E_K \\ & \tilde{\tilde{X}}_1 & \tilde{\tilde{X}}_2 & \cdots & \tilde{\tilde{X}}_K \\ C_1 & \tilde{\tilde{l}}_{11} & \tilde{\tilde{l}}_{12} & \cdots & \tilde{\tilde{l}}_{1K} \\ C_2 & \tilde{\tilde{l}}_{21} & \tilde{\tilde{l}}_{22} & \cdots & \tilde{\tilde{l}}_{2K} \\ \vdots & \vdots & \vdots & & \vdots \\ C_n & \tilde{\tilde{l}}_{n1} & \tilde{\tilde{l}}_{n2} & \cdots & \tilde{\tilde{l}}_{nK} \end{matrix} \quad (10)$$

Those opinions are expressed according to the scale accounted for in Table 4.

Table 4. IT2F scale.

Linguistic Variables	IT2F Numbers
Very Low (VL)	(0, 0, 0, 0.01; 1, 1) (0, 0, 0, 0.05; 0.9, 0.9)
Low (L)	(0, 0.1, 0.15, 0.3; 1, 1) (0.05, 0.1, 0.15, 0.2; 0.9, 0.9)
Medium Low (ML)	(0.1, 0.3, 0.35, 0.5; 1, 1) (0.2, 0.3, 0.35, 0.4; 0.9, 0.9)
Medium (M)	(0.3, 0.5, 0.55, 0.7; 1, 1) (0.4, 0.5, 0.55, 0.6; 0.9, 0.9)
Medium High (MH)	(0.5, 0.7, 0.75, 0.9; 1, 1) (0.6, 0.7, 0.75, 0.8; 0.9, 0.9)
High (H)	(0.7, 0.85, 0.9, 1; 1, 1) (0.8, 0.85, 0.9, 0.95; 0.9, 0.9)
Very High (VH)	(0.9, 1, 1, 1; 1, 1) (0.95, 1, 1, 1; 0.9, 0.9)

According to Gupta et al. [77], the average weight for each criterion ($\tilde{\tilde{w}}_{\delta j}$) could be determined by Equation (11).

$$\tilde{\tilde{w}}_{\delta j} = \frac{\sum_{k=1}^{K} \tilde{\tilde{X}}_k \otimes \tilde{\tilde{l}}_{jk}}{K} \quad (11)$$

The defuzzification of IT2F weights in this study is performed by the Center of Area (COA) method, providing the Best Nonfuzzy Performance (BNP) value, as suggested in [78], the BNP value can be obtained as in Equation (12).

$$w_{\delta j} = \frac{\int xu(x)dx}{\int u(x)dx} = \frac{-w_{\delta j1} \cdot w_{\delta j2} + w_{\delta j3} \cdot w_{\delta j4} + \frac{1}{3}\left(w_{\delta j4} - w_{\delta j3}\right)^2 - \frac{1}{3}\left(w_{\delta j2} - w_{\delta j1}\right)^2}{-w_{\delta j1} - w_{\delta j2} + w_{\delta j3} + w_{\delta j4}} \quad (12)$$

where $w_{j1,2,3 \text{ and } 4}$ represents the arithmetic mean of the upper and lower boundaries of the IT2F weight ($\tilde{\tilde{w}}_{\delta j}$).

The computation of the criterion minimum acceptable weight ($\tilde{\tilde{R}}_{\delta j}$) is determined by Equation (13), where R_k stands for the minimum acceptable criterion weight denoted as a percentage expressed by the kth expert. This variable can be defuzzified by Equation (12). The criteria whose weights ($w_{\delta j}$) are lower than the estimated minimum acceptable weight ($R_{\delta j}$) are omitted from the further evaluation procedure.

$$\tilde{\tilde{R}}_{\delta j} = \frac{\sum_{k=1}^{K} \tilde{\tilde{X}}_k \otimes R_k}{K} \quad (13)$$

3.4. *Interval Type-2 AHP Model*

The Analytical Hierarchy Process (AHP) method is a method of hierarchical weight decision analysis introduced by Saaty [79]. Based upon the pairwise comparison of a set of objects, the AHP is performed so as to elicit a corresponding priority vector that indicates preferences. The synthesis of the fuzzy set and the AHP method has successfully been applied in modeling diverse engineering and management problems, such as renewable energy project portfolio optimization [80], optimal maintenance strategy selection [55], green supplier selection [56], resilient supplier selection [81], partner selection in collaborative technological R&D projects [25], and supplier selection in the era of Industry 4.0 [82].

The first step within the criteria weighting process based upon the IT2F AHP methodology implies the establishment of a pairwise comparison matrix ($\overset{\approx}{A} = \left(\overset{\approx}{a}_{yj}\right)_{n\times n}$) among all the criteria $\left(C_j, j = 1,2,\ldots,n\right)$:

$$\overset{\approx}{A} = \left(\overset{\approx}{a}_{yj}\right)_{n\times n} = \begin{matrix} \\ C_1 \\ C_2 \\ \vdots \\ C_n \end{matrix} \begin{matrix} \begin{matrix} C_1 & C_2 & \cdots & C_n \end{matrix} \\ \begin{bmatrix} 1 & \overset{\approx}{a}_{12} & \cdots & \overset{\approx}{a}_{1n} \\ \overset{\approx}{a}_{21} & 1 & \cdots & \overset{\approx}{a}_{2n} \\ \vdots & \vdots & \ddots & \vdots \\ \overset{\approx}{a}_{n1} & \overset{\approx}{a}_{n2} & \cdots & 1 \end{bmatrix} \end{matrix} = \begin{bmatrix} 1 & \overset{\approx}{a}_{12} & \cdots & \overset{\approx}{a}_{1n} \\ \frac{1}{\overset{\approx}{a}_{12}} & 1 & \cdots & \overset{\approx}{a}_{2n} \\ \vdots & \vdots & \ddots & \vdots \\ \frac{1}{\overset{\approx}{a}_{1n}} & \frac{1}{\overset{\approx}{a}_{2n}} & \cdots & 1 \end{bmatrix} \tag{14}$$

The matrix elements ($\overset{\approx}{a}_{yj}$) refer to the preference of the criterion y over the criterion j determined by the experts involved in the prioritization process. In the first phase, preferential relationships are expressed by means of linguistic statements. In the next phase, it is converted to IT2F numbers by mapping on an IT2F scale. The IT2F numbers representing the linguistic statements used in the criteria weighting process in this study are introduced in Table 5 and graphically presented in Figure 3.

Table 5. IT2F scale for criteria weighting process.

Linguistic Statements	IT2F Numbers
Absolutely Strong (AS)	(7, 8, 9, 9; 1, 1) (7.2, 8.2, 8.8, 9; 0.8, 0.8)
Very Strong (VS)	(5, 6, 8, 9; 1, 1) (5.2, 6.2, 7.8, 8.8; 0.8, 0.8)
Fairly Strong (FS)	(3, 4, 6, 7; 1, 1) (3.2, 4.2, 5.8, 6.8; 0.8, 0.8)
Slightly Strong (SS)	(1, 2, 4, 5; 1, 1) (1.2, 2.2, 3.8, 4.8; 0.8, 0.8)
Exactly Equal (EE)	(1, 1, 1, 1; 1, 1) (1, 1, 1, 1; 1, 1)

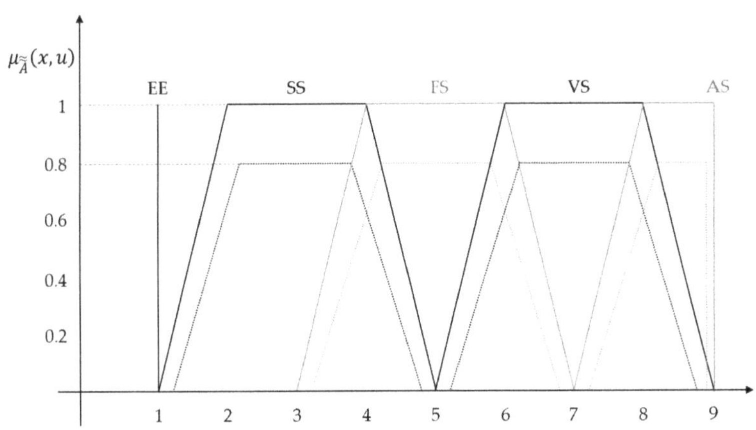

Figure 3. IT2F membership functions for linguistic statements used in criteria weighting process.

A consistency test is used to reveal the inconsistency within the established pairwise comparison matrix. In order to verify pairwise comparison matrix consistency, the consistency rate (CR) is introduced, and its value is obtained by Equation (15).

$$CR = \frac{CI}{RI} \tag{15}$$

where CI is a consistency index obtained by Equation (16)

$$CI = \frac{\lambda_{max} - n}{n - 1} \tag{16}$$

and λ_{max} is the largest eigenvalue of matrix $\overset{\approx}{A}$.

The RI is random index, and its value being dependent on the order of the matrix (see [79]).

If the $CR < 0.1$, then the pairwise comparison matrix is acceptable, otherwise, the matrix must be reformed.

If the criteria pairwise comparison is performed as a group process, then the individual pairwise comparison matrices can be aggregated using Equations (17) and (18), where $\overset{\approx k}{a}_{yj}$ is the criteria preferential relation expressed by the kth expert.

$$\overset{\approx}{a}_{yj} = \left(\prod_{k=1}^{K} \overset{\approx k}{a}_{yj}\right)^{\frac{1}{K}} = \left[\overset{\approx 1}{a}_{yj} \otimes \overset{\approx 2}{a}_{yj} \otimes \ldots \otimes \overset{\approx K}{a}_{yj}\right]^{\frac{1}{K}} \tag{17}$$

$$\sqrt[K]{\overset{\approx k}{a}_{yj}} = \begin{pmatrix} \sqrt[K]{a^U_{yj1}}, \sqrt[K]{a^U_{yj2}}, \sqrt[K]{a^U_{yj3}}, \sqrt[K]{a^U_{yj4}}; H^U_1(a_{yj}); H^U_2(a_{yj}), \\ \sqrt[K]{a^L_{yj1}}, \sqrt[K]{a^L_{yj2}}, \sqrt[K]{a^L_{yj3}}, \sqrt[K]{a^L_{yj4}}; H^L_1(a_{yj}); H^L_2(a_{yj}) \end{pmatrix} \tag{18}$$

There are several approaches to the generation of priorities from the pairwise comparison relations including the Least Squares (LS) method [83]; the geometric mean method [84]; the Logarithmic Least Squares (LLS) method [85]; the extent analysis method [86]; goal programming [87]; the Fuzzy Preference Programming (FPP) method [88]; the linear programming method [89]; the least deviation method [90]; the Weighted Least Square (WLS) and the quadratic programming methods [91].

A modification of the Buckley [84] fuzzy AHP model will be applied to generate criteria weights from the pairwise comparison matrix, whereby the modified model uses IT2F sets instead of the trapezoidal type-1 fuzzy set as it is more accurate in uncertainty modeling, due to the membership function, which is fuzzy by itself. This includes the generation of the fuzzy geometric mean ($\overset{\approx}{r}_y$) for each matrix row using the geometric mean technique as follows:

$$\overset{\approx}{r}_y = \left(\prod_{y=1}^{n} \overset{\approx}{a}_{yj}\right)^{\frac{1}{n}} = \left[\overset{\approx}{a}_{y1} \otimes \overset{\approx}{a}_{y2} \otimes \ldots \otimes \overset{\approx}{a}_{yn}\right]^{\frac{1}{n}} \tag{19}$$

The IT2F weights ($\overset{\approx}{w}_j$) are obtained by the fuzzy geometric mean ($\overset{\approx}{r}_y$) as follows:

$$\overset{\approx}{w}_j = \overset{\approx}{r}_y \otimes \left[\overset{\approx}{r}_1 \oplus \ldots \oplus \overset{\approx}{r}_y \oplus \ldots \overset{\approx}{r}_n\right]^{-1} \tag{20}$$

The non-fuzzy weights of C_j are obtained in the same manner as in Equation (12).

3.5. Interval Type-2 PROMETHEE Model

The Preference Ranking Organization METHod for Enrichment of Evaluations (PROMETHEE) [92] is a widely used outranking method that enables aggregation of the alternative evaluations established based on multiple, often conflicting criteria. The paper proposes using the extension of the PROMETHEE method in the context of the IT2F set. The majority of so-far-used IT2F MCDM models could be characterized as scoring or compromising models, whereas the extended outranking methods have not been thoroughly investigated. A literature review reveals only a few studies using the PROMETHEE method in the IT2F environment. For instance, Chen [93] established the PROMETHEE model that used signed distance-based generalized criteria and comprehensive preference indices in the IT2F set environment, while Wu et al. [66] used the IT2F PROMETHEE model to develop an investment decision-making framework. This model was also used as a part of the two-stage DM framework for the inland nuclear power plant site selection in synthesis with the GIS Wu et al. [94].

Let define the MCDM problem of m alternatives $(Z_i, i = 1, 2, \ldots, m)$ and n evaluation criteria $(C_j,\ j = 1, 2, \ldots, n)$. The fuzzy evaluation matrix is then defined as in Equation (21). The alternative performance $(\overset{\approx}{f}_{ij})$ is expressed using the scale provided in Table 4.

$$\overset{\approx}{F} = \left(\overset{\approx}{f}_{ij}\right)_{mxn} = \begin{matrix} \\ Z_1 \\ Z_2 \\ \vdots \\ Z_m \end{matrix} \begin{matrix} C_1 & C_2 & \cdots & C_n \\ \end{matrix} \begin{bmatrix} \overset{\approx}{f}_{11} & \overset{\approx}{f}_{12} & \cdots & \overset{\approx}{f}_{1n} \\ \overset{\approx}{f}_{21} & \overset{\approx}{f}_{22} & \cdots & \overset{\approx}{f}_{2n} \\ \vdots & \vdots & \vdots & \vdots \\ \overset{\approx}{f}_{m1} & \overset{\approx}{f}_{n1} & \cdots & \overset{\approx}{f}_{mn} \end{bmatrix} \tag{21}$$

$$\overset{\approx}{f}_{ij} = \left(\tilde{F}_{ij}^{U}, \tilde{F}_{ij}^{L}\right) = \left(f_{ij1}^{U}, f_{ij2}^{U}, f_{ij3}^{U}, f_{ij4}^{U}; H_1\left(\tilde{F}_{ij}^{U}\right), H_2\left(\tilde{F}_{ij}^{U}\right)\right)\left(f_{ij1}^{L}, f_{ij2}^{L}, f_{ij3}^{L}, f_{ij4}^{L}; H_1\left(\tilde{F}_{ij}^{L}\right), H_2\left(\tilde{F}_{ij}^{L}\right)\right) \tag{22}$$

In the next step, a normalized fuzzy evaluation matrix $(\overset{\approx}{D} = \left(\overset{\approx}{d}_{ij}\right)_{mxn})$ is established.

$$\overset{\approx}{d}_{ij} = \left(\tilde{D}_{ij}^{U}, \tilde{D}_{ij}^{L}\right) = \begin{cases} \left(\frac{f_{ij1}^{U}}{f_{max\,j}^{L}}, \frac{f_{ij2}^{U}}{f_{max\,j}^{L}}, \frac{f_{ij3}^{U}}{f_{max\,j}^{L}}, \frac{f_{ij4}^{U}}{f_{max\,j}^{L}}\right), \left(\frac{f_{ij1}^{L}}{f_{max\,j}^{L}}, \frac{f_{ij2}^{L}}{f_{max\,j}^{L}}, \frac{f_{ij3}^{L}}{f_{max\,j}^{L}}, \frac{f_{ij4}^{L}}{b_{max\,j}^{L}}\right), & if\ C_j \in C_I \\ \left(\frac{f_{min\,j}^{L}}{f_{ij4}^{U}}, \frac{f_{min\,j}^{L}}{f_{ij3}^{U}}, \frac{f_{min\,j}^{L}}{f_{ij2}^{U}}, \frac{f_{min\,j}^{L}}{f_{ij1}^{U}}\right), \left(\frac{f_{min\,j}^{L}}{f_{ij4}^{L}}, \frac{f_{min\,j}^{L}}{f_{ij3}^{L}}, \frac{f_{min\,j}^{L}}{f_{ij2}^{L}}, \frac{f_{min\,j}^{L}}{f_{ij1}^{L}}\right), & if\ C_j \in C_{II} \end{cases} \tag{23}$$

$$\begin{matrix} f_{max\,j}^{L} = max\left\{f_{ij}^{L} | i = 1, 2, \cdots, m\right\} \\ f_{min\,j}^{L} = min\left\{f_{ij}^{L} | i = 1, 2, \cdots, m\right\} \end{matrix} \quad C_I\ and\ C_{II}$$

C_I applies to benefit criteria and C_{II} to cost criteria.

The alternatives' outranking relations can be expressed by the preference function:

$$\overset{\approx}{P}_j(z_i, z_g) = \overset{\approx}{P}_j\left(\overset{\approx}{D}_{ij} - \overset{\approx}{D}_{gj}\right) \tag{24}$$

It is a non-falling function characterized as: $0 \leq \overset{\approx}{P}_j(z_i, z_g) \leq 1$ and $\overset{\approx}{P}_j(z_i, z_g) \neq \overset{\approx}{P}_j(z_g, z_i)$, therewith it acquiring the value 0 for $\overset{\approx}{D}_{ij} - \overset{\approx}{D}_{gj} < 0$. The preference function expresses the intensity of the preference of the alternative z_i over z_g, concerning the criterion C_j, which can be interpreted as:

- $\overset{\approx}{P}_j(z_i, z_g) = 0$—indifference—$\overset{\approx}{D}_{ij} = \overset{\approx}{D}_{gj}$
- $\overset{\approx}{P}_j(z_i, z_g) \sim 0$—weak preference—$\overset{\approx}{D}_{ij} > \overset{\approx}{D}_{gj}$
- $\overset{\approx}{P}_j(z_i, z_g) \sim 1$—strong preference—$\overset{\approx}{D}_{ij} >> \overset{\approx}{D}_{gj}$
- $\overset{\approx}{P}_j(z_i, z_g) = 1$—strict preference—$\overset{\approx}{D}_{ij} >>> \overset{\approx}{D}_{gj}$

The PROMETHEE method enables each decision criterion to be assigned a specific type of preference function by its characteristics, as well as the associated parameters that represent the intensity, limits, and speed of preference. The six different types of preference functions (Equations (25)–(30)) suggested by Brans and Vincke [92] can be used to express preferences in the majority of real-world problems.

Usual Criterion

$$\overset{\approx}{P}_j(z_i, z_g) = \begin{cases} 0, & \overset{\approx}{D}_{ij} - \overset{\approx}{D}_{gj} \leq 0 \\ 1, & \overset{\approx}{D}_{ij} - \overset{\approx}{D}_{gj} > 0 \end{cases} \tag{25}$$

U-shape Criterion

$$\overset{\approx}{P}_j(z_i, z_g) = \begin{cases} 0, & \overset{\approx}{D}_{ij} - \overset{\approx}{D}_{gj} \leq \overset{\approx}{q} \\ 1, & \overset{\approx}{D}_{ij} - \overset{\approx}{D}_{gj} > \overset{\approx}{q} \end{cases} \tag{26}$$

V-shape Criterion

$$\overset{\approx}{P}_j(z_i, z_g) = \begin{cases} 0, & \overset{\approx}{D}_{ij} - \overset{\approx}{D}_{gj} \leq 0 \\ \left(\overset{\approx}{D}_{ij} - \overset{\approx}{D}_{gj}\right)/\overset{\approx}{p}, & 0 < \overset{\approx}{D}_{ij} - \overset{\approx}{D}_{gj} \leq \overset{\approx}{p} \\ 1, & \overset{\approx}{D}_{ij} - \overset{\approx}{D}_{gj} > \overset{\approx}{p} \end{cases} \tag{27}$$

Level Criterion

$$\overset{\approx}{P}_j(z_i, z_g) = \begin{cases} 0, & \overset{\approx}{D}_{ij} - \overset{\approx}{D}_{gj} \leq \overset{\approx}{q} \\ 1/2, & \overset{\approx}{q} < \overset{\approx}{D}_{ij} - \overset{\approx}{D}_{gj} \leq \overset{\approx}{p} \\ 1, & \overset{\approx}{D}_{ij} - \overset{\approx}{D}_{gj} > \overset{\approx}{p} \end{cases} \tag{28}$$

V-shape with indifference Criterion

$$\overset{\approx}{P}_j(z_i, z_g) = \begin{cases} 0, & \overset{\approx}{D}_{ij} - \overset{\approx}{D}_{gj} \leq \overset{\approx}{q} \\ \left(\left(\overset{\approx}{D}_{ij} - \overset{\approx}{D}_{gj}\right) - \overset{\approx}{q}\right)/\left(\overset{\approx}{p} - \overset{\approx}{q}\right), & \overset{\approx}{q} < \overset{\approx}{D}_{ij} - \overset{\approx}{D}_{gj} \leq \overset{\approx}{p} \\ 1, & \overset{\approx}{D}_{ij} - \overset{\approx}{D}_{gj} > \overset{\approx}{p} \end{cases} \tag{29}$$

Gaussian Criterion

$$\overset{\approx}{P}_j(z_i, z_g) = \begin{cases} 0, & \overset{\approx}{D}_{ij} - \overset{\approx}{D}_{gj} \leq 0 \\ 1 - e^{-\left(\overset{\approx}{D}_{ij} - \overset{\approx}{D}_{gj}\right)^2/2\overset{\approx}{\delta}^2}, & \overset{\approx}{D}_{ij} - \overset{\approx}{D}_{gj} > 0 \end{cases} \tag{30}$$

where:

$$e^{\overset{\approx}{A}} = \left(\left(e^{a_1^U}, e^{a_2^U}, e^{a_3^U}, e^{a_4^U}; H_1\left(\tilde{A}^U\right), H_2\left(\tilde{A}^U\right)\right), \left(e^{a_1^L}, e^{a_2^L}, e^{a_3^L}, e^{a_4^L}; H_1\left(\tilde{A}^L\right), H_2\left(\tilde{A}^L\right)\right)\right) \tag{31}$$

In order to obtain the final alternatives rank, it is necessary to determine the preference index $\pi(z_i, z_g)$ which reflects the overall preference of the alternative z_i over the z_g concerning all the evaluation criteria, where $\overset{\approx}{w}_j$ is the relative weight of criterion C_j:

$$\pi(z_i, z_g) = \sum_{j=1}^{n} \overset{\approx}{w}_j \cdot \overset{\approx}{P}_j(z_i, z_g) \tag{32}$$

The exploitation of the obtained preferential relations to establish the rank of the alternatives includes the calculation of outgoing flows ($\varnothing^+(z_i)$) by Equation (33) and incoming flows ($\varnothing^{-}(z_i)$) using the Equation (34)—for partial ranking, and net flows ($\varnothing(z_i)$) by Equation (35)—for establishing the final rank of the alternative.

$$\varnothing^{+}(z_i) = \frac{1}{n-1}\sum_{z_x \in Z}^{n-1} \pi(z_i, z_x) \tag{33}$$

$$\varnothing^{-}(z_i) = \frac{1}{n-1}\sum_{z_x \in Z}^{n-1} \pi(z_x, z_i) \tag{34}$$

$$\varnothing(z_i) = \varnothing^{+}(z_i) - \varnothing^{-}(z_i) \tag{35}$$

4. Results

The case study has been conducted to verify the applicability and demonstrate the feasibility of the proposed hybrid IT2F MCDM model. The five-phase MCDM hybrid process has been conducted in accordance with the above-presented methodological framework (Figure 1). In this context, five alternative partners are identified to be evaluated. The decision-making process engaged a group of four experts: two university professors in the field of Engineering Management and two experts with experience in managing the OI projects.

4.1. Establishing the Evaluation Criteria List by IT2FD

At the first step, IT2F logic is used to assign weight factors ($\approx{X}_k$) to experts involved in the Delphi process (E_k). The experts' qualifications, experience designation, and IP phase they are involved in are used as evaluation criteria. This Delphi process included four experts, one from each of the following IP phases: idea generation, concept development, product development, and launch and market penetration phases.

The linguistic variables that describe these experts' experience, qualification, designation, and IP phase (given in Table 6) are quantified using IT2F numbers in accordance with Table 3. The arithmetic mean of the IT2F numbers assigned to the expert on this basis is used to determine its weight factor. The results are shown in Table 6.

Table 6. The experts' relative weights.

Expert	Experience	Qualification	Designation	IP Phase	IT2F Experts Weights ($\approx{X}_k$)
E_1	3	Master graduate	Specialist to Manager	Idea generation	(0.43, 0.58, 0.63, 0.75; 1, 1) (0.51, 0.58, 0.63, 0.68; 0.9, 0.9)
E_2	8	Postgraduate	Manager to GM	Product development	(0.75, 0.88, 0.89, 0.93; 1, 1) (0.81, 0.88, 0.89, 0.9; 0.9, 0.9)
E_3	15	Graduate	Specialist to Manager	Concept Development	(0.65, 0.8, 0.84, 0.93; 1, 1) (0.74, 0.8, 0.84, 0.88; 0.9, 0.9)
E_4	9	Graduate	Executive to Specialist	Launch and market penetration	(0.33, 0.49, 0.54, 0.68; 1, 1) (0.41, 0.49, 0.54, 0.59; 0.9, 0.9)

Each expert is asked through the questionnaire to evaluate the importance of each potential criterion from the list established in Section 3 (Table 2) by using the corresponding IT2F numbers. Those opinions are expressed according to the scale accounted for in Table 4, which resulted in an evaluation matrix (Table 7). The average criteria weights are generated from the established evaluation matrix, considering the experts' relative weights according to Equation (11). Along with the criteria filtering threshold, the minimum acceptable weight for each criterion is decided. The initial criteria are analyzed with regard to this threshold. According to the findings (Table 7), ten criteria are accepted, while the others are rejected since their significance weights are below the threshold.

Table 7. The results of the criteria evaluation.

Criteria	Aggregate Criteria Weight ($\tilde{\tilde{w}}_{\delta j}$)	Aggregate MAW ($\tilde{\tilde{R}}_{\delta j}$)	Criteria Weight	MAW	Selected Criteria
Technological innovation level	(0.46, 0.66, 0.71, 0.82; 1, 1) (0.57, 0.66, 0.71, 0.75; 0.9, 0.9)	(0.48, 0.61, 0.64, 0.73; 1, 1) (0.55, 0.61, 0.64, 0.67; 0.9, 0.9)	0.663	0.612	✓
Technological complementarity	(0.38, 0.58, 0.65, 0.82; 1, 1) (0.5, 0.58, 0.65, 0.72; 0.9, 0.9)	(0.43, 0.56, 0.59, 0.67; 1, 1) (0.5, 0.56, 0.59, 0.62; 0.9, 0.9)	0.608	0.561	✓
Product experience	(0.19, 0.39, 0.45, 0.64; 1, 1) (0.29, 0.39, 0.45, 0.52; 0.9, 0.9)	(0.33, 0.43, 0.45, 0.51; 1, 1) (0.38, 0.43, 0.45, 0.47; 0.9, 0.9)	0.416	0.430	
Number of patents held	(0.34, 0.54, 0.61, 0.78; 1, 1) (0.45, 0.54, 0.61, 0.67; 0.9, 0.9)	(0.39, 0.5, 0.53, 0.6; 1, 1) (0.45, 0.5, 0.53, 0.56; 0.9, 0.9)	0.566	0.507	✓
Expected capabilities of abstraction	(0.09, 0.26, 0.31, 0.48; 1, 1) (0.17, 0.26, 0.31, 0.37; 0.9, 0.9)	(0.26, 0.32, 0.34, 0.39; 1, 1) (0.29, 0.32, 0.34, 0.36; 0.9, 0.9)	0.280	0.327	
Technology transfer capability	(0.45, 0.65, 0.7, 0.82; 1, 1) (0.56, 0.65, 0.7, 0.75; 0.9, 0.9)	(0.47, 0.59, 0.63, 0.71; 1, 1) (0.54, 0.59, 0.63, 0.66; 0.9, 0.9)	0.655	0.601	✓
Overlapping knowledge base	(0.11, 0.27, 0.33, 0.49; 1, 1) (0.19, 0.27, 0.33, 0.38; 0.9, 0.9)	(0.26, 0.33, 0.35, 0.39; 1, 1) (0.3, 0.33, 0.35, 0.37; 0.9, 0.9)	0.295	0.333	
Product-specific knowledge	(0.4, 0.6, 0.67, 0.82; 1, 1) (0.51, 0.6, 0.67, 0.73; 0.9, 0.9)	(0.44, 0.56, 0.59, 0.67; 1, 1) (0.5, 0.56, 0.59, 0.62; 0.9, 0.9)	0.623	0.565	✓
Market knowledge complementarity	(0.27, 0.47, 0.53, 0.71; 1, 1) (0.38, 0.47, 0.53, 0.6; 0.9, 0.9)	(0.38, 0.49, 0.51, 0.58; 1, 1) (0.44, 0.49, 0.51, 0.54; 0.9, 0.9)	0.495	0.491	✓
Expected knowledge maturity	(0.22, 0.41, 0.47, 0.66; 1, 1) (0.31, 0.41, 0.47, 0.53; 0.9, 0.9)	(0.34, 0.43, 0.45, 0.52; 1, 1) (0.39, 0.43, 0.45, 0.48; 0.9, 0.9)	0.436	0.442	
Past experiences	(0.2, 0.39, 0.44, 0.62; 1, 1) (0.29, 0.39, 0.44, 0.5; 0.9, 0.9)	(0.35, 0.45, 0.47, 0.53; 1, 1) (0.4, 0.45, 0.47, 0.50; 0.9, 0.9)	0.408	0.451	
Financial assets	(0.06, 0.19, 0.24, 0.40; 1, 1) (0.12, 0.19, 0.24, 0.29; 0.9, 0.9)	(0.2, 0.26, 0.27, 0.31; 1, 1) (0.23, 0.26, 0.27, 0.29; 0.9, 0.9)	0.216	0.260	
Expected debt ratio and refund ability	(0.32, 0.52, 0.57, 0.72; 1, 1) (0.42, 0.52, 0.57, 0.62; 0.9, 0.9)	(0.39, 0.5, 0.53, 0.60; 1, 1) (0.45, 0.5, 0.53, 0.55; 0.9, 0.9)	0.528	0.503	✓
Financial resources demand of the project	(0.09, 0.26, 0.31, 0.48; 1, 1) (0.17, 0.26, 0.31, 0.37; 0.9, 0.9)	(0.23, 0.29, 0.31, 0.35; 1, 1) (0.27, 0.29, 0.31, 0.33; 0.9, 0.9)	0.280	0.297	
Return of investment	(0.16, 0.34, 0.4, 0.58; 1, 1) (0.25, 0.34, 0.4, 0.46; 0.9, 0.9)	(0.28, 0.36, 0.38, 0.44; 1, 1) (0.33, 0.36, 0.38, 0.41; 0.9, 0.9)	0.365	0.368	
Collaborative behavior	(0.43,0.62,0.67,0.80;1,1) (0.53,0.62,0.67,0.72;0.9,0.9)	(0.46,0.58,0.61,0.7;1,1) (0.53,0.58,0.61,0.65;0.9,0.9)	0.629	0.588	✓
Mutual trust and commitment	(0.21, 0.41, 0.47, 0.65; 1, 1) (0.31, 0.41, 0.47, 0.53; 0.9, 0.9)	(0.35, 0.44, 0.46, 0.52; 1, 1) (0.4, 0.44, 0.46, 0.49; 0.9, 0.9)	0.429	0.444	
Management and organizational culture	(0.24, 0.44, 0.5, 0.69; 1, 1) (0.33, 0.44, 0.5, 0.56; 0.9, 0.9)	(0.35, 0.45, 0.47, 0.54; 1, 1) (0.41, 0.45, 0.47, 0.5; 0.9, 0.9)	0.461	0.454	✓
Previous relationship	(0.03, 0.14, 0.18, 0.33; 1, 1) (0.08, 0.14, 0.18, 0.23; 0.9, 0.9)	(0.18, 0.23, 0.24, 0.27; 1, 1) (0.21, 0.23, 0.24, 0.25; 0.9, 0.9)	0.163	0.230	
Propensity to change	(0.04, 0.15, 0.19, 0.34; 1, 1) (0.09, 0.15, 0.19, 0.24; 0.9, 0.9)	(0.2, 0.24, 0.26, 0.29; 1, 1) (0.22, 0.24, 0.26, 0.27; 0.9, 0.9)	0.175	0.246	
Geographical proximity	(0.02, 0.11, 0.15, 0.29; 1, 1) (0.06, 0.11, 0.15, 0.20; 0.9, 0.9)	(0.17, 0.21, 0.23, 0.26; 1, 1) (0.19, 0.21, 0.23, 0.24; 0.9, 0.9)	0.136	0.218	
Symmetry of scale and scope	(0, 0.06, 0.09, 0.21; 1, 1) (0.03, 0.06, 0.09, 0.13; 0.9, 0.9)	(0.16, 0.2, 0.21, 0.23; 1, 1) (0.18, 0.2, 0.21, 0.22; 0.9, 0.9)	0.084	0.197	
Compatibility of corporation strategies	(0.18, 0.37, 0.43, 0.61; 1, 1) (0.27, 0.37, 0.43, 0.49; 0.9, 0.9)	(0.31, 0.4, 0.42, 0.48; 1, 1) (0.36, 0.4, 0.42, 0.44; 0.9, 0.9)	0.390	0.401	

Table 7. Cont.

Criteria	Aggregate Criteria Weight ($\tilde{\tilde{w}}_{\delta j}$)	Aggregate MAW ($\tilde{\tilde{R}}_{\delta j}$)	Criteria Weight	MAW	Selected Criteria
Convergence of expectations between partners	(0.36, 0.56, 0.63, 0.80; 1, 1) (0.47, 0.56, 0.63, 0.70; 0.9, 0.9)	(0.41, 0.53, 0.56, 0.64; 1, 1) (0.48, 0.53, 0.56, 0.59; 0.9, 0.9)	0.585	0.539	
Motivation and goal correspondence	(0.22, 0.41, 0.47, 0.66; 1, 1) (0.31, 0.41, 0.47, 0.53; 0.9, 0.9)	(0.33, 0.42, 0.44, 0.5; 1, 1) (0.38, 0.42, 0.44, 0.47; 0.9, 0.9)	0.436	0.446	
Strategic objectives of intellectual property management	(0.16,0.35,0.41,0.60;1,1) (0.25,0.35,0.41,0.47;0.9,0.9)	(0.31, 0.39, 0.41, 0.47; 1, 1) (0.35, 0.39, 0.41, 0.44; 0.9, 0.9)	0.375	0.395	
Market complementarity	(0.09, 0.25, 0.29, 0.46; 1, 1) (0.16, 0.25, 0.29, 0.35; 0.9, 0.9)	(0.23, 0.29, 0.31, 0.35; 1, 1) (0.26, 0.29, 0.31, 0.33; 0.9, 0.9)	0.265	0.295	

The IT2FD process results reinforce the significance of specific factors related to the partners' technological competencies, such as Technological innovation level, Technological complementarity, Number of patents held, and Technology transfer capability, highlighting the increasing technological complexity or necessity of technology fusion to innovate. In terms of the resource complementarity dimension, the evaluation process will incorporate Product-specific knowledge and Market knowledge complementarity criteria, with less emphasis on evaluating the financial terms of the corporation and strategic alignment.

4.2. Criteria Weighting by the IT2F AHP

The fuzzified linguistic variables defined in the previous section (Table 5) are now applied to describe the preference relations between the criteria considered by each expert. This has resulted in individual pairwise comparison matrices whose consistencies have been clarified and confirmed. Based on the established matrices and Equation (16) we can obtain CI: $CI_1 = 0.077, CI_2 = 0.087, CI_3 = 0.107, CI_4 = 0.106$; $RI = 1.49$, for $n = 10$. According to Equation (15) CR is obtained: $CR_1 = 0.0514\%$, $CR_2 = 0.0587\%$, $CR_3 = 0.0721\%$, $CR_4 = 0.0708\%$. Since $CR_{1,2,3,4} < 0.1$, established pairwise comparison matrices could be considered to be consistent.

The IT2F aggregated pairwise matrix is obtained by applying Equations (17) and (18). To generate criteria weights from these relations, the geometric mean is computed by Equation (19). Equation (20) is employed so as to establish IT2F criteria weights, and the composite criteria weights are available after defuzzifying by Equation (12) and normalizing.

The results (Table 8) emphasize that the criteria from the technological competencies dimension, specifically the technological innovation level (0.249) and the technology transfer capability (0.237), as well as the collaborative behavior criterion (0.14) as a represent of the cooperative capability dimension, are the most significant evaluation criteria according to the nature of the company's OI processes, the contextual conditions, and the innovation strategy at the corporate level.

Table 8. The results of the criteria weighting.

	Criteria	IT2F Criteria Geometric Means ($\tilde{\tilde{r}}_y$)	IT2F Criteria Weight ($\tilde{\tilde{w}}_j$)	Non-Fuzzy Normalized Weights (w_j)
C_1	Technological innovation level	(1.081, 2.353, 6.755, 11.827; 1, 1) (1.308, 2.657, 6.114, 10.414; 0.8, 0.8)	(0.238, 0.245, 0.253, 0.256; 1, 1) (0.24, 0.246, 0.253, 0.256; 0.8, 0.8)	0.249
C_2	Technological complementarity	(0.471, 0.856, 2.12, 3.755; 1, 1) (0.544, 0.943, 1.929, 3.275; 1, 1)	(0.104, 0.089, 0.079, 0.081; 1, 1) (0.1, 0.087, 0.08, 0.08; 1, 1)	0.088
C_3	Number of patents held	(0.258, 0.475, 1.399, 2.974; 1, 1) (0.297, 0.529, 1.241, 2.476; 0.8, 0.8)	(0.057, 0.049, 0.052, 0.064; 1, 1) (0.054, 0.049, 0.051, 0.061; 0.8, 0.8)	0.056

Table 8. *Cont.*

	Criteria	IT2F Criteria Geometric Means ($\tilde{\tilde{r}}_y$)	IT2F Criteria Weight ($\tilde{\tilde{w}}_j$)	Non-Fuzzy Normalized Weights (w_j)
C_4	Technology transfer capability	(0.995, 2.439, 6.777, 10.263; 1, 1) (1.251, 2.778, 6.226, 9.43; 0.8, 0.8)	(0.219, 0.254, 0.254, 0.222; 1, 1) (0.23, 0.257, 0.257, 0.231; 0.8, 0.8)	0.237
C_5	Product-specific knowledge base	(0.409, 0.85, 2.476, 4.616; 1, 1) (0.488, 0.956, 2.225, 3.995; 0.8, 0.8)	(0.09, 0.088, 0.093, 0.1; 1, 1) (0.089, 0.088, 0.092, 0.098; 0.8, 0.8)	0.093
C_6	Market knowledge complementarity	(0.135, 0.224, 0.626, 1.397; 1, 1) (0.151, 0.246, 0.554, 1.144; 0.8, 0.8)	(0.03, 0.023, 0.023, 0.03; 1, 1) (0.028, 0.023, 0.023, 0.028; 0.8, 0.8)	0.027
C_7	Expected debt ratio and refund ability	(0.098, 0.161, 0.459, 1.061; 1, 1) (0.109, 0.177, 0.404, 0.860; 0.8, 0.8)	(0.022, 0.017, 0.017, 0.023; 1, 1) (0.02, 0.016, 0.017, 0.021; 0.8, 0.8)	0.020
C_8	Collaborative behavior	(0.69, 1.404, 3.591, 5.825; 1, 1) (0.822, 1.567, 3.286, 5.227; 0.8, 0.8)	(0.152, 0.146, 0.135, 0.126; 1, 1) (0.151, 0.145, 0.136, 0.128; 0.8, 0.8)	0.140
C_9	Management and organizational culture	(0.048, 0.074, 0.222, 0.571; 1, 1) (0.052, 0.081, 0.193, 0.450; 0.8, 0.8)	(0.011, 0.008, 0.008, 0.012; 1, 1) (0.01, 0.007, 0.008, 0.011; 0.8, 0.8)	0.010
C_{10}	Convergence of expectations between partners	(0.353, 0.774, 2.254, 3.94; 1, 1) (0.428, 0.874, 2.036, 3.48; 0.8, 0.8)	(0.078, 0.08, 0.084, 0.085; 1, 1) (0.078, 0.081, 0.084, 0.085; 0.8, 0.8)	0.082

4.3. OI Partner Evaluation by IT2F PROMETHEE

The expert group has established a list of all potential OI partners, and after the initial screening, a total of five alternatives remained to be evaluated further. The experts reached a consistent evaluation of alternatives for each criterion based on the objective assessment of personalized knowledge and experience specific to a criterion using fuzzified linguistic terms. Those options are expressed according to the scale accounted in Table 4. As a result, the evaluation matrix was formed (Table 9). The matrix is normalized according to Equation (23).

Table 9. The IT2F evaluation matrix.

	Criteria	Alternative					Preference Function Type	Parameters	
		Z_1	Z_2	Z_3	Z_4	Z_5		p	q
C_1	Technological innovation level	L	MH	M	ML	MH	U-shape	-	-
C_2	Technological complementarity	M	VH	H	H	MH	V-shape	L	
C_3	Number of patents held	L	MH	ML	VL	M	V-shape	L	
C_4	Technology transfer capability	L	H	MH	M	ML	U-shape	L	
C_5	Product-specific knowledge base	H	MH	M	M	ML	V-shape	L	
C_6	Market knowledge complementarity	VH	VL	ML	H	L	V-shape with indifference	L	M
C_7	Expected debt ratio and refund ability	MH	ML	VL	ML	L	V-shape	ML	
C_8	Collaborative behavior	VH	L	M	H	MH	V-shape	ML	
C_9	Management and organizational culture	H	ML	MH	H	M	V-shape with indifference	VL	ML
C_{10}	Convergence of expectations between partners	ML	M	MH	MH	VL	V-shape with indifference	VL	ML

The preference functions for each criterion and the corresponding parameters according to the criteria characteristics are selected (Table 9). The level of technological innovation has a direct impact on the OI process's innovation performance; thus, the higher assessments strictly prefer their lower counterparts. It also proves to be the most important issue in OI partner selection for this company; therefore, the Usual Criterion type is designated for this criterion. The criteria C_2, C_3, C_4, C_5, C_7, and C_8 are considered superior only if the differences in the assessments reach a certain level. Before this level is reached, the higher level is linearly superior to the lower. It works for the criteria C_6, C_9, and C_{10} similarly, except for the fact that the difference cannot be made until the difference in the assessments has reached a certain level. Therefore, the corresponding preference function is designated as the V-shape with indifference Criterion. When it comes to criterion C_4, no difference can be made until the difference in the assessments reaches a certain level; thus, the U-shape Criterion is selected.

The preference index for each pair of alternatives, the outgoing flow, the incoming flow, and the net flow of each alternative OI partner are computed by Equations (33)–(35). For an easy ranking, the net flows are ultimately defuzzified by Equation (12). According to the calculation results presented in Table 10, the recommended ranking of the alternatives is: $Z_2 \to Z_3 \to Z_4 \to Z_5 \to Z_1$, so the alternative Z_2 is recommended to company as a compromise solution.

Table 10. The outgoing flow, the incoming flow, the net flow and the final ranking.

	Outgoing Flow ($\varnothing^+(z_i)$)	Incoming Flow ($\varnothing^-(z_i)$)	Net Flow ($\varnothing(z_i)$)	Non-Fuzzy Net Flow	Rank
Z_1	(0.141, 0.128, 0.119, 0.096; 1, 1) (0.140, 0.127, 0.118, 0.114; 0.8, 0.8)	(0.294, 0.302, 0.303, 0.260; 1, 1) (0.297, 0.303, 0.304, 0.270; 0.8, 0.8)	(−0.153, −0.174, −0.184, −0.164; 1, 1) (−0.158, −0.176, −0.185, −0.156; 0.8, 0.8)	−0.1700	5
Z_2	(0.3, 0.304, 0.271, 0.215; 1, 1) (0.302, 0.305, 0.271, 0.262; 0.8, 0.8)	(0.11, 0.099, 0.094, 0.077; 1, 1) (0.108, 0.099, 0.094, 0.088; 0.8, 0.8)	(0.191, 0.205, 0.177, 0.138; 1, 1) (0.194, 0.206, 0.177, 0.174; 0.8, 0.8)	0.1824	1
Z_3	(0.219, 0.224, 0.223, 0.172; 1, 1) (0.221, 0.224, 0.224, 0.189; 0.8, 0.8)	(0.188, 0.179, 0.146, 0.122; 1, 1) (0.188, 0.178, 0.145, 0.143)	(0.031, 0.045, 0.078, 0.05; 1, 1) (0.034, 0.046, 0.079, 0.046; 0.8, 0.8)	0.0523	2
Z_4	(0.203, 0.191, 0.187, 0.132; 1, 1) (0.203, 0.191, 0.187, 0.152; 0.8, 0.8)	(0.2, 0.2, 0.197, 0.133; 1, 1) (0.202, 0.2, 0.197, 0.165; 0.8, 0.8)	(0.003, −0.008, −0.01, −0.001; 1, 1) (0.001, −0.009, −0.009, −0.013; 0.8, 0.8)	−0.0050	3
Z_5	(0.169, 0.165, 0.164, 0.132; 1, 1) (0.170, 0.165, 0.164, 0.137; 0.8, 0.8)	(0.241, 0.232, 0.225, 0.156; 1, 1) (0.24, 0.232, 0.225, 0.188; 0.8, 0.8)	(−0.072, −0.067, −0.061, −0.024; 1, 1) (−0.07, −0.067, −0.062, −0.051; 0.8, 0.8)	−0.0581	4

5. Discussion

Selecting suitable partners capable of operationally and strategically contributing to the OI project goals is a strategic decision for companies striving to successfully transform the OI strategy into a sustainable competitive advantage.

Despite the extensive research on OI partner selection, there is still a notable research gap regarding establishing a comprehensive set of OI partner performance indicators. A comprehensive framework consisting of five critical dimensions (technological competencies, resource complementarity, financial terms of collaboration, cooperative capability, and strategic alignments), and twenty-seven indicators of OI partner performance has been developed in this study, which provides a holistic approach to OI partner selection, ensuring that the most critical technological, operational, and strategic aspects of OI partner evaluation were captured. This leads to a more integrated and coherent list of potential OI partner evaluation criteria.

In addition, this study aims to propose a hybrid MCDM methodology framework integrating Delphi, AHP, and PROMETHEE methods. In order to reflect the uncertainty, inherent to the decision-making process, in the best way, the theory of IT2F sets is employed. The proposed hybrid MCDM model represents a certain methodological advancement in identifying and evaluating OI partner performance indicators. The IT2F Delphi method enables the inclusion of expert subjective judgments, enabling a more comprehensive and inclusive approach to the identification of the most essential evaluation criteria. This results in a contingent and reliable set of OI partner performance indicators, enhancing the accuracy and practicality of the chosen evaluation criteria depending on the nature of the company's innovation processes as well as the contextual conditions and the corporate innovation strategy. The combination of IT2F AHP and IT2F PROMETHEE methods used in this paper yielded a more precise multi-criteria evaluation of the alternatives, under a high uncertainty level. Its main merits can be concluded from the following perspectives: criteria properties are treated in a proper way; it is suitable for solving complex decision-making problems when there are significant differences in participants' perceptions of the decision-making process; it has the ability to express preferences in a way similar to the human way of thinking; and it provides a mathematically simple synthesis of the obtained results for deriving criteria weights. In addition, the proposed approach is apt to incorporate imprecise data into the analysis using IT2F set theory. Namely, the classical MCDM methods that consider deterministic or random processes cannot effectively address OI partner selection problems due to their inherent fuzziness and imprecision. Hence, an IT2F MCDM algorithm is presented here to rectify the problem of vagueness and uncertainty in a more realistic way.

Furthermore, the study has some managerial implications. Based on the study findings, managers can establish specific management strategies, policies, and best practices to maximize the synergy of collaboration in OI projects in a systematic manner.

The proposed approach is illustrated through a case study of a high-tech company, to show the validity of the decision-making process. The study's results highlight the significance of specific factors related to the partners' technological competencies, such as Technological innovation level, Technological complementarity, Number of patents held, and Technology transfer capability, while the evaluation process will be less focused on the evaluation of financial terms of cooperation and strategic alignment. Based on the multi-criteria evaluation process, alternative Z_2 is selected as a compromise solution. We firmly believe that the underlying concept of this approach is both rational and comprehensible. The proposed hybrid IT2F MCDM model can be generalized and applied to other complex decision-making problems in the domains of engineering and management that encounter imprecise, indefinite, and subjective data or vague information.

The main limitations of the methodology presented in the paper are: (i) the decision-makers subjective assessments, which may influence the accuracy of the input data; (ii) the AHP method disregards potential dependences between evaluation criteria.

Future research can include the following: (i) employing methods that evaluate the interrelationships between the criteria, such as the Analytic Network Process (ANP) or DEMATEL; (ii) using other tools that accept uncertainty in decision-making, in particular, the proposed MCDM methods can be combined with other types of fuzzy sets for method extension to solve uncertainty in OI partner selection problems; and (iii) developing a comparative framework encompassing various MCDM methods that might highlight the optimal methods for selecting an OI partner.

Author Contributions: Conceptualization, S.P., J.V.V., D.D.M. and V.P.; methodology, S.P., J.V.V. and D.D.M.; validation, S.P., J.V.V. and D.D.M.; formal analysis, S.P., J.V.V., D.D.M. and V.P.; investigation, S.P., J.V.V. and D.D.M.; Data curation, S.P., J.V.V. and V.P.; Writing—original draft, S.P., J.V.V., D.D.M. and V.P.; writing—review & editing, J.V.V. and D.D.M.; visualization, S.P., J.V.V. and V.P.; supervision, J.V.V. and D.D.M. All authors have read and agreed to the published version of the manuscript.

Funding: This work was supported by the Ministry of Education, Science and Technological Development of the Republic of Serbia, under Grant 451-03-47/2023-01/200132 with University of Kragujevac—Faculty of Technical Sciences Čačak.

Data Availability Statement: Not applicable.

Conflicts of Interest: The authors declare no conflict of interest.

References

1. Zhang, X.; Chu, Z.; Ren, L.; Xing, J. Open Innovation and Sustainable Competitive Advantage: The Role of Organizational Learning. *Technol. Forecast. Soc. Chang.* **2023**, *186*, 122114. [CrossRef]
2. Barrett, G.; Dooley, L.; Bogue, J. Open Innovation within High-Tech SMEs: A Study of the Entrepreneurial Founder's Influence on Open Innovation Practices. *Technovation* **2021**, *103*, 102232. [CrossRef]
3. Kwon, S.; Motohashi, K. How Institutional Arrangements in the National Innovation System Affect Industrial Competitiveness: A Study of Japan and the US with Multiagent Simulation. *Technol. Forecast. Soc. Chang.* **2017**, *115*, 221–235. [CrossRef]
4. Tidd, J.; Bessant, J.R. *Managing Innovation: Integrating Technological, Market and Organizational Change*; John Wiley & Sons: Hoboken, NJ, USA, 2013.
5. Hung, K.P.; Chiang, Y.H. Open Innovation Proclivity, Entrepreneurial Orientation, and Perceived Firm Performance. *Int. J. Technol. Manag.* **2010**, *52*, 257–274. [CrossRef]
6. Chesbrough, H.; Bogers, M. Explicating Open Innovation: Clarifying an Emerging Paradigm for Understanding Innovation. In *New Frontiers in Open Innovation*; Oxford University Press, Forthcoming: Oxford, UK, 2014.

7. Becker, W.; Dietz, J. R&D Cooperation and Innovation Activities of Firms: Evidence for the German Manufacturing Industry. *Res. Policy* **2004**, *33*, 209–223.
8. Dubouloz, S.; Bocquet, R.; Equey Balzli, C.; Gardet, E.; Gandia, R. SMEs' Open Innovation: Applying a Barrier Approach. *Calif. Manag. Rev.* **2021**, *64*, 113–137. [CrossRef]
9. Bogers, M.; Chesbrough, H.; Moedas, C. Open Innovation: Research, Practices, and Policies. *Calif. Manag. Rev.* **2018**, *60*, 5–16. [CrossRef]
10. Hizam-Hanafiah, M.; Soomro, M.A. The Situation of Technology Companies in Industry 4.0 and the Open Innovation. *J. Open Innov. Technol. Mark. Complex.* **2021**, *7*, 34. [CrossRef]
11. Shmeleva, N.; Gamidullaeva, L.; Tolstykh, T.; Lazarenko, D. Challenges and Opportunities for Technology Transfer Networks in the Context of Open Innovation: Russian Experience. *J. Open Innov. Technol. Mark. Complex.* **2021**, *7*, 197. [CrossRef]
12. Carmona-Lavado, A.; Cuevas-Rodríguez, G.; Cabello-Medina, C.; Fedriani, E.M. Does Open Innovation Always Work? The Role of Complementary Assets. *Technol. Forecast. Soc. Chang.* **2021**, *162*, 120316. [CrossRef]
13. Wang, X.; Wang, Z.; Huang, Y.; Liu, Y.; Zhang, J.; Heng, X.; Zhu, D. Identifying R&D Partners through Subject-Action-Object Semantic Analysis in a Problem & Solution Pattern. *Technol. Anal. Strateg. Manag.* **2017**, *29*, 1167–1180.
14. Holle, M.; Elsesser, L.; Schuhmacher, M.; Lindemann, U. How to Motivate External Open Innovation-Partners: Identifying Suitable Measures. In Proceedings of the 2016 Portland International Conference on Management of Engineering and Technology (PICMET), Honolulu, HI, USA, 4–8 September 2016.
15. Du, J.; Leten, B.; Vanhaverbeke, W. Managing Open Innovation Projects with Science-based and Market-based Partners. *Res. Policy* **2014**, *43*, 828–840. [CrossRef]
16. Santos, R.S.; Soares, J.; Marques, P.C.; Navas, H.V.; Martins, J.M. Integrating Business, Social, and Environmental Goals in Open Innovation through Partner Selection. *Sustainability* **2021**, *13*, 12870. [CrossRef]
17. Wang, G.; Tian, X.; Hu, Y.; Evans, R.D.; Tian, M.; Wang, R. Manufacturing Process Innovation-Oriented Knowledge Evaluation Using MCDM and Fuzzy Linguistic Computing in an Open Innovation Environment. *Sustainability* **2017**, *9*, 1630. [CrossRef]
18. Nguyen, P.H.; Tran, L.C.; Nguyen, H.B.D.; Ho, T.P.T.; Duong, Q.A.; Tran, T.N. Unlocking the Potential of Open Innovation through Understanding the Interrelationship among Key Determinants of FDI Attractiveness. *J. Open Innov. Technol. Mark. Complex.* **2023**, *9*, 100021. [CrossRef]
19. Shahin, A.; Mahdian, E. Ranking the Indicators of Open Innovation Adoption based on NPD Factors. *Int. J. Bus. Innov. Res.* **2023**, *22*, 408–428. [CrossRef]
20. Ar, İ.M.; Peker, İ.; Birdoğan, B.A.K.İ. Determining the Appropriate Open Innovation Model for Logistics Firms Using An Integrated Fuzzy AHP-VIKOR Approach. *Int. J. Econ. Admin. Stud.* **2023**, *2023*, 135–148. [CrossRef]
21. Yildirim, E.; Ar, I.M.; Dabić, M.; Baki, B.; Peker, I. A Multi-Stage Decision Making Model for Determining a Suitable Innovation Structure Using an Open Innovation Approach. *J. Bus. Res.* **2022**, *147*, 379–391. [CrossRef]
22. Vakil Alroaia, Y. Open Innovation and SMEs: Providing a Model for Business Development (An Application on Iranian Industrial Park). *J. Appl. Res. Ind. Eng.* **2023**, *10*, 125–140.
23. Hakaki, A.; Nikabadi, M.S.; Heidarloo, M.A. An Optimized Model for Open Innovation Success in Manufacturing SMES. *RAIRO Oper. Res.* **2021**, *55*, 3339–3357. [CrossRef]
24. Mubarak, M.F.; Petraite, M.; Rasli, A.; Shabbir, M. Capability Framework to Support Supply Chain Open Innovation Networks. In *Blockchain Driven Supply Chain Management: A Multi-Dimensional Perspective*; Springer Nature: Singapore, 2023; pp. 119–134.
25. Vesic Vasovic, J.; Puzovic, S.; Paunovic, V. Selection of Partners in Collaborative Technological R&D Projects: An Approach to Criteria Prioritization. In *Operational Research in the Era of Digital Transformation and Business Analytics. BALCOR 2020. Springer Proceedings in Business and Economics*; Matsatsinis, N., Kitsios, F., Madas, M., Kamariotou, M., Eds.; Springer: Cham, Switzerland, 2023; pp. 77–87.
26. Aleksić, A.; Tadić, D. Industrial and Management Applications of Type-2 Multi-Attribute Decision-Making Techniques Extended with Type-2 Fuzzy Sets from 2013 to 2022. *Mathematics* **2023**, *11*, 2249. [CrossRef]
27. Yoon, B.; Song, B. A Systematic Approach of Partner Selection for Open Innovation. *Ind. Manag. Data Syst.* **2014**, *114*, 1068–1093. [CrossRef]
28. Manotungvorapun, N.; Gerdsri, N. Complementarity vs. Compatibility: What Really Matters for Partner Selection in Open Innovation? *Int. J. Transit. Innov. Syst.* **2016**, *5*, 122–139. [CrossRef]
29. Park, I.; Yoon, B. Identifying Potential Partnership for Open Innovation by Using Bibliographic Coupling and Keyword Vector Mapping. *Int. J. Comput. Electr. Autom. Contr. Inf. Eng.* **2013**, *7*, 206–211.
30. Wang, M.Y. Exploring Potential R&D Collaborators with Complementary Technologies: The Case of Biosensors. *Technol. Forecast. Soc. Chang.* **2012**, *79*, 862–874.
31. Jeon, J.; Lee, C.; Park, Y. How to Use Patent Information to Search Potential Technology Partners in Open Innovation. *J. Intellect. Prop. Rights* **2011**, *16*, 385–393.
32. Angue, K.; Ayerbe, C.; Mitkova, L. A Method Using Two Dimensions of the Patent Classification for Measuring the Technological Proximity: An Application in Identifying a Potential R&D Partner in Biotechnology. *J. Technol. Transf.* **2014**, *39*, 716–747.

33. Ades, C.; Figlioli, A.; Sbragia, R.; Porto, G.; Ary Plonski, G.; Celadon, K. Implementing Open Innovation: The Case of Natura, IBM and Siemens. *J. Technol. Manag. Innov.* **2013**, *8*, 57. [CrossRef]
34. Paixao Garcez, M.; Sbragia, R. The Selection of Partners in Technological Alliances Projects. *J. Technol. Manag. Innov.* **2013**, *8*, 49. [CrossRef]
35. Emden, Z.; Calantone, R.J.; Droge, C. Collaborating for New Product Development: Selecting the Partner with Maximum Potential to Create Value. *J. Prod. Innov. Manag.* **2006**, *23*, 330–341. [CrossRef]
36. Büyüközkan, G.; Feyzioğlu, O.; Nebol, E. Selection of the Strategic Alliance Partner in Logistics Value Chain. *Int. J. Prod. Econ.* **2008**, *113*, 148–158. [CrossRef]
37. Holmberg, S.R.; Cummings, J.L. Building Successful Strategic Alliances: Strategic Process and Analytical Tool for Selecting Partner Industries and Firms. *Long Range Plan.* **2009**, *42*, 164–193. [CrossRef]
38. Arranz, N.; de Arroyabe, J.C.F. The Choice of Partners in R&D Cooperation: An Empirical Analysis of Spanish Firms. *Technovation* **2008**, *28*, 88–100.
39. Shah, R.H.; Swaminathan, V. Factors Influencing Partner Selection in Strategic Alliances: The Moderating Role of Alliance Context. *Strateg. Manag. J.* **2008**, *29*, 471–494. [CrossRef]
40. Sarkar, M.B.; Echambadi, R.; Cavusgil, S.T.; Aulakh, P.S. The Influence of Complementarity, Compatibility, and Relationship Capital on Alliance Performance. *J. Acad. Mark. Sci.* **2001**, *29*, 358–373. [CrossRef]
41. Tian, Q.F.; Zhang, T.; Zhang, S.; Miao, D.D. Research on Elements Fusion Mechanism of Dual-use Science and Technology Collaborative Innovation. *J. Sci. Technol. Policy Manag.* **2020**, *37*, 136–145.
42. Harrigan, K.R. Joint Ventures and Competitive Strategy. *Strateg. Manag. J.* **1988**, *9*, 141–158. [CrossRef]
43. Guertler, M.R.; Haymerle, R.; Endres, F.; Lindemann, U. Identifying Open Innovation Partners: A Case Study in Plant Manufacturing. In Proceedings of the International Society for Professional Innovation Management (ISPIM) Innovation Summit, Brisbane, Australia, 1–18 December 2015.
44. Yin, J.; Tan, Q.M. A Research on the Evaluation of the Degree of Civil-military Integration and Its Optimization Measures. *Sci. Res. Manag.* **2020**, *41*, 90–97.
45. Geringer, J.M. Strategic Determinants of Partner Selection Criteria in International Joint Ventures. *J. Int. Bus. Stud.* **1991**, *22*, 41–62. [CrossRef]
46. Chen, M.K.; Wang, S.C. The Use of a Hybrid Fuzzy-Delphi-AHP Approach to Develop Global Business Intelligence for Information Service Firms. *Expert Syst. Appl.* **2010**, *37*, 7394–7407. [CrossRef]
47. Solesvik, M.; Westhead, P. Partner Selection for Strategic Alliances: Case Study Insights from the Maritime Industry. *Ind. Manag. Data Syst.* **2010**, *110*, 841–860. [CrossRef]
48. Kim, Y.; Lee, K. Technological Collaboration in the Korean Electronic Parts Industry: Patterns and Key Success Factors. *R&D Manag.* **2003**, *33*, 59–77.
49. Dacin, M.T.; Hitt, M.A.; Levitas, E. Selecting Partners for Successful International Alliances: Examination of US and Korean Firms. *J. World Bus.* **1997**, *32*, 3–16. [CrossRef]
50. Brouthers, K.D.; Brouthers, L.E.; Wilkinson, T.J. Strategic Alliances: Choose Your Partners. *Long Range Plan.* **1995**, *28*, 2–25. [CrossRef]
51. Zadeh, L.A. Fuzzy Sets. *Inf. Control* **1965**, *8*, 338–353. [CrossRef]
52. Liang, G.S. Fuzzy MCDM Based on Ideal and Anti-ideal Concepts. *Eur. J. Oper. Res.* **1999**, *112*, 682–691. [CrossRef]
53. Puzovic, S.; Vasovic Vasovic, J.; Radojicic, M.; Paunovic, V. An Integrated MCDM Approach to PLM Software Selection. *Acta Polytech. Hung.* **2019**, *16*, 45–65.
54. Aleksic, A.; Runic Ristic, M.; Komatina, N.; Tadic, D. Advanced Risk Assessment in Reverse Supply Chain Processes: A Case Study in Republic of Serbia. *Adv. Prod. Eng. Manag.* **2019**, *14*, 421–434. [CrossRef]
55. Mathew, M.; Chakrabortty, R.K.; Ryan, M.J. Selection of an Optimal Maintenance Strategy under Uncertain Conditions: An Interval Type-2 Fuzzy AHP-TOPSIS Method. *IEEE Trans. Eng. Manag.* **2020**, *69*, 1121–1134. [CrossRef]
56. Ecer, F. Multi-Criteria Decision Making for Green Supplier Selection Using Interval Type-2 Fuzzy AHP: A Case Study of a Home Appliance Manufacturer. *Oper. Res.* **2022**, *22*, 199–233. [CrossRef]
57. Jovcic, S.; Průša, P.; Dobrodolac, M.; Svadlenka, L. A Proposal for a Decision-Making Tool in Third-Party Logistics (3PL) Provider Selection based on Multi-Criteria Analysis and the Fuzzy Approach. *Sustainability* **2019**, *11*, 4236. [CrossRef]
58. Puška, A.; Stojanović, I. Fuzzy Multi-Criteria Analyses on Green Supplier Selection in an Agri-Food Company. *J. Intell. Manag. Decis.* **2022**, *1*, 2–16. [CrossRef]
59. Su, J.; Xu, B.; Li, L.; Wang, D.; Zhang, F. A Green Supply Chain Member Selection Method Considering Green Innovation Capability in a Hesitant Fuzzy Environment. *Axioms* **2023**, *12*, 188. [CrossRef]
60. Petrovic, T.; Vesic Vasovic, J.; Komatina, N.; Tadic, D.; Klipa, Đ.; Đuric, G. A Two-Stage Model based on EFQM, FBWM, and FMOORA for Business Excellence Evaluation in the Process of Manufacturing. *Axioms* **2022**, *11*, 704. [CrossRef]
61. Celik, E.; Gul, M.; Aydin, N.; Gumus, A.T.; Guneri, A.F. A Comprehensive Review of Multi Criteria Decision Making Approaches Based on Interval Type-2 Fuzzy Sets. *Knowl. Based Syst.* **2015**, *85*, 329–341. [CrossRef]

62. Shringi, A.; Arashpour, M.; Golafshani, E.M.; Dwyer, T.; Kalutara, P. Enhancing Safety Training Performance Using Extended Reality: A Hybrid Delphi–AHP Multi-Attribute Analysis in a Type-2 Fuzzy Environment. *Buildings* **2023**, *13*, 625. [CrossRef]
63. Mendel, J.M.; John, R.I.; Liu, F. Interval Type-2 Fuzzy Logic Systems Made Simple. *IEEE Trans. Fuzzy Syst.* **2006**, *14*, 808–821. [CrossRef]
64. Zadeh, L.A. The Concept of a Linguistic Variable and Its Application to Approximate Reasoning: I. *Inf. Sci.* **1975**, *8*, 199–249. [CrossRef]
65. Gölcük, İ. An Interval Type-2 Fuzzy Reasoning Model for Digital Transformation Project Risk Assessment. *Expert Syst. Appl.* **2020**, *159*, 113579. [CrossRef]
66. Wu, Y.; Zhang, T.; Yi, L. An Internal Type-2 Trapezoidal Fuzzy Sets-PROMETHEE-II based Investment Decision Framework of Compressed Air Energy Storage Project in China under the Perspective of Diffewt Investors. *J. Energy Storage* **2020**, *30*, 101548. [CrossRef]
67. Baykasoğlu, A.; Gölcük, İ. Development of an Interval Type-2 Fuzzy Sets based Hierarchical MADM Model by Combining DEMATEL and TOPSIS. *Expert Syst. Appl.* **2017**, *70*, 37–51. [CrossRef]
68. Boral, S.; Chakraborty, S. Failure Analysis of CNC Machines Due to Human Errors: An Integrated IT2F-MCDM-based FMEA Approach. *Eng. Fail. Anal.* **2021**, *130*, 105768. [CrossRef]
69. Bera, A.K.; Jana, D.K.; Banerjee, D.; Nandy, T. Supplier Selection Using Extended IT2 Fuzzy TOPSIS and IT2 Fuzzy MOORA Considering Subjective and Objective Factors. *Soft Comput.* **2020**, *24*, 8899–8915. [CrossRef]
70. Karagöz, S.; Deveci, M.; Simic, V.; Aydin, N. Interval Type-2 Fuzzy ARAS Method for Recycling Facility Location Problems. *Appl. Soft Comput.* **2021**, *102*, 107107.
71. Wei, Q.; Zhou, C.; Liu, Q.; Zhou, W.; Huang, J. A Barrier Evaluation Framework for Forest Carbon Sink Project Implementation in China Using an Integrated BWM-IT2F-PROMETHEE II method. *Expert Syst. Appl.* **2023**, *2023*, 120612.
72. Mendel, J.M.; John, R.B. Type-2 Fuzzy Sets Made Simple. *IEEE Trans. Fuzzy Syst.* **2002**, *10*, 117–127. [CrossRef]
73. Mabrouk, N. Green Supplier Selection Using Fuzzy Delphi Method for Developing Sustainable Supply Chain. *Decis. Sci. Lett.* **2021**, *10*, 63–70. [CrossRef]
74. Deveci, M.; Simic, V.; Karagoz, S.; Antucheviciene, J. An Interval Type-2 Fuzzy Sets based Delphi Approach to Evaluate Site Selection Indicators of Sustainable Vehicle Shredding Facilities. *Appl. Soft Comput.* **2022**, *118*, 108465.
75. Aleksić, A.; Nestić, S.; Huber, M.; Ljepava, N. The Assessment of the Key Competences for Lifelong Learning—The Fuzzy Model Approach for Sustainable Education. *Sustainability* **2022**, *14*, 2686. [CrossRef]
76. Ayyildiz, E.; Taskin Gumus, A.; Erkan, M. Individual Credit Ranking by an Integrated Interval Type-2 Trapezoidal Fuzzy ELECTRE Methodology. *Soft Comput.* **2020**, *24*, 16149–16163. [CrossRef]
77. Gupta, R.; Sachdeva, A.; Bhardwaj, A. Selection of 3pl Service Provider Using Integrated Fuzzy Delphi and Fuzzy TOPSIS. In Proceedings of the World Congress on Engineering and Computer Science, San Francisco, CA, USA, 20–22 October 2010.
78. Çolak, M.; Kaya, İ. Prioritization of Renewable Energy Alternatives by Using an Integrated Fuzzy MCDM Model: A Real Case Application for Turkey. *Renew. Sustain. Energy Rev.* **2017**, *80*, 840–853. [CrossRef]
79. Saaty, T.L. *The Analytic Hierarchy Process*; McGraw-Hill: New York, NY, USA, 1980.
80. Wu, Y.; Xu, C.; Ke, Y.; Tao, Y.; Li, X. Portfolio Optimization of Renewable Energy Projects Under Type-2 Fuzzy Environment with Sustainability Perspective. *Comput. Ind. Eng.* **2019**, *133*, 69–82.
81. Hoseini, S.A.; Hashemkhani Zolfani, S.; Skačkauskas, P.; Fallahpour, A.; Saberi, S. A Combined Interval Type-2 Fuzzy MCDM Framework for the Resilient Supplier Selection Problem. *Mathematics* **2021**, *10*, 44. [CrossRef]
82. Kaya, S.K.; Aycin, E. An Integrated Interval Type 2 Fuzzy AHP and COPRAS-G Methodologies for Supplier Selection in the Era of Industry 4.0. *Neural. Comput. Appl.* **2021**, *33*, 10515–10535. [CrossRef]
83. Saaty, T.L.; Vargas, L.G. Comparison of Eigenvalue, Logarithmic Least Squares and Least Squares Methods in Estimating Ratios. *Math. Model.* **1984**, *5*, 309–324. [CrossRef]
84. Buckley, J.J. Fuzzy Hierarchical Analysis. *Fuzzy Sets Syst.* **1985**, *17*, 233–247. [CrossRef]
85. Crawford, G.; Williams, C. A Note on the Analysis of Subjective Judgment Matrices. *J. Math. Psychol.* **1985**, *29*, 387–405. [CrossRef]
86. Chang, D.Y. Applications of the Extent Analysis Method on Fuzzy AHP. *Eur. J. Oper. Res.* **1996**, *95*, 649–655. [CrossRef]
87. Bryson, N. A Goal Programming Method for Generating Priority Vectors. *J. Oper. Res. Soc.* **1995**, *46*, 641–648. [CrossRef]
88. Mikhailov, L. A Fuzzy Programming Method for Deriving Priorities in the Analytic Hierarchy Process. *J. Oper. Res. Soc.* **2000**, *51*, 341–349. [CrossRef]
89. Chandran, B.; Golden, B.; Wasil, E. Linear Programming Models for Estimating Weights in the analytic Hierarchy Process. *Comput. Oper. Res.* **2005**, *32*, 2235–2254. [CrossRef]
90. Xu, Z.; Da, Q. A Least Deviation Method to Obtain a Priority Vector of a Fuzzy Preference Relation. *Eur. J. Oper. Res.* **2005**, *164*, 206–216. [CrossRef]
91. Ye-jun, X.U.; Qing-li, D.A. Weighted Least-Square Method and its Improvement for Priority of Incomplete Complementary Judgement Matrix. *Syst. Eng. Electron.* **2008**, *7*, 1273–1276.
92. Brans, J.P.; Vincke, P. Note—A Preference Ranking Organisation Method: (The PROMETHEE Method for Multiple Criteria Decision-Making). *Manag. Sci.* **1985**, *31*, 647–656. [CrossRef]

93. Chen, T.Y. A PROMETHEE-based Outranking Method for Multiple Criteria Decision Analysis with Interval Type-2 Fuzzy Sets. *Soft Comput.* **2014**, *18*, 923–940.
94. Wu, Y.; Liu, F.; Huang, Y.; Xu, C.; Zhang, B.; Ke, Y.; Jia, W. A Two-Stage Decision Framework for inland Nuclear Power Plant Site Selection based on GIS and Type-2 Fuzzy PROMETHEE II: Case Study in China. *Energy Sci. Eng.* **2020**, *8*, 1941–1961. [CrossRef]

mathematics

Article

Determining the Main Resilience Competencies by Applying Fuzzy Logic in Military Organization

Svajone Bekesiene [1,*], Oleksandr Nakonechnyi [2], Olena Kapustyan [2], Rasa Smaliukiene [1], Ramutė Vaičaitienė [1], Dalia Bagdžiūnienė [1] and Rosita Kanapeckaitė [1]

1 General Jonas Zemaitis Military Academy of Lithuania, Silo 5a, LT-10322 Vilnius, Lithuania
2 Taras Shevchenko National University of Kyiv, Volodymyrska St, 64/13, 01601 Kyiv, Ukraine
* Correspondence: svajone.bekesiene@lka.lt

Abstract: Military training programs have been developed to enhance soldier resilience competencies, which are necessary for soldiers to perform their duties effectively under stress. The ongoing military conflict in Ukraine and the experience of previous military missions abroad emphasize the need for effective training that helps soldiers recover quickly and continue their missions. However, selecting the most suitable resilience training program is challenging and the selection criteria need to be optimized to ensure the most needed competencies are considered. This study aimed to utilize a fuzzy MCDM method to establish the priority weight of decision-making criteria, identifying the core competencies necessary for soldier resilience training, and utilizing the fuzzy TOPSIS method to rank and select the most appropriate training program. The evaluation results were calculated using the MATLAB (R2020b) mathematical package developed by MathWorks. The application of the hierarchical MCDA model based on fuzzy sets theory indicated that mental agility is the most important competence in high-stress environments. The study found that the Mindfulness-Based Mind Fitness Training (MMFT) program, which is intended to regulate soldiers' emotions, had the highest rank among evaluated options according to the combined FAHP sub-factor fuzzy weights and alternatives evaluation conducted using FTOPSIS. The study provides valuable information on the selection of military resilience training programs.

Keywords: fuzzy logic; multi-criteria decision analysis; FAHP; FTOPSIS; resilience competencies; military organization

MSC: 03B52; 90B50; 90B90; 90C29; 90C31

Citation: Bekesiene, S.; Nakonechnyi, O.; Kapustyan, O.; Smaliukiene, R.; Vaičaitienė, R.; Bagdžiūnienė, D.; Kanapeckaitė, R. Determining the Main Resilience Competencies by Applying Fuzzy Logic in Military Organization. *Mathematics* **2023**, *11*, 2270. https://doi.org/10.3390/math11102270

Academic Editor: Aleksandar Aleksic

Received: 2 April 2023
Revised: 4 May 2023
Accepted: 9 May 2023
Published: 12 May 2023

1. Introduction

The development of resilience competence in the military is based on the premise that resilience is not a fixed state, but rather an ongoing process of learning and adapting [1]. Considering resilience as an individual's capacity to adapt positively to stressful situations creates both ambiguity and uncertainty when selecting the most ideal resilience training program and deciding which competencies need to be trained for resilience. As a solution to this challenge, fuzzy logic can be applied.

This study addresses the efficacy of resilience competence-building programs in the military when preparing for an actual conflict environment. The ongoing conflict in Ukraine, characterized by active hostilities (2014–2015), trench warfare (2016–2022) [2], and a large-scale Russian invasion of Ukraine (2022 and ongoing) [2,3], underscores the need for long-term resilience among soldiers and raises questions about the effectiveness of resilience competency-building programs based solely on data from military training and missions abroad. The motivation for this research comes from the shortage of studies that address the challenge of evaluating the soldiers' resilience competence development based on fuzzy logic rules. Furthermore, there are no established guidelines for military organizations to identify and prioritize those resilience competencies which require attention.

A set of factors and sub-factors related to soldiers' resilience were collected, and weights were assigned to each using the input of Ukrainian and Lithuanian military psychologists. A range of reports in the literature support the importance of dispositional attributes, beliefs, attitudes, and coping behaviors in resilience building [4]. Competencies such as self-awareness, self-regulation, optimism, mental agility, character strength, and connection have been identified as crucial contributors to resilience [5,6]. Experts agree that the unique composition of these competencies is essential for the resilience of soldiers in different military environments. When using a single expert decision-making approach, the basic homogeneous pairwise comparison within an MCDM framework is insufficient to accurately capture a decision maker's true perception with adequate effectiveness and precision; therefore, more advanced methods must be employed [7,8]. As such, experts' decisions regarding which competencies to include in resilience training programs can be considered applications of multi-criteria decision making (MCDM). The analytic hierarchy process (AHP) MCDM model is mostly useful for obtaining and using the weightings of factors and sub-factors in decision making [9]. Since pure AHP models do not deal with uncertainty, fuzzy techniques need to be integrated into the AHP to overcome inaccuracy in decision making [10]. The application of fuzzy set theory leads to higher accuracy in the analysis of human cognitive processes, converting linguistic judgments into fuzzy pairwise comparison matrixes [11]. The application of fuzzy set theory to the AHP to find the best solutions to compose training programs is increasing with research on curriculum development [12] and its use in distance learning [13,14]. Hybrid models continue to be researched, and techniques such as the TOPSIS (Technique for Order of Preference by Similarity to Ideal Solution) are used for explaining military problems. Several studies have used the AHP with the TOPSIS for multi-criteria decision making for military research purposes, including selecting new military personnel for the Indonesian Air Force [15], evaluating air combat effectiveness [16], and selecting military training aircrafts for the Spanish Air Force Academy [17]. However, while these studies have used the AHP, ANP, or TOPSIS to calculate the weights of criteria, these methods have not yet been applied in the context of soldiers' resilience competencies and skill selection. To address this gap, this study uses a fuzzy TOPSIS (Technique for Order of Preference by Similarity to Ideal Solution) technique, which is proven to lead to the best alternative that is closest to the optimal solution according to the criteria specified [18,19]. Therefore, the use of hybrid multi-criteria decision-making methodology in this study provides a more robust and comprehensive approach. By employing the fuzzy analytic hierarchy process (FAHP) to determine the fuzzy weights of factors and sub-factors, and the fuzzy TOPSIS to select the best training program among similar options, we intended to select the most appropriate training program for enhancing soldiers' resilience competencies and skills.

The primary objective of this study was to use a fuzzy MCDM method to identify the main competencies required for soldier resilience training, and the fuzzy TOPSIS method was applied to rank and select the most suitable training program. The application of fuzzy logic rules allows for a holistic approach and generates considerably clearer outcomes compared to traditional statistical models [8,20,21]. Aimed at assessing resilience competencies, the integration of fuzzy sets theory for resilience competencies assessment can be deemed as the goal of this research, which included: (a) assessing the relative importance of competencies and skills using FAHP; (b) modelling competencies and skills' values using fuzzy sets theory; (c) defining the overall index with fuzzy operators; and (d) identifying soldier resilience training programs that effectively promote soldiers' resilience in an actual combat environment by applying fuzzy TOPSIS. By applying the fuzzy MCDM methodology, objectivity was achieved in selecting which competencies to develop in army resilience training programs. The results of this study are crucial to developing more appropriate resilience building programs that lead to increased effectiveness among soldiers.

2. Preliminaries

2.1. The Fuzzy Sets Theory

Zadeh [22] first introduced the theory of the fuzzy set to deal with vagueness of human judgment. The fuzzy set (FS) is focused on the reasonableness of uncertainty due to inaccuracy or vagueness. The FS theory and fuzzy logic are now known as effective mathematical tools for multi decision criteria modeling and provide a major support for vague data analyses [23]. FSs allow partial membership and fuzzy numbers can be described by a specified interval of real numbers, each with a position of relationship between zero and one [24]. Concrete characterizations are used to describe fuzzy numbers. Typically, to explain fuzzy numbers, two definitions can be used.

Definition 1. *Let N ∈ F(R) be entitled a fuzzy number if* $x_0 \in R$ *exist such that* $\mu_N x_0 = 1$*, where for any* $\alpha \in [0, 1]$, $A_\alpha = x, \mu_{A_\alpha}(x) \geq \alpha$ *is a closed interval. All fuzzy numbers are characterized by F(R) sets, where R represents the set of real numbers.*

Definition 2. *A triangular fuzzy number (TFN) is designated following the specific design of number N = (l, m, u) and membership function* $\mu_N(x)$*: R →[0, 1]:*

$$\mu_N(x) = \begin{cases} 0, & x < l, \\ \frac{x}{m-l} - \frac{l}{m-l}, & x \in [l, m], \\ \frac{x}{m-u} - \frac{l}{m-u}, & x \in [m, u], \\ 0 & , x > u. \end{cases} \quad (1)$$

where $l \leq m \leq u$*, l is the lower and u is the upper value of the N, and m is the middle value of N. The set of elements* $\{x \in R | l < x < u\}$ *are supporting N. Therefore, by agreement, when l, m, and u are equal, N is a non-fuzzy number. In this study, the decision makers' assessments were collected by linguistic values, but for decision analysis, the triangular fuzzy numbers shown in Table 1 were used.*

Table 1. The triangular fuzzy numbers for pairwise comparison matrix.

Linguistic Value	Triangular Fuzzy Number	RTFN [1]
Elements are equally important (EI)	(1,1,1)	(1,1,1)
One element is equally moderately important to another (EMI)	(1/2,1,3/2)	(2/3,1,2)
One element is less important than another (WI)	(1,3/2,2)	(1/2,2/3,1)
One element is moderately more important than another (MI)	(3/2,2, 5/2)	(2/5,1/2,2/3)
One element is moderately more important than another (MSI)	(2,5/2,3)	(1/3,2/5,1/2)
One element is more important than another (SI)	(5/2,3,7/2)	(2/7,1/3,2/5)
One element is much more important than another (VSI)	(3,7/2,4)	(1/4,2/7,1/3)
One element is much, much more important than another (VS)	(7/2,4,9/2)	(2/9,1/4,2/7)
One element is entirely more important than another (ES)	(4,9/2,9/2)	(2/9,2/9,1/4)

[1] Notes: RTFN = reciprocal triangular fuzzy number.

To define two triangular fuzzy numbers $N_1 = (l_1, m_1, u_1)$ *and* $N_2 = (l_2, m_2, u_2)$, the main [18] operational laws can be used (see Table 2).

Table 2. The operational laws used with triangular fuzzy numbers.

Operations with Triangular Fuzzy Numbers	Operational Laws	
$N_1 \oplus N_2, when\ N_1 = (l_1, m_1, u_1)\ and\ N_2 = (l_2, m_2, u_2)$	$(l_1, m_1, u_1) \oplus (l_2, m_2, u_2) = (l_1 + l_2, m_1 + m_2, u_1 + u_2)$	(2)
$N_1 \otimes N_2, when\ N_1 = (l_1, m_1, u_1)\ and\ N_2 = (l_2, m_2, u_2)$	$(l_1, m_1, u_1) \otimes (l_2, m_2, u_2) \approx (l_1 \cdot l_2, m_1 \cdot m_2, u_1 \cdot u_2)$	(3)
$N_1 \otimes \lambda, when\ N_1 = (l_1, m_1, u_1)$	$(l_1, m_1, u_1) \otimes (\lambda, \lambda, \lambda) = (l_1 \cdot \lambda, m_1 \cdot \lambda, u_1 \cdot \lambda), \lambda > 0, \lambda \in \mathcal{R}$	(4)
$N_1^{-1}, when\ N_1 = (l_1, m_1, u_1)$	$(l_1, m_1, u_1)^{-1} \approx (1/u_1, 1/m_1, 1/l_1).$	(5)

Assessment in the pairwise judgement with triangular fuzzy numbers is typically represented by membership functions. For this study, triangular numbers (N_1, N_3, N_5, N_7, and N_9)

were used to characterize the judgement from "equally important" to "entirely more important", and $N_2, N_4, N_6,$ and N_8 were used as the middle opinion values. Therefore, the membership functions of the triangular fuzzy numbers are $N_i = (l_i, m_i, u_i)$ where $i = 1, 2, \cdots, 9$ and where l_i is lower, m_i is middle, and u_i is the upper limit of the N_i fuzzy number. A graphical view of the membership functions is shown in Figure 1.

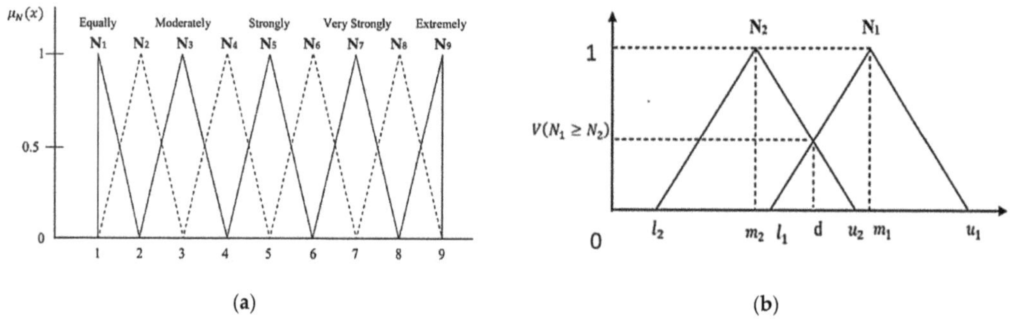

Figure 1. Scheme of triangular numbers description: (**a**) membership functions of the triangular numbers; (**b**) intersection between N_1 and N_2.

The fuzzy degree of judgment can be represented by Δ where $\Delta = u_i - l_i = l_i - u_i$. Moreover, if the value of Δ is larger, it characterizes an upper fuzzy point of judgment. Additionally, if $\Delta = 0$, the result is a non-fuzzy number. Scholars [25] suggest that the Δ value has to be greater than or equal to one half. In this study, we took into account these scholars' suggestions.

2.2. Fuzzy AHP Method

The need to prioritize different decision variables can be determined by applying the triangular fuzzy numbers. The extended analytical hierarchical process (AHP) method was employed to define the absolute priority of weights constructed on triangular fuzzy numbers. Based on the scope analysis method, each object can be used to perform the corresponding scope analysis for each objective. If we define $X = \{x_1, x_2, \cdots, x_m\}$ as an object set and $T = \{t_1, t_2, \cdots, t_m\}$ as a goal set, then the n extent values for each object can be established in the following way:

$$N^1_{z_i}, N^2_{z_i}, \cdots, N^m_{z_i}, \mathrm{I} = 1, 2, \cdots, \mathrm{m}, \tag{6}$$

where

$$N^j_{z_i} = \left(l^j_{z_i}, m^j_{z_i}, u^j_{z_i}\right), j = 1, 2, \cdots, n \tag{7}$$

are the triangular fuzzy numbers (TFN), and extended AHP analysis consists of the following steps:

Step 1: The valuation of fuzzy imitation extents for the i-th object are determined according to the methods of a previous work [26]:

$$S_i = \sum_{j=1}^{n} N^j_{z_i} \otimes \left[\sum_{i=1}^{m}\sum_{j=1}^{n} N^j_{z_i}\right]^{-1}. \tag{8}$$

To obtain the expression $\left[\sum_{i=1}^{m}\sum_{j=1}^{n} N^j_{z_i}\right]^{-1}$, we must complete additional fuzzy procedures with n values of the comprehensive analysis, which is represented by Equations (9) and (10):

$$\sum_{j=1}^{n} N^j_{z_i} = \left(\sum_{j=1}^{n} l_i, \sum_{j=1}^{n} m_i, \sum_{j=1}^{n} u_i\right) \tag{9}$$

$$\sum_{i=1}^{m}\sum_{j=1}^{n} N_{z_i}^{j}\left(\sum_{j=1}^{m} l_i, \sum_{j=1}^{m} m_i, \sum_{j=1}^{m} u_i\right) \tag{10}$$

Moreover, the inverse vector can be calculated by using Equation (11):

$$\left[\sum_{i=1}^{m}\sum_{j=1}^{n} N_{z_i}^{j}\right]^{-1} = \left(\frac{1}{\sum_{j=1}^{m} u_i}, \frac{1}{\sum_{j=1}^{m} m_i}, \frac{1}{\sum_{j=1}^{m} l_i}\right). \tag{11}$$

Step 2: The weight vector below every attribute, using the rule of the evaluation of fuzzy numbers, must be calculated. If we assume that $N_1 = (l_1, m_1, u_1)$ and $N_2 = (l_2, m_2, u_2)$ are the triangular fuzzy numbers, then the possibility that $N_1 \geq N_2$ is determined by

$$V(N_1 \geq N_2) = \sup_{x \geq y}\left[\min\left(\mu_{N_1}(x), \mu_{N_2}(y)\right)\right]. \tag{12}$$

When we can determine the pair (x, y) where $x \geq y$ and $\mu_{N_1}(x) = \mu_{N_2}(y) = 1$, the weight vector can be presented by the equation:

$$V(N_1 \geq N_2) = 1, \quad if\ n_1 \geq n_2. \tag{13}$$

When the pair $m_1 \leq m_2$ and $V(N_1 \geq N_2) = hzt(N_1 \cap N_2)$, the weight vector can be identified using Equation (14):

$$V(N_1 \geq N_2) = \mu(d) = \begin{cases} \frac{l_2 - u_1}{(m_1 - u_1) - (m_2 - l_2)}, \ l_2 \leq u_1, \\ otherwise, \\ 0. \end{cases} \tag{14}$$

In Equation (14), the value of d can be described as abscissa of the point D; that is, the maximum connection among N_1 *and* N_2 (see Figure 1b).

Step 3: We must determine vector weights. The possibility of a fuzzy number being bigger than z fuzzy numbers $N_i (i = 1, 2, \ldots, z)$ can be verified by Equation (15):

$$V(N \geq N_1, N_2, \ldots, N_3) = min V(N \geq N_i). \tag{15}$$

We suppose that abscissa d of the point D can be represented by

$$d'(A_i) = \min V(S_i \geq S_k), \tag{16}$$

where A_i is the i-th component of the k-th level and $k = 1, 2, \ldots, n; k \neq i$. Moreover, if we have an n number of components at the k-th level, the weight vector of the k-th level can be determined by Equation (17):

$$W' = \left(d'(A_1), d'(A_2), \ldots, d'(A_n)\right)^T \tag{17}$$

Step 4: The normalized weight vector after the normalization procedure is characterized by Equation (18):

$$W = \left(d(A_1), d(A_2), \ldots, d(A_n)\right)^T \tag{18}$$

where the weight vector W is representing a non-fuzzy number.

2.3. Consistency Checking

Mistakes made in representing preference relations in pairwise evaluations can lead to misleading judgments. Therefore, consistency checking and measurement of the lack are important themes in preference relations. In the conventional AHP, following scholars' suggestions [27], the consistency of the comparison matrix must be verified. However, many researchers [28–32] have not focused on the consistency examination procedure of the fuzzy AHP preference relation. This can be treated as a weakness of any FAHP that is developed or applied in these investigations. To address this drawback of the fuzzy AHP, other scholars [33,34] have resolved the consistency of fuzzy AHP priority relationships by substituting fuzzy position associations into their equivalent crisp multiplicative priority relationships, and then to check the consistency using Saaty's method. Here, the inconsistency rate of a matrix can be defined by the consistency index (CI) and consistency ratio (*CR*). According to the rule, if the CR is intolerable, the pairwise assessment should be reconsidered, and if *CR* < 0.1, the comparisons are acceptable, other than the inconsistent judgments, and the pairwise comparison should be revised [35]. The consistency ratio (*CR*) can be computed using Equation (19):

$$CR = \frac{CI}{RI} \tag{19}$$

where CI represents the consistency index, which shows the deviation from the stability, and RI represents the unplanned consistency index that can be obtained randomly from tables [36].

2.4. The Fuzzy TOPSIS Method

The Technique for Order Preference by Similarity to Ideal Situation (fuzzy TOPSIS) technique is an application of fuzzy logic and fuzzy sets. To conduct fuzzy TOPSIS analysis, the three main sets must first be determined: (1) alternatives $A = \{A_1, A_2, \ldots, A_m\}$, (2) evaluation criteria $C = \{C_1, C_2, \ldots, C_n\}$, and (3) decision makers $DM = \{DM_1, DM_2, \ldots, DM_i\}$. The consequence steps of the fuzzy TOPSIS algorithm can then be presented.

Step 1: Starting with the first step, we have to select the scale for linguistic variables and create the fuzzy rating of a decision matrix. Since the alternatives and criteria can be measured in linguistic terms, the triangular fuzzy numbers (TFN) characterized as linguistic terms by triangular scale can be chosen (see Table 3).

Table 3. Linguistic terms' connection with triangular fuzzy numbers' membership function.

FN [1]	Linguistic Terms	Triangular Scale (L, M, U) [1]
1	VL = Very Low	(1, 1, 3)
3	L = Low	(1, 3, 5)
5	M = Medium	(3, 5, 7)
7	H = High	(5, 7, 9)
9	VH = Very High	(7, 9, 9)

[1] Notes: FN = fuzzy number; L—lower value limit; M—middle value limit; U—upper value limit.

Step 2: Fuzzy linguistic assessments for the selections specified by decision makers and criteria weight must be given set fuzzy ratings by the *k*th decision maker for the *i*th alternative, and the *j*th criterion can be presented by the Equation (20):

$$\check{x}_{ij}^k = \left(a_{ij}^k, b_{ij}^k, c_{ij}^k\right) \tag{20}$$

w_j^k is the weight assigned by the *k*th decision maker to the *j*th criterion, which can be presented by the Equation (21):

$$w_j^k = \left\{w_{j1}^k, w_{j2}^k, w_{j3}^k\right\} \tag{21}$$

Step 3: The aggregated fuzzy scores for the alternatives can be calculated using Equation (22):

$$\check{x}_{ij} = \left(a_{ij}, b_{ij}, c_{ij}\right), a_{ij} = \min_k\left\{a_{ij}^k\right\}, b_{ij} = \frac{1}{l}\sum\nolimits_{k=1}^{l} b_{ij}^k, c_{ij} = \max_k\left\{c_{ij}^k\right\} \tag{22}$$

The aggregated fuzzy weight for a criterion can be presented by Equation (23):

$$w_{ij} = (a_{ij}, b_{ij}, c_{ij}), w_{ij} = \underset{k}{min}\left\{w_{ij}^{k}\right\}, w_{ij} = \frac{1}{l}\sum_{k=1}^{l} w_{j2}^{k}, w_{ij} = \underset{k}{max}\left\{w_{j3}^{k}\right\} \quad (23)$$

Step 4: The aggregated fuzzy decision matrix is built as specified below:

$$\tilde{D} = \begin{matrix} A_1 \\ A_2 \\ A_n \end{matrix} \begin{bmatrix} \tilde{x}_{11} & \cdots & \tilde{x}_{1n} \\ \vdots & \ddots & \vdots \\ \tilde{x}_{n1} & \cdots & \tilde{x}_{nn} \end{bmatrix} \quad (24)$$

where x_{ij} can be described as the aggregated fuzzy rating for the ith alternative.

Step 5: The fuzzy decision matrix normalization procedure is given below:

$$\tilde{R} = [\acute{r}_{ij}]_{m \times n} = \begin{bmatrix} \tilde{r}_{11} & \cdots & \tilde{r}_{1n} \\ \vdots & \ddots & \vdots \\ \tilde{r}_{m1} & \cdots & \tilde{x}_{mn} \end{bmatrix} \quad (25)$$

where

$$\tilde{r}_{ij} = \left(\frac{a_{ij}}{c_j^*}, \frac{b_{ij}}{c_j^*}, \frac{c_{ij}}{c_j^*}\right); \quad c_j^* = \underset{i}{max} c_{ij} \text{ (benefit criteria)} \quad (26)$$

$$\tilde{r}_{ij} = \left(\frac{a_j^-}{c_{ij}}, \frac{a_j^-}{b_{ij}}, \frac{a_j^-}{a_{ij}}\right); \quad a_j^- = \underset{i}{min} a_{ij} \text{ (cost criteria)} \quad (27)$$

After that, the linear scale transformation is used, and decision matrix is normalized. This procedure helps to change the TFN interval to [0, 1].

Step 6: The weighted normalized fuzzy decision matrix (WNFDM) is designed.

$$\tilde{V} = [\acute{v}_{ij}]_{m \times n} = [\acute{w}_j(.)\acute{r}_{ij}] = \begin{bmatrix} \acute{w}_1(.)\acute{r}_{11} & \cdots & \acute{w}_n(.)\acute{r}_{1n} \\ \vdots & \ddots & \vdots \\ \acute{w}_1(.)\acute{r}_{m1} & \cdots & \acute{w}_n(.)\acute{r}_{mn} \end{bmatrix} \quad (28)$$

Step 7: Now, the fuzzy positive ideal solution *(FPIS, A*)* and the fuzzy negative ideal solution ($FNIS, A^-$) can be computed. The computations of FPIS and FNIS can be completed using mathematical Equations (29) and (30):

$$A^* = \left\{\tilde{v}_1^*, \tilde{v}_2^*, \ldots, \tilde{v}_n^*\right\} = \left\{\left(\underset{j}{max} v_{ij} | i \in B\right), \left(\underset{j}{min} v_{ij} | i \in B\right)\right\}, (positive\ ideal\ solution) \quad (29)$$

$$A^- = \left\{\tilde{v}_1^-, \tilde{v}_2^-, \ldots, \tilde{v}_n^-\right\} = \left\{\left(\underset{j}{min} v_{ij} | i \in B\right), \left(\underset{j}{max} v_{ij} | i \in C\right)\right\}, (negative\ ideal\ solution) \quad (30)$$

where $\tilde{v}_i^*$ characterizes the *max* value of i across all judged alternatives, and $\tilde{v}_1^-$ is the *min* value of i across all the alternatives. B and C describe the positive and negative ideal justifications, respectively.

Step 8: The distances between alternatives are calculated. First, the fuzzy positive ideal result A^* and the distances between each alternative have to be calculated. Second, the distances between each alternative and fuzzy negative ideal result A^- are calculated. The distance between the FPIS and study alternatives and the distance between the FNIS and study alternatives can be computed by Equations (31) and (32), respectively:

$$d_i^* = \sum_{j=1}^{n} d(\tilde{v}_{ij}, \tilde{v}_j^*) \; i = 1, 2, \ldots, m; \quad (31)$$

$$d_i^- = \sum_{j=1}^{n} d(\tilde{v}_{ij}, \tilde{v}_j^-) \; i = 1, 2, \ldots, m; \quad (32)$$

where d represents the distance between two fuzzy numbers. When two triangular fuzzy numbers, (a_1, b_1, c_1) and (a_2, b_2, c_2), are specified, the value among the two criteria can be computed by the Equation (33):

$$d_v\left(\widetilde{N}_1, \widetilde{N}_2\right) = \sqrt{\frac{1}{3}[(a_1 - a_2)^2 + (b_1 - b_2)^2 + (c_1 - c_2)^2]} \quad (33)$$

It can be noted that $d\left(\widetilde{v}_{ij}, \widetilde{v}_j^*\right)$ and $d\left(\widetilde{v}_{ij}, \widetilde{v}_j^-\right)$ are crisp numbers.

Step 9: The calculation of the closeness coefficient (CC_i) of each alternative judged in the study can be recognized by the computation procedure represented by the equation:

$$CC_i = \frac{d_i^-}{d_i^+ + d_i^-} \quad (34)$$

Step 10: Finally, the alternatives must be ranked according to the CC_i. The great value of closeness index shows a good performance of the alternative [37].

3. Empirical Case Study Methodology

The methodology proposed in this study was basically made up of three steps. First, the decision-making problem for soldiers' resilience assessment were defined. Then, soldiers' resilience competencies and skills were decided. Finally, the judgement hierarchy was produced. The MATLAB (R2020b) mathematical package developed by MathWorks was used to calculate the evaluation results. A detailed evaluation framework for the study based on a hybrid methodology is illustrated in Figure 2.

3.1. Competencies and Skills That Affect Soldiers' Resilience

Research-based principles were applied to define groups of competencies and skills of soldier resilience training. After a comprehensive review of soldier resilience training programs, 14 main skills were chosen. Moreover, these fourteen skills that affect soldiers' resilience competencies were grouped by six characteristics: namely, Self-Awareness $\left(\widetilde{c}^1\right)$, Self-Regulation $\left(\widetilde{c}^2\right)$, Optimism $\left(\widetilde{c}^3\right)$, Mental Agility $\left(\widetilde{c}^4\right)$, Strength of Character $\left(\widetilde{c}^5\right)$, and Connection $\left(\widetilde{c}^6\right)$ [1,5,38]. The hierarchical structure of the soldiers' resilience-building model was designed to represent competencies as main factors, skills as sub-factors, and resilience training programs as alternatives (see Figure 3).

Self-awareness $\left(\widetilde{c}^1\right)$ can be represented by $\left(\widetilde{c}^{11}\right)$ and $\left(\widetilde{c}^{12}\right)$ [39]. $\left(\widetilde{c}^{11}\right)$ refers to a soldier's ability to identify the thoughts that arise in relation to a specific active event and the potential consequences of these thoughts: to separate an event from thoughts and consequences (emotions and reactions), in order to better understand their behavior (reactions) in a specific situation. $\left(\widetilde{c}^{12}\right)$ refers to a soldier's ability to identify beliefs and values that lead to overly strong emotions and reactions; to conduct an analysis of the negative, sad, depressing aspects of the situation; and to ask themselves: what can they do to change the situation, or is it necessary to discuss the situation with other people?

Self-regulation $\left(\widetilde{c}^2\right)$ is represented by three soldier skills, $\left(\widetilde{c}^{21}\right)$, $\left(\widetilde{c}^{22}\right)$, and $\left(\widetilde{c}^{23}\right)$ [40]. $\left(\widetilde{c}^{21}\right)$ concerns a soldier's ability to understand the components of the seven-step goal setting process and apply the skill in planning steps to achieve personal and career goals. $\left(\widetilde{c}^{22}\right)$ refers to a soldier's ability to control their physical state, focus on what is happening, control their breathing, try to relax, and work with their thoughts (ATC). $\left(\tilde{c}^{23}\right)$ represents how well a soldier can shift his attention away from unproductive thinking (interrupting) and focus more on the task at hand.

Optimism $\left(\tilde{c}^{3}\right)$ can be characterized by $\left(\tilde{c}^{31}\right)$ and $\left(\tilde{c}^{32}\right)$ [41,42]. First, it is important to notice positive events, evaluate why they led to positive emotions, what they mean, and what actions of the person or others caused positive occurrences $\left(\tilde{c}^{31}\right)$. Second, $\left(\tilde{c}^{32}\right)$ concerns a soldier's ability to stop catastrophic thoughts; reduce anxiety; find a solution to a problem by defining the best, worst, and most likely desired outcome; and develop a plan to help them to achieve the most likely desired outcome.

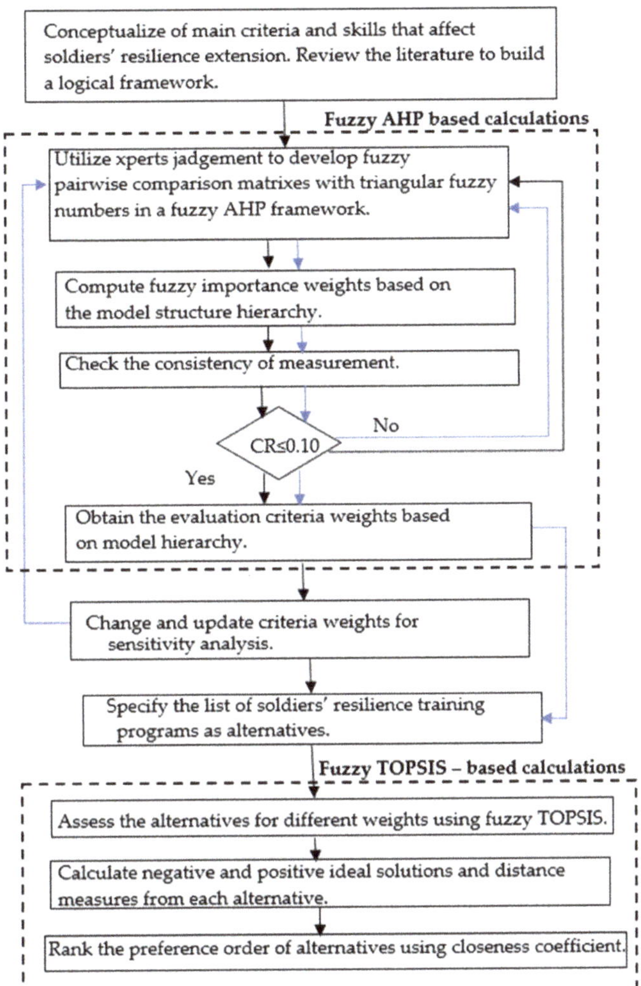

Figure 2. Assessment framework of soldiers' resilience building model.

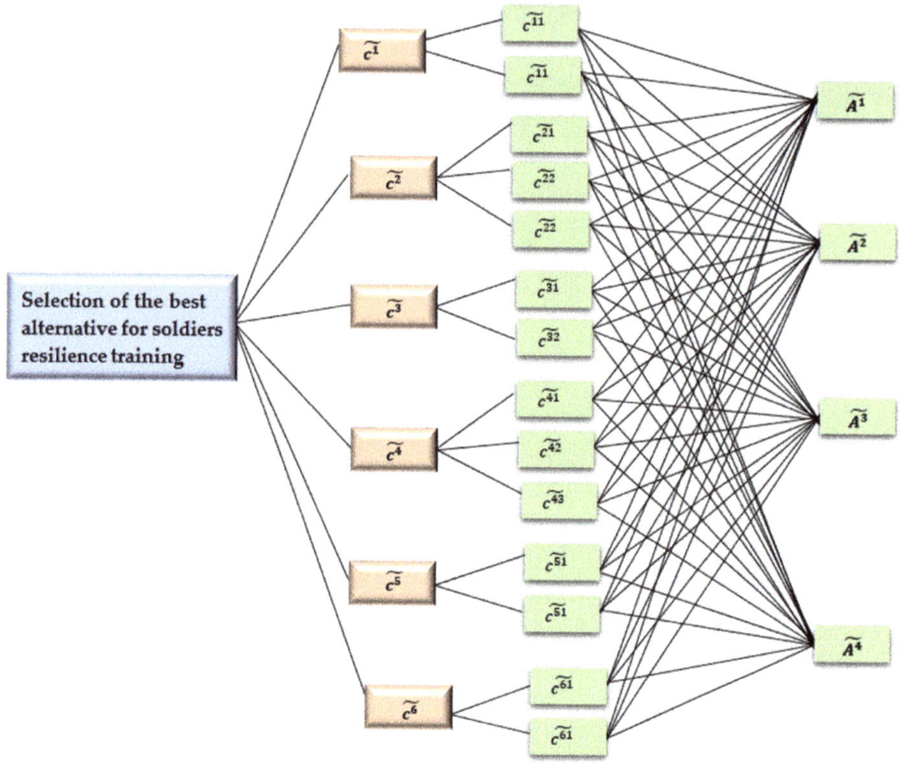

Figure 3. Hierarchical structure of soldiers' resilience-building model.

Mental agility $\left(\tilde{c}^{4}\right)$ can be represented by three skills, $\left(\tilde{c}^{41}\right)$, $\left(\tilde{c}^{42}\right)$, and $\left(\tilde{c}^{43}\right)$ [6,43,44]. $\left(\tilde{c}^{41}\right)$ defines the ability to identify and correct unproductive (fast, superficial) thinking (thoughts), apply mental cues (mental signs), and answer essential (critical) questions about the soldier's self to clarify information, determine what was omitted, etc. $\left(\tilde{c}^{42}\right)$ concerns a soldier's ability to carefully/thoroughly identify the cause of a problem and its solution strategies/methods, and how well a soldier can apply the six-step problem analysis and solution strategy, overcome the inertia of their thoughts, assess "confirmation bias" thoughts, and look at the situation "from the outside". $\left(\tilde{c}^{43}\right)$ is a soldier's ability to stop unproductive thinking (thoughts) in order to focus more on the task at hand, emphasizing the event, its positive aspects, and a positive perspective.

Strength of character $\left(\tilde{c}^{5}\right)$ can be represented by $\left(\tilde{c}^{51}\right)$ and $\left(\tilde{c}^{52}\right)$ [43,44]. $\left(\tilde{c}^{51}\right)$ is important because a soldier must act within a team. Therefore, soldiers have to be able to identify the strengths of one's own character and those of others, and the ways in which these strengths can be used to achieve personal effectiveness and strengthen positive relationships with others. Moreover, $\left(\tilde{c}^{52}\right)$ concerns how a soldier can identify their own and others' character strengths that help them to work in a team, overcome challenges, and be an effective leader.

Connection $\left(\tilde{c}^{6}\right)$ can be represented by $\left(\tilde{c}^{61}\right)$ and $\left(\tilde{c}^{62}\right)$ [6,43,44]. The $\left(\tilde{c}^{61}\right)$ skill represents a soldier's ability to communicate clearly and respectfully and apply the IDEAL model, which ensures trust, clarity and the ability to control the communication process. $\left(\tilde{c}^{62}\right)$ concerns a soldier's ability to give effective praise to promote excellence and motivate achievement, and to provide appropriate positive feedback to strengthen relationships with others. The structure of these competencies and skills is presented in Table 4.

Table 4. Competencies and skills as sub-factors that affect soldiers' resilience.

Competencies	Skills
1. **Self-awareness** $\left(\tilde{c}^{1}\right)$	**1.1.** ATC. Separate the A (activating Event) from their T (thoughts) and from the C (consequences: emotions and reactions) $\left(\tilde{c}^{11}\right)$. **1.2.** Detect icebergs $\left(\tilde{c}^{12}\right)$.
2. **Self-regulation** $\left(\tilde{c}^{2}\right)$	**2.1.** Goal setting $\left(\tilde{c}^{21}\right)$. **2.2.** Energy management $\left(\tilde{c}^{22}\right)$. **2.3.** Mental games $\left(\tilde{c}^{23}\right)$.
3. **Optimism** $\left(\tilde{c}^{3}\right)$	**3.1.** Hunt the good stuff $\left(\tilde{c}^{31}\right)$. **3.2.** Put it in perspective $\left(\tilde{c}^{32}\right)$.
4. **Mental agility** $\left(\tilde{c}^{4}\right)$	**4.1.** Avoid thinking traps $\left(\tilde{c}^{41}\right)$. **4.1.** Problem solving $\left(\tilde{c}^{42}\right)$. **4.2.** Real-time resilience $\left(\tilde{c}^{43}\right)$.
5. **Strengths of character** $\left(\tilde{c}^{5}\right)$	**5.1.** Identify character strengths in self and others $\left(\tilde{c}^{51}\right)$. **5.2.** Character strengths: challenges and leadership in themselves and in others $\left(\tilde{c}^{52}\right)$.
6. **Connection** $\left(\tilde{c}^{6}\right)$	**6.1.** Assertive communication $\left(\tilde{c}^{61}\right)$. **6.2.** Effective praise and active constructive responses $\left(\tilde{c}^{62}\right)$.

3.2. Training Programs for the Increasement of Soldiers' Resilience

In the present research, four different training programs focused on soldiers' resilience were chosen as alternatives. Descriptions of the four selected alternatives, whose primary target audience is soldiers, are presented below.

1. Army Center for Enhanced Performance $\left(\tilde{A}^{1}\right)$ [45]. The Army Center for Enhanced Performance (ACEP) strengthens the mind–body connection in addition to the development of psychological resilience. There are six components of training that lead to improved performance [45]: (1) mental skills' foundations, (2) building confidence, (3) goal setting, (4) attention control, (5) energy management, and (6) integrating imagery. This program is based on applied sport, health, and social psychology. Target audience—primarily soldiers.

2. Battlemind (also called Resiliency Training) $\left(\tilde{A}^2\right)$ [46]. Resilience training (RT) is designed to provide comprehensive mental training. It is designed to prepare soldiers to maintain good mental health despite the challenges of military life, combat, and transitioning once home. Resilience is developed as a soldier's inner strength, enabling him/her to face the challenges of his/her environment with courage and confidence. The program is based on a range of psychological theories, including cognitive restructuring, positive psychology, occupational health models, posttraumatic stress, mindfulness, etc.
3. Mindfulness-Based Mind Fitness Training $\left(\tilde{A}^3\right)$ [47]. This training consists of attention and concentration exercises for mindfulness, situational awareness, mental agility, emotion regulation, working memory, and more. These exercises change the structure and function of the brain. Training is carried out prior to deployment and is designed to protect the mental health of the soldiers in situations in which they are under stress. Studies have shown that the training program is beneficial and has reduced levels of PTSD, depression, and anxiety in soldiers upon return from deployment.
4. Master Resilience Training $\left(\tilde{A}^4\right)$ [48]. Master Resilience Training (MRT) is a standardized resilience training program. It is based on cognitive-behavioral and positive psychology methods. The program is based on Ellis' Adversity-Consequences-Beliefs (ABC) model and its effectiveness has been proven through empirical research.

4. Empirical Study Results

4.1. Data Collection Method

For this study, a cohort of 18 experts were interviewed using a pairwise comparison questionnaire. These experts were selected based on their professional competence, specifically their service experience in the field of resilience building, the length of their service in the military, and the completion of international missions. The 18 military psychologists involved in this study possessed extensive expertise in soldier resilience training, having continuously improved soldiers' resilience skills and post-deployment programming components. Eight of them were from Ukraine and ten from the Lithuania military area. Each expert conducted an independent evaluation by assessing six competencies and fourteen skills associated with soldier resilience through pairwise comparisons. In addition, these experts judged the four soldier resilience training programs. This study was conducted by researchers at the Military Academy of Lithuania in 2023.

4.2. Fuzzy AHP Analysis Results

The combined fuzzy analytic hierarchy process model was applied to measure the effect of critical resilience competencies in a two-level hierarchical structure. Accordingly, the fuzzy AHP model was performed using the following steps: (1) the Ukrainian and Lithuanian experts' verbal judgments were transformed into fuzzy weights that were connected with the triangular fuzzy number membership function specifications, as shown in Table 1; (2) the main criteria weighting was calculated; (3) the sub-criteria weighting was achieved. Following the analysis steps, the initial direct-relation matrixes were constructed for six resilience competencies and additionally for fourteen resilience skills. The Ukrainian and Lithuanian experts' opinions on the six main competencies presented in the initial direct-relation matrixes are shown in Table 5.

In the similar sequence, the matrixes of soldiers' resilience skills as sub-factors' of the resilience competencies were found and their corresponding fuzzy weights were computed. In addition, the consistency ratio (CR) coefficients were calculated to evaluate the consistency of the designed initial direct-relation matrixes.

Table 5. Experts' opinions on the six main competencies presented in the initial direct-relation matrix.

			DM1							*DM2*			
CC	$\tilde{c}^1$	$\tilde{c}^2$	$\tilde{c}^3$	$\tilde{c}^4$	$\tilde{c}^5$	$\tilde{c}^6$	CC	$\tilde{c}^1$	$\tilde{c}^2$	$\tilde{c}^3$	$\tilde{c}^4$	$\tilde{c}^5$	$\tilde{c}^6$
$\tilde{c}^1$	(1,1,1)	(1,1,1)	(1,1,1)	(1,1,1)	$\left(\frac{1}{2},1,\frac{3}{2}\right)$	$\left(3,\frac{7}{2},4\right)$	$\tilde{c}^1$	(1,1,1)	$\left(\frac{1}{2},1,\frac{3}{2}\right)$	$\left(\frac{3}{2},2,\frac{5}{2}\right)$	$\left(\frac{1}{2},1,\frac{3}{2}\right)$	$\left(3,\frac{7}{2},4\right)$	$\left(3,\frac{7}{2},4\right)$
$\tilde{c}^2$	(1,1,1)	(1,1,1)	(1,1,1)	(1,1,1)	$\left(1,\frac{3}{2},2\right)$	$\left(\frac{5}{2},3,\frac{7}{2}\right)$	$\tilde{c}^2$	$\left(\frac{2}{3},1,2\right)$	(1,1,1)	$\left(1,\frac{3}{2},2\right)$	$\left(\frac{1}{2},1,\frac{3}{2}\right)$	$\left(\frac{5}{2},3,\frac{7}{2}\right)$	$\left(\frac{3}{2},2,\frac{5}{2}\right)$
$\tilde{c}^3$	(1,1,1)	(1,1,1)	(1,1,1)	$\left(\frac{1}{2},1,\frac{3}{2}\right)$	$\left(\frac{3}{2},2,\frac{5}{2}\right)$	$\left(\frac{3}{2},2,\frac{5}{2}\right)$	$\tilde{c}^3$	$\left(\frac{2}{5},\frac{1}{2},\frac{2}{3}\right)$	$\left(\frac{1}{2},\frac{2}{3},1\right)$	(1,1,1)	$\left(\frac{1}{2},1,\frac{3}{2}\right)$	$\left(\frac{1}{2},1,\frac{3}{2}\right)$	$\left(\frac{1}{2},1,\frac{3}{2}\right)$
$\tilde{c}^4$	(1,1,1)	(1,1,1)	$\left(\frac{2}{3},1,2\right)$	(1,1,1)	$\left(\frac{5}{2},3,\frac{7}{2}\right)$	$\left(\frac{7}{2},4,\frac{9}{2}\right)$	$\tilde{c}^4$	$\left(\frac{2}{3},1,2\right)$	$\left(\frac{2}{3},1,2\right)$	$\left(\frac{2}{3},1,2\right)$	(1,1,1)	$\left(3,\frac{7}{2},4\right)$	$\left(\frac{5}{2},3,\frac{7}{2}\right)$
$\tilde{c}^5$	$\left(\frac{2}{3},1,2\right)$	$\left(\frac{1}{2},\frac{2}{3},1\right)$	$\left(\frac{2}{5},\frac{1}{2},\frac{2}{3}\right)$	$\left(\frac{2}{7},\frac{1}{3},\frac{2}{5}\right)$	(1,1,1)	(1,1,1)	$\tilde{c}^5$	$\left(\frac{1}{4},\frac{2}{7},\frac{1}{3}\right)$	$\left(\frac{2}{7},\frac{1}{3},\frac{2}{5}\right)$	$\left(\frac{2}{3},1,2\right)$	$\left(\frac{1}{4},\frac{2}{7},\frac{1}{3}\right)$	(1,1,1)	$\left(\frac{1}{2},1,\frac{3}{2}\right)$
$\tilde{c}^6$	$\left(\frac{1}{4},\frac{2}{7},\frac{1}{3}\right)$	$\left(\frac{2}{7},\frac{1}{3},\frac{2}{5}\right)$	$\left(\frac{2}{5},\frac{1}{2},\frac{2}{3}\right)$	$\left(\frac{2}{9},\frac{1}{4},\frac{2}{7}\right)$	(1,1,1)	(1,1,1)	$\tilde{c}^6$	$\left(\frac{1}{4},\frac{2}{7},\frac{1}{3}\right)$	$\left(\frac{2}{5},\frac{1}{2},\frac{2}{3}\right)$	$\left(\frac{2}{3},1,2\right)$	$\left(\frac{2}{7},\frac{1}{3},\frac{2}{5}\right)$	$\left(\frac{2}{3},1,2\right)$	(1,1,1)

Note: aggregated experts' opinions on six measurements, $\tilde{c}^1$ = Self-awareness, $\tilde{c}^2$ = Self-regulation, $\tilde{c}^3$ = Optimism, $\tilde{c}^4$ = Mental agility, $\tilde{c}^5$ = Strength of character, $\tilde{c}^6$ = Connection; DM1 = aggregated Ukrainian experts' assessment; DM2 = aggregated Lithuanian experts' assessment.

The global fuzzy weights of each of 14 skills were calculated using the value of a specific skill weight with the corresponding competence fuzzy weight. Calculated global fuzzy weights were used for ranking the best resilience training program by employing the fuzzy TOPSIS method. The global fuzzy weights based on the Ukrainian and Lithuanian experts' opinions are presented in Tables 6 and 7, correspondingly. Additionally, the ranks of the pairwise evaluation weights computed by fuzzy AHP for the six resilience competencies (see Table A1, Appendix A) and the fourteen resilience skills (see Table A2, Appendix A) were identified.

Table 6. Factor weight scores affecting soldiers' resilience levels based on Ukrainian experts' responses, established using the FAHP model.

Level 1	Level 2	*Global*
Competencies' Fuzzy Weight	*Skills' Fuzzy Weight*	*Fuzzy Weights*
W^1= (0.1397, 0.1907, 0.2404)	W^{11}= (0.5000, 0.5000, 0.5000)	$\hat{W}^{11}$= (0.0699, 0.0954, 0.1202)
	W^{12}= (0.5000, 0.5000, 0.5000)	$\hat{W}^{12}$= (0.0699, 0.0954, 0.1202)
W^2 = (0.1521, 0.1989, 0.2926)	W^{21}= (0.2602, 0.4357, 0.6597)	$\hat{W}^{21}$= (0.0396, 0.0663, 0.1003)
	W^{22}= (0.2659, 0.4100, 0.6897)	$\hat{W}^{22}$= (0.0404, 0.0624, 0.1049)
	W^{23}= (0.1103, 0.1543, 0.2220)	$\hat{W}^{23}$= (0.0168, 0.0235, 0.0338)
W^3 = (0.1331, 0.1950, 0.2590)	W^{31}= (0.5000, 0.5000, 0.5000)	$\hat{W}^{31}$= (0.0666, 0.0975, 0.1295)
	W^{32}= (0.5000, 0.5000, 0.5000)	$\hat{W}^{32}$= (0.0666, 0.0975, 0.1295)
W^4 = (0.1751, 0.2342, 0.3170)	W^{41}= (0.4898, 0.6253, 0.7732)	$\hat{W}^{41}$= (0.4898, 0.6253, 0.7732)
	W^{42}= (0.1348, 0.2056, 0.2950)	$\hat{W}^{42}$= (0.1348, 0.2056, 0.2950)
	W^{43}= (0.1296, 0.1690, 0.2577)	$\hat{W}^{43}$= (0.1296, 0.1690, 0.2577)
W^5 = (0.0757, 0.1073, 0.1606)	W^{51} = (0.5000, 0.5000, 0.5000)	$\hat{W}^{51}$= (0.0379, 0.0537, 0.0803)
	W^{52} = (0.5000, 0.5000, 0.5000)	$\hat{W}^{52}$= (0.0379, 0.0537, 0.0803)
W^6= (0.0562, 0.0740, 0.0967)	W^{61}= (0.5000, 0.5000, 0.5000)	$\hat{W}^{61}$= (0.0281, 0.0370, 0.0484)
	W^{62}= (0.5000, 0.5000, 0.5000)	$\hat{W}^{62}$= (0.0281, 0.0370, 0.0484)

Notes: W^1 = Self-Awareness, W^2 = Self-Regulation, W^3 = Optimism, W^4 = Mental agility, W^5 = Strength of character, W^6 = Connection.

4.3. Fuzzy TOPSIS Analysis Results

The calculated global fuzzy weights of the resilience skills were used as sub-factors weights for ranking the soldier resilience training programs using the fuzzy TOPSIS. To achieve the result, the following steps were performed: (1) developing the fuzzy decision matrix for the chosen alternatives and normalizing it; (2) computing the fuzzy positive and negative ideal solutions; (3) calculating the relative closeness and ranking the selected alternatives.

Table 7. Factor weight scores affecting soldiers' resilience level based on Lithuanian experts' responses, established using the FAHP model.

Level 1	Level 2	*Global*
Competencies' Fuzzy Weight	*Skills' Fuzzy Weight*	*Fuzzy Weights*
W^1 = (0.1377, 0.2578, 0.4361)	W^{11}= (0.4142, 0.5000, 0.8284) W^{12}= (0.2929, 0.5000, 0.5858)	$\hat{W}^{11}$= (0.0570, 0.1289, 0.3613) $\hat{W}^{12}$= (0.0403, 0.1289, 0.2555)
W^2 = (0.1167, 0.2182, 0.3877)	W^{21}= (0.2736, 0.4518, 0.6775) W^{22}= (0.2630, 0.4038, 0.6775) W^{23}= (0.1044, 0.1444, 0.2052)	$\hat{W}^{21}$= (0.0319, 0.0986, 0.2627) $\hat{W}^{22}$= (0.0307, 0.0881, 0.2627) $\hat{W}^{23}$= (0.0122, 0.0315, 0.0796)
W^3 = (0.0608, 0.1260, 0.2358)	W^{31}= (0.2679, 0.5000, 0.8038) W^{32}= (0.3094, 0.5000, 0.9282)	$\hat{W}^{31}$= (0.0163, 0.0630, 0.1896) $\hat{W}^{32}$= (0.1031, 0.0630, 0.2189)
W^4 = (0.1284, 0.2239, 0.4523)	W^{41}= (0.4542, 0.5772, 0.7079) W^{42}= (0.2320, 0.2989, 0.3980) W^{43}= (0.1050, 0.1238, 0.1580)	$\hat{W}^{41}$= (0.0583, 0.1292, 0.3201) $\hat{W}^{42}$= (0.0298, 0.0669, 0.1800) $\hat{W}^{43}$= (0.0135, 0.0277, 0.0714)
W^5 = (0.0479, 0.0830, 0.1472)	W^{51}= (0.6458, 0.7642, 0.8989) W^{52}= (0.2042, 0.2358, 0.2774)	$\hat{W}^{51}$= (0.0309, 0.0634, 0.1324) $\hat{W}^{52}$= (0.0098, 0.0196, 0.0408)
W^6= (0.0543, 0.0911, 0.1734)	W^{61}= (0.5798, 0.7143, 0.8697) W^{62}= (0.2367, 0.2857, 0.3551)	$\hat{W}^{61}$= (0.0315, 0.0651, 0.1508) $\hat{W}^{62}$= (0.0129, 0.0260, 0.0616)

Notes: W^1 = Self-Awareness, W^2 = Self-Regulation, W^3 = Optimism, W^4 = Mental agility, W^5 = Strength of character, W^6 = Connection.

As was mentioned before, in this study, we focused on four alternatives: (1) the Army Center for Enhanced Performance, (2) Resiliency Training, (3) Mindfulness-Based Mind Fitness Training, and (4) Master Resilience Training. Following TOPSIS methodology, the sub-factor $\tilde{c}^{21}$, which represents a soldier's ability to understand the components of the seven-step goal-setting process and apply this skill in planning steps to achieve personal and career goals, was marked as a non-beneficial attribute, whereas the other were beneficial. The conducted fuzzy TOPSIS analysis results are presented following the main analysis steps.

Fuzzy TOPSIS analysis began with defining the preference matrix of the four chosen soldier resilience training programs (alternatives) with respect to the 14 skills as sub-factors. For this, we used the linguistic values presented in Table 3. The experts' decision matrix on the four alternatives was constructed by following the linguistic preference of triangular fuzzy numbers characterized in Table 3. The preference matrix of the four alternatives for the 14 sub-factors expressed in linguistic terms is shown in Table 8.

Table 8. The preference matrix of the four alternatives for the 14 sub-factors, expressed in linguistic terms.

	$\tilde{c}^{11}$	$\tilde{c}^{12}$	$\tilde{c}^{21}$	$\tilde{c}^{22}$	$\tilde{c}^{23}$	$\tilde{c}^{31}$	$\tilde{c}^{32}$	$\tilde{c}^{41}$	$\tilde{c}^{42}$	$\tilde{c}^{43}$	$\tilde{c}^{51}$	$\tilde{c}^{52}$	$\tilde{c}^{61}$	$\tilde{c}^{62}$
$\tilde{A}^1$	M	VH	H	M	VH	H	M	VH	VH	VH	VH	H	VH	VH
$\tilde{A}^2$	M	M	VH	H	VH	M	H	M	M	M	VH	M	H	VH
$\tilde{A}^3$	VH	VH	M	VH	M	M	M	VH	VH	VH	L	L	L	M
$\tilde{A}^4$	M	M	M	M	M	M	M	M	M	M	M	VH	VH	M

Notes: descriptions of the linguistic terms are presented in Table 3.

Following the rules of the fuzzy TOPSIS method, the linguistic values were transformed into the corresponding triangular fuzzy numbers, and decision matrixes were normalized using Equations (25)–(27).

Next, Equation (25) was used to calculate the weighted, normalized fuzzy decision matrixes (WNFDM) for the two expert groups (Ukrainian and Lithuanian). The Ukrainian experts' data analysis result is presented in three tables according to the number of sub-factors (see Tables 9–11).

Table 9. Weighted normalized matrix for resilience competencies C^1 and C^2 by sub-factor.

DM1	c^{11}	c^{12}	c^{21}	c^{22}	c^{23}
A_1	0.023, 0.053, 0.09	0.054, 0.095, 0.120	0.013, 0.028, 0.060	0.013, 0.035, 0.082	0.013, 0.024, 0.034
A_2	0.023, 0.053, 0.093	0.023, 0.053, 0.093	0.013, 0.022, 0.043	0.022, 0.049, 0.105	0.013, 0.024, 0.034
A_3	0.054, 0.095, 0.120	0.054, 0.095, 0.120	0.017, 0.040, 0.100	0.031, 0.062, 0.105	0.006, 0.013, 0.026
A_4	0.023, 0.053, 0.093	0.023, 0.053, 0.093	0.017, 0.040, 0.100	0.013, 0.035, 0.082	0.006, 0.013, 0.026
DM2	c^{11}	c^{12}	c^{21}	c^{22}	c^{23}
A_1	0.019, 0.072, 0.281	0.031, 0.129, 0.256	0.011, 0.042, 0.158	0.010, 0.049, 0.204	0.009, 0.032, 0.080
A_2	0.019, 0.072, 0.281	0.013, 0.072, 0.199	0.011, 0.033, 0.113	0.017, 0.069, 0.263	0.009, 0.032, 0.080
A_3	0.044, 0.129, 0.361	0.031, 0.129, 0.256	0.014, 0.059, 0.263	0.024, 0.088, 0.263	0.004, 0.018, 0.062
A_4	0.019, 0.072, 0.281	0.013, 0.072, 0.199	0.014, 0.059, 0.263	0.010, 0.049, 0.204	0.004, 0.018, 0.062

Notes: DM1 = Ukrainian experts' assessment; DM2 = Lithuanian experts' assessment; C^1 = Self-awareness with sub-factors c^{11} and c^{12}, C^2 = Self-regulation with sub-factors c^{21}, c^{22}, and c^{23}.

Table 10. Weighted normalized matrix for resilience competencies C^3 and C^4 by sub-factor.

DM1	c^{31}	c^{32}	c^{41}	c^{42}	c^{43}
A_1	0.022, 0.042, 0.078	0.022, 0.054, 0.101	0.381, 0.625, 0.773	0.045, 0.069, 0.126	0.101, 0.169, 0.258
A_2	0.029, 0.059, 0.130	0.037, 0.076, 0.130	0.163, 0.347, 0.601	0.058, 0.123, 0.295	0.043, 0.094, 0.200
A_3	0.029, 0.059, 0.130	0.022, 0.054, 0.101	0.381, 0.625, 0.773	0.045, 0.069, 0.126	0.101, 0.169, 0.258
A_4	0.029, 0.059, 0.130	0.022, 0.054, 0.101	0.163, 0.347, 0.601	0.058, 0.123, 0.295	0.043, 0.094, 0.200
DM2	c^{31}	c^{32}	c^{41}	c^{42}	c^{43}
A_1	0.005, 0.027, 0.114	0.034, 0.035, 0.170	0.045, 0.129, 0.320	0.010, 0.022, 0.077	0.011, 0.028, 0.071
A_2	0.007, 0.038, 0.190	0.057, 0.049, 0.219	0.019, 0.072, 0.249	0.013, 0.040, 0.180	0.005, 0.015, 0.056
A_3	0.007, 0.038, 0.190	0.034, 0.035, 0.170	0.045, 0.129, 0.320	0.010, 0.022, 0.077	0.011, 0.028, 0.071
A_4	0.007, 0.038, 0.190	0.034, 0.035, 0.170	0.019, 0.072, 0.249	0.013, 0.040, 0.180	0.005, 0.015, 0.056

Notes: DM1 = Ukrainian experts' assessment; DM2 = Lithuanian experts' assessment; C^3 = Optimism with sub-factors c^{31} and c^{32}, C^4 = Mental agility with sub-factors c^{41}, c^{42}, and c^{43}.

Table 11. Weighted normalized matrix for resilience competencies C^5 and C^6 by sub-factor.

DM1	c^{51}	c^{52}	c^{61}	c^{62}
A_1	0.029, 0.054, 0.080	0.004, 0.008, 0.016	0.022, 0.037, 0.048	0.022, 0.037, 0.048
A_2	0.029, 0.054, 0.080	0.005, 0.011, 0.027	0.016, 0.029, 0.048	0.022, 0.037, 0.048
A_3	0.004, 0.018, 0.045	0.008, 0.018, 0.080	0.003, 0.012, 0.027	0.009, 0.021, 0.038
A_4	0.013, 0.030, 0.062	0.004, 0.006, 0.011	0.022, 0.037, 0.048	0.009, 0.021, 0.038
A_4	0.013, 0.030, 0.062	0.004, 0.006, 0.011	0.022, 0.037, 0.048	0.009, 0.021, 0.038
DM2	c^{51}	c^{52}	c^{61}	c^{62}
A_1	0.024, 0.063, 0.132	0.001, 0.003, 0.008	0.025, 0.065, 0.151	0.010, 0.026, 0.062
A_2	0.024, 0.063, 0.132	0.001, 0.004, 0.014	0.018, 0.051, 0.151	0.010, 0.026, 0.062
A_3	0.003, 0.021, 0.074	0.002, 0.007, 0.041	0.004, 0.022, 0.084	0.004, 0.014, 0.048
A_4	0.010, 0.035, 0.103	0.001, 0.002, 0.006	0.025, 0.065, 0.151	0.004, 0.014, 0.048

Notes: DM1 = Ukrainian experts' assessment; DM2 = Lithuanian experts' assessment; C^5 = Strength of character with sub-factors c^{51} and c^{52}, C^6 = Self-regulation with sub-factors c^{61} and c^{62}.

Consequently, we computed the *(FPIS, A*)* to assess the fuzzy positive ideal solution and the $(FNIS, A^-)$ as fuzzy negative ideal solution. The FPIS, A* represents the maximum value of $\tilde{v}_i^*$ for soldier resilience training programs which were included in this study, while $\tilde{v}_1^-$ is the minimum value (FNIS). The FPIS, A* and FNIS, A^- were calculated using Equations (29) and (30). Due to the number of sub-factors (14 skills) and page layout, the calculated results are separated and presented in three tables. The investigation results of the Ukrainian (DM1) and Lithuanian (DM2) expert groups are shown in Tables 12–14.

Table 12. Fuzzy positive ideal solution (FPIS, A*) and fuzzy negative ideal solution (FNIS, A^-) for the resilience competencies C^1 and C^2 by sub-factor.

DM1	c^{11}	c^{12}	c^{21}	c^{22}	c^{23}
FPIS, A*	0.0544, 0.0954, 0.1202	0.0544, 0.0954, 0.1202	0.0170, 0.0398, 0.1003	0.0314, 0.0624, 0.1049	0.0131, 0.0235, 0.0338
FNIS, A^-	0.0233, 0.0530, 0.0935	0.0233, 0.0530, 0.0935	0.0132, 0.0221, 0.0430	0.0135, 0.0347, 0.0816	0.0056, 0.0131, 0.0263
DM2	c^{11}	c^{12}	c^{21}	c^{22}	c^{23}
FPIS, A*	0.0443, 0.1289, 0.3613	0.0313, 0.1289, 0.2555	0.0137, 0.0592, 0.2627	0.0239, 0.0881, 0.2627	0.0095, 0.0315, 0.0796
FNIS, A^-	0.0190, 0.0716, 0.2810	0.0134, 0.0716, 0.1987	0.0106, 0.0329, 0.1126	0.0102, 0.0489, 0.2043	0.0041, 0.0175, 0.0619

Notes: DM1 = Ukrainian experts' assessment; DM2 = Lithuanian experts' assessment; C^1 = Self-awareness, C^1 = Self-regulation; FPIS, A*= Fuzzy positive ideal solution; FNIS, A^- = fuzzy negative ideal solution.

Table 13. Fuzzy positive ideal solution (FPIS, A*) and fuzzy negative ideal solution (FNIS, A^-) for the resilience competencies C^3 and C^4 by sub-factor.

DM1	c^{31}	c^{32}	c^{41}	c^{42}	c^{43}
FPIS, A*	0.0285, 0.0585, 0.1295	0.0370, 0.0758, 0.1295	0.3810, 0.6253, 0.7732	0.0578, 0.1234, 0.2950	0.1008, 0.1690, 0.2577
FNIS, A^-	0.0222, 0.0418, 0.0777	0.0222, 0.0542, 0.1007	0.1633, 0.3474, 0.6014	0.0449, 0.0685, 0.1264	0.0432, 0.0939, 0.2004
DM2	c^{31}	c^{32}	c^{41}	c^{42}	c^{43}
FPIS, A*	0.0070, 0.0378, 0.1896	0.0573, 0.0490, 0.2189	0.0453, 0.1292, 0.3201	0.0128, 0.0401, 0.1800	0.0105, 0.0277, 0.0714
FNIS, A^-	0.0054, 0.0270, 0.1138	0.0344, 0.0350, 0.1703	0.0194, 0.0718, 0.2490	0.0099, 0.0223, 0.0771	0.0045, 0.0154, 0.0555

Notes: DM1 = Ukrainian experts' assessment; DM2 = Lithuanian experts' assessment; C^3 = Optimism, C^4 = Mental agility. FPIS, A*= Fuzzy positive ideal solution; FNIS, A^- = fuzzy negative ideal solution.

Table 14. Fuzzy positive ideal solution (FPIS, A*) and fuzzy negative ideal solution (FNIS, A^-) for the resilience competencies C^5 and C^6 by sub-factor.

DM1	c^{51}	c^{52}	c^{61}	c^{62}
FPIS, A *	0.0295, 0.0537, 0.0803	0.0076, 0.0179, 0.0803	0.0219, 0.0370, 0.0484	0.0219, 0.0370, 0.0484
FNIS, A^-	0.0042, 0.0179, 0.0446	0.0042, 0.0060, 0.0115	0.0031, 0.0123, 0.0269	0.0094, 0.0206, 0.0376
DM2	c^{51}	c^{52}	c^{61}	c^{62}
FPIS, A *	0.0240, 0.0634, 0.1324	0.0020, 0.0065, 0.0408	0.0245, 0.0651, 0.1508	0.0100, 0.0260, 0.0616
FNIS, A^-	0.0034, 0.0211, 0.0736	0.0011, 0.0022, 0.0058	0.0035, 0.0217, 0.0838	0.0043, 0.0144, 0.0479

Notes: DM1 = Ukrainian experts' assessment; DM2 = Lithuanian experts' assessment; C^5 = Strength of character, C^6 = Self-regulation. FPIS, A*= Fuzzy positive ideal solution; FNIS, A^- = fuzzy negative ideal solution.

The distances between the four resilience training programs were assessed. The positive and negative ideal solutions and final ranking of each resilience training program are presented in the table below (see Table 15).

Table 15. Identified distances from positive FPIS to negative FNIS ideal solutions and final rankings of alternatives.

		DM1					*DM2*		
Alternative	S_i^+	S_i^-	$^1CC_i=\frac{d_i^-}{d_i^++d_i^-}$	Rank	Alternative	S_i^+	S_i^-	$^1CC_i=\frac{d_i^-}{d_i^++d_i^-}$	Rank
A1	0.2759	0.4144	0.6003	2	**A1**	0.3172	0.2580	0.449	4
A2	0.4400	0.2544	0.3663	3	**A2**	0.2985	0.2876	0.491	2
A3	0.2015	0.4887	0.7080	1	**A3**	0.2078	0.3674	0.639	1
A4	0.4866	0.2040	0.2954	4	**A4**	0.3163	0.2595	0.451	3

Notes: DM1 = Ukrainian experts' assessment; DM2 = Lithuanian experts' assessment. d_i^+ = distance between alternative and FPIS, S_i^+; d_i^-= distance between alternative and FNIS, S_i^-; CC_i= closeness coefficient of each resilience training program judged as an alternative.

To illustrate the soldier resilience training programs' rankings, the closeness coefficient was chosen (CC_i), and the graphical results are presented in Figure 4. The larger values of CC_i indicate the most preferred alternatives, and the alternative A3 marks the maximum (0.7080) value of closeness coefficient, whereas the alternative A4 marks the lowest value of 0.2954.

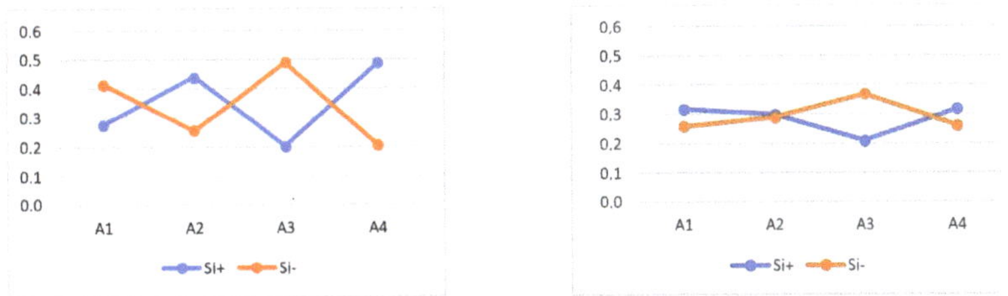

(**a**) (**b**)

Figure 4. The geometric distances from FPIS (di*) and FNIS (di⁻) of four alternatives reached as study result: (**a**) Ukrainian experts' opinion; (**b**) Lithuanian experts' opinion. A detailed description of presented values is shown in Table 15.

Based on the conducted research on soldier resilience training and the opinion analysis of Ukrainian and Lithuanian experts, a ranking of military resilience training programs was compiled. Lastly, a sensitivity examination was performed to evaluate the effect of the changed weights on dissimilar resilience competencies as factors, and between resilience skills as sub-factors.

5. Sensitivity Analysis

The results of the final rankings let us identify that the best soldier resilience training program choice was A3 according to the Ukrainian and Lithuanian experts' judgement (see Table 16). Therefore, a sensitivity exploration was performed to evaluate the effect of sub-factors' weights on the best resilience training program choice. Consequently, the different sub-factors were eliminated, and dissimilar cases of analysis were conducted. The sensitivity evaluation analysis showed the dissimilar rankings for soldier resilience training programs. The investigation outcomes are presented in Figure 5 and Table 16.

Table 16. Resilience training programs' rankings identified under different cases analysis.

Alternative	*DM1*			Alternative	*DM2*		
	Case (1)	Case (2)	Case (3)		Case (1)	Case (2)	Case (3)
A1	2	3	4	A1	4	2	4
A2	3	2	2	A2	1	4	2
A3	1	1	1	A3	3	1	1
A4	4	4	3	A4	2	3	3

Notes: DM1 = Ukrainian experts' assessment; DM2 = Lithuanian experts' assessment.

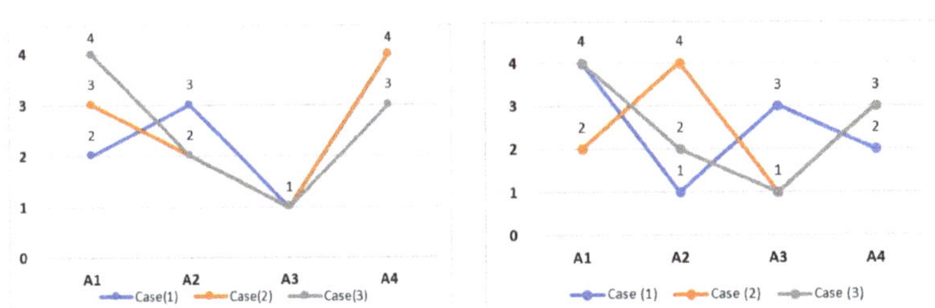

(**a**) (**b**)

Figure 5. A graphical presentation of the sensitivity analysis results: (**a**) analysis result for the evaluation of the Ukrainian experts' opinion; (**b**) analysis result for the evaluation of the Lithuanian experts' opinion.

The cases of sensitivity analysis were performed following a rank analysis using skills as sub-factors. The Ukrainian experts pointed out that the three mental agility competence skills were their top priorities (see Table A1, Appendix A). Therefore, we used these vital sub-factors and conducted three different case analyses:

- Case 1. The sub-factors 'Hunt the good stuff' $\left(\tilde{c}^{31}\right)$ and 'Put it in perspective' $\left(\tilde{c}^{32}\right)$ were eliminated, and the obtained ranking showed the altered result.
- Case 2. The sub-factors 'Avoid thinking traps' $\left(\tilde{c}^{41}\right)$, 'Problem solving' $\left(\tilde{c}^{42}\right)$, and 'Real-time resilience' $\left(\tilde{c}^{43}\right)$ were eliminated, as these skills (sub-factors) were identified as vital for Ukrainian soldier resilience training. The elimination of the mental agility training part produced the different rankings of the resilience training programs.
- Case 3. The sub-factor 'Avoid thinking traps' $\left(\tilde{c}^{41}\right)$ was eliminated, and consequently, a different ranking was achieved, because this skill was ranked as a top interest.

The Lithuanian experts pointed out that three skills, two of which belonged to the Self-awareness competence $\left(\tilde{c}^{1}\right)$ and one of which represented the mental agility competence $\left(\tilde{c}^{4}\right)$, were their top priorities. Consequently, we used these essential sub-factors and conducted three different case analyses:

- Case 1. The two sub-factors 'Separate the A (activating Event) from their T (thoughts) and from the C (consequences: emotions and reactions)' $\left(\mathrm{ATC}, \tilde{c}^{11}\right)$ and 'Detect icebergs' $\left(\tilde{c}^{12}\right)$ were eliminated and the obtained ranking showed different ranking results for the four training programs.
- Case 2. The sub-factors 'Hunt the good stuff' $\left(\tilde{c}^{31}\right)$ and 'Put it in perspective' $\left(\tilde{c}^{32}\right)$ were eliminated and the obtained rankings showed different results.
- Case 3. The sub-factor 'Avoid thinking traps' $\left(\tilde{c}^{41}\right)$ was eliminated, and consequently, a different ranking was achieved, because the mental agility competence $\tilde{c}^{41}$ sub-factor was ranked as a skill of top importance.

Sensitivity analysis showed that the different weights of the resilience competencies and the skills used as sub-factors in this study lead to changes in the resilience training programs' rankings. The sensitivity analyses completed using different scenarios clearly illustrate the sensitivity of ideal rankings established based on military psychologists' opinions. After taking into account the differences in the specifics of the military services of today's Ukrainian and Lithuanian soldiers and the fact that several resilience training programs were chosen across different scenarios, the analysis showed the differences in assessment, program selection, and ratings.

In addition, to confirm the success of the assessment model proven in this study, the outcomes were matched with the ranking results of the traditional Technique for Order of Preference by Similarity to Ideal Solution (TOPSIS) method and the grey relational analysis (GRA) method. The conducted calculation outcomes are presented in Table A3 (Appendix A). The ranking results of the soldier resilience training programs achieved by the model suggested in this study are not meaningfully dissimilar from the ranking outcomes of the traditional TOPSIS and GRA methods in general, for both DM1 and DM2. Table A3 shows that the results of the model developed in this study are close to those obtained by the traditional method, which indicates that the model developed in this study is useful and correct.

6. Discussion

The present study built upon existing research on competence training for resilience in militaries by utilizing a hierarchy MCDA model based on fuzzy sets theory. This model helped determine which competencies and skills should be included in resilience training programs to make them the most effective in a contemporary military environment. Previous studies [1,5,46,48–50] have identified sets of competencies and skills that are essential for solders to develop during the pre-deployment period to allow them to recover from stressful situations quickly and efficiently. Insufficient resilience training has been linked to a variety of negative outcomes, such as disturbed sleep habits, low energy, headaches, and other disorders [51].

The current study indicated that, according to experts from Ukraine and Lithuania, the MMFT (A3) is the most effective training program for fostering resilience in the pre-deployment period, which can help protect against harmful levels of stress. The validity and effectiveness of this program have been proven through rigorous research in neuroscience and stress physiology [52]. Previous studies have shown that this program increases an individual's tolerance in high-stress contexts, and that after MMFT training, soldiers' attention, memory, and sleep quality obviously improved [53]. In this study, the impact of MMFT as a form of resilience training was found to be significant by both Ukrainian and Lithuanian psychologists. The MMFT's maximum relative closeness coefficient was determined by Ukraine experts as follows: MMFT (A3)> ACEP (A1)>RT (A2)> MRT (A4). Furthermore, the maximum relative closeness coefficients of the resilience training programs were also identified according to the opinion of the Lithuanian experts, whose rankings arranged the resilience training programs in the following order: MMFT (A3)>RT (A2)> MRT (A4)> ACEP (A1).

The results of the current study contribute to the identification of the most important competencies that are relevant in today's military environment. Using sensitivity analysis on the Ukrainian experts' data, it was determined that mental agility competence, including the three skills 'Avoid thinking traps' $\left(\tilde{c}^{41}\right)$, 'Problem solving' $\left(\tilde{c}^{42}\right)$, and 'Real-time resilience' $\left(\tilde{c}^{41}\right)$, contributes most to solders' resilience in a real combat environment. Meanwhile, according to the assessment of Lithuanian military psychologists, who participate in training soldiers for military missions and training, the most contributing skills are 'ATC' $\left(\tilde{c}^{11}\right)$, 'Detect icebergs' $\left(\tilde{c}^{12}\right)$, and 'Avoid thinking traps' $\left(\tilde{c}^{41}\right)$. As can be seen from the results of this study, the necessary competencies and skills for resilience training are not identical between those for soldiers participating in conventional war and those intended for soldiers participating in military training and missions.

Following previous research that demonstrated the efficacy of MMFT in mediating stress reduction and improving psychological functioning [54], this study predicted that modifying the MMFT program could result in even greater effects on the development of resilience of participants.

The present study has several limitations that need to be considered when interpreting the findings. First, the experts selected for this study were from Ukraine and Lithuania, with very different geopolitical situations; therefore, the opinions of the experts may have been affected by these circumstances. Second, the study only examined resilience training programs based on different concepts, which may not cover all available military resilience programs. Third, the subjectivity of the experts' opinions introduces a level of uncertainty, which could be reduced by including more experts in the analysis. Given these limitations, caution is necessary when interpreting the findings. To better understand the quality assessment process, additional analyses could be conducted with a broader range of resilience competencies and more alternative resilience training programs.

7. Conclusions

Imprecise qualitative decisions regarding what the core competencies are for solders' residence building, and regarding what the most effective training program is, can be improved by applying the FAHP and the FTOPSIS as MCDM techniques. Apparently, the application of fuzzy AHP weights in fuzzy TOPSIS helps to reach farther realistic and reliable results. This study found that there were differences in preferences regarding resilience criteria and skills among well-known soldier resilience training programs. Therefore, the choice to apply MCDM techniques can be effectively used while ranking the best training program, given the existence of complex and imprecise constraints.

The sensitivity analysis carried out in this study provided valuable information on the impact of the soldier resilience training programs on the ranking process. Based on the results of the FTOPSIS and the sensitivity analysis, it was concluded that MMFT (A3) is the most effective alternative to resilience training for Ukrainian and Lithuanian soldiers. Overall, the findings of this study suggest that MMFT could be an effective tool for building soldiers' resilience and preparing them for the demands of contemporary military environments. In addition, according to the sensitivity analysis, MMFT attained the highest ranking according to the evaluation results of Ukrainian experts and was ranked top or third according to the opinion of Lithuanian experts.

The present study contributes to the existing literature by presenting MCDM-based solutions to address the issue of insufficient accuracy in capturing experts' decision making when selecting the optimal resilience training program for military personnel in high-stress environments. The MCDA model, based on fuzzy sets theory, highlights mental agility as the most critical competence. Since the current research focused solely on conventional resilience competencies included in the resilience training program, the list of competencies used here is non-exhaustive. Some of the competencies might be weakly expressed and underestimated by experts despite their relevance to resilience. Therefore, future research should incorporate additional resilience competencies in the evaluation, and alternative MCDM techniques such as PROMETHEE with fuzzy logic, interval numbers, or hesitant fuzzy sets should be explored.

Author Contributions: Conceptualization by R.K. and S.B.; methodology by R.K. and S.B.; software by S.B.; validation by O.N. and S.B.; formal analysis by S.B., resources by S.B.; writing for original draft preparation by R.S., R.V., O.N., D.B. and S.B.; writing for review and editing by S.B., O.K. and R.S.; visualization by S.B.; supervision by S.B.; project administration by S.B.; and funding acquisition by S.B. All authors have read and agreed to the published version of the manuscript.

Funding: This research was funded by the Research Council of Lithuania (LMTLT) under project agreement No S-LU-22-9; the principal investigator of the grant was Svajone Bekesiene.

Institutional Review Board Statement: The study was approved by the General Jonas Zemaitis Military Academy, Protocol No. PR-1815.

Informed Consent Statement: Informed consent was obtained from all subjects involved in the study.

Data Availability Statement: Not applicable.

Conflicts of Interest: The authors declare no conflict of interest. The funder had no role in the design of the study; in the collection, analyses, or interpretation of data; in the writing of the manuscript, or in the decision to publish the results.

Appendix A

Table A1. FAHP criteria importance weight scores presented by the two expert groups.

[1] Sub-Criteria	Ukrainian Experts		[1] Sub-Criteria	Lithuanian Experts	
	[2] FAHP	Rank		[2] FAHP	Rank
$\tilde{C}^1$	0.1842	4	$\tilde{C}^1$	**0.2462**	**2**
$\tilde{C}^2$	**0.2077**	**2**	$\tilde{C}^2$	**0.2139**	**3**
$\tilde{C}^3$	**0.1895**	**3**	$\tilde{C}^3$	0.1251	4
$\tilde{C}^4$	**0.2344**	**1**	$\tilde{C}^4$	**0.2382**	**1**
$\tilde{C}^5$	0.1109	5	$\tilde{C}^5$	0.0823	6
$\tilde{C}^6$	0.0732	6	$\tilde{C}^6$	0.0944	5

Notes: [1] Criteria are presented in Table 3 as skills. [2] FAHP = pairwise comparison weights calculated for the criteria using the fuzzy analytic hierarchy process. Bold numbers represent three important ranks.

Table A2. The FAHP sub-criteria importance weight scores presented by the two expert groups.

[1] Sub-Criteria	Ukrainian Experts		[1] Sub-Criteria	Lithuanian Experts	
	[2] FAHP	Rank		[2] FAHP	Rank
$\hat{c}^{11}$	0.0539	5	$\hat{c}^{11}$	**0.1346**	**1**
$\hat{c}^{12}$	0.0539	5	$\hat{c}^{12}$	**0.1045**	**3**
$\hat{c}^{21}$	0.0389	6	$\hat{c}^{21}$	0.0967	4
$\hat{c}^{22}$	0.0392	6	$\hat{c}^{22}$	0.0938	6
$\hat{c}^{23}$	0.0140	9	$\hat{c}^{23}$	0.0303	11
$\hat{c}^{31}$	0.0554	4	$\hat{c}^{31}$	0.0661	8
$\hat{c}^{32}$	0.0554	4	$\hat{c}^{32}$	0.0947	5
$\hat{c}^{41}$	**0.3565**	**1**	$\hat{c}^{41}$	**0.1249**	**2**
$\hat{c}^{42}$	**0.1200**	**2**	$\hat{c}^{42}$	0.0681	7
$\hat{c}^{43}$	**0.1050**	**3**	$\hat{c}^{43}$	0.0277	12
$\hat{c}^{51}$	0.0325	7	$\hat{c}^{51}$	0.0558	10
$\hat{c}^{52}$	0.0325	7	$\hat{c}^{52}$	0.0173	14
$\hat{c}^{61}$	0.0214	8	$\hat{c}^{61}$	0.0609	9
$\hat{c}^{62}$	0.0214	8	$\hat{c}^{62}$	0.0247	13

Notes: [1] Sub-criteria are presented in Table 3 as skills. [2] FAHP = pairwise comparison weights calculated for the sub-criteria using the fuzzy analytic hierarchy process. Bold numbers represent three important ranks.

Table A3. Comparison of ranking results of different models.

Alternative	Fuzzy TOPSIS		TOPSIS		Grey Relational Analysis Method	
	Distance Closeness	Rank	Distance Closeness	Rank	Distance Closeness	Rank
	DM1					
A1	0.6003	2	0.7303	2	0.7375	2
A2	0.3663	3	0.3857	3	0.5881	3
A3	0.7080	1	0.7543	1	0.7557	1
A4	0.2954	4	0.1464	4	0.4700	4
	DM2					
A1	0.4486	4	0.3609	4	0.6145	4
A2	0.6387	1	0.7153	1	0.7102	1
A3	0.4907	2	0.6380	2	0.7047	2
A4	0.4506	3	0.4773	3	0.6490	3

Notes: DM1 = Ukrainian experts' assessment; DM2 = Lithuanian experts' assessment.

References

1. Howard, J.D.; Kupczynski, L.; Groff, S.L.; Gibson, A.M. A Quantitative Analysis of Resilience Training's Influence on Army National Guardsmen Resilience and Performance. *Adv. J. Educ. Soc. Sci.* **2022**, *7*, 26–46.
2. EU Response to Russia's Invasion of Ukraine–Consilium. Available online: https://www.consilium.europa.eu/en/policies/eu-response-ukraine-invasion/ (accessed on 21 March 2023).
3. Ukraine: Civilian Casualty Update 20 March 2023. OHCHR. Available online: https://www.ohchr.org/en/news/2023/03/ukraine-civilian-casualty-update-20-march-2023 (accessed on 21 March 2023).
4. Gottschall, S.; Guérin, E. Organizational and Non-Organizational Risk and Resilience Factors Associated with Mental Health and Well-Being in the Royal Canadian Navy. *Curr. Psychol.* **2021**, *42*, 6179–6193. [CrossRef]
5. Reivich, K.J.; Seligman, M.E.P.; McBride, S. Master Resilience Training in the U.S. Army. *Am. Psychol.* **2011**, *66*, 25–34. [CrossRef] [PubMed]
6. Lester, P.B.; Harms, P.D.; Herian, M.N.; Krasikova, D.V.; Beal, S.J. Report #3: Longitudinal Analysis of the Impact of Master Resilience Training on Self-Reported Resilience and Psychological Health Data. *Compr. Soldier Fit. Program Eval.* **2011**, *3*. [CrossRef]
7. Chatterjee, K.; Kar, S. Unified Granular-Number-Based AHP-VIKOR Multi-Criteria Decision Framework. *Granul. Comput.* **2017**, *2*, 199–221. [CrossRef]
8. Chatterjee, K.; Kar, S. Multi-Criteria Analysis of Supply Chain Risk Management Using Interval Valued Fuzzy TOPSIS. *Opsearch* **2016**, *53*, 474–499. [CrossRef]
9. Fang, J.; Partovi, F.Y. Criteria Determination of Analytic Hierarchy Process Using a Topic Model. *Expert Syst. Appl.* **2021**, *169*, 114306. [CrossRef]
10. Yariyan, P.; Zabihi, H.; Wolf, I.D.; Karami, M.; Amiriyan, S. Earthquake Risk Assessment Using an Integrated Fuzzy Analytic Hierarchy Process with Artificial Neural Networks Based on GIS: A Case Study of Sanandaj in Iran. *Int. J. Disaster Risk Reduct.* **2020**, *50*, 101705. [CrossRef]
11. Mujiya Ulkhaq, M.; Nartadhi, R.L.; Akshinta, P.Y. Evaluating Service Quality of Korean Restaurants: A Fuzzy Analytic Hierarchy Approach. *Ind. Eng. Manag. Syst.* **2016**, *15*, 77–91. [CrossRef]
12. Sharma, D.; Puppala, H.; Asthana, R.; Uddin, Z. Fuzzy Analytical Hierarchy Process to Evaluate the Curriculum of an Undergraduate Program with a Vision to Design an Industry-Ready Undergraduate Engineering Program. *J. MESA* **2022**, *13*, 703–713.
13. Bekesiene, S.; Vasiliauskas, A.V.; Hošková-Mayerová, Š.; Vasilienė-Vasiliauskienė, V. Comprehensive Assessment of Distance Learning Modules by Fuzzy AHP-TOPSIS Method. *Mathematics* **2021**, *9*, 409. [CrossRef]
14. Yilmaz, H.; Karadayi-Usta, S.; Yanık, S. A Novel Neutrosophic AHP-Copeland Approach for Distance Education: Towards Sustainability. *Interact. Learn. Environ.* **2022**, 1–23. [CrossRef]
15. Wibowo, A.S.; Permanasari, A.E.; Fauziati, S. Combat Aircraft Effectiveness Assessment Using Hybrid Multi-Criteria Decision Making Methodology. In Proceedings of the 2016 2nd International Conference on Science and Technology-Computer, Yogyakarta, Indonesia, 27–28 October 2016.
16. Wang, J.; Fan, K.; Su, Y.; Liang, S.; Wang, W. Air Combat Effectiveness Assessment of Military Aircraft Using a Fuzzy AHP and TOPSIS Methodology. In Proceedings of the 2008 Asia Simulation Conference-7th International Conference on System Simulation and Scientific Computing, Beijing, China, 10–12 October 2008.
17. Sánchez-Lozano, J.M.; Rodríguez, O.N. Application of Fuzzy Reference Ideal Method (FRIM) to the Military Advanced Training Aircraft Selection. *Appl. Soft Comput. J.* **2020**, *88*, 106061. [CrossRef]
18. Ghorui, N.; Ghosh, A.; Algehyne, E.A.; Mondal, S.P.; Saha, A.K. AHP-TOPSIS Inspired Shopping Mall Site Selection Problem with Fuzzy Data. *Mathematics* **2020**, *8*, 1380. [CrossRef]
19. Wang, C.-N.; Nguyen, T.-L.; Dang, T.-T. Two-Stage Fuzzy MCDM for Green Supplier Selection in Steel Industry. *Intell. Autom. Soft Comput.* **2022**, *33*, 1245–1260. [CrossRef]
20. Basilio, M.; Pereira, V.; Costa, H.; Santos, M.; Ghosh, A. A Systematic Review of the Applications of Multi-Criteria Decision Aid Methods (1977–2022). *Electronics* **2022**, *11*, 1720. [CrossRef]
21. Jiang, Y.; Yang, C.; Ma, H. A Review of Fuzzy Logic and Neural Network Based Intelligent Control Design for Discrete-Time Systems. *Discret. Dyn. Nat. Soc.* **2016**, *2016*, 7217364. [CrossRef]
22. Zadeh, L.A. Fuzzy Sets. *Inf. Control* **1965**, *8*, 338–353. [CrossRef]
23. Chen, G.; Pham, T.T. *Introduction to Fuzzy Sets, Fuzzy Logic, and Fuzzy Control Systems*; CRC Press: Boca Raton, FL, USA, 2019; ISBN 9780367397883.
24. Chang, C.-W.; Wu, C.-R.; Lin, H.-L. Applying Fuzzy Hierarchy Multiple Attributes to Construct an Expert Decision Making Process. *Expert Syst. Appl.* **2009**, *36*, 7363–7368. [CrossRef]
25. Zhu, K.J.; Jing, Y.; Chang, D.Y. A discussion on extent analysis method and applications of fuzzy AHP. *Eur. J. Oper. Res.* **1999**, *116*, 450–456. [CrossRef]
26. Chang, D.Y. Applications of the Extent Analysis Method on Fuzzy AHP. *Eur. J. Oper. Res.* **1996**, *95*, 649–655. [CrossRef]
27. Saaty, T.L. A scaling method for priorities in a hierarchical structure. *J. Math. Psychol.* **1977**, *15*, 234–281. [CrossRef]
28. Ishizaka, A.; Labib, A. Analytic Hierarchy Process and Expert Choice: Benefits and Limitations. *OR Insight* **2009**, *22*, 201–220. [CrossRef]

29. Kulak, O.; Kahraman, C. Fuzzy Multi-Attribute Selection among Transportation Companies Using Axiomatic Design and Analytic Hierarchy Process. *Inf. Sci.* **2005**, *170*, 191–210. [CrossRef]
30. Erensal, Y.C.; Öncan, T.; Demircan, M.L. Determining Key Capabilities in Technology Management Using Fuzzy Analytic Hierarchy Process: A Case Study of Turkey. *Inf. Sci.* **2006**, *176*, 2755–2770. [CrossRef]
31. Bozbura, F.T.; Beskese, A.; Kahraman, C. Prioritization of Human Capital Measurement Indicators Using Fuzzy AHP. *Expert Syst. Appl.* **2007**, *32*, 1100–1112. [CrossRef]
32. Dağdeviren, M.; Yüksel, I. Developing a Fuzzy Analytic Hierarchy Process (AHP) Model for Behavior-Based Safety Management. *Inf. Sci.* **2008**, *178*, 1717–1733. [CrossRef]
33. Kwong, C.K.; Bai, H. Determining the Importance Weights for the Customer Requirements in QFD Using a Fuzzy AHF with an Extent Analysis Approach. *IIE Trans. Inst. Ind. Eng.* **2003**, *35*, 619–626. [CrossRef]
34. Chan, F.T.S.; Kumar, N. Global Supplier Development Considering Risk Factors Using Fuzzy Extended AHP-Based Approach. *Omega* **2007**, *35*, 417–431. [CrossRef]
35. Alavi, I. Fuzzy Ahp Method for Plant Species Selection in Mine Reclamation Plans: Case Study Sungun Copper Mine. *Iran. J. Fuzzy Syst.* **2014**, *11*, 23–38.
36. Pant, S.; Kumar, A.; Ram, M.; Klochkov, Y.; Sharma, H.K. Consistency Indices in Analytic Hierarchy Process: A Review. *Mathematics* **2022**, *10*, 1206. [CrossRef]
37. Shih, H.S.; Shyur, H.J.; Lee, E.S. An Extension of TOPSIS for Group Decision Making. *Math. Comput. Model.* **2007**, *45*, 801–813. [CrossRef]
38. Griffith, J.; West, C. Master Resilience Training and Its Relationship to Individual Well-Being and Stress Buffering among Army National Guard Soldiers. *J. Behav. Health Serv. Res.* **2013**, *40*, 140–155. [CrossRef]
39. Montgomery, W.H. *Beyond Words: Leader Self-Awareness and Interpersonal Skills*; U.S. Army War College: Carlisle, UK, 2007.
40. Mitchell, M.S.; Greenbaum, R.L.; Vogel, R.M.; Mawritz, M.B.; Keating, D.J. Can You Handle the Pressure? The Effect of Performance Pressure on Stress Appraisals, Self-Regulation, and Behavior. *Acad. Manag. J.* **2019**, *62*, 531–552. [CrossRef]
41. Rozek, D.C.; Keane, C.; Sippel, L.M.; Stein, J.Y.; Rollo-Carlson, C.; Bryan, C.J. Short-Term Effects of Crisis Response Planning on Optimism in a U.S. Army Sample. *Early Interv. Psychiatry* **2019**, *13*, 682–685. [CrossRef]
42. Hassett, A.L.; Fisher, J.A.; Vie, L.L.; Kelley, W.L.; Clauw, D.J.; Seligman, M.E.P. Association between Predeployment Optimism and Onset of Postdeployment Pain in US Army Soldiers. *JAMA Netw. Open* **2019**, *2*, e188076. [CrossRef]
43. Chopik, W.J.; Kelley, W.L.; Vie, L.L.; Oh, J.; Bonett, D.G.; Lucas, R.E.; Seligman, M.E.P. Development of Character Strengths across the Deployment Cycle among U.S. Army Soldiers. *J. Pers.* **2021**, *89*, 23–34. [CrossRef]
44. MRT Skills Overview; The Trustees of the University of Pennsylvania: Philadelphia, PA, USA, 2014. Available online: https://drum.armymwr.com/application/files/9914/9332/0727/Drum_Resilience_Training_Skills_Overview.pdf (accessed on 21 March 2023).
45. Dibble, J.J. Army Center for Enhanced Performance. *Prof. Forum.* 2015, pp. 22–25. Available online: https://www.benning.army.mil/infantry/magazine/issues/2015/Apr-Jun/pdfs/Dibble_HTML.pdf (accessed on 21 March 2023).
46. Castro, C.A.; Hoge, C.W.; Cox, A.L. Battlemind Training: Building Soldier Resiliency. Human Dimensions in Military Operations—Military Leaders' Strategies for Addressing Stress and Psychological Support; Meeting Proceedings RTO-MP-HFM-134; 2006; (42), pp. 1–6. Available online: http://www.rto.nato.int/abstracts.asp (accessed on 21 March 2023).
47. Stanley, E.A. Mindfulness-Based Mind Fitness Training: An Approach for Enhancing Performance and Building Resilience in High-Stress Contexts. In *The Wiley Blackwell Handbook of Mindfulness*; Wiley: Hoboken, NJ, USA, 2014; pp. 964–985.
48. Master Resilience Training & Its Mental Health Benefits. Available online: https://www.asymca.org/blog/master-resilience-training (accessed on 27 March 2023).
49. McInerney, S.A.; Waldrep, E.; Benight, C.C. Resilience Enhancing Programs in the U.S. Military: An Exploration of Theory and Applied Practice. *Mil. Psychol.* **2022**, 1–12. [CrossRef]
50. Smaliukienė, R.; Bekesiene, S.; Mažeikienė, A.; Larsson, G.; Karčiauskaitė, D.; Mazgelytė, E.; Vaičaitienė, R. Hair Cortisol, Perceived Stress, and the Effect of Group Dynamics: A Longitudinal Study of Young Men during Compulsory Military Training in Lithuania. *Int. J. Environ. Res. Public Health* **2022**, *19*, 1663. [CrossRef]
51. Fordjour, G.A.; Chan, A.P.C.; Fordjour, A.A. Exploring Potential Predictors of Psychological Distress among Employees: A Systematic Review. *Int. J. Psychiatry Res.* **2020**, *3*, 1–11. [CrossRef]
52. Thomas, K.H.; Albright, D.L.; Shields, M.M.; Kaufman, E.; Michaud, C.; Taylor, S.P.; Hamner, K. Predictors of Depression Diagnoses and Symptoms in United States Female Veterans: Results from a National Survey and Implications for Programming. *J. Mil. Veterans Health* **2016**, *24*, 6–16.
53. Larsen, K.L.; Stanley, E.A. Mindfulness-Based Mind Fitness Training (MMFT). In *Handbook of Mindfulness-Based Programs*; Routledge: Hoboken, NJ, USA, 2019; pp. 53–63.
54. Carmody, J.; Baer, R.A. Relationships between Mindfulness Practice and Levels of Mindfulness, Medical and Psychological Symptoms and Well-Being in a Mindfulness-Based Stress Reduction Program. *J. Behav. Med.* **2008**, *31*, 23–33. [CrossRef] [PubMed]

mathematics

MDPI

Article

Analysis of the Relationship between the Organizational Resilience Factors and Key Performance Indicators' Recovery Time in Uncertain Environments in Industrial Enterprises

Michael Huber [1], Nikola Komatina [1], Vladan Paunović [2] and Snežana Nestić [1,*]

[1] Faculty of Engineering, University of Kragujevac, 34000 Kragujevac, Serbia; michael.huber@kg.ac.rs (M.H.); nkomatina@kg.ac.rs (N.K.)
[2] Faculty of Technical Sciences, University of Kragujevac, 34000 Kragujevac, Serbia; vladan.paunovic@ftn.kg.ac.rs
* Correspondence: s.nestic@kg.ac.rs

Abstract: In terms of uncertain business conditions, the ability of an enterprise to bounce back after severe disruptions, or simply resilience, may be seen as one of the major features needed to sustain successful business operations. This research has the objective of proposing an algorithm for the organizational resilience assessment in industrial companies and conducting an analysis of the relationship between the organizational Resilience Factors and Key Performance Indicators recovery times. As the variables that are an integral part of the research are exposed to a high degree of uncertainty, they are modeled using fuzzy set theory. The methodology used for the research is an enhanced fuzzy Delphi, where the fuzzy geometric mean is employed as an aggregation operator. The relationship between the organizational resilience factors and Key Performance Indicators' recovery time is based on the correlation analysis. The proposed model is based on real data from one complex industrial enterprise. The main finding of the research is that calculations indicate a significant negative correlation between treated variables.

Keywords: organizational resilience; key performance indicators; recovery time; fuzzy delphi; fuzzy sets theory

MSC: 03E72

Citation: Huber, M.; Komatina, N.; Paunović, V.; Nestić, S. Analysis of the Relationship between the Organizational Resilience Factors and Key Performance Indicators' Recovery Time in Uncertain Environments in Industrial Enterprises. *Mathematics* **2023**, *11*, 3075. https://doi.org/10.3390/math11143075

Academic Editor: Georgios Tsekouras

Received: 14 June 2023
Revised: 4 July 2023
Accepted: 8 July 2023
Published: 12 July 2023

1. Introduction

Over the previous decades, resilience-scoped research has been conducted from different perspectives: Resistance and recovery, adaptation, and anticipation [1]. Also, as research interest has grown over the years, there is little consensus about what resilience means or how it is designed [2]. During a period of stable business conditions, organizational performance indices do not have significant oscillations. On the other hand, if severe disruptions occur, a sudden drop in performance might happen [3]. In practice, performances such as quality, cost, productivity, innovativeness, time, etc. need to be managed by companies [4] to make their business activities successful. As performance represents a complex variable, in practice it is measured and managed through Key Performance Indicators (KPIs) [5]. Common sense implies that more resilient organizations will recover their performance faster compared to those that are not so resilient.

It may be assumed that organizational resilience models are complex, which implies that their evaluation cannot be performed directly; assessment models that rely on the judgments of decision-makers could be applied. This assumption is important since many management problems demand this approach to assessment, which induces a certain degree of uncertainty.

The motivation for this research stems from the various uncertainties of the above-mentioned business context. Companies can be affected by various factors such as competition, changes in the market, political instability, and natural disasters. COVID-19 has shown how quickly an uncertain situation can develop and how companies must respond.

Uncertainties for companies have been omnipresent not only since pandemics, unstable geopolitical situations, or endangered supply chains. Entrepreneurial resilience is therefore seen as an important capability for companies to cope with these very changes and crises [6]. In a rapidly changing world, it is thus crucial that companies be resilient as a precondition for success. Therefore, both uncertainty and resilience are closely related concepts that are of significant importance to companies. As organizational resilience models, as well as business processes, are complex in nature, their evaluation cannot be performed directly. This implies that they require evaluation models based on the judgments of decision-makers. This feature is important because, in a variety of management problems, it is not possible to directly measure the variables of interest. This is because those variables are subject to a certain degree of uncertainty. At the same time, it is closer to human thinking to use linguistic variables for assessment. Different mathematical theories support the quantitative description of linguistic expressions [7]. Many mathematical theories support modeling linguistic expressions in a quantitative way. The theory of fuzzy sets [7,8] is used in many research areas to describe uncertainty quantitatively.

The fuzzy Delphi method will be used to assess the organizational resilience of the company. Here, the decision-making method is based on a consensus of expert opinions and uncertain information on a particular topic or issue. The overall objective of this research is the analysis of the dependency between organizational resilience factors (RFs) and KPIs' recovery times. To achieve the defined research goal, organizational resilience should be assessed as well as the recovery time of KPIs in the treated company.

Furthermore, the following chapters are organized as follows: Section 2 provides an overview of the relevant literature. Section 3 presents the proposed model. Section 4 presents the case study in a corporate context, and Section 5 provides a critical discussion and conclusion.

2. Literature Review

The literature supports the use of type one fuzzy sets for modeling existing uncertainties [9,10]. Type one fuzzy sets are used for the research. The features of type one fuzzy sets are the triangular membership function, granulation, and domain. The granulation is often chosen in accordance with the nature of the problem being solved. The domain might be chosen according to the DM assessment or following the literature guidelines [11].

A significant number of scholars support the application of type one fuzzy sets since they provide a solid base for calculations embracing uncertainties with a reasonable number of mathematical operations.

Considering all the issues raised, methods such as Delphi with type one fuzzy sets are used to solve fuzzy group decision-making problems [12,13]. The aggregation of DMs' opinions into unique opinions can be obtained by applying the different aggregation operators [14,15]. Mostly, in the domain of solving real business problems in the presence of uncertainty, fuzzy arithmetic mean [16–19] and fuzzy geometric mean [20–22] are applied.

This section embraces the analysis of the Fuzzy Delphi technique compared with the fuzzy Delphi technique enhanced with type one fuzzy numbers and applied to solving similar problems in management. The comparative analysis is presented in Table 1.

Table 1. Comparative analysis of the proposed Delphi technique with type one fuzzy numbers.

Authors	The Number of DMs	Membership Function Shape/Granulation/Domain	The Aggregation Operator/Defuzzification Procedure/the Distance between Two Fuzzy Numbers/Checking the Consensus of Decision-Makers Assessments
Chen and Lee [23]	-	TFN/5/[0–1]	the proposed aggregation method/simple gravity method/-/the proposed threshold value [23]
Habibi et al. [12]	-	TFN/5/[0–1] TFN/7/[0–1]	the proposed aggregation procedure/center gravity method/-/the usually used threshold [24]
Liu and Chu [25]	-	TrFN/3/[0–10]	the proposed aggregation procedure/-/-/the proposed procedure by Horng et al. [24]
Kumar et al. [26]	-	TFN/9/[0.1–0.9]	the proposed aggregation procedure [26]/center of gravity method/-/-
Jani et al. [16]	12	TFN/7/[0–1]	fuzzy arithmetic mean/-/Euclidean distance/threshold value defined by Mahmoudi et al. [27]
Singh and Sarkar [28]	15	TFN/5/[0.1–0.9]	the proposed aggregation procedure/center of gravity method/-/the proposed procedure based on a threshold value defined by Kumar et al. [29]
Bui et al. [20]	-	TFN/5/[0–1]	fuzzy geometric mean/method of the maximum possibility/-/the proposed procedure for establishing equilibrium across the fundamental judgments among the expert group [7]
Khan et al. [21]	12	TFN/5/[0–1]	fuzzy geometric mean/center of gravity/-/procedure defined by Horng et al. [24]
Abdollahi et al. [17]	15	TrFN/5/[0–9]	fuzzy arithmetic mean/-/the defuzzification procedure [30]/distance between two consecutive rounds [27]
Tsai et al. [18]	14	TFN/5/[0–1]	fuzzy arithmetic mean/center of gravity method/-/-
Dawood et al. [19]	-	TFN/5/[0–1]	fuzzy arithmetic mean/center of gravity method/Euclidean distance/The consensus must be higher than or equal to 75% to declare an acceptable agreement amongst the experts [31]; defined threshold value; distance between two consecutive rounds [27]
Mabrouk [13]	-	TFN/5/[0–1]	the proposed aggregation model/the proposed defuzzification method/-/defined the filtering threshold for the critical attributes
Aleksić et al. [22]	5	TFN/7/[1–9]	fuzzy geometric mean/-/Hamming distance/combining the Graded Mean Integration Representation and Average Percent of Majority Opinions Cut-off Rate [32]
The proposed model	9	TFN/5/[0–10]	fuzzy square mean/-/Euclidean distance/intraclass correlation coefficient [33]

Authors should discuss the results and how they can be interpreted from the perspective of previous studies and the working hypotheses. The findings and their implications should be discussed in the broadest possible context. Future research directions may also be highlighted.

In the analyzed papers, almost all authors use TFNs to describe the assessment of DMs. Up to now, there is no recommendation in the literature on how to determine the granulation and the domains of the employed fuzzy numbers in the realization of Delphi studies. The number of linguistic variables is most influenced by the complexity of the problem as well as the number of decision-makers included in the Delphi study. Having in mind the stated, it is worth mentioning that most scholars [13,17–21,23,28] employ five linguistic expressions for describing uncertainties in their research.

In the analyzed Delphi studies, the majority of authors [12,13,16–21,23] suggested that the domain should be defined on the set of real lines belonging to the interval [0–1].

In this research, the triangular membership function is used for modeling RF value estimates on sub-processes of the manufacturing process, as in almost all analyzed works. In the literature, many authors suggest that TFNs can capture uncertainties and inaccuracies adequately, and on the other hand, their usage does not require complex computations [7]. The number of pre-defined linguistic terms used to describe the considered uncertainty is five, as in the majority of analyzed papers. The domain of TFNs defined in this research belongs to the interval [0–10], as suggested by Liu and Chu [25].

The aggregation of DMs' assessments into a single assessment is based on the use of different operators. The selection of aggregation methods for DMs' estimates can be acknowledged as a problem in itself.

In this research, the authors suggest a fuzzy quadratic mean operator, which represents the difference between the presented research and papers that can be found in the relevant literature and is presented in Table 1.

The linguistic expression representing the result of the previous round is obtained from the condition of the minimum distance of pre-defined linguistic expressions and TFN, which describe the aggregate value of the DMs' assessment. Euclidean distance is most often used in a variety of research [16,19], as in this one particular study. Some scholars use Hamming distance as well [22].

Checking the consistency of DMs' assessments is based on different procedures [34]. In this research, the procedure for checking if the consensus of DMs' opinions is reached is performed by using an intraclass correlation coefficient [35]. It can be concluded that it is necessary to perform a sensitivity analysis of the results of the consensus check obtained by applying different methods.

3. Methodology

This section proposes the three-stage fuzzy model, which represents the core of this research. Simultaneously, a literature review is provided. In the first stage, the level of RFs is determined at the level of the product delivery process within the analyzed company by applying the proposed fuzzy Delphi technique. The second stage of the proposed model is used to determine the weighted aggregated fuzzy value of RFs at the level of each KPI as well as the scatterplot dependency between RFs and KPIs. The proposed two-stage model is presented in Figure 1.

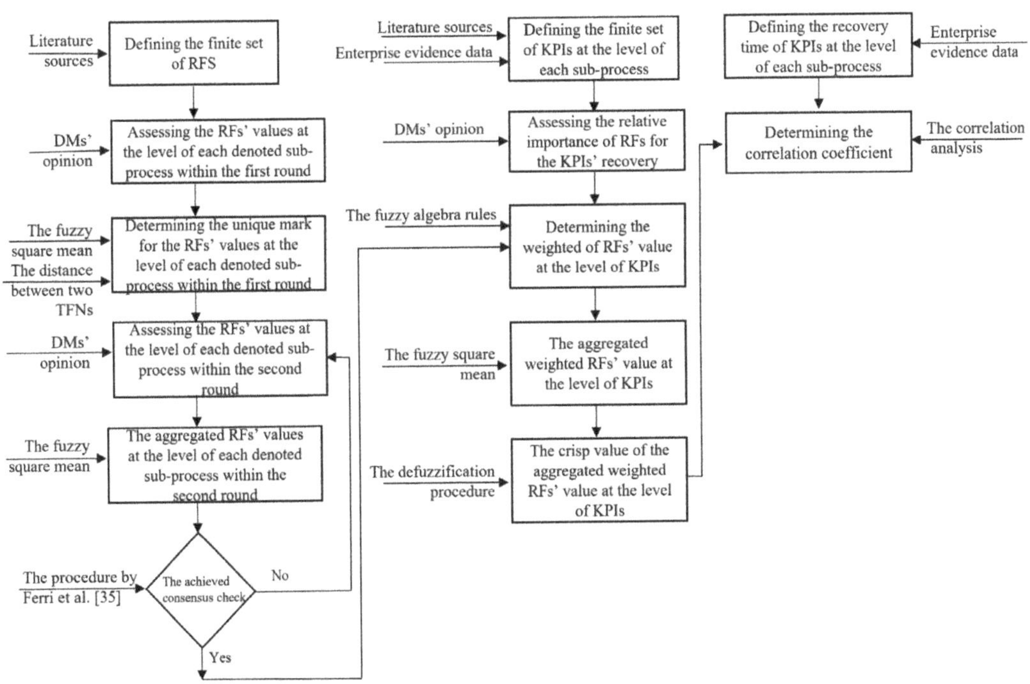

Figure 1. The proposed fuzzy model for assessment and analysis of the relationship between the weighted aggregated RFs' values and KPIs' recovery time.

To execute the proposed research, the following steps should be performed in the corporate context: (1) the definition of a finite set of RFs (see Section 3.1.1); (2) the definition of the main processes (MP) and sub-processes (SP) of the company (see Section 3.1.2); (3) the definition of the KPIs that are managed at the level of identified subprocesses (see Section 3.1.3); (4) the identification of a group of experts who have in-depth knowledge and experience related to enterprise resilience (see Section 3.1.4); (5) the execution of the proposed Delphi method to reach a consensus opinion of the experts related to the RFs' level in the treated company (see Section 3.3); (6) the calculation of the aggregated weighted RFs value at the level of each identified KPI; (7) the assessment of the KPIs' recovery time; (8) The scatter plot analysis of RFs values and KPIs' recovery time values; (9) analysis of the results to identify weaknesses and opportunities for improvement.

3.1. Defining the Finite Set of Input Variables

3.1.1. Defining the Finite Set Resilience Factors

Formally, the list of proposed RFs is represented by a formal set: $\{1,..,j,..,J\}$. The total number of considered RFs is denoted as J. The index of RF is marked as $j, j = 1,..,J$. In this research, the set of RFs is defined according to the referent literature [36]. The considered RFs that are significant for a production company are: management commitment $(j = 1)$, reporting culture $(j = 2)$, learning $(j = 3)$, awareness $(j = 4)$, preparedness $(j = 5)$, flexibility $(j = 6)$, self-organization $(j = 7)$, teamwork $(j = 8)$, redundancy $(j = 9)$, and fault-tolerance $(j = 10)$.

3.1.2. Defining the Finite Set of Business Sub-Processes

The classification of the business process and its' corresponding subprocesses is determined in compliance with the APQC framework [37]. Within this research, a process entitled "Deliver Physical Products" is analyzed. Its' subprocesses can be formally represented by a set of indices: $\{1,..,p,..,P\}$. The finite number of subprocesses is denoted as P, and $p, p = 1,..,P$ represents the index of the subprocess. The sub-processes of Deliver Physical Products are: planning for and aligning supply chain resources $(p = 1)$, procuring materials and services $(p = 2)$, produce/assemble/test product $(p = 3)$, and managing logistics and warehousing $(p = 4)$.

3.1.3. Defining the Managed KPIs

There is no specific recommendation on which KPIs should be managed in different companies, so it is their responsibility to choose adequate KPIs based on their size, business domain, and other features. For this research, the set of KPIs is defined in compliance with the APQC framework to provide generality, and at the same time, it is adjusted to the company that is analyzed to provide expediency.

The set of considered KPIs is presented by a set of indices: $\{1,..,i,..,I\}$. The total number of the considered KPIs is denoted as I. The index of the KPI is marked as $i, i = 1,..,I$. In this research, these KPIs are [37]: Total cost of quality per $100,000 in revenue $(i = 1)$, employee retention rate $(i = 2)$, percentage of sales orders scheduled to customer requests $(i = 3)$, total cost to perform the procurement process group per purchase order $(i = 4)$, average procure-to-pay cycle time in days $(i = 5)$, percentage of unique suppliers who are active suppliers $(i = 6)$, scrap and rework costs as a percentage of cost of goods sold $(i = 7)$, total cost to manufacture per $1000 revenue $(i = 8)$, percentage of defective parts per million $(i = 9)$, average cycle time in calendar days from delivery order to successful completion of delivery and disposal of back-hauled goods $(i = 10)$, perfect order performance $(i = 11)$, percentage of supplier on-time delivery $(i = 12)$.

3.1.4. Defining the Set of DMs' Team

The assessment of the level of each RF, $j, j = 1,..,J$ at the level of each business process should be presented as a fuzzy group decision-making problem. DMs should be aware of the RF level so they can manage and enhance it continuously.

In this research, a set of DMs should be presented by a set of indices: $\{1,..,e,..,E\}$. Ukupan broj DMs are denoted as E. The index of DM is marked as $e,.,e = 1,..,E$.

The DMs team consists of the highest-ranking employees in the company structure, which enables wide insight into the enterprise's functioning and experience in the decision-making process. In the analyzed company, the DMs' team consists of the Chief Executive Officer ($e = 1$), Operations Manager ($e = 2$), Management System Manager ($e = 3$), Global Supply Chain Manager ($e = 4$), Human Resource Manager ($e = 5$), Marketing Manager ($e = 6$), Service and Sale Manager ($e = 7$), Chief Information Officer ($e = 8$), Research and Development Manager ($e = 9$). It should be noted that the DMs' team is responsible for all assessments that are proposed by this research.

3.2. The Selection of Linguistic Variables for the Existing Uncertainties' Description

In this research, existing uncertainties are: (1) the values of RFs at the level of sub-processes of the delivery product business process; and (2) the relative importance of RFs for the KPIs' recovery.

The assessment of RFs' values at a level for each considered sub-process is performed by using five linguistic expressions, which are modeled by TFNs. These linguistic variables and their corresponding TFNs are given:

Very low value (*B*1)—(0, 1.5, 3)
Low value (*B*2)—(1, 2.5, 4)
Medium value (*B*3)—(3, 5, 7)
High value (*B*4)—(6, 7.5, 9)
Very high value (*B*5)—(7, 8.5, 10)

The domains of TFNs that are used for the quantitative description of RFs' level in the analyzed company within the interval [0–10]. The values 0 and 10 denote that RF has the lowest value or the highest value, respectively.

The assessment of the relative importance of RFs for the KPIs' recovery is described by the seven linguistic expressions modeled by TFNs:

Extremely low importance (*A*1)—(0, 0, 2.5)
Low importance (*A*2)—(0.5, 2, 3.5)
Fairly low importance (*A*3)—(1.5, 3.5, 5.5)
Medium importance (*A*4)—(3, 5, 7)
Fairly high importance (*A*5)—(5, 6.5, 8)
High importance (*A*6)—(6.5, 8, 9.5)
Extremely high importance (*A*7)—(7.5, 10, 10)

The domains of these TFNs are defined on a real line belonging to the interval [0–10]. Values 0 and 10 denote that RF has no relative importance for the KPIs' recovery or has extremely high importance, respectively.

3.3. The Assessment of RFs Values' Level by the Proposed Fuzzy Delphi Technique

The Delphi technique is one of the most popular qualitative methods of group decision-making. The simplest explanation of the Delphi technique can be interpreted as the collection and processing of data, which is realized through several rounds.

Within the execution of the technique, one of the most important questions is how to determine the optimal set of DMs. There are no recommendations or guidelines on how to determine the optimal number of DMs. Some scholars [38,39] suggest that there should be between five and ten DMs that provide the assessment. It may be suggested that, through the analysis of the research context, an optimal number of DMs may be determined.

According to best practice, it is assumed that the DMs participating in the Delphi study have a precise perception of the identified problem or that they have in-depth knowledge of the treated area(s). At the same time, the experience level of DMs can vary, and they can be ranked within various levels in the company hierarchy. An important issue during the realization is that the anonymity of DMs must be provided during the execution of the technique so individual biases and personal thoughts do not impact other participants. In this research, the DMs have been selected according to their importance for the company's operations, considering their knowledge and competence.

The Delphi method is realized in several rounds. During the first round, DMs express their assessment regarding the treated problem. The mapping of DMs' assessments into a single assessment can be executed by applying different aggregation operators. The average value of the DMs' assessments is submitted in writing to the DMs again, who should adjust their assessments in the second round according to that value. By applying the different procedures [34], it can be determined if the DMs have reached a consensus. If they are, the average value of the estimates obtained in the second round is accepted as the decision. Otherwise, the described process of data collection and processing is repeated. It should be noticed that the DMs' team is delivering individual assessments to determine the RFs' value. This is because their competence covers several aspects of business activities, and all the uncertainties should be considered.

The questionnaire is adapted taking into account the verified research [22] and it is introduced to each DM with explanations of the different resilience levels within the enterprise. The questionnaire contains guidelines with linguistic expressions defining the level of organizational resilience for each RF as follows:

There are no blueprints or plans for the construction of organizational resilience, there is no awareness of organizational resilience—B1;

There are drafts of activities for securing organizational resilience—B2;

There are clear plans and activities for securing organizational resilience, and the competencies of all employees in the field of organizational resilience management are ensured—B3;

Competencies of all employees in the field of organizational resilience management are ensured, and there is a partially developed awareness of organizational resilience—B4;

All needed competences are ensured, and there is the absolute commitment of management and all employees regarding organizational resilience management—B5.

The proposed fuzzy Delphi technique is realized in the following steps:

Step 1. During the first round, each DM $e, e = 1, \ldots, E$ assesses the level of RFs $j, j = 1, .., J$ at the level of each sub-process $p, p = 1, .., P$ by using one of the five pre-defined linguistic expressions, $\tilde{v}_{jp}^{1e} = \left(l_{jp}^{1e}, m_{jp}^{1e}, u_{jp}^{1e}\right)$.

Step 2. Let us determine the aggregated value of the DMs' assessment in the first round, $\tilde{b}_{jp}^{1}$ by applying the operator of the square mean:

$$\tilde{v}_{jp}^{1} = \left(\sqrt{\frac{1}{E} \cdot \sum_{e=1,..E} \left(l_{jp}^{1e}\right)^2}, \sqrt{\frac{1}{E} \cdot \sum_{e=1,..E} \left(m_{jp}^{1e}\right)^2}, \sqrt{\frac{1}{E} \cdot \sum_{e=1,..E} \left(u_{jp}^{1e}\right)^2}\right) \quad (1)$$

So that:

$$\tilde{v}_{jp}^{1} = \left(l_{jp}^{1}, m_{jp}^{1}, u_{jp}^{1}\right), \quad (2)$$

Step 3. Let us calculate the distance between $\tilde{v}_{jp}^{1}$ and TFNs that correspond to the pre-defined linguistic expressions $Bk, k = 1, .., 5$, $d\left(\tilde{v}_{jp}, Bk\right)$.

Step 4. To each RF $j, j = 1, .., J$ at the level of sub-process, $p = 1, .., P$, should be adjoined one of the pre-defined linguistic expressions $Bk, k = 1, .., K$ according to the expression:

$$\min_{k=1,..,K} d\left(\tilde{v}_{jp}, Bk\right) = B_{jp}^{*}, \quad (3)$$

Step 5. During the second round, DMs adjust their assessment according to the average value of B_{jp}^{*}. Let the DMs' assessments in the second round be denoted as $\tilde{v}_{jp}^{2e}$.

Step 6. Let us check the correlation degree between the DMs assessment in the first, $\tilde{v}_{jp}^{1e}$, and the second round, $\tilde{v}_{jp}^{2e}$. If the degree of correlation is higher than or equal to 0.5, it can be considered that a consensus of DMs has been reached according to the developed procedure [35]. If there is no statistical dependency between DMs' assessments in the first and second rounds, it is necessary to perform the second round of the assessment.

3.4. The Calculation of the Aggregated Weighted RFs Value at the Level of Each Identified KPI

This part of the research contains the steps for determining the relative importance of RFs for the KPIs' recovery that is managed under the business process, which is entitled Deliver Physical Products. The assessment of the relative importance of RFs for the KPIs' recovery is treated as a problem itself, with the assessment in the form of consensus.

The DMs have seven linguistic expressions at their disposal. The guidelines with the linguistic expressions defining the importance of RFs for the treated KPI's recovery time are as follows:

The treated RF has extremely low importance for the treated KPI's recovery time—A1;
The treated RF has low importance for the treated KPI's recovery time—A2;
The treated RF has fairly low importance for the treated KPI's recovery time—A3;
The treated RF has medium importance for the treated KPI's recovery time—A4;
The treated RF has fairly high importance for the treated KPI's recovery time—A5;
The treated RF has high importance for the treated KPI's recovery time—A6;
The treated RF has extremely high importance for the treated KPI's recovery time—A7.

After this, the determination of the weighted *aggregated RFs' value* at the level of each denoted KPI is performed by applying the operator of the fuzzy square mean. The proposed procedure is realized as follows:

Step 1. The assessment of RFs $j, j = 1, .., J$ relative importance of RFs for the KPIs' $i, i = 1, .., I$ recovery time is denoted by TFN $\tilde{\varphi}_{ji}$.

Step 2. Let us calculate the weighted value of each RF $j, j = 1, .., J$ at the level of each denoted KPI:

$$\tilde{\theta}_{ji} = \tilde{v}_{jp} \cdot \tilde{\varphi}_{ji}, \quad (4)$$

Step 3. Let us determine the weighted aggregated fuzzy value of RFs at the level of each KPI $i, i = 1, .., I$, $\tilde{\theta}_i$ by applying the operator of the fuzzy geometric mean.

3.5. The Proposed Procedure for Analysis of the Relationship between the Weighted Aggregated RFs' Values and KPIs' Recovery Time

The KPIs for recovery time, t_i, are obtained from the enterprise records. It is worth mentioning that this research does not take into consideration KPI management but rather follows the sudden drop in KPI values. The time needed for the complete recovery of the KPI's values is denoted as the recovery time. It is presented in months.

Here, an assumption is introduced: there is a linear correlation between KPIs' recovery time and the weighted aggregated fuzzy value of RFs at the level of each KPI. This assumption will be checked based on the determination of the coefficient of correlation between the named variables.

The final steps of the research represent the analysis of the relationship between the weighted aggregated RFs' value and KPIs' recovery time. This should be executed as follows:

Step 1. Let us determine the representative scalar TFN $\tilde{\theta}_i$, Δ_i by applying the simple gravity method.

Step 2. Let us determine the correlation coefficient between the KPIs' recovery time, t_i and the weighted aggregated value of RFs at the level of each considered KPI, Δ_i.

4. A Case Study in a Complex Production Company

The analyzed enterprise follows a decentralized organizational structure. Here, all business units are mapped within a matrix organization, which acts autonomously in the global supply chain of precise industry components. Nevertheless, all organizational units should interact closely with each other without limiting their independence, flexibility, and agility in the market. To meet the challenges mentioned, it is essential for almost all organizational units and employees—regardless of the specific company size, characteristics, form, and maturity—to maintain a management system. The analyzed enterprise has a well-structured business process in compliance with the ISO 9001 and ISO 14001 standards, so it is possible to propose a similar business process framework, such as APQC.

The DMs were engaged in the first round, and after the calculations, which are presented in Section 4.1, they also participated in the second round of the fuzzy Delphi.

4.1. Application of Applying the Proposed Fuzzy Delphi Technique

The defined team of DMs has received an email containing the relevant data for assessing the level of RF values, as explained in Section 3.3. The input data for fuzzy Delphi are assessed values (RFs) at the level of the business process of Delivering Physical Products. This data is presented in Appendix A for round one and in Appendix B for the second round.

The proposed fuzzy Delphi technique is illustrated in the example of determining the value RF $j = 1$ at the level of sub-process alignment of supply chain resources ($p = 1$).

During the first round, the DMs assessed the values of the treated RF in the following manner:

B5, B4, B4, B4, B4, B4, B4, B4, B3, B2

The aggregated value of the DMs' assessment in the first round, $\tilde{v}_{11}^{1}$ is obtained by applying the operator of the fuzzy square mean:

$$\tilde{v}_{11}^{1} = \begin{pmatrix} \sqrt{\frac{7^2+6^2+6^2+6^2+6^2+6^2+6^2+3^2+1^2}{9}}, \\ \sqrt{\frac{8.5^2+7.5^2+7.5^2+7.5^2+7.5^2+7.5^2+7.5^2+5^2+2.5^2}{9}}, \\ \sqrt{\frac{10^2+9^2+9^2+9^2+9^2+9^2+9^2+7^2+4^2}{9}} \end{pmatrix} = (5.53, 7, 8.50)$$

Let us determine the distance of TFN $\tilde{v}_{11}$ from $B1$:

$$d(\tilde{v}_{11}^{1}, B1) = \sqrt{\frac{1}{3} \cdot \left((5.53 - 0)^2 + (7 - 1.5)^2 + (8.50 - 3)^2 \right)} = 5.411$$

In a similar manner, the distance of TFN $\tilde{v}_{11}$ from the rest of the pre-defined linguistic expressions is calculated:

$$d(\tilde{v}_{11}^{1}, B2) = 4.510$$
$$d(\tilde{v}_{11}^{1}, B3) = 2.054$$
$$d(\tilde{v}_{11}^{1}, B4) = 0.490$$
$$d(\tilde{v}_{11}^{1}, B5) = 4.219$$

Let us determine a linguistic expression that can be used to describe the aggregated value of the DMs' assessment in the first round according to the expression:

$$min(5.411; 4.510; 2.054; 0.490; 4.219) = 0.490 \rightarrow B4$$

The DMs' assessments in the second round are B4, B3, B3, B2, B3, B3, B3, B2, and B2.

The aggregated value of the DMs' assessments in the second round is obtained by using the operator of the fuzzy square mean.

$$\tilde{v}_{11}^{2} = (3.06, 4.71, 6.45)$$

The check of the consistency of the DMs' assessment is delivered according to the developed procedure [35].

By using further calculations, the value of the correlation coefficient can be obtained. The value of the correlation coefficient between the assessment of the value RF $j = 1$ in the first round and in the second round is 0.8. The obtained value of the correlation coefficient shows that there is a strong positive relationship between the estimates of DMs in the first and second rounds, so it can be concluded that the obtained value of RF $j = 1$ j in the second round can be considered the final value.

Similarly, the aggregated values of RFs were determined at the level of each sub-process of the considered business process and presented in Appendix B.

Based on the obtained values of the correlation coefficients, it can be concluded that the values of RFs obtained in the second round can be accepted as the final values of RFs at the level of each sub-process.

4.2. The Calculation of the Aggregated Weighted RFs Value at the Level of Each Identified KPI

The assessment of the relative importance of RFs for the KPIs' recovery is performed for each denoted KPI at the level of each treated sub-process (Table 2). The DMs have performed this activity within the scope of the panel discussion that is executed after the second round of fuzzy Delphi. The assessment itself was based on the guidelines explained in Section 3.4. The panel discussion took place at the company headquarters with all DMs that participated in previous activities.

Table 2. The relative importance of RFs for the KPIs' recovery at the level of each treated sub-process.

RFs	$i = 1$	$i=2$	$i = 3$	$i = 4$	$i = 5$	$i = 6$	$i = 7$	$i = 8$	$i = 9$	$i = 10$	$i = 11$	$i = 12$
$j = 1$	A3	A7	A5	A6	A5	A7	A3	A7	A4	A5	A7	A7
$j = 2$	A5	A2	A6	A7	A5	A3	A3	A6	A4	A6	A6	A7
$j = 3$	A5	A6	A5	A5	A2	A2	A6	A5	A4	A4	A6	A5
$j = 4$	A6	A6	A6	A5	A7	A6	A6	A6	A5	A4	A5	A6
$j = 5$	A5	A6	A5	A5	A5	A3	A5	A4	A6	A6	A6	A6
$j = 6$	A4	A2	A4	A6	A6	A3	A4	A5	A2	A4	A4	A5
$j = 7$	A2	A4	A4	A4	A3	A3	A4	A3	A3	A3	A4	A4
$j = 8$	A3	A7	A4	A4	A3	A4	A4	A3	A3	A4	A5	A4
$j = 9$	A5	A2	A2	A3	A2	A1	A6	A5	A5	A5	A4	A3
$j = 10$	A7	A2	A3	A3	A3	A5	A5	A4	A5	A4	A4	A5

Let us determine the aggregate weighted value of the RF (j = 1) at the level of the KPI $(i = 1)$ $\tilde{z}_{11}$:

$$\tilde{z}_{11} = \tilde{v}_{11} \cdot A3 = (3.06, 4.71, 6.45) \cdot (1.5, 3.5, 5.5) = (4.59, 16.49, 35.48)$$

The other aggregated weighted values of RFs are calculated similarly to those presented in Appendix C.

4.3. The Determination of the Relationship between the Weighted Aggregated RFs' Values and KPIs' Recovery Time

The recovery time is taken from the company records, as explained in Section 3.5. The representative scalars of the resilience at the level of KPIs, as well as the recovery time of each KPI, are given in Table 3.

The input data for the correlation analysis are the representative scalars of the total aggregated weighted values of RFs and the recovery time expressed in months. The obtained value of the correlation coefficient is presented as follows (Table 4).

Table 3. The total aggregated weighted crisp values of RFs and KPIs' recovery time in months.

KPIs	z_i	t_i
$i = 1$	31.10	10
$i = 2$	36.76	7
$i = 3$	32.42	5
$i = 4$	37.07	7
$i = 5$	31.50	4
$i = 6$	32.41	6
$i = 7$	27.61	10
$i = 8$	27.81	6
$i = 9$	23.92	7
$i = 10$	22.84	9
$i = 11$	27.09	7
$i = 12$	25.61	8

Table 4. Impact of the aggregated weighted values of RFs on KPIs' recovery time.

	The Weighted Aggregated RFs' Value at the Level of Each KPI	The Recovery Time of Each KPI
The weighted aggregated RFs' value at the level of each KPI	1	
The recovery time of each KPI	−0.73857	1

Based on the obtained value of the correlation coefficient, it can be concluded that there is a statistically significant influence of the values of RFs on the recovery time of KPIs. The value of the coefficient is negative, which indicates that if the value of RFs increases, the recovery time decreases.

5. Discussion and Conclusions

After the execution of the proposed fuzzy Delphi technique, the value of RFs is obtained at the level of each denoted subprocess. In the next step, the relative importance of the RFs for the recovery of each KPI is obtained through a direct assessment. The weighted value of the RFs is obtained through the multiplication of the previously defined variables. The weighted aggregated fuzzy value of each RF is obtained by applying the aggregation operator to the fuzzy square mean. By applying the Simple Gravity Method, the representative scalar of the weighted aggregated fuzzy value of each RF is determined.

The output of the research is the analysis of the relationship between the weighted aggregated value of each RF and the recovery time of each KPI. From the presented calculations considering correlation analysis, it is shown that the introduced assumption of a negative correlation is confirmed. There is a negative statistical dependence between the RFs and the time needed for KPIs' recovery.

Comparing the results with the already presented research, the following may be concluded: The domains where the aggregation of resilience is conducted may be presented as follows: military service [40], social resilience measurement [41], and quantification of operational supply chain resilience [42]. Each of the mentioned papers considers their own set of resilience indicators/factors, so it can be concluded that there is no unique list of RFs. In the mentioned papers, resilience indicators/factors are presented with crisp values compared to the proposed research, which is done by using linguistic variables. It may be concluded that there are different approaches to aggregate resilience indicators/factors. The aggregated value may be determined in an exact manner by applying multi-attribute decision-making techniques, such as the analytical hierarchy process [40], or by applying simple aggregation operators [41,42]. In the presented research, the aggregated value is obtained through the application of the fuzzy Delphi technique and fuzzy square mean operator.

Improving the overall resilience of companies and their decision-makers requires a holistic approach that takes various aspects into account. Based on the case study conducted, the authors of this paper, together with the DMs that provided input data for the case study, derive the following general recommendations for increasing resilience:

(1) Establish strong risk management practices: companies should implement a comprehensive risk management system that identifies potential risks, evaluates them, and takes appropriate measures to address them. Such an approach makes it possible to respond to potential threats at an early stage and minimize damage.
(2) Diversification of business activities: companies should reduce their dependence on individual products, markets, or suppliers. A broader base enables them to respond better to changes in the market and cushion potential risks more effectively.
(3) Promote flexibility and adaptability: companies should develop a corporate culture that promotes flexibility and adaptability. This includes fostering a spirit of innovation, a willingness to change, and the development of agile structures and processes.
(4) Empowering leaders: decision-makers should have a high level of resilience Companies should support their leaders by providing them with the necessary resources, training, and coaching to deal with challenging situations.
(5) Continuous training and learning: companies should ensure that their employees are continuously trained to keep up with changing demands and challenges. This includes both technical and generic competencies, such as problem-solving skills, communication, and teamwork.
(6) Build a strong network: companies should build and maintain relationships with relevant stakeholders, including customers, suppliers, partners, and regulators. A strong network can be invaluable in times of crisis to gain support and find solutions together.
(7) Leverage technology and digital transformation: companies should take advantage of modern technologies to make their processes more efficient and improve their resilience. This can include the use of data analytics, artificial intelligence, and other technologies to identify risks early and make informed decisions.

These recommendations serve as a starting point to improve the resilience of companies and their decision-makers. Companies must consider their challenges and needs and develop tailored solutions accordingly. The other approach that may be combined with the proposed measures may include the ranking of the proposed RFs to identify those ranked last, so the DMs may propose more concrete measures to improve those and sustain the values of those ranked first.

The main contribution of the research may be summarized as follows: There are just a few papers that treat the problem in a similar manner, defining interconnections between RFs and KPIs. All the uncertainties that exist in the model are described by using linguistic variables modeled by fuzzy sets theory. The fuzzy values of RFs at the level of delivery of physical product sub-processes are obtained by using the enhanced fuzzy Delphi method. The weighted aggregated fuzzy value of resilience at the level of a KPI is determined in an exact manner by applying fuzzy algebra rules.

The main constraint of the research is the selection of the DM team, which consists of the top management representatives, considering their knowledge, skills, and experience related to overall business operations, strategy, organizational state, and functioning.

On the other hand, it may be considered that the proposed model is flexible in terms of changing the number of KPIs and RFs. Also, the number of DMs can be changed due to the nature of the treated organization.

Future research should cover the extension of the Delphi method by using some other method for checking the consensus, developing a new method, and comparing the obtained results. For resilience management benchmarking, it can be assumed that this model should be used in some business processes and other branches of industry and the economy. Also, it would be useful to test the proposed model with different types of fuzzy numbers to determine their suitability for embracing the existing uncertainties.

Author Contributions: Conceptualization, M.H. and S.N.; investigation, N.K. and S.N.; methodology, M.H. and V.P., supervision, S.N., validation, M.H., N.K. and V.P.; visualization, M.H. and N.K.; writing—original draft, M.H., S.N., N.K. and V.P. All authors have read and agreed to the published version of the manuscript.

Funding: This research received no external funding.

Data Availability Statement: Not applicable.

Conflicts of Interest: The authors declare no conflict of interest.

Appendix A. The First Round of the Proposed Fuzzy Delphi

Table A1. The assessment of the DMs in the first round.

Sub-Processes	RFs	*e*=1	*e*=2	*e*=3	*e*=4	*e*=5	*e*=6	*e*=7	*e*=8	*e*=9
$p = 1$	$j = 1$	B5	B5	B5	B5	B5	B5	B5	B4	B3
	$j = 2$	B5	B5	B4	B4	B3	B5	B4	B5	B3
	$j = 3$	B5	B5	B4	B5	B4	B4	B4	B5	B3
	$j = 4$	B5	B4	B5	B5	B4	B3	B2	B3	B3
	$j = 5$	B5	B5	B5	B5	B3	B2	B3	B3	B4
	$j = 6$	B5	B5	B5	B5	B5	B3	B4	B5	B3
	$j = 7$	B4	B5	B3	B5	B4	B3	B2	B5	B5
	$j = 8$	B5	B5	B5	B5	B5	B5	B5	B5	B5
	$j = 9$	B5	B3	B2	B4	B4	B2	B2	B4	B3
	$j = 10$	B5	B5	B5	B4	B4	B5	B5	B5	B5
$p = 2$	$j = 1$	B5	B5	B5	B5	B5	B4	B5	B4	B3
	$j = 2$	B5	B5	B5	B3	B3	B4	B5	B4	B3
	$j = 3$	B5	B5	B4	B4	B4	B4	B4	B5	B3
	$j = 4$	B5	B4	B5	B5	B4	B3	B2	B3	B3
	$j = 5$	B5	B5	B5	B5	B3	B2	B3	B3	B4
	$j = 6$	B5	B5	B5	B5	B5	B3	B3	B5	B3
	$j = 7$	B4	B5	B3	B5	B4	B3	B2	B5	B5
	$j = 8$	B5	B5	B5	B5	B5	B5	B5	B5	B5
	$j = 9$	B5	B3	B3	B4	B4	B2	B2	B4	B3
	$j = 10$	B5	B5	B5	B5	B4	B5	B5	B5	B5
$p = 3$	$j = 1$	B5	B5	B4	B4	B5	B5	B4	B5	B5
	$j = 2$	B5	B4	B2	B4	B3	B4	B3	B3	B3
	$j = 3$	B5	B5	B4	B4	B4	B5	B4	B5	B3
	$j = 4$	B5	B3	B4	B4	B4	B4	B2	B3	B4
	$j = 5$	B5	B5	B4	B4	B4	B2	B2	B2	B3
	$j = 6$	B5	B5	B4	B4	B5	B3	B4	B5	B4
	$j = 7$	B4	B5	B2	B4	B4	B3	B2	B4	B3
	$j = 8$	B5	B5	B5	B5	B5	B5	B5	B5	B5
	$j = 9$	B5	B3	B3	B3	B4	B4	B2	B4	B3
	$j = 10$	B5	B5	B5	B4	B4	B4	B5	B5	B5
$p = 4$	$j = 1$	B5	B4	B4	B3	B5	B5	B3	B3	B5
	$j = 2$	B5	B3	B2	B4	B2	B4	B3	B3	B3
	$j = 3$	B3	B4	B4	B3	B4	B3	B4	B4	B3
	$j = 4$	B4	B3	B2	B3	B4	B4	B2	B4	B4
	$j = 5$	B4	B4	B3	B4	B4	B3	B2	B2	B2
	$j = 6$	B5	B5	B3	B4	B5	B3	B3	B3	B3
	$j = 7$	B3	B3	B2	B4	B4	B2	B2	B4	B4
	$j = 8$	B5	B5	B5	B5	B5	B5	B5	B5	B5
	$j = 9$	B5	B3	B2	B3	B4	B3	B2	B4	B2
	$j = 10$	B4	B3	B5	B3	B3	B4	B5	B4	B4

Table A2. The aggregated values of RFs at the level of sub-process Align supply chain resources ($p = 1$).

RFs	The Aggregated Value in the First Round	The Linguistic Expression
$j = 1$	(5.53,7,8.50)	B4
$j = 2$	(4.70,6.19,7.75)	B4
$j = 3$	(4.48,6.07,7.72)	B3
$j = 4$	(3.97,5.33,6.83)	B3
$j = 5$	(4.33,5.65,7.06)	B3
$j = 6$	(5.30,6.72,8.20)	B4
$j = 7$	(4.27,5.67,7.17)	B3
$j = 8$	(7,8.50,10)	B5
$j = 9$	(2.69,4.06,5.65)	B3
$j = 10$	(5.61,7.15,8.72)	B4

Table A3. The aggregated values of RFs at the level of sub-process Procure materials and services ($p = 2$).

RFs	The Aggregated Value in the First Round	The Linguistic Expression
$j = 1$	(5.53,7,8.50)	B4
$j = 2$	(4.45,5.87,7.37)	B3
$j = 3$	(4.14,5.77,7.48)	B3
$j = 4$	(3.97,5.33,6.81)	B3
$j = 5$	(4.33,5.65,7.06)	B3
$j = 6$	(5.08,6.43,7.84)	B4
$j = 7$	(4.27,5.67,7.17)	B3
$j = 8$	(7,8.5,10)	B5
$j = 9$	(2.71,4.14,5.72)	B3
$j = 10$	(5.99,7.51,9.04)	B4

Table A4. The aggregated values of RFs at the level of sub-process Test product, ($p = 3$).

RFs	The Aggregated Value in the First Round	The Linguistic Expression
$j = 1$	(5.47,7.03,8.63)	B4
$j = 2$	(2.73,4.20,5.78)	B3
$j = 3$	(4.64,6.21,7.85)	B3
$j = 4$	(3.04,4.67,6.39)	B3
$j = 5$	(3.33,4.72,6.25)	B3
$j = 6$	(4.64,6.21,7.85)	B4
$j = 7$	(2.87,4.39,6.03)	B3
$j = 8$	(6.90,8.39,9.89)	B5
$j = 9$	(2.73,4.20,5.78)	B3
$j = 10$	(5.33,6.90,8.51)	B4

Table A5. The aggregated values of RFs at the level of sub-process Manage logistics and warehousing, ($p = 4$).

RFs	The Aggregated Value in the First Round	The Linguistic Expression
$j = 1$	(4.28,5.71,7.23)	B3
$j = 2$	(2.54,3.88,5.39)	B3
$j = 3$	(2.33,4.08,5.86)	B3
$j = 4$	(2.29,3.97,5.73)	B3
$j = 5$	(2.05,3.64,5.32)	B3
$j = 6$	(3.68,5,6.43)	B3
$j = 7$	(2.69,4.09,5.65)	B3
$j = 8$	(6.90,8.39,9.89)	B5
$j = 9$	(2.52,3.83,5.31)	B3
$j = 10$	(3.51,5.07,6.72)	B3

Appendix B. The Second Round of the Proposed Fuzzy Delphi

Table A6. The assessment of the DMs in the second round.

Sub-Processes	RFs	$e = 1$	$e = 2$	$e = 3$	$e = 4$	$e = 5$	$e = 6$	$e = 7$	$e = 8$	$e = 9$
$p = 1$	$j = 1$	B4	B3	B3	B2	B3	B3	B3	B2	B2
	$j = 2$	B4	B4	B3	B3	B2	B3	B3	B4	B2
	$j = 3$	B4	B2	B2	B3	B3	B2	B3	B3	B2
	$j = 4$	B4	B3	B3	B3	B2	B2	B1	B2	B2
	$j = 5$	B4	B5	B3	B3	B2	B1	B2	B2	B2
	$j = 6$	B5	B4	B2	B4	B3	B2	B2	B4	B2
	$j = 7$	B3	B3	B1	B3	B2	B2	B1	B3	B3
	$j = 8$	B5	B5	B4	B4	B5	B5	B5	B5	B3
	$j = 9$	B3	B2	B1	B2	B2	B1	B1	B1	B1
	$j = 10$	B4	B3	B3	B2	B2	B3	B3	B4	B4

Table A6. *Cont.*

Sub-Processes	RFs	$e = 1$	$e = 2$	$e = 3$	$e = 4$	$e = 5$	$e = 6$	$e = 7$	$e = 8$	$e = 9$
$p = 2$	$j = 1$	B5	B4	B4	B4	B3	B4	B4	B4	B3
	$j = 2$	B4	B3	B3	B2	B3	B3	B3	B3	B3
	$j = 3$	B3	B3	B3	B2	B2	B3	B3	B3	B2
	$j = 4$	B4	B3	B3	B3	B2	B2	B1	B2	B2
	$j = 5$	B4	B4	B4	B3	B2	B1	B2	B2	B2
	$j = 6$	B4	B4	B2	B3	B3	B2	B2	B3	B2
	$j = 7$	B3	B3	B1	B3	B2	B2	B1	B3	B3
	$j = 8$	B5	B5	B4	B4	B5	B5	B5	B5	B3
	$j = 9$	B3	B1	B1	B2	B2	B1	B1	B1	B2
	$j = 10$	B4	B3	B3	B2	B2	B3	B3	B3	B3
$p = 3$	$j = 1$	B3	B2	B2	B2	B2	B3	B2	B3	B3
	$j = 2$	B3	B2	B1	B2	B1	B3	B2	B2	B3
	$j = 3$	B3	B3	B2	B2	B2	B2	B3	B4	B2
	$j = 4$	B3	B2	B3	B3	B2	B2	B1	B2	B2
	$j = 5$	B3	B4	B3	B2	B2	B1	B1	B1	B2
	$j = 6$	B4	B3	B3	B3	B4	B2	B2	B3	B3
	$j = 7$	B2	B3	B1	B3	B3	B2	B1	B3	B2
	$j = 8$	B5	B4	B5	B4	B5	B5	B5	B5	B4
	$j = 9$	B3	B1	B1	B1	B2	B2	B1	B2	B2
	$j = 10$	B3	B3	B3	B2	B2	B2	B4	B3	B2
$p = 4$	$j = 1$	B3	B1	B2	B1	B2	B2	B1	B1	B2
	$j = 2$	B3	B1	B1	B2	B1	B2	B1	B1	B1
	$j = 3$	B2	B2	B2	B1	B2	B2	B2	B2	B1
	$j = 4$	B3	B2	B1	B2	B2	B2	B1	B2	B2
	$j = 5$	B3	B2	B1	B3	B2	B2	B1	B1	B1
	$j = 6$	B3	B4	B2	B2	B3	B2	B2	B2	B2
	$j = 7$	B2	B2	B1	B2	B2	B1	B1	B2	B2
	$j = 8$	B5	B5	B4	B4	B5	B5	B5	B5	B5
	$j = 9$	B3	B2	B1	B1	B2	B2	B1	B1	B1
	$j = 10$	B2	B1	B2	B1	B1	B2	B3	B3	B3

Table A7. The aggregated values of RFs at the level of sub-process Align supply chain resource ($\boldsymbol{p} = 1$).

RFs	The Aggregated Value in the Second Round	The Measure of Achieved Consensus
$j = 1$	(3.06,4.71,6.45)	0.8
$j = 2$	(4.03,5.59,7.29)	0.94
$j = 3$	(2.91,4.49,6.16)	0.50
$j = 4$	(2.73,4.20,5.78)	0.90
$j = 5$	(3.45,4.78,6.25)	0.88
$j = 6$	(4.35,5.69,7.12)	0.76
$j = 7$	(2.29,3.97,5.73)	0.91
$j = 8$	(6.45,7.97,9.49)	1
$j = 9$	(1.15,2.47,3.97)	0.79
$j = 10$	(4.03,5.59,7.23)	0.85

Table A8. The aggregated values impact RFs at the level of sub-processes Procure materials and services, ($\boldsymbol{p} = 2$).

RFs	The Aggregated Value in the Second Round	The Measure of Achieved Consensus
$j = 1$	(5.61,7.15,8.72)	0.59
$j = 2$	(3.33,5.14,6.99)	0.61
$j = 3$	(2.69,4.56,6.45)	0.62
$j = 4$	(2.73,4.20,5.78)	0.90
$j = 5$	(3.67,4.96,6.37)	0.92
$j = 6$	(3.38,4.86,6.44)	0.74
$j = 7$	(2.29,3.97,5.73)	0.91
$j = 8$	(6.45,7.97,9.49)	1
$j = 9$	(1.15,2.47,3.97)	0.79
$j = 10$	(3.20,4.93,6.72)	0.71

Table A9. The aggregated values of RFs at the level of sub-process Test product, ($p = 3$).

RFs	The Aggregated Value in the Second Round	The Measure of Achieved Consensus
$j = 1$	(2.13,3.82,5.54)	0.71
$j = 2$	(1.18,3.41,5.04)	0.55
$j = 3$	(2.75,4.25,5.84)	0.56
$j = 4$	(1.18,3.47,5.12)	0.70
$j = 5$	(2.52,3.83,5.31)	0.86
$j = 6$	(3.64,5.27,6.98)	0.78
$j = 7$	(2.08,3.70,5.40)	0.84
$j = 8$	(6.68,8.18,9.68)	0.50
$j = 9$	(1.20,2.56,4.07)	0.88
$j = 10$	(2.91,4.49,6.16)	0.69

Table A10. The aggregated values of RFs at the level of sub-process Manage logistics and warehousing, ($p = 4$).

RFs	The Aggregated Value in the Second Round	The Measure of Achieved Consensus
$j = 1$	(1.20,2.56,4.07)	0.68
$j = 2$	(0.94,2.41,3.90)	0.94
$j = 3$	(1.60,3.09,4.67)	0.60
$j = 4$	(1.29,2.73,4.26)	0.58
$j = 5$	(1.53,2.94,4.50)	0.77
$j = 6$	(2.58,4,5.53)	0.87
$j = 7$	(0.82,2.22,3.70)	0.72
$j = 8$	(6.79,8.29,9.79)	0.66
$j = 9$	(0.67,2.01,3.48)	0.50
$j = 10$	(2.05,3.64,5.32)	0.59

Appendix C. The Weighted Aggregated Fuzzy Value of RFs at the Level of KPI

Table A11. The weighted aggregated fuzzy value of RFs at the level of KPI in the scope of sub-process Align supply chain resources ($p = 1$).

RFs	$i = 1$	$i = 2$	$i = 3$
$j = 1$	(4.59,16.49,35.48)	(22.95,47.10,64.50)	(15.30,30.62,51.60)
$j = 2$	(20.15,36.34,58.32)	(2.02,11.18,25.52)	(30.23,44.72,69.26)
$j = 3$	(14.55,29.19,49.28)	(21.83,35.92,58.52)	(14.55,29.19,49.28)
$j = 4$	(20.48,33.60,54.91)	(20.48,33.60,54.91)	(20.48,33.60,54.91)
$j = 5$	(17.25,31.07,50)	(25.88,38.24,59.38)	(17.25,31.07,50)
$j = 6$	(13.05,28.45,49.84)	(2.18,11.38,24.92)	(13.05,28.45,49.84)
$j = 7$	(1.15,7.94,20.06)	(6.87,19.85,40.11)	(6.87,19.85,40.11)
$j = 8$	(9.68,27.90,52.20)	(48.38,79.70,94.90)	(19.35,39.85,66.43)
$j = 9$	(5.75,16.06,31.76)	(0.58,4.94,13.90)	(0.58,4.94,13.90)
$j = 10$	(30.23,55.90,72.30)	(2.02,11.18,25.31)	(6.05,19.57,39.77)
Weighted aggregated fuzzy value of RFs	(12.90,30.94,49.45)	(22.05,36.40,51.81)	(16.45,30.14,50.68)

Table A12. The weighted aggregated fuzzy value of RFs at the level of KPI in the scope of sub-process procurement materials and services ($p = 2$).

RFs	$i = 4$	$i = 5$	$i = 6$
$j = 1$	(42.08,57.20,82.84)	(28.05,46.48,69.76)	(42.08,71.50,87.20)
$j = 2$	(24.98,51.40,69.90)	(16.65,33.41,55.92)	(5,17.99,38.45)
$j = 3$	(13.45,29.64,51.60)	(1.35,9.12,22.58)	(1.35,9.12,22.58)
$j = 4$	(13.65,27.30,46.24)	(20.48,42,57.80)	(20.48,33.60,54.91)
$j = 5$	(18.35,32.24,50.46)	(18.35,32.24,50.46)	(5.51,17.36,35.04)
$j = 6$	(25.35,38.88,61.18)	(25.35,38.88,61.18)	(5.07,17.01,35.42)
$j = 7$	(6.87,19.85,40.11)	(3.44,13.90,31.52)	(3.44,13.90,31.52)
$j = 8$	(19.35,39.85,66.43)	(9.68,27.90,52.20)	(19.35,39.85,66.43)
$j = 9$	(1.73,8.65,21.84)	(0.58,4.94,13.90)	(0,0,9.93)
$j = 10$	(4.80,17.26,36.98)	(4.80,17.26,36.98)	(16,32.05,53.76)
Weighted aggregated fuzzy value of RFs	(20.47,35.28,55.46)	(16.10,29.95,48.44)	(17.07,31.70,48.46)

Table A13. The weighted aggregated fuzzy value of RFs at the level of KPI in the scope of the sub-process Test product ($p = 3$).

RFs	$i = 7$	$i = 8$	$i = 9$
$j = 1$	(3.20,13.37,30.47)	(15.98,38.20,55.40)	(6.39,19.10,38.70)
$j = 2$	(1.77,11.94,27.72)	(8.85,27.28,47.88)	(3.54,17.05,35.28)
$j = 3$	(20.63,34,55.48)	(13.75,27.63,46.72)	(8.25,21.25,40.88)
$j = 4$	(8.85,27.76,48.64)	(8.85,27.76,48.64)	(5.90,22.56,40.96)
$j = 5$	(12.60,24.90,42.48)	(7.56,19.15,37.17)	(18.90,30.64,50.45)
$j = 6$	(10.92,26.35,48.86)	(18.20,34.26,55.84)	(1.82,10.54,24.43)
$j = 7$	(6.24,18.50,37.80)	(3.12,12.95,29.70)	(3.12,12.95,29.70)
$j = 8$	(20.04,40.90,67.76)	(10.02,28.63,53.24)	(10.02,28.63,53.24)
$j = 9$	(9,20.48,38.67)	(6,16.64,32.56)	(6,16.64,32.56)
$j = 10$	(14.55,29.19,49.28)	(8.73,22.45,43.12)	(14.55,29.19,49.28)
Weighted aggregated fuzzy value of RFs	(10.53,26.17,46.13)	(11.01,26.54,45.88)	(9.35,21.87,40.54)

Table A14. The weighted aggregated fuzzy value of RFs at the level of a KPI in the scope of a sub-process Manage logistics and warehousing ($p = 4$).

RFs	$i = 10$	$i = 11$	$i = 12$
$j = 1$	(6,16.64,32.58)	(9,25.60,40.70)	(9,25.60,40.70)
$j = 2$	(7.05,19.28,37.05)	(7.05,19.28,37.05)	(7.05,24.10,39)
$j = 3$	(4.80,15.45,32.69)	(12,24.72,44.37)	(8,20.09,37.36)
$j = 4$	(3.87,13.65,29.82)	(6.45,17.75,34.08)	(9.68,21.84,40.47)
$j = 5$	(11.48,23.52,42.75)	(11.48,23.52,42.75)	(11.48,23.52,42.75)
$j = 6$	(7.74,22,38.71)	(7.74,22,38.71)	(12.90,28.60,44.24)
$j = 7$	(1.23,7.77,20.35)	(2.46,11.10,25.90)	(2.46,11.10,25.90)
$j = 8$	(20.37,41.45,68.53)	(33.95,53.98,78.32)	(20.37,41.45,68.53)
$j = 9$	(3.35,13.07,27.84)	(2.01,10.05,24.36)	(1.01,7.04,19.14)
$j = 10$	(6.15,18.35,37.24)	(6.15,18.35,37.24)	(10.25,23.86,42.56)
Weighted aggregated fuzzy value of RFs	(8.84,20.97,38.72)	(13.11,25.41,42.74)	(10.57,24.39,41.88)

References

1. Duchek, S. Organizational Resilience: A Capability-Based Conceptualization. *Bus. Res.* **2020**, *13*, 215–246. [CrossRef]
2. Hepfer, M.; Lawrence, T.B. The Heterogeneity of Organizational Resilience: Exploring Functional, Operational and Strategic Resilience. *Organ. Theory* **2022**, *3*, 26317877221074700. [CrossRef]
3. Beuren, I.M.; dos Santos, V.; Theiss, V. Organizational Resilience, Job Satisfaction and Business Performance. *Int. J. Product. Perform. Manag.* **2021**, *71*, 2262–2279. [CrossRef]
4. Ishaq Bhatti, M.; Awan, H.M.; Razaq, Z. The Key Performance Indicators (KPIs) and Their Impact on Overall Organizational Performance. *Qual. Quant.* **2014**, *48*, 3127–3143. [CrossRef]
5. Závadský, J.; Korenková, V.; Závadská, Z.; Kadárová, J.; Tuček, D. Competences in the Quality Management System Evaluation Based on the Most Worldwide Used Key Performance Indicators. *Calitatea* **2019**, *20*, 29–41.
6. Fatoki, O. The Impact of Entrepreneurial Resilience on the Success of Small and Medium Enterprises in South Africa. *Sustainability* **2018**, *10*, 2527. [CrossRef]
7. Zimmermann, H.-J. *Fuzzy Set Theory—And Its Applications*; Springer Science & Business Media: Berlin/Heidelberg, Germany, 2011; ISBN 978-94-010-0646-0.
8. Dubois, D.; Prade, H. Systems of Linear Fuzzy Constraints. *Fuzzy Sets Syst.* **1980**, *3*, 37–48. [CrossRef]
9. Arsovski, S.; Putnik, G.; Arsovski, Z.; Tadic, D.; Aleksic, A.; Djordjevic, A.; Moljevic, S. Modelling and Enhancement of Organizational Resilience Potential in Process Industry SMEs. *Sustainability* **2015**, *7*, 16483–16497. [CrossRef]
10. Akram, M.; Zahid, K.; Deveci, M. Multi-Criteria Group Decision-Making for Optimal Management of Water Supply with Fuzzy ELECTRE-Based Outranking Method. *Appl. Soft Comput.* **2023**, *143*, 110403. [CrossRef]
11. Jin, L. Uncertain Probability, Regular Probability Interval and Relative Proximity. *Fuzzy Sets Syst.* **2023**, *467*, 108579. [CrossRef]
12. Habibi, A.; Jahantigh, F.F.; Sarafrazi, A. Fuzzy Delphi Technique for Forecasting and Screening Items. *Asian J. Res. Bus. Econ. Manag.* **2015**, *5*, 130. [CrossRef]
13. Mabrouk, N. Green Supplier Selection Using Fuzzy Delphi Method for Developing Sustainable Supply Chain. *Decis. Sci. Lett.* **2021**, *10*, 63–70. [CrossRef]
14. Grabisch, M.; Marichal, J.-L.; Mesiar, R.; Pap, E. *Aggregation Functions*; Illustrated Edition; Cambridge University Press: Cambridge, UK, 2009; ISBN 978-0-521-51926-7.
15. Calvo, T.; Mayor, G.; Mesiar, R. *Aggregation Operators: New Trends and Applications*; Studies in Fuzziness and Soft Computing; Physica-Verlag HD: Berlin/Heidelberg, Germany, 2002; Volume 97, ISBN 978-3-662-00319-0.

16. Jani, N.M.; Zakaria, M.H.; Maksom, Z.; Haniff, M.S.M.; Mustapha, R. Validating Antecedents of Customer Engagement in Social Networking Sites Using Fuzzy Delphi Analysis. *Int. J. Adv. Comput. Sci. Appl.* **2018**, *9*, 294–304. [CrossRef]
17. Abdollahi, S.M.; Ranjbarian, B.; Kazemi, A. An Investigation of the Antecedents of Consumers' Confusion in Purchasing an Outbound Package Tour in the City of Isfahan by Fuzzy Delphi Method. *Iran. J. Manag. Stud.* **2020**, *13*, 527–564. [CrossRef]
18. Tsai, H.-C.; Lee, A.-S.; Lee, H.-N.; Chen, C.-N.; Liu, Y.-C. An Application of the Fuzzy Delphi Method and Fuzzy AHP on the Discussion of Training Indicators for the Regional Competition, Taiwan National Skills Competition, in the Trade of Joinery. *Sustainability* **2020**, *12*, 4290. [CrossRef]
19. Dawood, K.A.; Sharif, K.Y.; Ghani, A.A.; Zulzalil, H.; Zaidan, A.A.; Zaidan, B.B. Towards a Unified Criteria Model for Usability Evaluation in the Context of Open Source Software Based on a Fuzzy Delphi Method. *Inf. Softw. Technol.* **2021**, *130*, 106453. [CrossRef]
20. Bui, T.D.; Tsai, F.M.; Tseng, M.-L.; Ali, M.H. Identifying Sustainable Solid Waste Management Barriers in Practice Using the Fuzzy Delphi Method. *Resour. Conserv. Recycl.* **2020**, *154*, 104625. [CrossRef]
21. Khan, S.; Haleem, A.; Khan, M.I. Risk Management in Halal Supply Chain: An Integrated Fuzzy Delphi and DEMATEL Approach. *J. Model. Manag.* **2020**, *16*, 172–214. [CrossRef]
22. Aleksic, A.; Nestic, S.; Tadic, D.; Komatina, N. Determination of Organizational Resilience Level within Business Processes in Production Companies. In Proceedings of the 6th International Scientific Conference COMET-a, Jahorina, Bosnia and Herzegovina, 11–19 November 2022; University of East Sarajevo, Faculty of Mechanical Engineering: Jahorina, Bosnia and Herzegovina, 2022; pp. 750–757.
23. Chen, C.-A.; Lee, S.-R. Developing the Country Brand of Taiwan from the Perspective of Exports. *Asian J. Empir. Res.* **2013**, *3*, 1223–1236.
24. Horng, J.-S.; Chou, S.-F.; Liu, C.-H.; Tsai, C.-Y. Creativity, Aesthetics and Eco-Friendliness: A Physical Dining Environment Design Synthetic Assessment Model of Innovative Restaurants. *Tour. Manag.* **2013**, *36*, 15–25. [CrossRef]
25. Liu, C.-H.S.; Chou, S.-F. Tourism Strategy Development and Facilitation of Integrative Processes among Brand Equity, Marketing and Motivation. *Tour. Manag.* **2016**, *54*, 298–308. [CrossRef]
26. Kumar, A.; Mangla, S.K.; Luthra, S.; Rana, N.P.; Dwivedi, Y.K. Predicting Changing Pattern: Building Model for Consumer Decision Making in Digital Market. *J. Enterp. Inf. Manag.* **2018**, *31*, 674–703. [CrossRef]
27. Mahmoudi, S.; Ranjbarian, B.; Fathi, S. Factors Influencing on Iran's Image as a Tourism Destination. *Int. J. Serv. Oper. Manag.* **2017**, *26*, 186–210. [CrossRef]
28. Singh, P.K.; Sarkar, P. A Framework Based on Fuzzy Delphi and DEMATEL for Sustainable Product Development: A Case of Indian Automotive Industry. *J. Clean. Prod.* **2020**, *246*, 118991. [CrossRef]
29. Kumar, A.; Dash, M.K. Causal Modelling and Analysis Evaluation of Online Reputation Management Using Fuzzy Delphi and DEMATEL. *Int. J. Strateg. Decis. Sci.* **2017**, *8*, 27–45. [CrossRef]
30. Farhadian, T.; Shahgholian, K. A Fuzzy Delphi Method for Identifying Effective Indexes in Absorption Banking Resources in Iran and Pathology He Performance of a Bank with Fan P-Swot (Case Study: Mehr Bank of Iran). *Int. J. Biol. Pharm. Allied Sci.* **2015**, *4*, 557–568.
31. Chu, H.-C.; Hwang, G.-J. A Delphi-Based Approach to Developing Expert Systems with the Cooperation of Multiple Experts. *Expert Syst. Appl.* **2008**, *34*, 2826–2840. [CrossRef]
32. Islam, D.M.; Dinwoodie, J.; Roe, M. Promoting Development Through Multimodal Freight Transport in Bangladesh. *Transp. Rev.* **2006**, *26*, 571–591. [CrossRef]
33. Buck, A.J.; Gross, M.; Hakim, S.; Weinblatt, J. Using the Delphi Process to Analyze Social Policy Implementation: A Post Hoc Case from Vocational Rehabilitation. *Policy Sci.* **1993**, *26*, 271–288. [CrossRef]
34. von der Gracht, H.A. Consensus Measurement in Delphi Studies: Review and Implications for Future Quality Assurance. *Technol. Forecast. Soc. Change* **2012**, *79*, 1525–1536. [CrossRef]
35. Ferri, C.P.; Prince, M.; Brayne, C.; Brodaty, H.; Fratiglioni, L.; Ganguli, M.; Hall, K.; Hasegawa, K.; Hendrie, H.; Huang, Y.; et al. Global Prevalence of Dementia: A Delphi Consensus Study. *Lancet* **2005**, *366*, 2112–2117. [CrossRef] [PubMed]
36. Macuzić, I.; Tadić, D.; Aleksić, A.; Stefanović, M. A Two Step Fuzzy Model for the Assessment and Ranking of Organizational Resilience Factors in the Process Industry. *J. Loss Prev. Process Ind.* **2016**, *40*, 122. [CrossRef]
37. APQC. *Process Classification Framework V6.11*; APQC: Houston, TX, USA, 2015.
38. Somerville, J.A. Effective Use of the Delphi Process in Research: Its Characteristics, Strengths and Limitations. Unpublished Ph.D. Thesis, Oregon State University, Corvallís, OR, USA, 2008.
39. Malone, D.C.; Abarca, J.; Hansten, P.D.; Grizzle, A.J.; Armstrong, E.P.; Van Bergen, R.C.; Duncan-Edgar, B.S.; Solomon, S.L.; Lipton, R.B. Identification of Serious Drug-Drug Interactions: Results of the Partnership to Prevent Drug-Drug Interactions. *J. Am. Pharm. Assoc.* **2004**, *44*, 142–151. [CrossRef]
40. Hodicky, J.; Özkan, G.; Özdemir, H.; Stodola, P.; Drozd, J.; Buck, W. Analytic Hierarchy Process (AHP)-Based Aggregation Mechanism for Resilience Measurement: NATO Aggregated Resilience Decision Support Model. *Entropy* **2020**, *22*, 1037. [CrossRef] [PubMed]

41. Copeland, S.; Comes, T.; Bach, S.; Nagenborg, M.; Schulte, Y.; Doorn, N. Measuring Social Resilience: Trade-Offs, Challenges and Opportunities for Indicator Models in Transforming Societies. *Int. J. Disaster Risk Reduct.* **2020**, *51*, 101799. [CrossRef]
42. Munoz, A.; Dunbar, M. On the Quantification of Operational Supply Chain Resilience. *Int. J. Prod. Res.* **2015**, *53*, 6736–6751. [CrossRef]

Σ mathematics

MDPI

Article

Enhancing Sustainability in Belize's Ecotourism Sector: A Fuzzy Delphi and Fuzzy DEMATEL Investigation of Key Indicators

Marvin Ruano [1], Chien-Yi Huang [2], Phi-Hung Nguyen [3,*], Lan-Anh Thi Nguyen [3], Hong-Quan Le [3] and Linh-Chi Tran [3]

1 Department of Management, National Taipei University of Technology, Taipei 10608, Taiwan; marvin.ruano100@gmail.com
2 Department of Industrial Engineering and Management, National Taipei University of Technology, Taipei 10608, Taiwan; jayhuang@mail.ntut.edu.tw
3 Research Center of Applied Sciences, Faculty of Business, FPT University, Hanoi 100000, Vietnam; anhntl84@fe.edu.vn (L.-A.T.N.); quanlhhs150282@fpt.edu.vn (H.-Q.L.); chitlhs150498@fpt.edu.vn (L.-C.T.)
* Correspondence: hungnp30@fe.edu.vn

Abstract: Sustainable ecotourism has become a strategy to balance tourism growth with environmental and sociocultural considerations. This study aims to propose an integrated approach of the Delphi technique and the decision-making trial and evaluation laboratory (DEMATEL) based on fuzzy set theory to investigate sustainable ecotourism indicators in Belize. The study covers six dimensions: environmental, social, cultural, economic, political, and intrinsic. Firstly, the Fuzzy Delphi technique constructs a comprehensive set of indicators with expert consensus, resulting in 51 relevant and representative indicators out of the initial 63. Secondly, the Fuzzy DEMATEL approach is then applied to analyze the interdependencies among indicators and identify their causal relationships, providing insights into the complex dynamics of sustainable ecotourism in Belize. The results provide a structured decision-making framework to prioritize actions, allocate resources effectively, and promote sustainable practices in the ecotourism sector. Therefore, these findings enhance the understanding of indicator interconnections across dimensions, enabling informed decision making for policymakers, industry practitioners, and researchers. Policymakers can develop policies and regulations that foster sustainable practices, while industry practitioners can enhance visitor experiences, engage with local communities, and ensure the industry's long-term viability. Researchers can further investigate specific dimensions and indicators to advance the knowledge and implementation of sustainable ecotourism. Finally, this investigation supports the goal of achieving a harmonious and sustainable balance between tourism development and environmental preservation in Belize. By safeguarding the natural and cultural heritage of the region, sustainable ecotourism can benefit present and future generations.

Keywords: Belize; ecotourism; sustainable; MCDM; fuzzy set theory; Delphi; DEMATEL

MSC: 97M30; 91B02; 62P05; 91B84

Citation: Ruano, M.; Huang, C.-Y.; Nguyen, P.-H.; Nguyen, L.-A.T.; Le, H.-Q.; Tran, L.-C. Enhancing Sustainability in Belize's Ecotourism Sector: A Fuzzy Delphi and Fuzzy DEMATEL Investigation of Key Indicators. *Mathematics* **2023**, *11*, 2816. https://doi.org/10.3390/math11132816

Academic Editor: Aleksandar Aleksić

Received: 31 May 2023
Revised: 18 June 2023
Accepted: 21 June 2023
Published: 23 June 2023

1. Introduction

Tourism is a rapidly expanding global sector that has significant implications for the environment, economy, and society [1]. In response, ecotourism has emerged as a sustainable tourism model that promotes responsible travel, environmentally friendly practices, and economic benefits for local communities [2–4]. Belize, a nation in Central America, has prioritized ecotourism to enhance environmental protection efforts and generate income for its citizens. The country boasts a diverse cultural and natural heritage, including the Belize Barrier Reef, the second-largest coral reef system globally, and numerous protected areas such as national parks, wildlife reserves, and marine reserves. Visitors worldwide come to Belize to experience its unique biodiversity and cultural heritage [5,6]. However, the growth of ecotourism in Belize presents challenges, particularly in ensuring its long-term sustainability [7]. The expansion of tourism activities carries the risk of overuse and degradation of natural resources, displacement of local communities, and the commodification of culture [8]. Therefore, it is crucial to identify and monitor indicators of sustainable ecotourism in Belize to ensure that tourism contributes to the country's economic development and conservation efforts while minimizing negative impacts on the environment and society.

Sustainable ecotourism has gained prominence in recent years as the tourism industry's growth places increasing pressure on natural resources and ecosystems. Ecotourism offers a solution by promoting conservation and sustainable development through responsible travel to natural areas, benefiting both the environment and local communities. Key aspects of sustainable ecotourism include community involvement and empowerment, conservation management, environmental education, and economic benefits for local communities [8,9]. Methods to promote sustainable ecotourism encompass certification programs, stakeholder engagement, sustainable tourism planning, and monitoring and evaluation of tourism impacts [10,11]. Additionally, applying technologies such as geographic information systems (GISs) and remote sensing facilitates the assessment and management of tourism sites [12–14]. Multi-criteria decision-making (MCDM) methods play a vital role in sustainable ecotourism research, allowing the evaluation and comparison of alternatives based on multiple criteria [15]. These methods consider various environmental, economic, social, and cultural factors that impact ecotourism sustainability [16]. MCDM methods enable decision makers to prioritize criteria, identify trade-offs, and generate comprehensive rankings of alternatives [17,18]. Various tools, such as the analytic hierarchy process (AHP), the Technique for Order of Preference by Similarity to Ideal Solution (TOPSIS), and the Preference Ranking Organization Method for Enrichment of Evaluations (PROMETHEE), have been utilized in ecotourism research to identify the most sustainable alternatives based on different criteria, assisting policymakers and stakeholders in making informed decisions to promote sustainable ecotourism development [19].

Belize, renowned for its abundant biodiversity and cultural heritage, has emerged as a sought-after ecotourism destination. Ecotourism, which emphasizes preserving and celebrating the natural environment and local culture, has gained popularity among Belize travelers. In this study, our objective is to explore sustainable ecotourism indicators specific to Belize. When it comes to identifying a comprehensive set of indicators for sustainable ecotourism, the integrated Fuzzy Delphi and Fuzzy DEMATEL approach provides distinct advantages over other methods. This approach stands out because it addresses the inherent uncertainties and subjectivities associated with indicator selection. By incorporating the Fuzzy Delphi method, expert opinions and subjective judgments can seamlessly integrate into the process, ensuring a well-rounded and inclusive perspective.

Moreover, the Fuzzy DEMATEL method enables an assessment of the interrelationships and dependencies among the identified indicators. This aspect is crucial in the complex context of ecotourism, where numerous factors interact and influence sustainability outcomes. By uncovering these intricate relationships, decision makers gain a holistic understanding of how indicators interplay, leading to a more integrated and coherent set of indicators that reflect the multidimensional nature of sustainable ecotourism in Belize. The integration of the Fuzzy Delphi and Fuzzy DEMATEL approaches also addresses the limitations of both individual methods. The Fuzzy Delphi method's potential biases in expert opinions and the Fuzzy DEMATEL method's sensitivity to threshold values are effectively mitigated by combining the two approaches. This integration results in a robust and reliable set of indicators, enhancing the accuracy and practicality of the chosen indicators for guiding sustainable ecotourism practices in Belize.

With the growing awareness of the detrimental effects of tourism on the environment and local communities, the notion of sustainability has gained significant prominence in the tourism industry. In line with this, the present study aims to address two specific research questions in the context of Belize:

(i) What are the key sustainable ecotourism indicators in the context of Belize?

(ii) What is the interdependence among these indicators?

By answering these questions, this study seeks to contribute to the understanding and promotion of sustainable practices within the ecotourism sector in Belize as per the following perspectives:

(i) This study contributes to the field of sustainable ecotourism in Belize by developing a comprehensive set of indicators using the integrated Fuzzy Delphi and Fuzzy DEMATEL approach. These indicators cover dimensions such as environmental conservation, community involvement, cultural preservation, and economic benefits.

(ii) The proposed method represents a methodological advancement in identifying and evaluating sustainable ecotourism indicators. The Fuzzy Delphi method enables the incorporation of expert opinions and subjective judgments, allowing for a more comprehensive and inclusive approach to indicator selection. On the other hand, the Fuzzy DEMATEL method facilitates the assessment of interrelationships and dependencies among the identified indicators, providing insights into their complex interactions.

(iii) By identifying the key sustainable ecotourism indicators, this research offers valuable guidance for decision-making processes, policy formulation, and the development of sustainable tourism practices. Stakeholders can utilize the results of this study to prioritize their efforts, allocate resources effectively, and implement strategies that ensure the long-term sustainability of ecotourism in Belize. Furthermore, the comprehensive set of indicators can serve as a benchmark for monitoring and evaluating the performance and progress of ecotourism initiatives, facilitating continuous improvement and adaptation.

The following is the structure of this study. Section 2 is devoted to a review of the literature. Section 3 describes the research process and methods. The discussions and results of empirical analysis are presented in Section 4. Finally, Section 5 summarizes the conclusion, implications, limitations and furture research.

2. Literature Review

2.1. Literature Review on Sustainable Ecotourism Indicators

Ecotourism has been widely recognized as a sustainable form of tourism that supports conservation efforts, promotes environmental awareness, and provides socioeconomic benefits to local communities. However, various ecotourism activities have negatively impacted the local environment, culture, and society. Therefore, identifying and implementing sustainable ecotourism indicators are essential to ensuring that ecotourism activities are sustainable in the long run.

Several studies have been conducted to assess and evaluate sustainable ecotourism indicators, employing various methods and approaches.

In a study by Ocampo et al. [20], the Fuzzy Delphi method was utilized to identify sustainable ecotourism indicators. Initially, they started with 666 tourism indicators and then tailored them specifically to ecotourism, resulting in 59 indicators. A final set of 39 indicators was derived through further refinement, aligning with the specific context of Philippine ecotourism. This research highlights the effectiveness of the Fuzzy Delphi method in narrowing down and customizing indicators to suit the requirements of sustainable ecotourism practices. Sobhani et al. [21] conducted a notable study assessing sustainable ecotourism indicators in Tehran, Iran. Their research encompassed three main categories: environmental–physical, demographic–social, and economic–institutional. Within these categories, a total of 38 environmental–physical indicators, 42 demographic–social indicators, and 30 economic–institutional indicators were identified.

In a study focused on monitoring ecotourism sustainability in the northern forests of Iran, Godratollah Barzekar et al. [22] used the Delphi process to achieve consensus on nine criteria and identified a total of 61 indicators. These indicators cover various dimensions, including ecology, economy, society, culture, and institutions, specific to the northern forests of Iran. Azlizam Aziz et al. [23] utilized the Delphi method to determine criteria and indicators for monitoring ecotourism sustainability. Through a consensus-based approach, they identified 21 environmental factors, 8 economic factors, 6 cultural factors, 21 societal factors, and 5 institutional factors. This study highlights the multidimensional nature of ecotourism sustainability assessment. Meanwhile, Asadpourian et al. [24] conducted a comprehensive analysis in Lorestan Province, Iran, using a combination of the SWOT analysis and AHP. Their integrated approach resulted in a framework consisting of 30 indicators across three dimensions: economic, social, and environmental. This study demonstrates the importance of considering multiple dimensions and employing robust analytical processes in the identification of indicators for sustainable ecotourism.

These above-mentioned studies collectively contribute to the understanding and evaluation of sustainable ecotourism indicators by employing different methodologies and approaches. They provide valuable insights into indicator selection, customization, and assessment across various dimensions, supporting the development and implementation of sustainable ecotourism practices in different regions. After comprehensively summarizing the literature on sustainable ecotourism indicators, it is evident that a wide range of approaches and methodologies have been employed to assess and evaluate the sustainability of ecotourism ventures. Within this body of research, the authors have identified six distinct groups of indicators utilized in different studies. These indicator groups encompass various dimensions and aspects of sustainable ecotourism, providing a holistic perspective on the environmental, social, cultural, economic, political, and intrinsic dimensions. The following sections delve into each group, highlighting their respective characteristics and contributions to assessing sustainable ecotourism in Table 1.

Table 1. Potential list of sustainable ecotourism indicators.

Main Dimensions	Code	Indicators
Environmental [25–28]	EN 1	Number of endangered species and protection/conservation of flora and fauna.
	EN 2	Cleanliness and quality of tourism facilities, and access to drainage and wastewater treatment systems.
	EN 3	Climate/weather and quality of air, water, and land.
	EN 4	Environmental codes of conduct, awareness, and education for tourists.
	EN 5	Environmental emergency action plans.
	EN 6	Environmental laws and sites' rules and regulations.
	EN 7	Environmentally responsible suppliers.
	EN 8	Negative impacts of tourism on the environment.
	EN 9	Proper use and consumption of water.
	EN 10	Proper use of electrical power.
	EN 11	Recycling, reduction, and reuse of waste.
	EN 12	Respect of ecosystem and proper use of coastal land and forest with every new development project.
	EN 13	Restoration and reduction of damage caused by tourism.
	EN 14	Use of biodegradable products.
	EN 15	Water, land, and air pollution.
	EN 16	Workshops and discussions on environmentally friendly management techniques.

Table 1. *Cont.*

Main Dimensions	Code	Indicators
Social [29–33]	SO 1	Disability laws.
	SO 2	Exploitation of employees, child labor, or sex tourism.
	SO 3	Fair compensation and compliance with labor laws.
	SO 4	Impacts of tourists on local issues.
	SO 5	Improvement of well-being, quality of life, and safety of local community.
	SO 6	Level of interaction between locals and tourists.
	SO 7	Local business support.
	SO 8	Organizational structure inclusivity.
	SO 9	Poverty risk and social exclusion.
	SO 10	Professional development and education of locals.
	SO 11	Protection of minority groups.
	SO 12	Social equity in all organizational practices.
	SO 13	Stakeholder rights regarding tourist activities.
	SO 14	Training and promotion of qualified employees within the tourism industry.
Cultural [24,33–36]	CU 1	Authenticity of local products and services.
	CU 2	Availability and accessibility of information about local culture.
	CU 3	Cultural codes of conduct for tourists.
	CU 4	Illegal trade of artifacts.
	CU 5	Negative effects of development projects on cultural identities.
	CU 6	Promotion and protection of local culture.
Economic [25,37–40]	EC 1	Ability to attract more investment.
	EC 2	Availability and accessibility of medical services, transportation, and recreational facilities.
	EC 3	Conservation of local economy pace.
	EC 4	Cost management of tourism operations.
	EC 5	Domestic spending, reinvestment, and business expansion.
	EC 6	Economic and financial development of stakeholders.
	EC 7	Employment opportunities, financial subsidization, and compensation for locals.
	EC 8	Fair trade practices and principles.
	EC 9	Implementation of green design technology.
	EC 10	Risk management and production stabilization.
	EC 11	Satisfactory goods and services.
	EC 12	Support and contribution towards development of local goods, services, and infrastructure.
	EC 13	Support for local suppliers and subsidization of local production and manufacturing.
	EC 14	Taxes on land, buildings, and other structures.
	EC 15	Tourist expenditure and annual gross income in tourism jobs.

Table 1. *Cont.*

Main Dimensions	Code	Indicators
Political [25,41–43]	PO 1	Foreign involvement and ownership in local business.
	PO 2	Common organizational goals and employee loyalty and job security.
	PO 3	Democratic organizational cultures, group management, autonomy, flexibility, freedom of speech, and participatory decision making.
	PO 4	Ethical, moral, and transparent organizational structures.
	PO 5	Monitorization of operational, management, and financial results.
	PO 6	Political prejudices, bias, and discrimination.
	PO 7	Reflection of sustainability values on business practices.
Intrinsic [37,44]	IN 1	Attitude of locals toward satisfaction, service quality, and training mechanisms.
	IN 2	Average length of stay per tourist.
	IN 3	Crime rates, accidents, visitor safety and security, and legal compliance (prosecutions, fines, etc.).
	IN 4	Overall service quality of local businesses and potential businesses.
	IN 5	Tourist satisfaction with related activities and the volume of tourists, returning tourists, and seasonality.

The environmental dimension is a vital aspect of sustainable ecotourism, aiming to promote the conservation and protection of the environment [25–27]. Evaluating the environmental impact of ecotourism activities requires using indicators that assess sustainability. Several critical indicators have been identified in the literature. These indicators include the protection and conservation of endangered species, the cleanliness and quality of tourism facilities, climate and weather conditions, environmental codes of conduct and tourist awareness, environmental emergency action plans, and adherence to environmental laws and regulations. Additionally, the use of environmentally responsible suppliers and the implementation of restoration measures to reduce damage caused by tourism are essential factors. By implementing sustainable practices, such as ecotourism, it is possible to mitigate the negative impact of tourism activities on the environment and contribute to environmental sustainability. Monitoring and evaluating these indicators are essential for the successful development of sustainable ecotourism.

The social dimension of sustainable ecotourism encompasses a range of indicators that focus on the well-being of local communities, human rights, and social equity [28–30]. It recognizes both the potential benefits of tourism for local people and the risks of negative impacts on their livelihoods, culture, and identity. Key social indicators include compliance with disability laws to ensure accessibility and inclusivity, prevention of exploitation of employees, consideration of local issues and tensions between tourists and communities, support for local businesses and entrepreneurship, inclusive organizational structures, poverty reduction and avoidance of social exclusion, professional development and education opportunities for locals, protection of minority groups and cultural heritage, social equity in all organizational practices, recognition of stakeholder rights, and training and promotion of qualified employees. These indicators highlight the importance of fostering positive social impacts, respecting local rights and traditions, and promoting equitable participation and benefits for all stakeholders involved in ecotourism.

The cultural dimension is a critical aspect of sustainable ecotourism, encompassing indicators focusing on preserving and promoting local cultures [24,31]. Cultural indicators play a significant role in creating authentic tourist experiences, fostering economic growth for local businesses, and respecting the values and beliefs of the host community. Key cultural indicators include the authenticity of local products and services, availability and accessibility of information about local culture, cultural codes of conduct for tourists, addressing the issue of the illegal trade of artefacts, mitigating negative impacts of development projects on cultural identities, and promoting and protecting local culture through festivals, handicrafts, and cuisine. These indicators emphasize the importance of preserving cultural heritage, preventing misrepresentation, and ensuring sustainable development that respects the cultural identity of the host community.

The economic dimension is a vital aspect of sustainable ecotourism, focusing on indicators that assess the financial sustainability and contribution of ecotourism to the local economy [25,32]. Economic indicators evaluate factors such as the ability to attract investment, availability and accessibility of essential services and facilities, conservation of the local economy's pace, cost management of tourism operations, domestic spending and business expansion, economic and financial development of stakeholders, employment opportunities and compensation for locals, fair trade practices, implementation of green design technology, and risk management and production stabilization. These indicators help measure the economic benefits, employment generation, financial well-being, and the region's long-term viability. Effective risk management practices and production stabilization are crucial for managing potential risks and maintaining consistent quality, ultimately enhancing customer satisfaction and attracting more visitors. Monitoring and evaluating economic indicators enable ecotourism operators to make informed decisions and improve their economic sustainability.

The political dimension is a vital component in assessing the sustainability of ecotourism, as it examines the political climate of the destination and its impact on the industry [33–35]. Political indicators play a significant role in evaluating various aspects such as foreign involvement and ownership in local businesses, common organizational goals and employee loyalty, democratic organizational cultures and participatory decision making, ethical and transparent organizational structures, monitoring of operational and financial results, political prejudices and discrimination, and the reflection of sustainability values in business practices. These indicators help identify areas for improvement, promote sustainable practices, and ensure the industry's long-term viability. By addressing political factors and fostering an environment of collaboration, transparency, and ethical responsibility, ecotourism can thrive while benefiting the local community and preserving the natural and cultural heritage of the destination.

The intrinsic dimension of sustainable ecotourism indicators focuses on measuring aspects inherent to the tourism industry, providing valuable information about service quality and tourist satisfaction [36,37]. These indicators encompass various factors such as the attitude of locals towards service quality and training mechanisms, the average length of stay per tourist, crime rates and visitor safety, the overall service quality of local businesses, and tourist satisfaction with related activities. By assessing these intrinsic indicators, stakeholders can gain insights into the effectiveness of training programs, marketing campaigns, safety measures, and service quality. This information can guide efforts to enhance the tourism experience, attract more visitors, and foster repeat visits, ultimately contributing to ecotourism destinations' long-term sustainability and success.

2.2. Literature Review on Established Methods

In order to effectively assess and analyze the multitude of variables and factors involved in sustainable ecotourism, a robust and systematic method is essential to filter and prioritize these variables. This ensures that the indicators used for evaluation are relevant, comprehensive, and representative of the sustainability dimensions. Focusing on different aspects of methodology in the field of environmental modeling and sustainability assessment, Pianosi et al. [45] discussed the concept of sensitivity analysis (SA) and its applications in environmental modeling. Andria et al. [46] presented a fuzzy approach for assessing the sustainability of tourist destinations, addressing the limitations of the traditional carrying capacity method. In another study, Andria et al. [47] emphasized the increasing importance of "smartness" and "sustainability" in decision-making processes for practitioners and policymakers. Finally, Andria et al. [48] presented a method for ranking tourist destinations and evaluating their sustainability performance. They employed a fuzzy multiple-criteria decision-making method to determine sustainability performance values and rank destinations accordingly. These studies collectively highlight the significance of incorporating fuzzy approaches and sensitivity analysis techniques for robust sustainability assessment in various domains, including environmental modeling and tourist destinations.

Of the numerous methods utilized in academic literature, the Delphi technique has gained prominence as a widely accepted and efficient method for achieving consensus on a particular subject. This method enables the systematic collection of expert opinions on a particular subject through sequentially applied feedback questionnaires interspersed with summary data on earlier responses [49]. The Delphi method, recognized as a dependable qualitative research strategy, can address challenges, enhance decision making, and foster consensus among groups in diverse domains [50]. It can be identified by four key characteristics: participant anonymity, iterative rounds of feedback and opinion revision, controlled feedback that informs participants about the perspectives of others, and the provision for Delphi participants to clarify or modify their views. Furthermore, the method enables quantitative analysis and interpretation of data through statistical group responses [51]. A loss of individual knowledge occurs as a result of the Delphi method's requirement that experts adjust their judgments to reflect the average worth of all expert opinions. Furthermore, the Delphi technique does not consider data imprecision and uncertainty. Therefore, using a defuzzifyng function based on questionnaires, the conventional Delphi approach is combined with fuzzy sets to validate essential elements and choose assessment indicators [52–55]. In order to deal with uncertainty and imprecision, fuzzy theory is a mathematical framework that permits variables to have partial membership in a set. This implies that fuzzy theory allows for more nuanced and probabilistic explanations of variables as opposed to utilizing binary true/false values, which can be advantageous when working with complicated or ambiguous concepts. When dealing with linguistic variables, such as "high" or "low" levels of a given factor, fuzzy theory is constructive since it enables varying degrees of membership in a collection rather than requiring a binary classification [55]. Evaluating the relationship between indicators is crucial in assessing sustainable ecotourism, as it allows for a deeper understanding of the complex interactions and dependencies among different indicators. The Delphi technique and the DEMATEL approach have both been widely used to study the cause-and-effect linkages between indicators and provide a systematic framework for assessing their interdependencies [56,57]. The Delphi method helps generate expert consensus and identify relevant indicators, while DEMATEL offers a quantitative analysis of the relationships between these indicators. Additionally, Fuzzy DEMATEL, an extension of DEMATEL that incorporates fuzzy logic, has been utilized to address uncertainties and vagueness in the assessment process [58,59]. These two methods have been effective in various fields, allowing researchers to gain valuable insights into the complex relationships among indicators and their implications for sustainable development [60–62]. Therefore, considering the significance of these methods, applying Fuzzy Delphi–DEMATEL in the context of sustainable ecotourism can provide valuable insights for decision making and policy formulation.

2.3. Research Gaps

With its rich biodiversity and significant ecotourism potential, Belize presents a unique case for exploring the relationships between indicators and identifying priority areas for intervention. Despite the extensive research on sustainable ecotourism indicators, there is still a notable research gap regarding a comprehensive assessment of variables and their interrelationships. Moreover, a limited number of studies have explicitly focused on the context of Belize, an ecologically diverse and significant ecotourism destination. Therefore, this work aims to address these research gaps by applying the combined methods of Fuzzy Delphi and Fuzzy DEMATEL in the context of sustainable ecotourism in Belize. This study seeks to comprehensively analyze indicators, evaluate their interdependencies, and offer valuable insights for sustainable ecotourism development in Belize by utilizing these methods. This research's findings can contribute to the body of knowledge and offer practical guidance for policymakers, stakeholders, and ecotourism professionals in Belize and elsewhere.

3. Methodology

3.1. Research Process

The research process involves three phases, as presented in Figure 1 as follows:

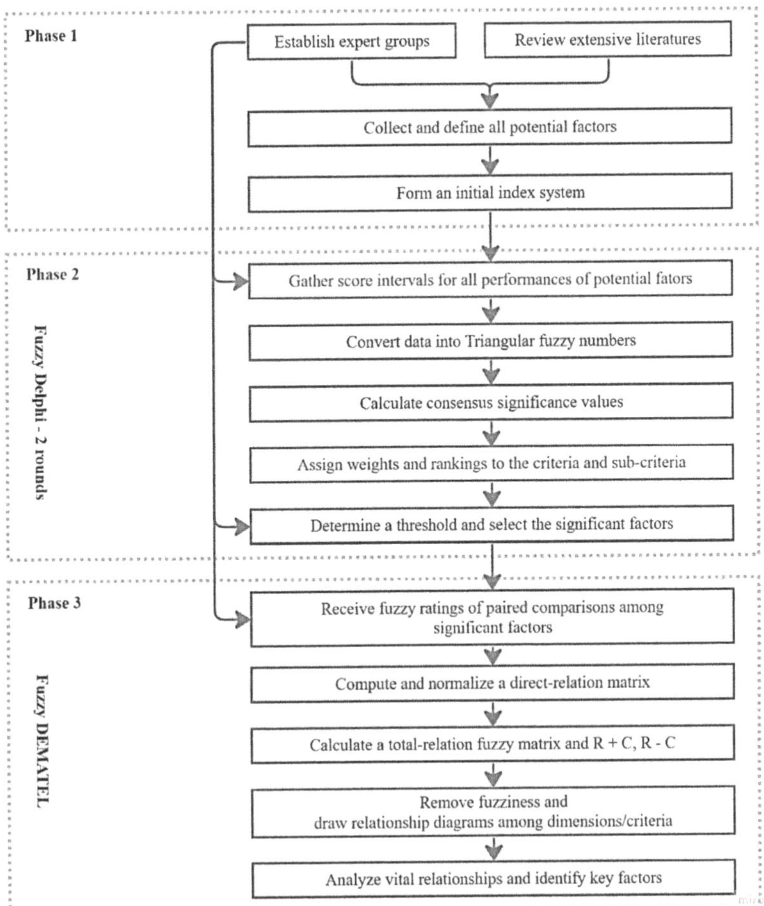

Figure 1. Research framework.

In the first phase, potential indicators are collected from literature review and experts' opinions to build an initial index system. This phase focuses on gathering relevant information and establishing a foundation for the subsequent analysis.

In the second phase, the Fuzzy Delphi method is employed to recognize significant factors. Expert groups are surveyed through a questionnaire, and consensus significance values are calculated to validate critical factors. The significance of factors is determined based on their level of influence. It is important to note that qualitative data are transformed into quantitative data using linguistic terms transformation (refer to Table 2). Two rounds of Fuzzy Delphi are conducted to enhance the reliability and accuracy of the results.

Table 2. TFNs corresponding to linguistic terms of Fuzzy Delphi method.

Linguistic Terms	TFNs
Equal	(0, 0, 0.25)
Moderate	(0, 0.25, 0.5)
Strong	(0.25, 0.5, 0.75)
Demonstrated	(0.5, 0.75, 1)
Extreme	(0.75, 1, 1)

In the third phase, a causal structure model is developed using the Fuzzy DEMATEL method to identify critical factors. This involves analyzing the causal relationships between the significant factors. The Fuzzy DEMATEL method allows for a comprehensive understanding of the interdependencies and correlations among the identified factors.

This hybrid framework offers several advantages, including the incorporation of vague judgments into quantitative values, integration of expert comments, and the ability to explore correlations between factors under uncertain circumstances. It is well suited for addressing real-world decision-making issues, particularly in the context of identifying indicators for sustainable ecotourism in Belize.

3.2. Fuzzy Delphi Method

The integration of the Delphi method with fuzzy set theory, known as Fuzzy Delphi, offers several advantages, including facilitating consensus among diverse perspectives, saving time and cost, and reducing the number of rounds required for opinion collection [63]. In this study, to better capture and represent expert knowledge, triangular fuzzy numbers (TFNs) are employed. TFNs are characterized by three actual numbers (l, m, u), where l represents the lower limit, m represents the maximum, and u represents the upper limit. The use of TFNs enhances decision-making capabilities in complex problem-solving scenarios [64].

Step 1: Given P experts and Q attributes, where expert i (P_i = 1, 2, 3, ... , n) concludes that characteristic j (Q_j = 1, 2, 3, ... , m) can be represented as a TFN, $F_{ij} = (l_{ij}; m_{ij}; u_{ij})$. In this representation, l_{ij} denotes the lower limit, m_{ij} denotes the modal value, and u_{ij} denotes the upper limit of the TFN. As a consequence, linguistic values are produced utilizing linguistic words and TFNs, as illustrated in Table 2.

The weight of attribute j then refers to $F_{ij} = (l_{ij}; m_{ij}; u_{ij})$ where $\left[l_{ij} = min\left(l_{ij}\right); m_{ij} = \sqrt[n]{\prod_1^n m_{ij}}; u_{ij} = max\left(u_{ij}\right)\right]$.

Step 2: The convex combination value and the alpha cut value are distinct approaches for summarizing and interpreting TFNs in fuzzy systems [64]. The convex combination value of a TFN involves calculating a weighted average that considers the membership levels at different points along its range. It provides a single representative value that captures the general information of the TFN, taking into account its entire shape. This method enables a more nuanced representation of experts' opinions, facilitating decision making and analysis. In contrast, the alpha cut value of a TFN focuses on a specific level of membership or confidence. It determines the crisp value at which the TFN possesses a certain degree of membership. By identifying the point or interval along the x-axis where the membership function exceeds a predefined threshold (α), the alpha cut value provides a crisp value suitable for crisp decisions or comparisons [65].

In this study, the convex combination value of TFNs is employed as a method to merge multiple TFNs into a unified value. To determine the convex combination value, a parameter λ is introduced, where λ is adjusted between 0 and 1 based on the experts' perceptions, whether they are positive or negative, and in accordance with the average judgments of the expert group. This adjustment ensures that the resulting value aligns with the collective opinions and reflects the level of consensus among the experts. By utilizing a parameter λ (in this study, λ = 0.5), the convex combination value $D_b(\alpha_b, \beta_b)$ is obtained using Equation (1):

$$\begin{aligned} \alpha_b &= u_b - \lambda(u_b - m_b) \\ \beta_b &= l_b - \lambda(m_b - l_b) \end{aligned} \quad (1)$$

Step 3: Next, the precise value of $D_b(\alpha_b, \beta_b)$ is calculated using Equation (2):

$$D_b = \int (\alpha_b, \beta_b) = \lambda[\alpha_b + (1-\lambda)\beta_b] \quad (2)$$

Step 4: The threshold for the valid attributes is generated using Equation (3):

$$Therehold(\sigma) = \sum_{a=1}^{n} \frac{D_b}{n} \quad (3)$$

The attribute b is acceptable if $D_b > \sigma$, but it is refused if $D_b < \sigma$, according to [66].

3.3. Fuzzy DEMATEL Method

DEMATEL is a reliable technique for analyzing causal correlations and significant effects among attributes [67]. This method incorporates expert opinions, which are initially expressed qualitatively and then converted into fuzzy numbers to eliminate ambiguity and achieve a shared perspective. Notably, the Fuzzy DEMATEL approach utilizes the total-relation matrix to identify linkages between criteria and subcriteria, as well as cause-and-effect relationships [68,69]. One of the key advantages of this method is its reliance on pairwise comparisons, enabling the consideration of relationships during the decision-making process [70,71].

The experts express their judgments regarding the relationships between attributes on a 5-point linguistic scale (Table 3).

Table 3. TFNs corresponding to linguistic terms of Fuzzy DEMATEL method.

Linguistic Terms	TFNs
No effect	(0, 0, 0.25)
Extremely weak effect	(0, 0.25, 0.5)
Weak effect	(0.25, 0.5, 0.75)
Strong effect	(0.5, 0.75, 1)
Extremely strong effect	(0.75, 1, 1)

Step 1: The fuzzy weight (E_{kij}), assigned by the kth expert, represents the level of influence of the ith attribute on the jth attribute in a decision committee based on a 5-point linguistic scale (Table 3). It is expressed using TFNs or linguistic terms to capture the expert's subjective perception. These fuzzy TFNs quantify the expert's opinion and contribute to the decision-making process by considering their expertise and knowledge of the assessed attributes using Equations (4) and (5).

$$E_{kij} = \left(l_{ij}^k, m_{ij}^k, u_{ij}^k\right) \quad (4)$$

$$E_{kij} = \left[\frac{l_{ij}^k - min\, l_{ij}^k}{max\, u_{ij}^k - min\, l_{ij}^k}, \frac{m_{ij}^k - min\, m_{ij}^k}{max\, u_{ij}^k - min\, l_{ij}^k}, \frac{u_{ij}^k - min\, u_{ij}^k}{max\, u_{ij}^k - min\, l_{ij}^k},\right] \quad (5)$$

Step 2: The left (Lv) and right (Rv) values are transformed into normalized values, as indicated in Equation (6). These normalized values are subsequently utilized to calculate the total normalized crisp values (Cv), as illustrated in Equation (7).

$$(Lv_{ij}, Rv_{ij}) = \left(\frac{m_{ij}^{k}}{1 + m_{ij}^{k} - l_{ij}^{k}}, \frac{u_{ij}^{k}}{1 + u_{ij}^{k} - m_{ij}^{k}} \right) \tag{6}$$

$$Cv_{ij}^{k} = \frac{\left[Lv_{ij}\left(1 - Lv_{ij}\right) + Rv_{ij} \right]}{\left(1 - Lv_{ij} + Rv_{ij}\right)} \tag{7}$$

Step 3: A synthetic value is obtained to calculate the individual judgment of each expert using Equations (8)–(10):

$$X_{ij}^{k} = min\ Rv_{ij} - Cv_{ij}^{k}\left(max\ u_{ij}^{k} - min l_{ij}^{k} \right) \tag{8}$$

$$Z_{ij}^{k} = \frac{\left(X_{ij}^{1} + X_{ij}^{2} + X_{ij}^{3} + \ldots + X_{ij}^{n} \right)}{k} \tag{9}$$

$$Z_{ij}^{k} = \left(l_{ij}^{Z}, m_{ij}^{Z}, u_{ij}^{Z} \right) \tag{10}$$

Step 4: The direct-relation fuzzy matrix is normalized using Equations (11) and (12):

$$r = max \sum_{j=1}^{k} u_{ij}^{Z} \tag{11}$$

$$\widetilde{H}_{ij} = \frac{Z_{ij}^{k}}{r} = \left(\frac{l_{ij}^{Z}}{r}, \frac{m_{ij}^{Z}}{r}, \frac{u_{ij}^{Z}}{r} \right) = \left(l''_{ij}, m''_{ij}, u''_{ij} \right) \tag{12}$$

Step 5: The total-relation fuzzy matrix (T) is determined using Equations (13)–(17):

$$T = \lim_{k \to \infty} \left(\widetilde{H}^{1}, \widetilde{H}^{2}, \widetilde{H}^{3} \right) \tag{13}$$

$$\widetilde{t}_{ij} = \left(l_{ij}^{t}, m_{ij}^{t}, u_{ij}^{t} \right) \tag{14}$$

$$l_{ij}^{t} = H_{l} \times \left(I - H_{l} \right)^{-1} \tag{15}$$

$$m_{ij}^{t} = H_{m} \times \left(I - H_{m} \right)^{-1} \tag{16}$$

$$u_{ij}^{t} = H_{u} \times \left(I - H_{u} \right)^{-1} \tag{17}$$

Step 6: The total-relation fuzzy matrix (TM) is defuzzified using Equation (18).

$$t_{ij} = \frac{l_{ij}^{t} + 2m_{ij}^{t} + u_{ij}^{t}}{4} \tag{18}$$

Step 7: The R value and C value are calculated using variables retrieved from the total-relation defuzzified matrix using Equations (19) and (20).

$$\mathrm{R_j} = \sum_{\mathrm{j=1}}^{\mathrm{n}} \mathrm{t_{ij}} (\mathrm{j} = 1, 2, 3, \ldots, \mathrm{n}) \tag{19}$$

$$C_i = \sum_{i=1}^{n} t_{ij} (i = 1, 2, 3, \ldots, n) \tag{20}$$

The number of elements in each row (R_j) shows how much one component impacts other factors in the system. On the other hand, the number of components in each column (C_i) reveals how much other systemic factors influence a factor.

Step 8: Drawing cause/effect interrelationship:

The cause/effect interrelationships can be visualized by plotting the values of ($R_j + C_i$) and ($R_j - C_i$) on a Cartesian coordinate system. The ($R_j + C_i$) values represent the degree of interaction between a specific factor and other factors in the system, with higher values indicating stronger interactions. Conversely, the ($R_j - C_i$) values indicate the strength of the causal relationship, with positive values indicating the factor as a cause and negative values indicating it as an effect.

To create an influential relation map (IRM), the defuzzified values are used in Equations (21) and (22) to determine the influence level among the different aspects.

$$p = \frac{\sum_{n=1}^{n} t_{ij}}{TM^2} \tag{21}$$

where p is the threshold to filter the influence between two aspects. If $t_{ij} > p$, then there is interaction between two aspects, and the influence level is t_{ij}.

$$\text{influence level} = \begin{cases} t_{ij} < p : weak \\ weak < t_{ij} < \frac{p_1 + p_2}{2} : medium \\ t_{ij} > p : strong \end{cases} \tag{22}$$

where $p_1 = \frac{\sum_{n=1}^{n} t_{ij}}{count(t_{ij} > p)}$ and $p_2 = \max(t_{ij})$.

4. Results

4.1. Expert Panel

The selection of participants is a crucial step in implementing the Delphi technique. Niederberger et al. [69] emphasized the importance of creating a well-balanced panel by exercising judgment to include experts from diverse backgrounds. The individuals invited to participate should possess a deep understanding of the topic at hand. The number of respondents should be neither too small, as this may limit the breadth of evaluation, nor too large, as this may become challenging to coordinate. A sample size of 10 to 20 experts is generally considered sufficient to generate meaningful outcomes [53]. Nguyen et al. [70] further highlighted that a Delphi group instills greater confidence when comprising at least 10 experts.

The authors emphasize the importance of gathering credible opinions from a powerful group of experts in order to develop a framework for addressing indicators related to sustainable ecotourism. To ensure the accuracy of the data and research results, a diverse range of experts, including government officials, tourism organizations, SMEs, ecologists, tourism specialists, and sustainable development practitioners, should be involved in the decision-making process. The methodology presented in this paragraph demonstrates a collaborative and interdisciplinary approach to addressing various factors promoting sustainable ecotourism. The authors invited 20 qualified respondents from various backgrounds to participate in a communication meeting, providing their opinions on the importance and relationships of indicators in sustainable ecotourism. By incorporating a wide range of perspectives and expertise, the researchers aim to develop a comprehensive framework that considers the needs and priorities of all stakeholders involved in sustainable ecotourism. Table 4 summarizes the profiles of the experts.

Table 4. The general information of 20 respondents.

Information	Item	Frequency	Percentage
Age	From 25 to 40	3	15
	From 40 to 60	9	45
	Over 60	8	40
Gender	Male	8	40
	Female	12	60
Education	Bachelor	3	15
	Master	8	40
	Doctor	8	40
Position occupation	Scholar	8	40
	Policymaker	7	35
	Manager	5	25
Experience	5–10 years	8	40
	10–20 years	6	30
	Over 20 years	6	30

4.2. Fuzzy Delphi Results

The Fuzzy Delphi analysis is conducted in two rounds. The importance of each factor is represented by the absolute mean of the experts' agreement, as displayed in Table 5. In the first round, with a threshold of $\delta = 0.301$, 9 out of 63 elements were eliminated. A threshold of $\delta = 0.304$ was used in the second round, removing 3 additional elements from the initial set of 63.

Table 5. Fuzzy Delphi method results.

Dimensions	Criteria	Round 1		Round 2		Accepted Criteria
		Weight	Validate	Weight	Validate	
Environmental	EN 1	0.3235	Accept	0.3387	Accept	EN 1
	EN 2	0.2210	Reject			
	EN 3	0.3226	Accept	0.3178	Accept	EN 3
	EN 4	0.3120	Accept	0.3090	Accept	EN 4
	EN 5	0.2222	Reject			
	EN 6	0.3246	Accept	0.3051	Accept	EN 6
	EN 7	0.3283	Accept	0.3087	Accept	EN 7
	EN 8	0.3018	Accept	0.1563	Reject	
	EN 9	0.3225	Accept	0.3267	Accept	EN 9
	EN 10	0.3231	Accept	0.3164	Accept	EN 10
	EN 11	0.3073	Accept	0.3374	Accept	EN 11
	EN 12	0.3220	Accept	0.3196	Accept	EN 12
	EN 13	0.3136	Accept	0.3107	Accept	EN 13
	EN 14	0.3145	Accept	0.3216	Accept	EN 14
	EN 15	0.3261	Accept	0.3225	Accept	EN 15
	EN 16	0.3272	Accept	0.3145	Accept	EN 16
Social	SO 1	0.1250	Reject			
	SO 2	0.3029	Accept	0.2210	Reject	
	SO 3	0.3149	Accept	0.3257	Accept	SO 3
	SO 4	0.3162	Accept	0.3337	Accept	SO 4
	SO 5	0.3252	Accept	0.3193	Accept	SO 5
	SO 6	0.3220	Accept	0.3178	Accept	SO 6
	SO 7	0.3188	Accept	0.3155	Accept	SO 7
	SO 8	0.3192	Accept	0.3059	Accept	SO 8
	SO 9	0.3192	Accept	0.3274	Accept	SO 9
	SO 10	0.3246	Accept	0.3047	Accept	SO 10
	SO 11	0.2500	Reject			
	SO 12	0.3323	Accept	0.3248	Accept	SO 12
	SO 13	0.3299	Accept	0.3196	Accept	SO 13
	SO 14	0.3182	Accept	0.3082	Accept	SO 14

Table 5. *Cont.*

Dimensions	Criteria	Round 1		Round 2		Accepted Criteria
		Weight	Validate	Weight	Validate	
Cultural	CU 1	0.3178	Accept	0.3267	Accept	CU 1
	CU 2	0.3159	Accept	0.3176	Accept	CU 2
	CU 3	0.3094	Accept	0.3216	Accept	CU 3
	CU 4	0.3283	Accept	0.3192	Accept	CU 4
	CU 5	0.3257	Accept	0.3094	Accept	CU 5
	CU 6	0.3127	Accept	0.3090	Accept	CU 6
Economic	EC 1	0.3067	Accept	0.3172	Accept	EC 1
	EC 2	0.1250	Reject			
	EC 3	0.3263	Accept	0.3257	Accept	EC 3
	EC 4	0.3036	Accept	0.1250	Reject	
	EC 5	0.3054	Accept	0.3323	Accept	EC 5
	EC 6	0.3000	Reject			
	EC 7	0.3051	Accept	0.3111	Accept	EC 7
	EC 8	0.3129	Accept	0.3210	Accept	EC 8
	EC 9	0.1250	Reject			
	EC 10	0.3111	Accept	0.3220	Accept	EC 10
	EC 11	0.3003	Reject			
	EC 12	0.3320	Accept	0.3290	Accept	EC 12
	EC 13	0.3349	Accept	0.3192	Accept	EC 13
	EC 14	0.1250	Reject			
	EC 15	0.3014	Accept	0.3114	Accept	EC 15
Political	PO 1	0.3288	Accept	0.3212	Accept	PO 1
	PO 2	0.3202	Accept	0.3075	Accept	PO 2
	PO 3	0.3123	Accept	0.3212	Accept	PO 3
	PO 4	0.3155	Accept	0.3222	Accept	PO 4
	PO 5	0.3029	Accept	0.3164	Accept	PO 5
	PO 6	0.3078	Accept	0.3220	Accept	PO 6
	PO 7	0.3131	Accept	0.3248	Accept	PO 7
Intrinsic	IN 1	0.3272	Accept	0.3226	Accept	IN 1
	IN 2	0.3267	Accept	0.3257	Accept	IN 2
	IN 3	0.3188	Accept	0.3174	Accept	IN 3
	IN 4	0.3214	Accept	0.3127	Accept	IN 4
	IN 5	0.3059	Accept	0.3198	Accept	IN 5
		Threshold = 0.3009		Threshold = 0.3041		

In the context of sustainable ecotourism, certain indicators were eliminated from consideration in the environmental, social, and economic dimensions. Specifically, in the environmental dimension, the indicators cleanliness and quality of tourism facilities, access to drainage and wastewater treatment systems (EN2), environmental emergency action plans (EN5), and negative impacts of tourism on the environment (EN8) were excluded. In the social dimension, the indicators disability laws (SO1), exploitation of employees, child labor or sex tourism (SO2), and protection of minority groups (SO11) were disqualified. In the economic dimension, most factors were excluded, including the indicators availability and accessibility of medical services, transportation and recreational facilities (EC2), cost management of tourism operations (EC4), economic and financial development of stakeholders (EC6), implementation of green design technology (EC9), satisfactory goods and services (EC11), and taxes on land, buildings, and other structures (EC14). However, indicators in the cultural, political, and intrinsic dimensions were preserved for consideration in the sustainable ecotourism framework. In conclusion, a total of 51 elements were deemed suitable for use in the next stage of the Fuzzy DEMATEL analysis.

4.3. Fuzzy DEMATEL Results

The interrelationships among the six dimensions, namely environmental (EN), social (SO), cultural (CU), economic (EC), political (PO), and intrinsic (IN), as well as the sub-criteria within each dimension, were explained using the Fuzzy DEMATEL method. To illustrate the computational procedure for the dimensions, experts provided their opinions with fuzzy ratings based on Table 3.

In the next step, linguistic ratings were collected, and an initial integrated direct-causal-relationships fuzzy matrix was derived using Equation (4). Subsequently, the fuzzy matrix was normalized using Equations (8) and (9). The fuzzy total-relation matrix was obtained based on Equations (10)–(14). The total-relation matrix and the normalized direct-relation matrix were then generated. The R value was computed by summing the variables in each row, while the C value was calculated by summing the variables in each column. The difference between R and C represents the net influence levels, where positive values indicate that one dimension has a more significant influence on other dimensions than they have on it. Negative values indicate that the dimension is more likely to be affected by others. The R + C value represents the correlation intensity among dimensions, with higher values indicating more significant importance. The calculation process described above was applied to the main dimensions, and the results are presented in Table 6.

Table 6. The crisp total-relation defuzzified matrix.

Main Dimensions	EN	SO	CU	EC	PO	IN	R	C	R + C	R − C	Relation
EN	0.69	0.81	0.86	0.84	0.85	0.74	4.79	4.76	9.55	0.03	Cause
SO	0.78	0.69	0.80	0.87	0.84	0.71	4.69	7.91	12.60	−3.22	Effect
CU	0.78	0.80	0.71	0.82	0.86	0.73	4.70	4.47	9.17	0.23	Cause
EC	0.87	0.89	0.90	0.78	0.93	0.78	5.14	4.84	9.98	0.30	Cause
PO	0.76	0.77	0.80	0.80	0.70	0.69	4.51	8.00	12.51	−3.49	Effect
IN	0.88	0.87	0.89	0.88	0.91	0.67	5.11	4.53	9.65	0.58	Cause

The influence level of the six dimensions in the case of sustainable ecotourism in Belize can be prioritized as social (SO) > political (PO) > economic (EC) > intrinsic (IN) > environmental (EN) > cultural (CU), based on the D+R values. Based on the D−R values, environmental (EN), cultural (CU), economic (EC), and intrinsic (IN) are net causes, while the remaining dimensions are net effects.

A threshold value was determined by calculating the average of all elements in the total-relation matrix to identify significant influence relationships among the dimensions and criteria. If an element in the full influence matrix exceeded this threshold value, it indicated a higher relevance. Conversely, if the value fell below the threshold, indicating low relevance, it was removed and set to 0 in the matrix. In this specific case study, the threshold value was determined to be 0.804. Taking the second row of Table 6 as an example, the values for the economic dimension (0.87) and political dimension (0.84) surpass the threshold value. Thus, the social dimension influences both the economic dimension and the political dimension. Similarly, the influential relation map (IRM) among the dimensions was identified and is presented in Table 7. Figure 2 sets the rules for the intensity levels of the relationships, which combine three intensity levels (strong, medium, and weak). Moreover, the interdependencies and relationships among the dimensions are visually depicted in Figure 3.

Table 7. Influential relation map among dimensions.

Main Dimensions	EN	SO	CU	EC	PO	IN
EN	0.00	0.81	0.86	0.84	0.85	0.00
SO	0.00	0.00	0.00	0.87	0.84	0.00
CU	0.00	0.00	0.00	0.82	0.86	0.00
EC	0.87	0.89	0.90	0.00	0.93	0.00
PO	0.00	0.00	0.00	0.80	0.00	0.00
IN	0.88	0.87	0.89	0.88	0.91	0.00

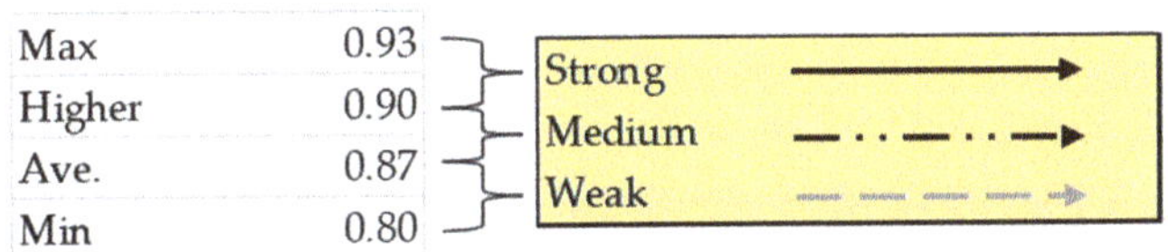

Figure 2. Intensity level of the relationships.

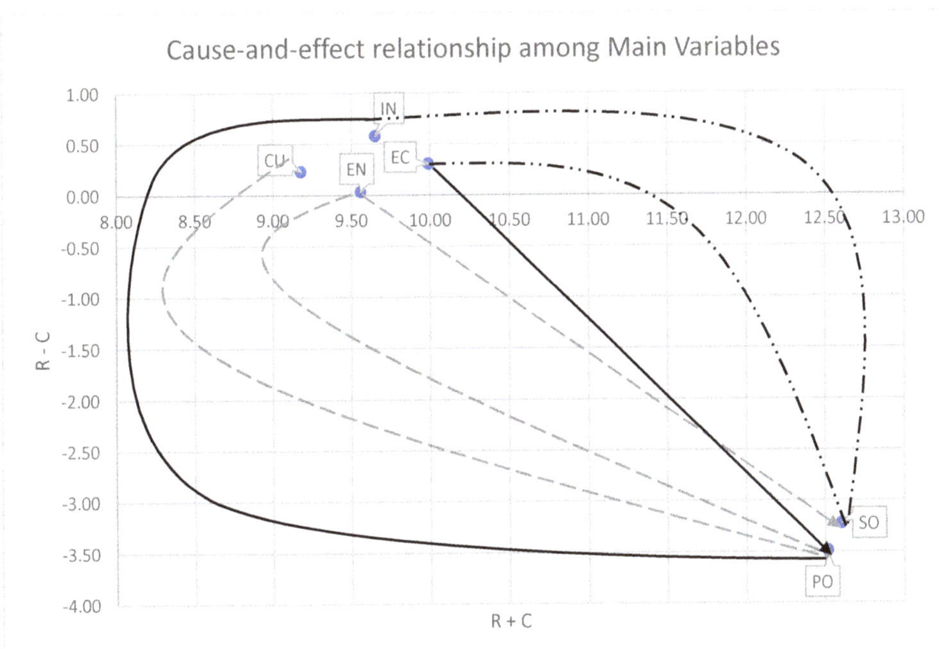

Figure 3. Significant cause/effect relationship diagram among main dimensions.

The direction of the arrow should describe the matrix. If the figure is "0", this means there is no relationship between the dimensions. For example, EN (horizontal) has relationships with SO and PO, and has no interrelationship with IN. Furthermore, dimensions under the exact cause or effect system group are not considered causality relationships, which means they are deleted. To be more specific, IN and EC substantially affect PO while having a medium effect on SO as denoted in Figures 2 and 3.

The same computational procedure of the dimensions was applied to each dimension group to show the relationships between the subcriteria within. Figures 4–9 show that the impact–relationship map in the net format of criteria under six dimensions can be plotted. The findings and analysis of the subcriteria within each dimension are outlined below according to rules in Figure 2.

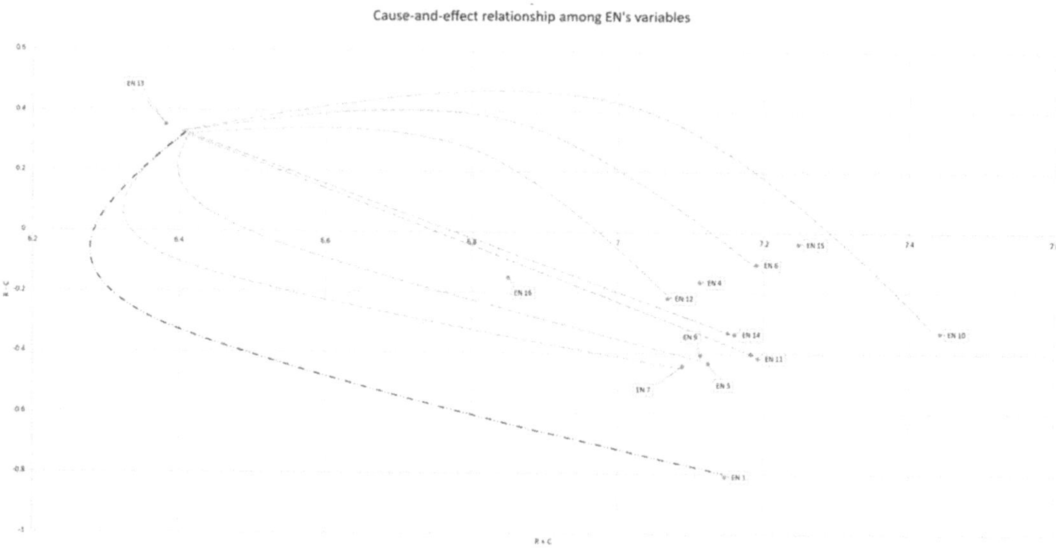

Figure 4. Impact–relation map of environmental dimension.

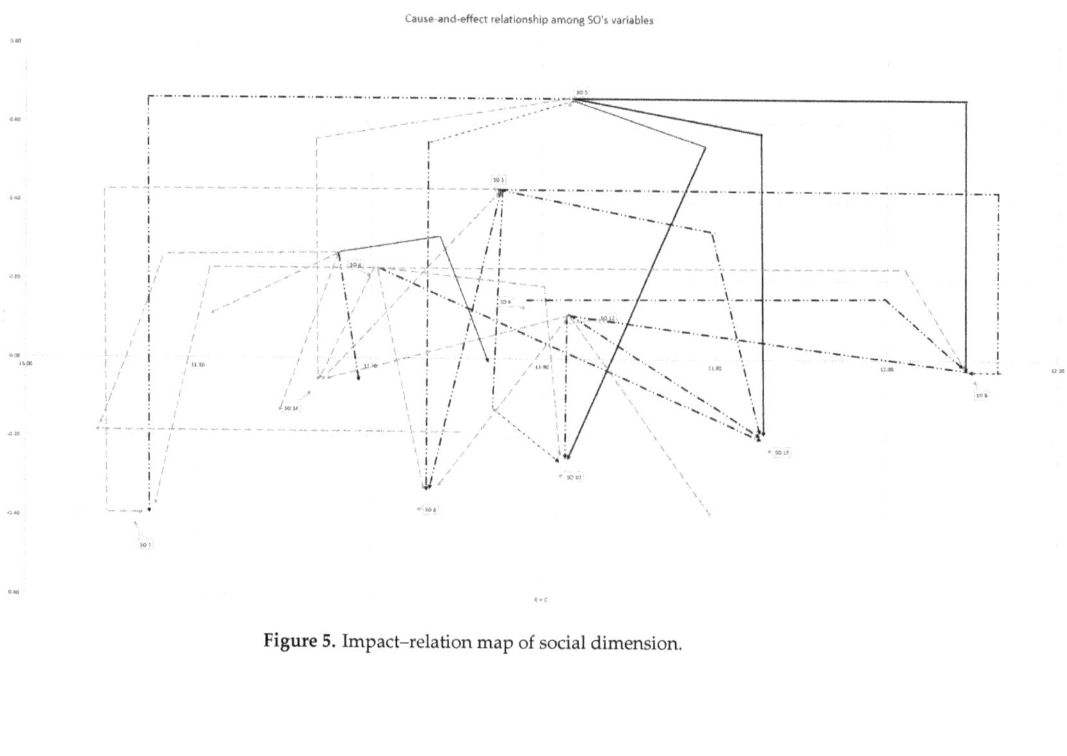

Figure 5. Impact–relation map of social dimension.

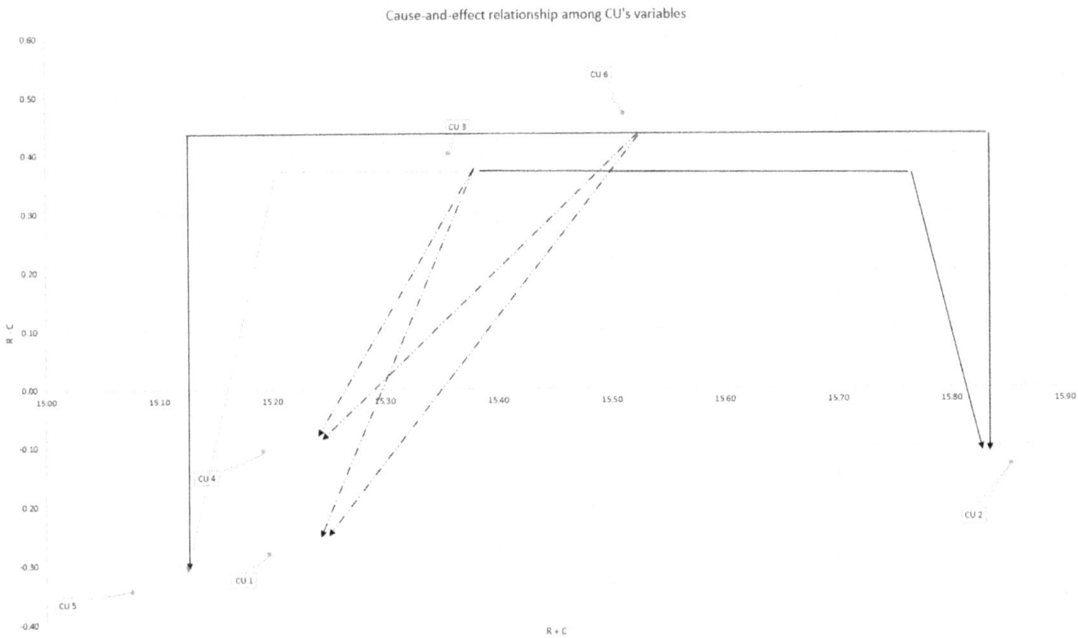

Figure 6. Impact–relation map of cultural dimension.

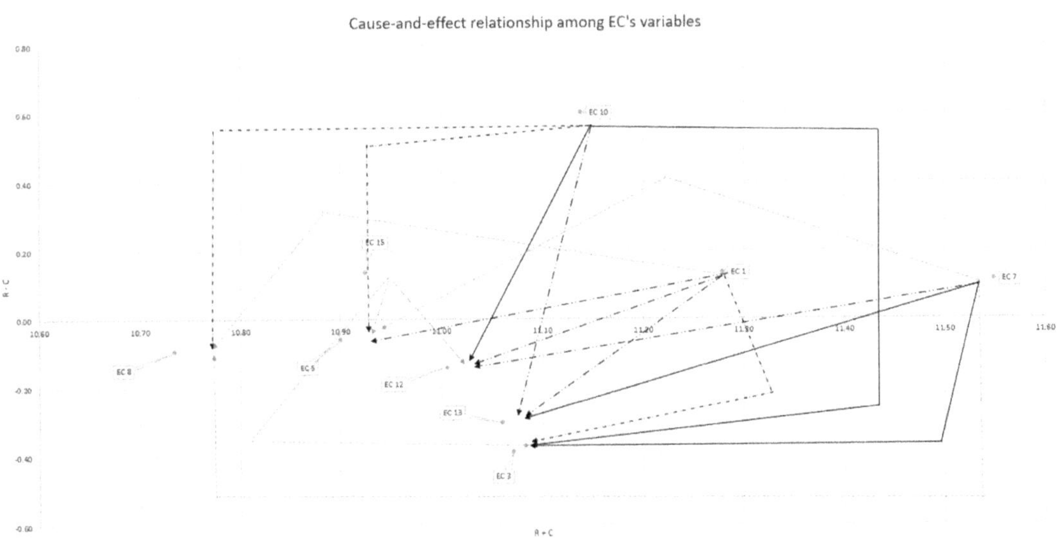

Figure 7. Impact–relation map of economic dimension.

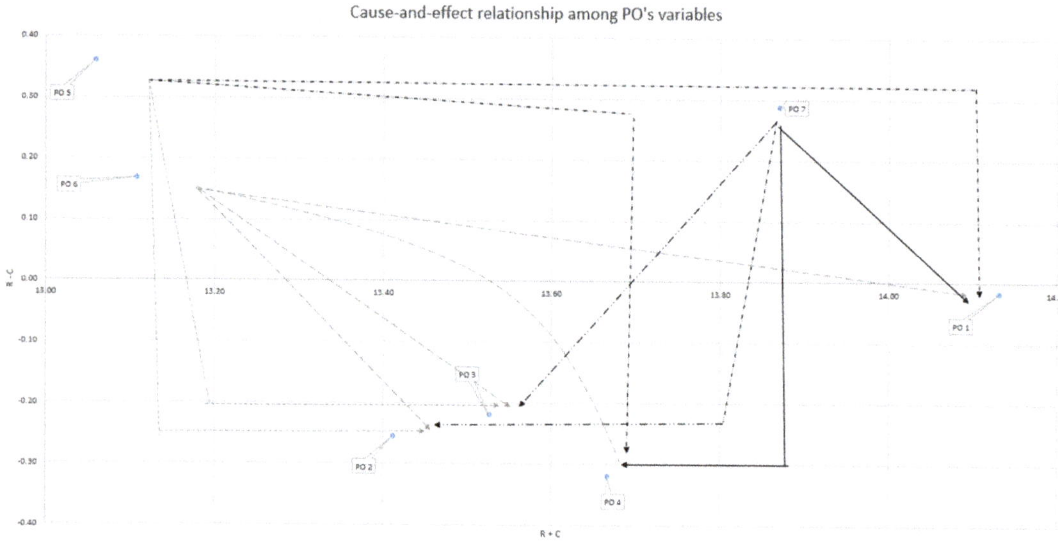

Figure 8. Impact–relation map of political dimension.

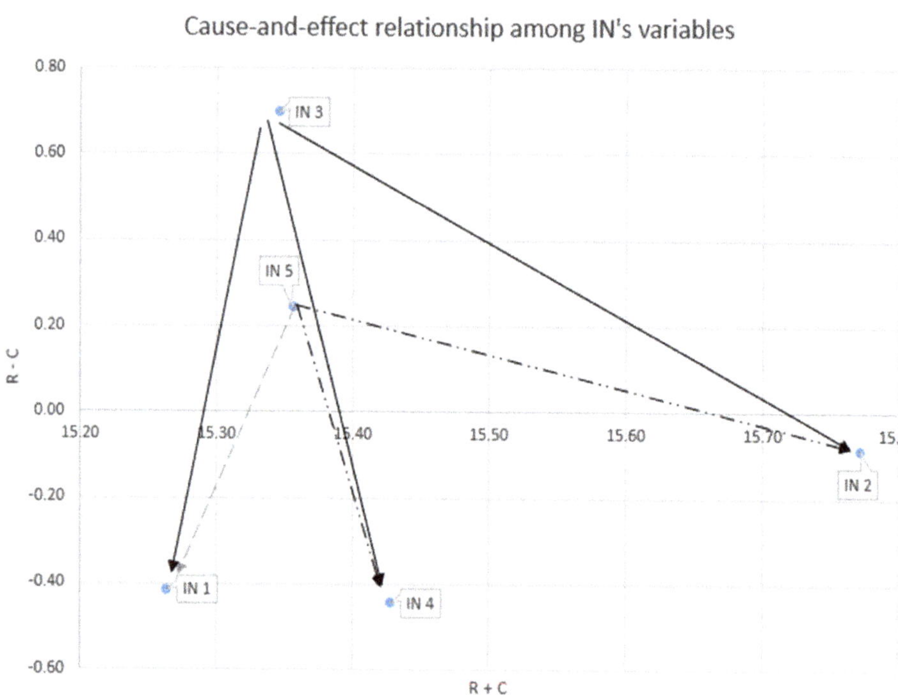

Figure 9. Impact–relation map of intrinsic dimension.

Regarding the environmental dimension in Figure 4, the group of EN variables is unique in that only EN13 was identified as a cause factor, and most of the time the relationship between cause factors and effect factors in this group is relatively weak. The impact of EN13 on EN1 has medium influence; the rest have weak influence.

Regarding the social dimension in Figure 5, this is the most complicated group of variables for the SO variable group. There are five factors that are said to be cause factors and six factors that are said to be effect factors. Among the cause variables, SO5 has the highest position as well as having the most "strong influence" curves, which means that SO5 has the most significant impact on the effect factors. In contrast, among the effect variables, SO13 is the variable most affected by the cause factors.

With respect to the cultural dimension in Figure 6, for the group of CU variables, there are two variables (CU3 and CU6) that are considered as cause factors while the remaining four are effect factors. Similar to the argument regarding SO, CU6 has the most substantial impact on the effect factors and CU2 is the most affected by the cause factors.

Regarding the economic dimension (Figure 7), political dimension (Figure 8), and intrinsic dimension (Figure 9), similar comments can be made on these three remaining groups of variables. EC10, PO7, and IN3 are the cause factors that have the most decisive impact on the effect factors. In contrast, EC3, PO1, and IN2 are the most affected effect factors.

4.4. Discussion

In the existing literature review, numerous studies have explored the field of sustainable ecotourism using various quantitative and qualitative methodologies. For instance, studies have investigated the ecotourism suitability of Babol in Iran [71], the development of ecotourism in Thailand [72], and the site selection of ecotourism in Zhejiang province [73]. However, this study stands out due to its unique integration of the Fuzzy Delphi and Fuzzy DEMATEL approaches, which has not been previously attempted. By utilizing this integrated methodology, we are able to effectively identify and analyze the key variables in sustainable ecotourism, thereby gaining a comprehensive understanding of their interrelationships. This knowledge is invaluable for prioritizing and implementing initiatives or interventions that promote the long-term developmentw of ecotourism in Belize. The findings of this research contribute significantly to the existing body of knowledge and provide practical insights for sustainable ecotourism management in the region.

The findings of our study support the notion that all six dimensions—environmental, social, cultural, economic, political, and intrinsic—serve as qualified indicators for evaluating sustainable ecotourism, which aligns with previous research in this field. Previous studies have also emphasized the significance of considering multiple dimensions when assessing sustainable tourism practices. For instance, Carpio et al. [74] identified several dimensions—environmental, economic, social, and institutional—as important indicators of sustainable tourism. Similarly, Janusz et al. [27] acknowledged three dimensions—economic, environmental, and social—as crucial for evaluating sustainable tourism.

The application of the Fuzzy DEMATEL method in our study allowed us to investigate the cause/effect relationships among the six dimensions of sustainable ecotourism. Our findings reveal that the environmental, cultural, economic, and intrinsic dimensions serve as net causes of sustainable ecotourism, indicating that they have a direct positive impact on other dimensions. Conversely, the political and social dimensions emerge as net effects of sustainable ecotourism, suggesting that they are influenced more by the other dimensions. Specifically, our results highlight the strong influence of the intrinsic and economic dimensions on the political dimension. The intrinsic factors, encompassing personal growth and spiritual well-being, may shape the values and attitudes of policymakers, leading to their support for sustainable practices. Likewise, economic factors such as local economic benefits and resource efficiency can incentivize policymakers to prioritize sustainability in tourism development. This finding aligns with previous research that emphasizes the importance of stakeholder engagement and participation in fostering sustainable tourism [75,76]. Additionally, our findings underscore the significance of the cultural dimension as a net cause of sustainable ecotourism. This emphasizes the importance of preserving local heritage, promoting cultural authenticity, and involving local communities in ecotourism development [77].

Overall, our findings reveal the complex interplay among the different dimensions of sustainable ecotourism and emphasize the need for a multidimensional approach to sustainable tourism development. Policies and practices that prioritize the environmental, cultural, economic, and intrinsic dimensions, while considering their impact on other dimensions, are likely to be the most effective in promoting sustainable ecotourism. By understanding these cause/effect relationships, policymakers and stakeholders can make informed decisions and implement strategies that address the interconnected nature of sustainable tourism and contribute to the long-term well-being of both the environment and local communities.

In this study, the inclusion of environmental indicators in the assessment of sustainable ecotourism provides a comprehensive framework for evaluating the impact of tourism on the environment. Our findings, which align with a previous study in Belize [78], suggest that EN2, EN5, and EN8 may be crucial in assessing the environmental impact of sustainable ecotourism. Furthermore, the results reveal that among the environmental indicators, only EN13 (recovery and mitigation of tourism damage) serves as a net cause, while the other indicators act as net effects. This implies that focusing on the restoration and reduction of tourism damage can have a positive influence on other environmental indicators. Therefore, it is crucial to prioritize recovery efforts and minimize the negative impacts caused by tourism activities in order to enhance the overall environmental sustainability of ecotourism. This finding is consistent with the study on local people's perceptions of ecotourism in Belize conducted by Holladay et al. [79].

The results of the Fuzzy Delphi process show that the social aspect of sustainable ecotourism can be assessed using 11 indicators: SO3, SO4, SO5, SO6, SO7, SO8, SO9, SO10, SO12, SO13, and SO14. SO1, SO2, and SO11 were eliminated based on expert assessment and consensus in the Fuzzy Delphi process. One possible reason SO1 was dropped is that it may not be considered an essential issue in the particular context of the study. In addition, disability law may be seen as a legal obligation rather than a social responsibility of the tourism industry in some regions, which may explain why it was dropped. SO2 was likely eliminated because it overlaps with other indicators such as fair treatment, compliance with labor laws, and protection of minorities. Finally, SO11 may have been dropped due to the difficulty of quantifying and measuring the extent to which minority groups are protected.

These results indicate that the cultural dimension of sustainable ecotourism is mainly influenced by the promotion and protection of local culture (CU6) and cultural codes of conduct for tourists (CU3), which have substantial effects on the availability and accessibility of information about local culture (CU2). According to Lonely Planet, the cultural legacy of Belize is closely connected to its natural surroundings [80]. Tourists have the opportunity to participate in genuine cultural activities, such as acquiring knowledge about ancient Maya agricultural methods, joining a drumming workshop conducted by the Garifuna community, or getting to know the Kriol culture in Gales Point Manatee village. This implies that well-informed tourists are more likely to respect and appreciate local cultures and have clear guidelines for interacting with local communities. Additionally, CU6 strongly affects the adverse effects of development projects on cultural identities (CU5), suggesting that protecting and promoting local culture can mitigate the potential negative impacts of tourism-related development on cultural heritage. Therefore, ecotourism stakeholders need to prioritize promoting and protecting local culture and establish clear guidelines for cultural interaction and respect to ensure the sustainability of the cultural dimension of ecotourism.

After the Fuzzy Delphi process, six metrics, namely EC2, EC4, EC6, EC9, EC11, and EC14, were removed for various reasons, but mainly because these variables are small factors that can be included in other variables; for example, EC2 can be included under EC12 and EC4 can be included under EC1 or EC7. The results of the Fuzzy DEMATEL analysis in this respect show that EC10, a measure of risk management and production stability, has a substantial impact on both EC3 and EC12. EC3 refers to keeping the local economy up to speed, while EC12 refers to measures to support and contribute to developing local goods, services, and infrastructure. This suggests that effective risk management and production stabilization can help sustain the local economy by promoting the growth of local goods and services while maintaining a steady economic pace. This is also mentioned in *Tourism, Local Economic Development, and Poverty* [81] and the book titled *Regional Economic Development* [82]. In addition, the analysis suggests that EC7, which measures employment opportunities, financial subsidies, and compensation to local people, significantly affects both EC3 and EC13. This suggests that supporting local employment and providing reasonable compensation and financial assistance can contribute to sustaining the local economy's growth while promoting entrepreneurs' development as local suppliers and manufacturers. These results suggest that the sustainable tourism industry should prioritize effective risk management and support local employment to promote conservation and economic development.

Through Fuzzy Delphi, no variable in the PO dimension is excluded. This proves that experts agree that all variables are significant for sustainable ecotourism. In addition, the Fuzzy DEMATEL results show that PO7, which measures the reflection of sustainability values on business operations, significantly impacts PO4, which measures ethical organizational structures, ethics, and transparency, and PO1, which measures external organizational structures involved in and owning local businesses. PO7 is also considered the cause factor with the most significant impact on the effect factor in this dimension. In contrast, PO4 is the effect factor most affected by the cause factor. The results suggest that focusing on sustainable values in the tourism industry can help promote an ethical and transparent organizational structure while reducing the negative impact of participation and foreign ownership on local businesses.

Like the PO dimension, the IN dimension also had no variable removed using Fuzzy Delphi. The results of the Fuzzy DEMATEL method indicate that IN3 has a significant impact on IN1, IN2, and IN4, indicating that safety and security are essential factors affecting the intrinsic aspect of sustainable tourism. Tourists prioritize safety and security when choosing a travel destination, and these factors directly affect tourists' overall satisfaction. Therefore, IN3 (crime, accident rate, and visitor safety and security) affects IN1 (local people's attitude towards satisfaction, service quality, and training mechanisms), IN2 (average duration of stay per visitor), and IN4 (overall service quality of local businesses and potential businesses), as they all have a relationship.

5. Conclusions, Implications, Limitations and Future Work

5.1. Conclusions

In this study, we make significant contributions to the field of sustainable tourism performance evaluation. The authors have developed a comprehensive framework consisting of six dimensions and 63 indicators, which provides a holistic approach to assessing the sustainability of tourism activities. The framework was developed through a rigorous process that involved expert consultation using the Fuzzy Delphi method, ensuring that the most critical aspects of sustainable tourism were captured. Our findings confirm our hypothesis regarding the importance of different dimensions in sustainable tourism performance. The environmental dimension emerged as the most critical dimension, highlighting the significance of environmental preservation and minimizing the negative impact of tourism activities. This was followed by the economic, social, cultural, and political dimensions, all of which play essential roles in sustainable tourism development. To further explore the interrelationships between these dimensions, we conducted an analysis using the Fuzzy DEMATEL method. This analysis revealed several indicators that exerted a strong influence on other indicators within the framework. Identifying these influential indicators is crucial for prioritizing efforts and resources in sustainable tourism development. To summarize, this study emphasizes the need for a multidimensional and integrated approach to sustainable tourism performance evaluation. By considering the environmental, economic, social, cultural, and political dimensions, policymakers, managers, and stakeholders can gain a comprehensive understanding of the sustainability of tourism activities and make informed decisions to enhance sustainability.

5.2. Theoretical Implications

The theoretical implications of this study are of significant importance to both academics and practitioners in the field of sustainable ecotourism, not only in Belize but also in the broader travel and tourism sector. This study contributes to the theoretical understanding of sustainable ecotourism in the following ways: Firstly, this study identifies and emphasizes six crucial dimensions—environmental, social, cultural, economic, political, and intrinsic factors—which have a direct impact on sustainable ecotourism. By focusing on these key areas, sustainable development efforts can be more targeted and effective, addressing the specific challenges and opportunities within the realm of sustainable ecotourism. This framework provides a comprehensive understanding of the multifaceted nature of sustainable ecotourism, enabling stakeholders to develop strategies and policies that align with these dimensions. Secondly, this study reinforces the significance of specific factors that contribute to sustainable ecotourism, such as the restoration and reduction of damage caused by tourism (EN13), the improvement of the well-being, quality of life, and safety of the local community (SO5), and the promotion and protection of local culture (CU6). While previous studies have recognized the importance of these factors, this study provides additional evidence and reaffirms their critical role in the ecotourism industry. These findings serve as a valuable reference for future research and can guide practitioners in implementing measures to enhance sustainable business growth. Lastly, this study uncovers the causal relationships between the variables within each dimension, offering insights into the underlying mechanisms and interdependencies that shape sustainable ecotourism. By understanding these causal connections, stakeholders can identify areas for improvement, identify root causes of issues, and develop effective interventions to address any existing shortcomings. This knowledge empowers decision makers to prioritize their efforts and allocate resources efficiently, fostering the long-term sustainability of ecotourism initiatives.

5.3. Managerial Implications

The conclusions drawn from this study have important managerial implications for the tourism industry. Managers and policymakers should focus on four key aspects—economy (EC), intrinsic dimension (IN), environment (EN), and culture (CU)—as they are considered crucial cause factors for the development and improvement of ecotourism. Firstly, in terms of the economic dimension, authorities need to prioritize job opportunities and implement favorable financial policies for the local communities in tourism-exploited areas. Additionally, effective risk management and production stabilization measures should be implemented. Attracting investment resources is also vital in enhancing the overall quality of the tourism industry and attracting more visitors. Secondly, in the intrinsic dimension, special attention should be given to tourist satisfaction with related activities and the volume of tourists, including returning visitors and seasonality. Meeting tourists' needs and providing valuable experiences are key factors in determining whether they will return to a destination. It is also important to address crime rates, visitor safety, and security concerns to ensure a safe and secure environment for tourists. Thirdly, in the environmental dimension, this study emphasizes the importance of the restoration and reduction of damage caused by tourism. Policies and practical actions aimed at recovering and mitigating the negative impacts of tourism are essential for sustainable ecotourism. These efforts help preserve the unique characteristics of the destination, differentiating it from other places and serving as a decisive factor in attracting tourists. Lastly, the findings related to the cultural dimension highlight two key factors that require attention: cultural codes of conduct for tourists and the promotion and protection of local culture. Emphasizing and showcasing the cultural heritage of the area can be an effective strategy to attract tourists and promote sustainable ecotourism. Encouraging visitors to learn about and respect local culture and traditions is crucial in preserving cultural identities and fostering a positive cultural exchange between tourists and local communities.

5.4. Limitations and Future Work

The limitations of this study should be acknowledged to provide a clearer understanding of the study's scope and potential areas for improvement. Firstly, the framework developed in this study was focused on sustainable tourism in Belize, and its applicability to other regions or countries may require further examination. Future research should aim to validate and refine the framework in different geographical contexts to enhance its generalizability and ensure its effectiveness in diverse settings. Secondly, although the framework included a comprehensive set of dimensions and indicators, there may be additional factors that were not considered in this study. Future research could explore the inclusion of other dimensions, such as technological advancements, governance structures, or community engagement, which may play significant roles in sustainable tourism performance. Another limitation is that the data used in this study relied on expert consultation and may be subjective to some extent. Future research could incorporate primary data collection from various stakeholders, including tourists, local communities, and industry practitioners, to provide a more comprehensive and objective assessment of sustainable tourism performance. Furthermore, the analysis in this study focused on identifying causal relationships between dimensions using the Fuzzy DEMATEL method. Future research could expand on this by conducting quantitative analyses, such as structural equation modeling or regression analysis, to examine the complex interplay between dimensions and their impact on overall sustainable tourism performance. Additionally, while the defuzzification procedure used in this study facilitated the interpretation and analysis of cause/effect relationships, it may not capture the full complexity of fuzzy algebra rules. Exploring the application of fuzzy algebra rules for determining cause/effect relationships could be a valuable extension in future research. Moreover, this study primarily focused on the evaluation of sustainable tourism performance rather than providing detailed strategies or interventions for improvement. Future research could delve deeper into developing specific management strategies, policies, and best practices based on the findings of this study, to guide practitioners and policymakers in implementing sustainable tourism initiatives effectively. Lastly, the temporal aspect of sustainable tourism was not explicitly addressed in this study. Future research could explore the dynamics of sustainable tourism performance over time and investigate the long-term impacts of various interventions and policies on sustainability outcomes. By recognizing and addressing these limitations, future research can build upon the findings of this study and contribute to the advancement of knowledge in the field of sustainable tourism, ultimately leading to more effective and impactful practices and policies.

Author Contributions: Conceptualization, L.-A.T.N., P.-H.N. and M.R.; methodology, L.-A.T.N., P.-H.N. and M.R.; software, H.-Q.L. and L.-C.T.; validation, L.-A.T.N., P.-H.N., M.R. and C.-Y.H.; formal analysis, L.-A.T.N., P.-H.N. and M.R.; investigation, M.R.; resources, H.-Q.L. and L.-C.T.; data curation, P.-H.N. and M.R.; writing—original draft preparation, P.-H.N. and M.R.; writing—review and editing, P.-H.N., L.-A.T.N., H.-Q.L., L.-C.T. and M.R.; visualization, L.-A.T.N., P.-H.N. and M.R.; supervision, C.-Y.H.; project administration, P.-H.N. and M.R.; funding acquisition, P.-H.N. and M.R. All authors have read and agreed to the published version of the manuscript.

Funding: This research received no external funding.

Data Availability Statement: All the data generated or analyzed during this study are available via the following link: http://dx.doi.org/10.17632/6n3xn4wpsj.1 (accessed on 1 June 2023).

Acknowledgments: The authors would like to extend their sincere gratitude to the experts and policymakers in Belize for their invaluable contributions to this study. Their expertise, insights, and support have played a crucial role in the development and execution of this research. Their willingness to share their knowledge and collaborate with the research team is deeply appreciated. Without their active participation and cooperation, this study would not have been possible. The authors are grateful for their time, guidance, and commitment to sustainable tourism in Belize, which have greatly enriched the findings and implications of this study.

Conflicts of Interest: The authors declare no conflict of interest.

References

1. Why Tourism? Available online: https://www.unwto.org/why-tourism (accessed on 17 May 2023).
2. Sustainable Tourism. Available online: https://sdgs.un.org/topics/sustainable-tourism (accessed on 17 May 2023).
3. How Ecotourism Benefits the Environment and Local Communities. Available online: https://www.worldpackers.com/articles/ecotourism-benefits (accessed on 17 May 2023).
4. Kiper, T. Role of Ecotourism in Sustainable Development. In *Advances in Landscape Architecture*; InTech: London, UK, 2013.
5. Resilience and Conservation in a Changing Climate: The Case of Belize. Available online: https://www.worldbank.org/en/news/feature/2021/03/24/resilience-and-conservation-in-a-changing-climate-the-case-of-belize (accessed on 17 May 2023).
6. Saving Belize's Magnificent and Endangered Barrier Reef. Available online: https://www.worldwildlife.org/stories/saving-belize-s-magnificent-and-endangered-barrier-reef (accessed on 17 May 2023).
7. Is Ecotourism Truly 'Eco' Friendly in Belize? Available online: https://www.sanpedrosun.com/conservation/2019/09/25/is-ecotourism-truly-eco-friendly-in-belize/ (accessed on 17 May 2023).
8. Baloch, Q.B.; Shah, S.N.; Iqbal, N.; Sheeraz, M.; Asadullah, M.; Mahar, S.; Khan, A.U. Impact of Tourism Development upon Environmental Sustainability: A Suggested Framework for Sustainable Ecotourism. *Environ. Sci. Pollut. Res.* **2023**, *30*, 5917–5930. [CrossRef] [PubMed]
9. Stronza, A.L.; Hunt, C.A.; Fitzgerald, L.A. Ecotourism for Conservation? *Annu. Rev. Environ. Resour.* **2019**, *44*, 229–253. [CrossRef]
10. Diamantis, D. Stakeholder Ecotourism Management: Exchanges, Coordination's and Adaptations. *J. Ecotourism* **2018**, *17*, 203–205. [CrossRef]
11. Kim, M.; Xie, Y.; Cirella, G.T. Sustainable Transformative Economy: Community-Based Ecotourism. *Sustainability* **2019**, *11*, 4977. [CrossRef]
12. Bunruamkaew, K.; Murayam, Y. Site Suitability Evaluation for Ecotourism Using GIS & AHP: A Case Study of Surat Thani Province, Thailand. *Procedia Soc. Behav. Sci.* **2011**, *21*, 269–278. [CrossRef]
13. Masoum, M.; Nasiri, H.; Hosseini, A.; Rafii, Y. Ecotourism Planning Using Remote Sensing and GIS. In *Current Issues in Hospitality and Tourism*; CRC Press: Boca Raton, FL, USA, 2012; pp. 225–229.
14. Akbarian Ronizi, S.R.; Mokarram, M.; Negahban, S. Investigation of Sustainable Rural Tourism Activities With Different Risk: A GIS-MCDM Case in Isfahan, Iran. *Earth Space Sci.* **2023**, *10*, e2021EA002153. [CrossRef]
15. Garabinović, D.; Papić, M.; Kostić, M. Multi-Criteria Decision Making Trends in Ecotourism and Sustainable Tourism. *Ekon. Poljopr.* **2021**, *68*, 321–340. [CrossRef]
16. Kheybari, S. Adjusting Trade-Offs in Multi-Criteria Decision-Making Problems. *Int. J. Inf. Technol. Decis. Mak.* **2021**, *20*, 1499–1517. [CrossRef]
17. Mardani, A.; Jusoh, A.; MD Nor, K.; Khalifah, Z.; Zakwan, N.; Valipour, A. Multiple Criteria Decision-Making Techniques and Their Applications—A Review of the Literature from 2000 to 2014. *Econ. Res.-Ekon. Istraživanja* **2015**, *28*, 516–571. [CrossRef]
18. Mulliner, E.; Malys, N.; Maliene, V. Comparative Analysis of MCDM Methods for the Assessment of Sustainable Housing Affordability. *Omega (Westport)* **2016**, *59*, 146–156. [CrossRef]
19. Abdel-Basset, M.; Gamal, A.; Chakrabortty, R.K.; Ryan, M. A New Hybrid Multi-Criteria Decision-Making Approach for Location Selection of Sustainable Offshore Wind Energy Stations: A Case Study. *J. Clean. Prod.* **2021**, *280*, 124462. [CrossRef]
20. Ocampo, L.; Ebisa, J.A.; Ombe, J.; Geen Escoto, M. Sustainable Ecotourism Indicators with Fuzzy Delphi Method—A Philippine Perspective. *Ecol. Indic.* **2018**, *93*, 874–888. [CrossRef]
21. Sobhani, P.; Esmaeilzadeh, H.; Sadeghi, S.M.M.; Marcu, M.V.; Wolf, I.D. Evaluating Ecotourism Sustainability Indicators for Protected Areas in Tehran, Iran. *Forests* **2022**, *13*, 740. [CrossRef]
22. Barzekar, G.; Aziz, A.; Mariapan, M.; Ismail, M.H. Delphi Technique for Generating Criteria and Indicators in Monitoring Ecotourism Sustainability in Northern Forests of Iran: Case Study on Dohezar and Sehezar Watersheds. 2011. Available online: https://depot.ceon.pl/handle/123456789/5389 (accessed on 20 June 2023).
23. Aziz, A.; Barzekar, G.; Ajuhari, Z.; Idris, N.H. Criteria & Indicators for Monitoring Ecotourism Sustainability in a Protected Watershed: A Delphi Consensus. *IOSR J. Environ. Sci. Toxicol. Food Technol.* **2015**, *10*, 105–111.
24. Asadpourian, Z.; Rahimian, M.; Gholamrezai, S. SWOT-AHP-TOWS Analysis for Sustainable Ecotourism Development in the Best Area in Lorestan Province, Iran. *Soc. Indic. Res.* **2020**, *152*, 289–315. [CrossRef]
25. Xu, L.; Ao, C.; Liu, B.; Cai, Z. Ecotourism and Sustainable Development: A Scientometric Review of Global Research Trends. *Environ. Dev. Sustain.* **2023**, *25*, 2977–3003. [CrossRef]
26. Sonuç, N. Environment, Tourism and Sustainability (Ecotourism Management, Environment and Sustainable Tourism). In *Encyclopedia of Sustainable Management*; Springer International Publishing: Cham, Switerland, 2020; pp. 1–6.
27. Janusz, G.K.; Bajdor, P. Towards to Sustainable Tourism–Framework, Activities and Dimensions. *Procedia Econ. Financ.* **2013**, *6*, 523–529. [CrossRef]

28. West, P.; Igoe, J.; Brockington, D. Parks and Peoples: The Social Impact of Protected Areas. *Annu. Rev. Anthr.* **2006**, *35*, 251–277. [CrossRef]
29. Oldekop, J.A.; Holmes, G.; Harris, W.E.; Evans, K.L. A Global Assessment of the Social and Conservation Outcomes of Protected Areas. *Conserv. Biol.* **2016**, *30*, 133–141. [CrossRef]
30. Pasape, L.; Anderson, W.; Lindi, G. Assessment of Indicators of Sustainable Ecotourism in Tanzania. *Anatolia* **2015**, *26*, 73–84. [CrossRef]
31. Fallah, M.; Ocampo, L. The Use of the Delphi Method with Non-Parametric Analysis for Identifying Sustainability Criteria and Indicators in Evaluating Ecotourism Management: The Case of Penang National Park (Malaysia). *Env. Syst. Decis.* **2021**, *41*, 45–62. [CrossRef]
32. Pasape, L.; Anderson, W.; Lindi, G. Good Governance Strategies for Sustainable Ecotourism in Tanzania. *J. Ecotourism* **2015**, *14*, 145–165. [CrossRef]
33. Li, W. Environmental Management Indicators for Ecotourism in China's Nature Reserves: A Case Study in Tianmushan Nature Reserve. *Tour. Manag.* **2004**, *25*, 559–564. [CrossRef]
34. Forje, G.W.; Tchamba, M.N. Ecotourism Governance and Protected Areas Sustainability in Cameroon: The Case of Campo Ma'an National Park. *Curr. Res. Environ. Sustain.* **2022**, *4*, 100172. [CrossRef]
35. Mearns, K.F. Lessons from the Application of Sustainability Indicators to Community-Based Ecotourism Ventures in Southern Africa. *Afr. J. Bus. Manag.* **2012**, *6*, 7851. [CrossRef]
36. Vereczi, G. Sustainability Indicators for Ecotourism Destinations and Operations. In *Quality Assurance and Certification in Ecotourism*; CABI: Wallingford, UK, 2007; pp. 101–115.
37. Tsaur, S.-H.; Lin, Y.-C.; Lin, J.-H. Evaluating Ecotourism Sustainability from the Integrated Perspective of Resource, Community and Tourism. *Tour. Manag.* **2006**, *27*, 640–653. [CrossRef]
38. Ashok, S.; Tewari, H.R.; Behera, M.D.; Majumdar, A. Development of Ecotourism Sustainability Assessment Framework Employing Delphi, C&I and Participatory Methods: A Case Study of KBR, West Sikkim, India. *Tour. Manag. Perspect.* **2017**, *21*, 24–41. [CrossRef]
39. Bulatović, J.; Rajović, G. Applying Sustainable Tourism Indicators to Community-Based Ecotourism Tourist Village Eco-Katun Štavna. *Eur. J. Econ. Stud.* **2016**, *16*, 309–330. [CrossRef]
40. Bhuiyan, M.A.H.; Siwar, C.; Ismail, S.M. Sustainability Measurement for Ecotourism Destination in Malaysia: A Study on Lake Kenyir, Terengganu. *Soc. Indic. Res.* **2016**, *128*, 1029–1045. [CrossRef]
41. Bunruamkaew, K.; Murayama, Y. Land Use and Natural Resources Planning for Sustainable Ecotourism Using GIS in Surat Thani, Thailand. *Sustainability* **2012**, *4*, 412–429. [CrossRef]
42. Wondirad, A.; Tolkach, D.; King, B. Stakeholder Collaboration as a Major Factor for Sustainable Ecotourism Development in Developing Countries. *Tour. Manag.* **2020**, *78*, 104024. [CrossRef]
43. Andarani, P.; Lestari, D.F.; Rezagama, A.; Sariffuddin, S. Sustainable Ecotourism Development Based on Participatory Rural Appraisal: A Case Study of Thekelan Village, Central Java, Indonesia. *E3S Web Conf.* **2018**, *73*, 02019. [CrossRef]
44. Aydin, I.Z.; Öztürk, A. Identifying, Monitoring, and Evaluating Sustainable Ecotourism Management Criteria and Indicators for Protected Areas in Türkiye: The Case of Camili Biosphere Reserve. *Sustainability* **2023**, *15*, 2933. [CrossRef]
45. Pianosi, F.; Beven, K.; Freer, J.; Hall, J.W.; Rougier, J.; Stephenson, D.B.; Wagener, T. Sensitivity Analysis of Environmental Models: A Systematic Review with Practical Workflow. *Environ. Model. Softw.* **2016**, *79*, 214–232. [CrossRef]
46. Andria, J.; di Tollo, G.; Pesenti, R. Andria, J.; di Tollo, G.; Pesenti, R. A Fuzzy Evaluation of Tourism Sustainability. In *Business and Consumer Analytics: New Ideas*; Springer International Publishing: Cham, Switzerland, 2019; pp. 911–932.
47. Andria, J.; di Tollo, G.; Pesenti, R. A Heuristic Fuzzy Algorithm for Assessing and Managing Tourism Sustainability. *Soft Comput.* **2020**, *24*, 4027–4040. [CrossRef]
48. Andria, J.; di Tollo, G.; Pesenti, R. Fuzzy Multi-Criteria Decision-Making: An Entropy-Based Approach to Assess Tourism Sustainability. *Tour. Econ.* **2021**, *27*, 168–186. [CrossRef]
49. Grime, M.M.; Wright, G. Delphi Method. In *Wiley StatsRef: Statistics Reference Online*; Wiley: Hoboken, NJ, USA, 2016; pp. 1–6.
50. Fletcher, A.J.; Marchildon, G.P. Using the Delphi Method for Qualitative, Participatory Action Research in Health Leadership. *Int. J. Qual. Methods* **2014**, *13*, 1–18. [CrossRef]
51. Goodman, C.M. The Delphi Technique: A Critique. *J. Adv. Nurs.* **1987**, *12*, 729–734. [CrossRef]
52. Saffie, N.A.M.; Shukor, N.M.; Rasmani, K.A. Fuzzy Delphi Method: Issues and Challenges. In Proceedings of the 2016 International Conference on Logistics, Informatics and Service Sciences (LISS), Sydney, Australia, 24–27 July 2016; pp. 1–7.
53. Nguyen, P.-H.; Nguyen, T.-L.; Le, H.-Q.; Pham, T.-Q.; Nguyen, H.-A.; Pham, C.-V. How Does the Competitiveness Index Promote Foreign Direct Investment at the Provincial Level in Vietnam? An Integrated Grey Delphi–DEA Model Approach. *Mathematics* **2023**, *11*, 1500. [CrossRef]
54. Quiñones, R.S.; Caladcad, J.A.A.; Himang, C.M.; Quiñones, H.G.; Castro, C.J.; Caballes, S.A.A.; Abellana, D.P.M.; Jabilles, E.M.Y.; Ocampo, L.A. Using Delphi and Fuzzy DEMATEL for Analyzing the Intertwined Relationships of the Barriers of University Technology Transfer: Evidence from a Developing Economy. *Int. J. Innov. Stud.* **2020**, *4*, 85–104. [CrossRef]
55. Amirghodsi, S.; Naeini, A.B.; Makui, A. An Integrated Delphi-DEMATEL-ELECTRE Method on Gray Numbers to Rank Technology Providers. *IEEE Trans. Eng. Manag.* **2022**, *69*, 1348–1364. [CrossRef]

56. Muhammad, M.N.; Cavus, N. Fuzzy DEMATEL Method for Identifying LMS Evaluation Criteria. *Procedia Comput. Sci.* **2017**, *120*, 742–749. [CrossRef]
57. Suzan, V.; Yavuzer, H. A Fuzzy Dematel Method To Evaluate The Most Common Diseases In Internal Medicine. *Int. J. Fuzzy Syst.* **2020**, *22*, 2385–2395. [CrossRef]
58. Singh, P.K.; Sarkar, P. A Framework Based on Fuzzy Delphi and DEMATEL for Sustainable Product Development: A Case of Indian Automotive Industry. *J. Clean. Prod.* **2020**, *246*, 118991. [CrossRef]
59. Ghag, N.; Acharya, P.; Khanapuri, V. Analyzing the Sustainable International Competitiveness Factors of SMEs by Fuzzy Delphi and Neutrosophic DEMATEL. *Bus. Strategy Dev.* **2023**, *2023*, 1–17. [CrossRef]
60. Mohammadfam, I.; Mirzaei Aliabadi, M.; Soltanian, A.R.; Tabibzadeh, M.; Mahdinia, M. Investigating Interactions among Vital Variables Affecting Situation Awareness Based on Fuzzy DEMATEL Method. *Int. J. Ind. Erg.* **2019**, *74*, 102842. [CrossRef]
61. Wang, F. Preference Degree of Triangular Fuzzy Numbers and Its Application to Multi-Attribute Group Decision Making. *Expert Syst. Appl.* **2021**, *178*, 114982. [CrossRef]
62. Cheng, C.-H.; Lin, Y. Evaluating the Best Main Battle Tank Using Fuzzy Decision Theory with Linguistic Criteria Evaluation. *Eur. J. Oper. Res.* **2002**, *142*, 174–186. [CrossRef]
63. Bui, T.D.; Tsai, F.M.; Tseng, M.-L.; Ali, M.H. Identifying Sustainable Solid Waste Management Barriers in Practice Using the Fuzzy Delphi Method. *Resour. Conserv. Recycl.* **2020**, *154*, 104625. [CrossRef]
64. Saraswathi, A. A Fuzzy-Trapezoidal DEMATEL Approach Method for Solving Decision Making Problems under Uncertainty. *AIP Conf. Proc.* **2019**, *2112*, 020076.
65. Wu, W.-W.; Lee, Y.-T. Developing Global Managers' Competencies Using the Fuzzy DEMATEL Method. *Expert Syst. Appl.* **2007**, *32*, 499–507. [CrossRef]
66. Li, Y.; Hu, Y.; Zhang, X.; Deng, Y.; Mahadevan, S. An Evidential DEMATEL Method to Identify Critical Success Factors in Emergency Management. *Appl. Soft Comput.* **2014**, *22*, 504–510. [CrossRef]
67. Yazdi, M.; Nedjati, A.; Zarei, E.; Abbassi, R. A Novel Extension of DEMATEL Approach for Probabilistic Safety Analysis in Process Systems. *Saf. Sci.* **2020**, *121*, 119–136. [CrossRef]
68. Li, R.-J. Fuzzy Method in Group Decision Making. *Comput. Math. Appl.* **1999**, *38*, 91–101. [CrossRef]
69. Niederberger, M.; Köberich, S. Coming to Consensus: The Delphi Technique. *Eur. J. Cardiovasc. Nurs.* **2021**, *20*, 692–695. [CrossRef]
70. Nguyen, P.-H.; Tran, L.-C.; Nguyen, H.B.-D.; Ho, T.P.-T.; Duong, Q.-A.; Tran, T.-N. Unlocking the Potential of Open Innovation through Understanding the Interrelationship among Key Determinants of FDI Attractiveness. *J. Open Innov. Technol. Mark. Complex.* **2023**, *9*, 100021. [CrossRef]
71. Zabihi, H.; Alizadeh, M.; Wolf, I.D.; Karami, M.; Ahmad, A.; Salamian, H. A GIS-Based Fuzzy-Analytic Hierarchy Process (F-AHP) for Ecotourism Suitability Decision Making: A Case Study of Babol in Iran. *Tour. Manag. Perspect.* **2020**, *36*, 100726. [CrossRef]
72. Tseng, M.L.; Lin, C.; Remen Lin, C.W.; Wu, K.J.; Sriphon, T. Ecotourism Development in Thailand: Community Participation Leads to the Value of Attractions Using Linguistic Preferences. *J. Clean. Prod.* **2019**, *231*, 1319–1329. [CrossRef]
73. Fang, Y. Site Selection of Ecotourism: A Case Study of Zhejiang Province. *IJISET-Int. J. Innov. Sci. Eng. Technol.* **2017**, *4*, 321–326.
74. Carpio, A.A.; Chamen, C.K.; Deonio, A.F.; Ferrer, R.J.; Ilagan, E.A.; Rasco, M.D.; Royena, A.; Saguin, A.J.; Tesiorna, M.F.; Amparo, J.M. The Perceived Critical Sustainability Dimensions of Ecotourism among Human Ecology Students. *J. Hum. Ecol.* **2019**, *8*, 68–90.
75. Byrd, E.T. Stakeholders in Sustainable Tourism Development and Their Roles: Applying Stakeholder Theory to Sustainable Tourism Development. *Tour. Rev.* **2007**, *62*, 6–13. [CrossRef]
76. Kiryluk, H.; Glińska, E.; Ryciuk, U.; Vierikko, K.; Rollnik-Sadowska, E. Stakeholders Engagement for Solving Mobility Problems in Touristic Remote Areas from the Baltic Sea Region. *PLoS ONE* **2021**, *16*, e0253166. [CrossRef]
77. Yang, L.; Hu, X.; Lee, H.M.; Zhang, Y. The Impacts of Ecotourists' Perceived Authenticity and Perceived Values on Their Behaviors: Evidence from Huangshan World Natural and Cultural Heritage Site. *Sustainability* **2023**, *15*, 1551. [CrossRef]
78. Nuenninghoff, S.; Lemay, M.; Rogers, C. Sustainable Tourism in Belize. Inter-American Development Bank: Washington, DC, USA, 2013.
79. Holladay, P.J.; Ormsby, A.A. A Comparative Study of Local Perceptions of Ecotourism and Conservation at Five Blues Lake National Park, Belize. *J. Ecotourism* **2011**, *10*, 118–134. [CrossRef]
80. Going Local: Explore Cultural Tourism in Belize. Available online: https://www.lonelyplanet.com/articles/going-local-explore-cultural-tourism-in-belize (accessed on 26 May 2023).

81. Goodwin, H. Tourism, Local Economic Development, and Poverty Reduction. *Appl. Res. Econ. Dev.* **2008**, *5*, 55–64.
82. Regional Economic Development: Analysis and Planning Strategy-Robert J. Stimson, Roger R. Stough, Brian H. Roberts-Google Sách. Available online: https://books.google.com.vn/books?hl=vi&lr=&id=UwGY2VG4JDYC&oi=fnd&pg=PR5&dq=The+Impact+of+Risk+Management+on+Local+Economic+Development&ots=v3UZZeIerp&sig=P0WZ1X6tQSMoK06jTxYiaA1C7k4&redir_esc=y#v=onepage&q=The%20Impact%20of%20Risk%20Management%20on%20Local%20Economic%20Development&f=false (accessed on 22 May 2023).

mathematics

Article

Determining the Risk Level in Client Analysis by Applying Fuzzy Logic in Insurance Sector

Jelena Lukić [1,*], Mirjana Misita [2], Dragan D. Milanović [2], Ankica Borota-Tišma [3] and Aleksandra Janković [4]

1 Danube Insurance Company a.d.o, Makedonska 4, 11000 Belgrade, Serbia
2 Faculty of Mechanical Engineering, University of Belgrade, Kraljice Marije 16, 11000 Belgrade, Serbia
3 Belgrade Business and Arts Academy of Applied Studies, Kraljice Marije 73, 11000 Belgrade, Serbia
4 The Academy of Applied Technical Studies Belgrade, Katarine Ambrozić 3, 11000 Belgrade, Serbia
* Correspondence: jrusov@gmail.com; Tel.: +38-1641917479

Abstract: The aim of the paper is to determine the risk level of a contract extension with the existing policyholders, which is further propagated to the business effectiveness and long-term sustainability of the company. The uncertainties in the relative importance of risk factors, their values, and risk levels are described by the linguistic forms, which are modeled by using the fuzzy sets theory. The evaluations of the relative importance of risk factors are stated as a fuzzy group decision-making problem. The weights of risk factors are obtained by using a fuzzy analytic hierarchy process. The determination of production rules for the assessment of the risk level is based on fuzzy IF-THAN rules. The verification of the model is performed by using real-life data originating from the insurance company which operates in the Republic of Serbia.

Keywords: risk level; fuzzy data; FAHP; fuzzy logic; production rules

MSC: 97M30

Citation: Lukić, J.; Misita, M.; Milanović, D.D.; Borota-Tišma, A.; Janković, A. Determining the Risk Level in Client Analysis by Applying Fuzzy Logic in Insurance Sector. *Mathematics* **2022**, *10*, 3268. https://doi.org/10.3390/math10183268

Academic Editor: Michael Voskoglou

Received: 4 August 2022
Accepted: 3 September 2022
Published: 8 September 2022

1. Introduction

Changes in the business world, primarily in the domain of politics and economy, might lead to business uncertainties in all organizations, especially insurance companies. The enhancement of business efficiency is one of the most important tasks for operational and strategic management. To determine whether there will be an extension of the insurance contract for each insured client, it is necessary to anticipate the possibility of further damage to the insurer. Insurance companies have different policies and business strategies, which are based either on high levels of management or acceptable risk levels; it is necessary to analyze the evidence of risk factor values (RFs) and then decide to extend contracts with clients. In practice, it can be seen, that these two RFs need to be analyzed: the amount of money and the number of claims. It is also worth considering that these two RFs, alongside with receivables ratio responsible for measuring premium payments, have a significant impact on clients. Uncertainties in the relative importance of the RFs and their values cannot be accurately determined when the conditions persistently change, due to difficulty in determining the complexity involved in the risk of extension of the contract with the insured. Different types of vagueness, imprecision, and uncertainties are described by linguistic forms that are assigned with different numerical values as a certain degree of affiliation [1–3]. The development in some areas of mathematics such as fuzzy sets theory [4,5] allows uncertainties to be quantitatively represented in a fairly proper way. The basic characteristic of a fuzzy number is a membership function which can take different forms. In the literature, the triangular and trapezoidal membership functions are widely used [6] because they do not require complex calculations; it should be emphasized that the domains of fuzzy numbers are defined on a real line with respect to the nature of linguistic expressions and estimates of DMs.

According to the experience of the best practice, it is known that decision makers have different assessments of the relative importance of RFs. Therefore, it is fully justified to introduce the assumption that determining their relative importance should be posed as a fuzzy group decision-making problem. Decision makers can make significantly better judgments if they look at each pair of RFs separately, by analogy [6]. In accordance with the introduced assumptions, the weight vector of RFs is given by using the Fuzzy Analytic Hierarchy Process (FAHP).

Complex problems, such as those that can determine the level of risk, can be successfully solved by applying fuzzy logic [4,5]. The theory of fuzzy logic has been usefully proven when it is necessary to decide based on experience, intuition, and subjective assessments of individual parameters by decision-makers. Zadeh [2] emphasized the use of fuzzy logic: (1) the mathematical concept is very simple, (2) it is flexible, (3) fuzzy logic tolerates the imprecise data, (4) it can incorporate into the decisions the experience of DM who know and understand the problem, (5) Fuzzy logic is based on the native language, which presents the best way for communication. In addition, this concept has certain shortcomings, such as a large number of production rules. Thus, one of the basic goals is to reduce the number of rules so that they can be effectively used in solving real problems.

The assessment of the level of risk in practice is performed in relation to these two RF: the amount and the frequency of claims; it is believed that RFs are equally important. The motive of this research can be defined as the extent to which the business of the insured, as an individual, may affect the risk level assessment of the insurance company. The given fuzzy logic model is for determining whether it is a risky business for the insurance company to extend the contract with existing clients.

The motivation for this research comes from the fact that there are no research papers that treat the problem of determining the level of business risk based on fuzzy logic rules. In addition, there are no guides or developed methodologies for the company to assess the level of risk of doing business with a client before signing a contract with him.

The research challenges, motivations and the scientific research area is the application of a fuzzy model in the field of determining critical workflow processes to improve business management and risk transfer; this research investigates risk forecasting and management by forming a fuzzy model to determine whether it is risky for an insurance company to extend contracts with existing policyholders based on the flow, the amount, and the number of their claims. Predicting the financial result gives the basic concept of development and business characteristics of the insurance company; it achieves the stabilization of the company's business, and then the growth, development, and improvement of the insurance market, as well as full protection of the interests of policyholders.

Decision-making on the extension of contracts with existing clients in the Republic of Serbia is greatly influenced by changes in the business world, especially in the domain of politics. In addition, the decision makers responsible for the extension of contractual obligation with the specific insured, being also managers within insurance companies are often described as inexperienced, relatively incompetent and dependent, with a lack of tendency to take risks. Weaknesses and failures of managers can lead to wrong business decisions, because of which there can be immediate and long-term consequences for the business and positioning of the insurer in the market.

Many authors believe that the basis of the problem of insolvency of insurance companies lies in low-quality and unprofessional management, while insufficient premium is the ultimate manifestation of this problem. Quality management, i.e., management of DM is of essential importance for the stability of each DM. In addition, lately increasing competition in the insurance field has put focus on the use of new methodologies based on fuzzy logic [7].

These are the goals and main reasons why the authors developed a mathematical model that would be employed to determine the exact level of risk for each client. The model gives significantly better results compared to the used risk matrices in an insurance company. In this regard, there is a possibility of real application of this model because its great importance can be seen in increasing the objectivity of management for decision-making. The obtained results are important for determining a more precise strategy, which leads to an increase in business efficiency.

Integration of the fuzzy sets theory and risk assessment approach can be marked as the aim of this research: (a) the assessment of the relative importance of RFs by using FAHP; (b) modeling of RFs values by fuzzy sets theory; (c) determining the overall index by using the fuzzy operators; (d) defining production rules that can easily and simply determine the level of business risk for each client, in the field of non-life insurance.

The paper is organized as follows: In Section 2, there is a detailed literature review related to the applied fuzzy sets theory for modeling uncertainties and fuzzy logic for the determination of production rules. The proposed methodology is presented in Section 3. In Section 4, the proposed model is illustrated by real-life data which comes from domestic insurance companies which exist in the Republic of Serbia. The discussion of the obtained results and Conclusion is given in Section 5.

2. Literature Review

For the purposeful presentation of the literature review, this section is divided into 3 sub-sections: (1) basic consideration of management problems in the field of insurance, (2) modeling of uncertainties into the relative importance of RFs and their values as well as risk levels, and (3) determining of production rules.

2.1. Some Management Problems in the Insurance Domain

In the literature, many papers consider the problem of assessing the level of the business risk of an insurance company. Many authors suggest that it is necessary to combine risk level assessment methods with fuzzy sets theory as there are a lot of uncertainties in the considered problem. Shapiro [8] analyzed and discussed the benefits of applying fuzzy sets theory and fuzzy logic in solving management problems in the field of insurance.

The problems of investment management in the field of insurance, scheduling of liabilities, as well as cash flow management, are solved by applying the fuzzy logic by Shapiro [9]. Determining the time [10] structure of interest rates, in the field of life insurance, is given by using the fuzzy regression analysis in [10,11]. Berry-Stölzle, et al. [12] suggest that the assessment of the required solvency in property insurance can be successfully performed by using fuzzy regression analysis. Shapiro [13] suggests that annuity damage modeling should be based on the fuzzy set theory. Abul-Haggad and Barakat [14] have developed a fuzzy risk matrix combined with the Mamdani method. In this way, it is possible to accept and process expert knowledge in a much more intuitive way that is closer to human thinking. Markowski and Mannan [15] propose the procedure for determining three types of fuzzy risk matrices (low-cost, standard, and high-cost) that can be used for different safety analyses in the chemical industry. The problem of determining the identification of an insurance company can be successfully solved by applying fuzzy logic rules, according to the opinion of Zapa and Cogollo [16].

2.2. Modelling of Existing Uncertainties

The uncertain and imprecise data (in this paper these are the relative importance of RFs and their values as well as the risk level) can be adequately represented by linguistic expressions. The choice of membership function can be considered a problem in itself. Triangular or trapezoidal membership functions are most often used because they do not require great computation complexity. Furthermore, there is no guideline or recommendation in the relevant literature for the determination of the bounds in the domain of fuzzy numbers. Hence, it can be said that the number and type of linguistic variables are determined by DMs, depending on the type and complexity of the problem; it should be mentioned that linguistic variables can be modeled by self-confidence interval (LIT) [17].

The relative importance of considered RFs is not equal and does not change over time. The assessment of relative importance depends on the knowledge and experience of DMs.

In general, the relative importance of items can be determined: (a) in a direct way [18] in the direct method of processing, decision-makers associate pre-defined linguistic terms to each RFs that describe their weight, and (b) by setting up a fuzzy pair-wise comparison matrix [19–21]; it is considered that in this way DMs can make a better and more accurate assessment. The fuzzy rating of the relative importance of items is, more or less, burdened by DMs' errors. Therefore, it is necessary to check the extent to which these errors affect the accuracy of the vector weights. In conventional AHP [22], the consistency estimate of DMs is based on applying the Eigenvector method. There are many procedures for handling FAHP, their advantages, and disadvantages, which are analyzed by Kahraman, et al. [23]. The similarities and differences between the proposed FAHP are presented in Table 1.

Table 1. FAHP.

Author's	Type Variable/ Granularity/Domain	Group Decision-Making Problem/ Aggregation Method	Pair-Wise Comparison Matrix/Consistency Checking	Handling of Uncertainties in FAHP	The Weights Vector	Application Domain
Chen, et al. [24]	TFNs/5/[1–3.5]	Yes/the proposed procedure	Concept equal possibilities/ Eigenvector [22]	Extent analysis [25]	crisp	Evaluation performance in the education domain
Sultana, et al. [26]	TFNs/5/[1–9]	-	Yes	Extent analysis [25]	crisp	
Sirisawat and Kiatcharoenpol [27]	TFNs/9/[1–10]	-	-	Extent analysis [25]	crisp	
Jakšić, et al. [20]	TFNs/5/[1–5]	-	-	Extent analysis [25]	crisp	Ranking of banks
Banduka, et al. [28]	TFNs/5/[1–5]	-	-	Extent analysis [25]	crisp	Extension of FMEA in automotive industry
Lyu, et al. [29]	Defined procedure for determination of fuzzy elements of fuzzy pair-wise comparison matrix/the elements of pair-wise comparison matrix is given by applying the ranking of fuzzy numbers	-	Eigenvector [22]	Extent analysis [25]	crisp	Risk assessment in civil engineering
Bakır and Atalık [21]	TFNs/9/[1–9]	Yes/fuzzy geometric mean	-	The proposed method by Buckley [30]	Crisp is given by applying the center of area method [4] and linear normalization procedure TFNs/5 [31]	Assessment of quality in the air industry
Calabrese, et al. [32]	TFNs/5/[1–3.5]	-	The defuzzification procedure for TFNs [33]/ Eigenvector [22]	Extent analysis [25]	crisp	ranking of ISO sustainability subjects
The proposed model	TFNs/3/[1–5]	Yes/the proposed procedure	The center of area method [4]/ Eigenvector [22]	Extent analysis [25]	crisp	Assessment of RFs in the insurance sector

By comparing papers that deal with the proposed procedure certain similarities can be noticed. In the analyzed papers, elements of fuzzy pair-wise comparison matrix are described by TFNs, as in this research. As it is noted, the granulation of used fuzzy numbers depends on the size and complexity of the considered problem. There are no recommendations on how to determine granulation. The nine-point scale has been proposed by [21,27]. Most authors suggest a five-point scale [20,24,26,29,33]. The point scale is introduced in this research which represents one of the differences between this and the other analyzed papers.

In the literature, many authors are determining the relative importance of fuzzy group decision-making problems [21,24] as in this research. The authors are of the opinion that a more accurate assessment of the relative importance can be more accurately determined if more DMs participate in the decision-making process. The aggregation of the opinions of DMs into unique marks can be performed by applying the different aggregation operators such as: fuzzy geometric mean [21], and (ii) the proposed procedure [24], as in this research.

In conventional AHP [22] it was emphasized that it is necessary to check the consistency of estimates of DMs. A fuzzy pair-wise comparison matrix can be considered consistent if the corresponding crisp matrix is consistent. The fuzzy pair-wise comparison matrix can be transformed into a correspondent pair-wise comparison matrix by using different defuzzification procedures, such as: (i) simple defuzzification is applied in [33], (ii) α cut level applied in [24], and (iii) the center of the area method which is applied in this research. There are many developed methods for checking the consistency of a pair-wise comparison matrix [25]. One of the most widely used methods is the Eigenvector used in the analyzed papers (see Table 1), as in this research.

The determination of the weights vector can be based on: (i) the method proposed by Buckey [30] and (ii) the method of extended analysis [25]. Some authors [21] consider that the method proposed by Buckey [30] has certain advantages over the method of extended analysis. On the other hand, the method of extended analysis [25] is easy to understand, and therefore, the method of extended analysis is widely used for handling FAHP (see Table 1) as well as in this research.

2.3. Fuzzy Production Rules

Assume that the output variable depends on several input variables that have different values. The number of possible values that can be assigned to an output variable is equal to the number of combinations with repetition. The solution obtained in this way is not applicable in practice. Therefore, the application of IF-THEN logic rules can lead to solutions that practitioners can easily understand and apply. The problem becomes significantly more complex if there are multiple input variables whose values can be described using several linguistic terms. The solution to such complex problems can be successfully obtained through the experience and knowledge of DMs, which are formalized by fuzzy IF-THAN rules (see Table 2).

Table 2. Fuzzy IF-THAN rules.

Author's	Number of Input Variables	Number, Type and Domain of Linguistic Terms	Normalized Input Variable Values/the Weights Vector of Input Variables/the Weighted Input Variables Values	The Overall Index/Defuzzification	Number of Decision Rules and its Type	Application Domain
Sii, et al. [34]	2	6/TrFNs/[0–10]	-	-	4/Fuzzy rules made by experts	safety of the marine system
Gentile, et al. [35]	3	5/Gausian and TFNs/[100–500] and [0–1]	-	Fuzzy union/moment method	5/TFNs and TrFNS	safety principles to plant design and operating plants
Tadić, et al. [36]	15	7/TFNs/[0–1]	-	Fuzzy union/moment method	7/TrFNs	Customer satisfaction with banking service quality
Aleksić, et al. [37]	7	5/TFNs/[0–1]	The linear normalization procedure/FAHP combined with FOWA and fuzzy union/product of fuzzy numbers	Arithmetic mean	5/TrFNs	Assessment of organization's vulnerability
Tadić, et al. [38]	3	5/TFNs/[0–1]	-/FAHP/dilatation operator	Fuzzy cut/moment method	5/TFNs	Inherent safety index for food industry
The proposed model	3	4/TFNs/different measurement scales	Yes/linear normalization procedure-	Fuzzy union of the TFNs describing the weighted normalized input variable values/moment method	4/TrFNs	Assessment risk level in insurance companies

According to the overall index value, production rules can be defined. In this way, the number of production rules is significantly reduced, and at the same time, the effectiveness of solving complex problems is significantly increased. A brief retrospective of these papers is given below. The similarities and differences between the proposed fuzzy IF-THAN rules are presented in Table 2.

The authors have used a different number of input variables whose membership functions have different shapes. In the analyzed papers, the authors assumed that input, as well as output variables, are described by uncertain numbers whose domains are defined on real lines into intervals [0–1]. If the values of the input variables are defined on other measurement scales (as in this paper), then it is necessary to perform their normalization. In this research, the linear normalization procedure is performed. There are many papers in which the overall index depends on the values and weights of input variables [38,39]. With respect to the results published in the literature in the field of risk analysis, it can be considered that the risk level can describe with not less than 3 and not more than 5 linguistic terms. Basically, all authors (see Table 2) discuss those linguistic expressions that describe the values of output variables that can be modeled with sufficient accuracy by using TrFNs, as in this research. As is well known, TrFNs capture uncertainty better than TFNs. In that case, it is necessary to normalize the values of input variables in Aleksić, et al. [37] developed a procedure for determining the overall index, which is described by precise numbers. [38] or fuzzy union in the rest analyzed papers, as in this research. By applying the defuzzification procedure, the fuzzy overall index value is presented by a precise value. There are many defuzzification procedures that can be found in the relevant literature [5]. The widely used defuzzification procedure is the current method in this research.

The results obtained by analyzing the relevant literature show that the determination of the risk level of a contract extension with the clients in the field of insurance is based on: (i) respecting two RFs (the claim amount and the claim frequency) which have the same relative importance and (ii) subjective assessment of the DMs. The best practice experience shows that it is necessary to consider the Claims ratio, which is included as the third RF in this research; it is assumed that the considered RFs do not have equal importance and that it is determined in an exact manner. In the analyzed literature, the RF values are described as crisp. Due to the significant economic and political changes which are happening in the region, it can be said that describing RF values by precise numbers, and especially the claim amount, is not appropriate; it considerably makes it difficult for DMs to estimate the risk of a contract extension with the clients. In this research, the RF values are modeled by using the fuzzy sets theory which allows for them to be described in a sufficient enough manner. Determination of the risk level is based on the proposed model which significantly decreases the subjectivity of the DMs.

3. The Proposed Model

One of the important problems in any insurance company is the risk level analysis of business due to the extension of contracts with clients. The solution to this problem greatly affects the achievement of business goals, primarily the survival and the development of the insurance company.

It is known that the risk level of insurance policy extensions is affected by numerous RFs that can be formally represented by a set of indexes $\{1, \ldots, i, \ldots, I\}$ where I is the total number of RFs and i, $i = 1, \ldots, I$ is the index of RF. The number and the type of RFs are determined by DMs according to their experience and knowledge as well as the results of the best practice. In this research, an insurance extension risk assessment is considered with respect to three RFs: the amount of claims $(i = 1)$, the number of claims $(i = 2)$, and the claims ratio $(i = 3)$.

In this research, based on the results from the most successful insurance companies, we have introduced RFs that do not have the same relative importance. The relative importance of RFs is assessed by DMs. The DMs are presented by sets of indices $\{1, \ldots, e, \ldots, E\}$. The total number of DMs is denoted as E and e, $e = 1, \ldots, E$ is the index of DM. In this paper, DMs are underwriters, as one of the most important functions of insurers in the decision-making process. The underwriter needs to match earned premium with the claims with an eye on profitability. If the premium is not sufficient to cover the claims, the insurer is confronted with the probability of loss, and the underwriting risk arises; this risk could include the underestimated liabilities arising from unpaid business written in previous years, for example. The relative importance of RFs is stated by a fuzzy pair-wise comparison matrix at the level of each DM.

In general, RFs values can be adequately described by using K different linguistic expressions which are modeled by TFNs, $\widetilde{v}_{ji}, i = 1, \ldots I; j = 1, \ldots J$. The domains of these TFNs belong to different intervals on the real line and have different measurement units. Interval limits are determined according to DMs estimates; they base their estimates on evidence data and experience. The weighted normalized RFs values are given by using the fuzzy algebra rules.

The procedure for determining the level of risk of contract renewal for each client separately, considering RFs values as well as their weights, is further briefly described. Firstly, the fuzzy overall risk index values, as well as their representative scalars are calculated by using fuzzy algebra rules.

The total number of output rules N is given according to the following logistic rule $N = J^K$. In this manuscript, respecting the introduced assumption $N = 3^4 = 81$; it can be clearly concluded that the use of the principle of approximate reasoning is not justified in practice. Reducing the number of output variables, which at the same time leads to increased decision-making efficiency, can be achieved by using fuzzy IF-THAN rules based on the Mamdani method [18,40], as well as fuzzy preference relation (LIT 2) [41]. Mamdani method is widely accepted for identifying the level of risk in insurance for the collection of expert knowledge because it allows expertise in a more intuitive way [8]. Each rule can be represented by a classical implication where the logical operator cut in the logic phase is replaced by taking the minimum value under certain conditions. Because of that, it can be said that the system is simplified by discarding the least significant rules. In this way, the practical applicability of the developed model is significantly increased and at the same time, sufficient inference accuracy is achieved. Representation of the considered problem can be presented in black box form, as it is shown on Figure 1.

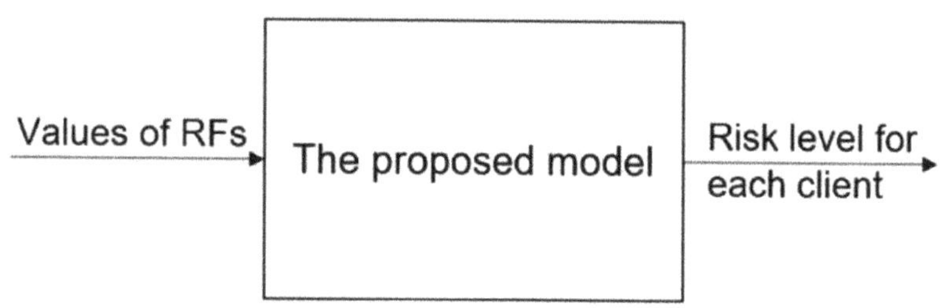

Figure 1. Black box form of the considered problem.

A flowchart of the research methodology is presented in Figure 2.

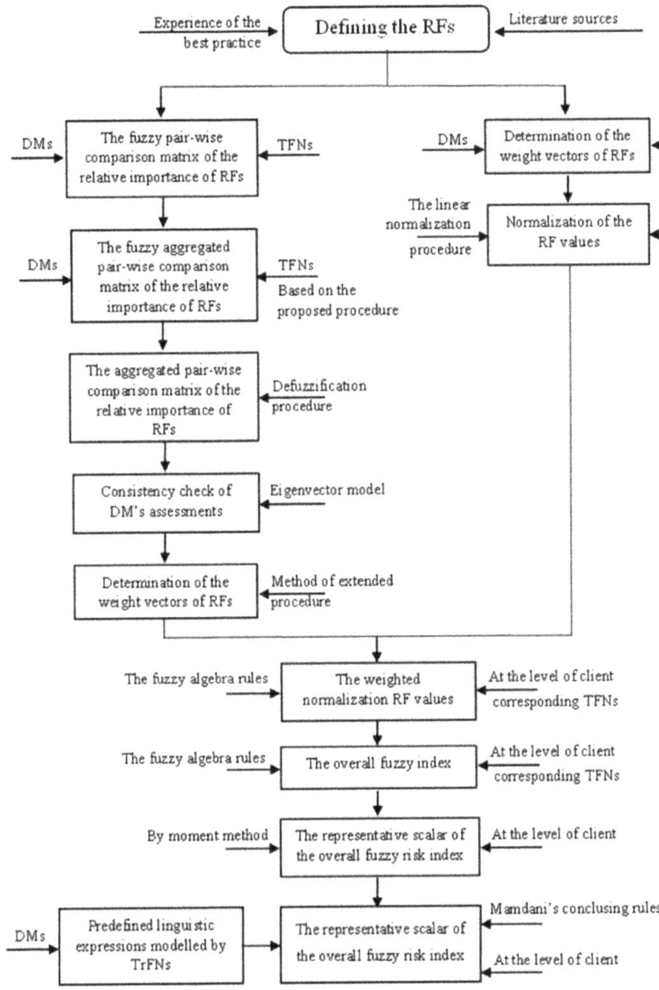

Figure 2. The proposed methodology.

3.1. The Modelling of the Relative Importance of the RFs

The relative importance of RFs is not equal, and it does not change over time; they involve a high degree of subjective judgments, knowledge, and experience of DMs; they use pre-defined linguistic expressions which are modeled by TFNs, $\widetilde{W}_{ii'}^{e} = \left(l_{ii'}^{e}, m_{ii'}^{e}, u_{ii'}^{e}\right)$ with the lower and upper bounds $l_{ii'}^{e}, u_{ii'}^{e}$ and modal value $m_{ii'}^{e}$, respectively. Values in the domain of these TFNs belong to a real set within the interval [1–5]. A value of 1 or 5 means that the relative importance of RF over RF is very small, or extremely large, respectively.

If the strong relative importance RF i' over RF i holds, then the pair-wise comparison scale can be represented by the TFN

$$\widetilde{W}_{ii'}^{e} = \left(\widetilde{W}_{i'i}^{e}\right)^{-1} = \left(\frac{1}{u_{ii'}^{e}}, \frac{1}{m_{ii'}^{e}}, \frac{1}{l_{ii'}^{e}}\right)$$

If $i = i'$ then the relative importance of RF i over RF i' is represented by a single point 1, which is a TFN (1, 1, 1).

These TFNs are given in the following way and presented in Figure 3:

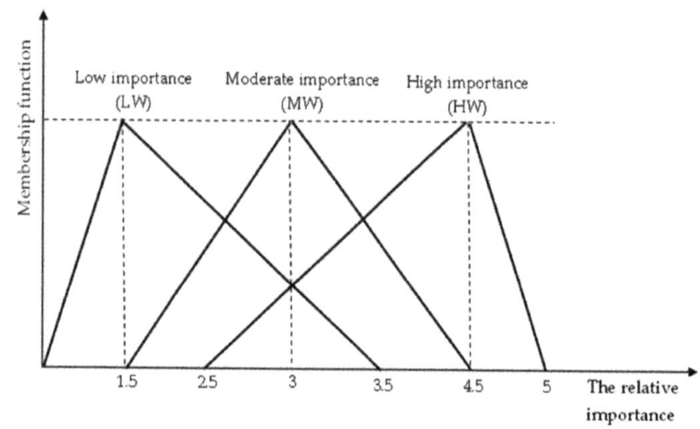

Figure 3. The TFNs describing the relative importance of RFs.

low importance (LW)—(1, 1.5, 3.5)
moderate importance (MW)—(1.5, 3, 4.5)
high importance (HW)—(2.5, 4.5, 5)

Evaluation and relative importance of RFs is based on the consideration of the probability of the possible event, the outcome (amount of the claim incurred), and frequency (number of claims over the period of time). Decision makers, managers, will manage by analyzing the relative importance of RFs, frequency and the size of a claim for one client. The linguistic domain scales are determined by the number and amount of liquidated damage claims for the clients in the insurance company. The domain of the TFN is made on the basis of an analysis of the total number and amount of claims, the control environment, the inherent risks and the measurement in terms of impact and probability.

3.2. The Modeling of RFs Values and Risk Levels

DMs have defined linguistic expressions that can be used to describe the values of treated RFs; these linguistic expressions and their corresponding TFNs are presented in the following.

RF- the claim amount: Claims covered by property insurance (things) are, as a rule, only pecuniary damage claims, incurred on an insured thing or object, which can be partial or total considering the claim intensity. Based on the actual database, it is known that the amount of claim (expressed in thousands of monetary units) is neither less than zero nor more than 300. Values of these RFs can be described by four linguistic forms: *Small (L1), Medium (L2), Large (L3), and Total (L4)*; these linguistic forms are modeled by using TFNs (see Figure 4); it should be noted that the values in the domain were determined by DMs based on their experience. Figure 4b shows these normalized TFNs.

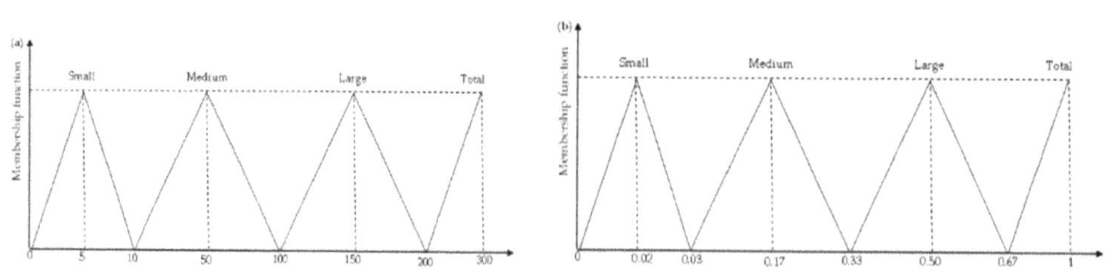

Figure 4. Linguistic forms for describing: (**a**) values of settled claims and (**b**) corresponding the normalized TFNs.

Property insurance claims, considering the amount of claim, can be partial or total. The domain of the TFN for the amount of claims is determined by defining the limit values of the claim. Limited amounts are defined by expert experience according to the actual movement of the amount of claims all clients in the insurance company in the case of property insurance. Based on expert experience, the limit values are defined by the domain of the TFN based on the average claim amount for all insureds in the case of property insurance; it therefore seems that the claims exceeding RSD 300,000 are considered to be total, and all under this amount of partial claims. Fuzzy sets that describe the input variable 'claim amount' do not cover equal intervals, as a result of the fact that these fuzzy sets are defined on the basis of the empirical dana of the real insurance company portfolio.

RF- the claim frequency: The number of incurred claims by one client in the observed period defines the frequency of claims. The value of this RF can be described by four linguistic forms joined by TFNs correspondents: *Negligible (M1), Moderate (M2), High (M3), and Extremely high (M4)* (see Figure 5). The values in the domain of these linguistic expressions are determined by respecting the number of incurred adverse events of each client under each contract. By using the normalization procedure, the normalized TFNs are presented in Figure 5b.

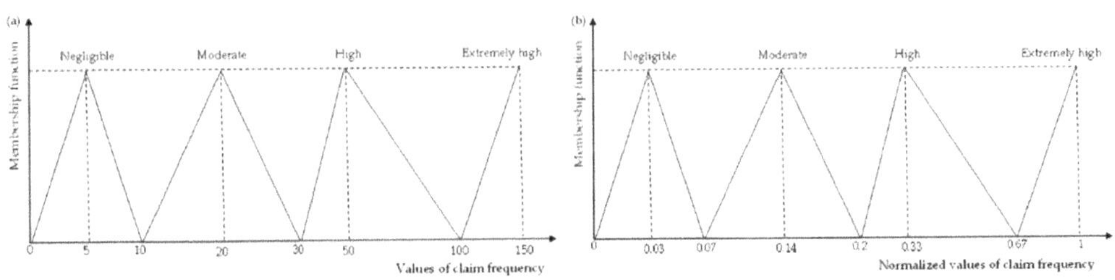

Figure 5. Linguistic forms for describing (**a**) the claim frequency and (**b**) corresponding the normalized TFNs.

The values of RF number of claims depend on the claim frequency of one client. The input variable claim frequency refers to the number of claims caused by one client over the observed period. As well as with the first input variable, it is necessary to determine its domain. Based on the actual movement of the number of claims in the case of property insurance for all clients in the insurance company, the experts will define the interval to which the linguistic scales belong. The second input variable, claim frequency, refers to the number of claims incurred by one client during the insurance period. The parameter assessment is defined on the basis of expert experience and the actual database about number of claims incurred by each client individually in the insurance company for property insurance. The claim frequency is estimated according to the number of harmful events caused by one client, observed through all his property insurance contracts. Because of the ranking of frequency, which is based on the real database expertise, this input variable is shown by 4 fuzzy sets that do not cover equal intervals.

RF- Claim ratio: The claim ratio can be defined as the ratio of incurred claims and earned premiums in the observed period (year) and it is considered the simplest measure of premium adequacy in the field of non-life insurance; it can be mentioned that the value of this RF may affect the company's profit for the entire insurance period of the observed client. As a percentage share of incurred claims in the earned premiums, it is necessary to consider the claim ratio when determining the risk level in the insurance company; it is an indicator of the sufficiency of the premium to cover insurance liabilities. Based on the value of this RF, it can be determined whether the premium is sufficient to cover policy liabilities. Based on the best practice experience, the values of RF can be described by four linguistic forms which are modeled by TFNs: *Optimal (K1), Very good (K2), Conditionally acceptable (K3), and Unacceptable (K4),* which are presented in Figure 6.

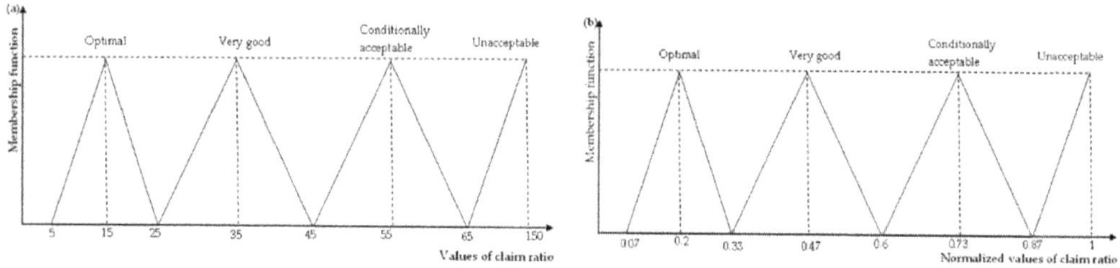

Figure 6. Linguistic forms for describing (**a**) claim ratio and (**b**) corresponding the normalized TFNs.

Fuzzy sets that describe the input variable 'claim amount' do not cover equal intervals, as a result of the fact that these fuzzy sets are defined on the basis of empirical data of the real portfolio of the insurance company. Earned premium depends on each contract with each client. The linguistic variables corresponding to the TFN ratio claims this input variable shown by 4 fuzzy sets that do not cover equal intervals, because the ratio claims represent the percentage share of the damage in the premium.

3.3. Risk Levels

The management of insurance companies may define different levels of risk. For instance, a risk level may refer to the maximum percentage of change given the worst-case level of RFs values. Based on the best practice results from the insurance domain, the risk level can be modeled by one of the four predetermined linguistic terms which are modeled by trapezoidal fuzzy numbers (TrFNs) which are presented in Figure 7:

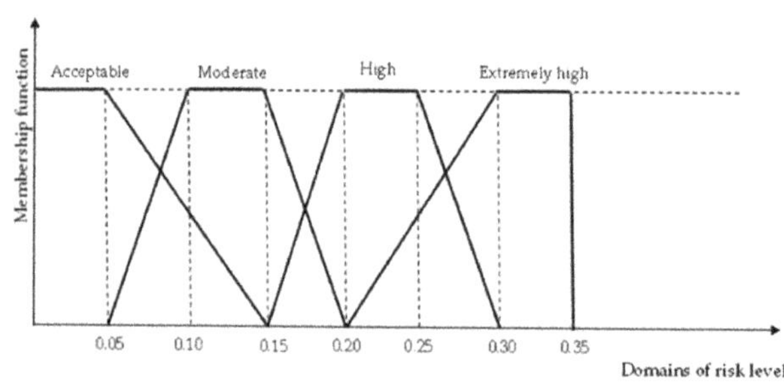

Figure 7. Linguistic expressions and corresponding TrFNs for describing risk levels.

Acceptable (Q1)—(0, 0, 0.05, 0.15)
Moderate (Q2)—(0.05, 0.1, 0.15, 0.2)
High (Q3)—(0.15, 0.2, 0.25, 0.3)
Extremly high (Q4)—(0.2, 0.3, 0.35, 0.35)

The domains of these TFNs are defined into a set of real line intervals [0–0.35]. The upper bound of this interval was determined by the assumption that the considered RFs have different weights. If the overlap from one TrFNs to the other TrFNs is very high, it obviously indicates that there is a lack of knowledge about the risk level or a lack of sufficient partitioning. The proposed values of the defined risk level represent the initial draft assessed by DMs' opinion in the insurance companies in the Republic of Serbia.

3.4. The Proposed Algorithm

In this Section the proposed Algorithm is presented and carried out in the following steps:

Step1. Fuzzy rating of the relative importance of each paiI If RI, $i = 1, \ldots, I$ is performed by DMs, so that:

$$\widetilde{W}^{e}_{ii'}, i, i' = 1, \ldots, I; i \neq i'$$

Step 2. Fuzzy aggregated pair-wise comparison matrix of the relative importance of RFs is $\left[\widetilde{W}_{ii'}\right]_{I\times I}$.

where:

$$l_{ii'} = \min_{e=1,\dots,E} l^e_{ii'},\ m_{ii'} = \sqrt[E]{m^e_{ii'}},\ u_{ii'} = \max_{e=1,\dots,E} u^e_{ii'},$$

Step 3. Transform the fuzzy pair-wise comparison matrix into the pair-wise comparison matrix of the relative importance RFs:

$$[\theta_{ii'}]_{I\times I}$$

where:

$\theta_{ii'}$ is the representative scalar of the TFN of $\widetilde{W}_{ii'}$, which is obtained by the moment method [4].

The consistency of the pairwise comparison matrix, is verified by applying the eigenvector method [22].

Step 4. Calculate the normalized weights vector, of treated RFs by using the method of extended analysis [25]:

$$[\omega_i]_{I\times 1}$$

Step 5. Each RF can be described by using K_j predefined linguistic forms modeled by TFN,

$$\widetilde{v}_{ij}, i = 1,\dots I;\ j = 1,\dots J$$

Step 6. Normalized RFs values, $\widetilde{r}_{ji}, i = 1,\dots I;\ j = 1,\dots J$, were obtained by linear normalization procedure [38].

Step 7. The weighted normalized RFs values are given by using fuzzy algebra rules [9]:

$$\widetilde{z}_{ij} = \omega_i \cdot \widetilde{r}_{ij}\ ,\ i = 1,\dots I;\ j = 1,\dots J$$

Step 8. Determine the overall fuzzy risk index, $\widetilde{\rho}_j$:

$$\widetilde{\rho}_j = \bigcup_i \widetilde{z}_{ji}$$

where I is the overall output variables (in this case the overall number of RFs).

Step 9. The representative scalar of the TFN, $\widetilde{\rho}_j, \rho_j$ is calculated by the moment method [9]:

$$\rho_j = defuzz\ \widetilde{\rho}_j$$

Step 10. There are several manners for determining the IF-THEN rules. In this paper, rules are built from the DMs' knowledge and experience by analogy to Mamdani's concluding rules. There are four production rules modeled by the TFNs $\widetilde{s}_q, q = 1,\dots,4$.

The region of risk in the observed insurance company can be defined according to the rule:

IF the value of "the overall risk index value" equals ρ_j, THEN the region of risk is described by the linguistic form where

$$\max_{q=1,\dots,4} \mu_{\widetilde{s}_q}(\rho_j) = \mu_{\widetilde{s}^*_q}$$

In this way, the fuzzy risk matrix is constructed.

Step 11. The proposed model is verified by real life data.

4. Illustrative Example

The proposed model is tested on real-life data obtained in the period from 2009 to 2019 and comes from the domain of property insurance, one of the most common types of non-life insurance, from one of the largest insurance companies in the Republic of Serbia. In insurance companies, risk management is the responsibility of actuaries and underwriters, so the assessment of the importance of the treated RF is obtained from the actuaries and underwriters. By applying the interview method, we have obtained a fuzzy rating of actuaries and underwriters. RF values at the level of each client can be based on data evidence. Validation of the model has been performed on a sample of 100 clients from the group of clients who were observed over the period of 10 years and had claims for at least 7 years. The sample was determined randomly without repetition.

The insured in advance pays the premium, and the insurance company pays off compensation to the client, if, and when an insured adverse event occurs. If DMs do not determine the level of the client's risk from the aspect of a contract extension for the next period, that could lead to the inability of the company to settle its obligations to other policyholders. The consequence of a bad or insufficiently good decision of DMs may jeopardize the liquidity and survival of the insurance company.

The procedure of the proposed Algorithm is shown below.

To reduce the number of calculations, the example was formed on decisions of three DMs, respectively, the assumption is E = 3.

According to the proposed algorithm (Step 1) fuzzy pair-wise comparison matrix at the level of each DM is constructed

$$\begin{bmatrix} (1,1,1) & LW, 1/MW, (1,1,1) & HW, MW, MW \\ & (1,1,1) & MW, HW, LW \\ & & (1,1,1) \end{bmatrix}$$

The proposed process of aggregation is illustrated by the following example (Step 2 of the proposed Algorithm):

$$l_{12} = \min_{e=1,\ldots,E}(1,\ 0.22,\ 1) = 0.22$$

$$m_{12} = \sqrt[3]{(1.5 \cdot 0.33 \cdot 1)} = 0.79$$

$$u_{12} = \max_{e=1,\ldots,3}(3.5,\ 0.67,\ 1) = 3.50$$

The aggregate values of the other elements of the unclear aggregate comparison matrix are determined in a similar way (Step 2 of the proposed Algorithm), so that:

$$\begin{bmatrix} (1,1,1) & (0.22,\ 0.79,\ 3.5) & (1.5,\ 3.39,\ 5) \\ (0.29,\ 1.27,\ 4.55) & (1,1,1) & (1,\ 2.69,\ 5) \\ (0.2,\ 0.29,\ 0.67) & (0.2,\ 0.37,\ 1) & (1,1,1) \end{bmatrix}$$

The pair-wise comparison matrix of the relative importance of RFs (Step 3 of the proposed algorithm):

$$\begin{bmatrix} 1 & 0.79 & 3.39 \\ 1.27 & 1 & 2.69 \\ 0.29 & 0.37 & 1 \end{bmatrix},\ C.I. = 0.064$$

By applying the concept of extent analysis (Step 4 of the proposed Algorithm), the weights vector is calculated:

$$W_p = (1,\ 0.97,\ 0.5)$$

The normalized weights vector ω:

$$\omega = \begin{pmatrix} 0.41 & 0.39 & 0.20 \end{pmatrix}$$

The proposed procedure (Step 5 to Step 9 of the proposed Algorithm) is illustrated in the next example. Let the considered RFs be described by the linguistic characterization: *Medium (L1), High (M3) and Unacceptable (K4)*. The weighted normalized values of these RFs are shown in Figure 8.

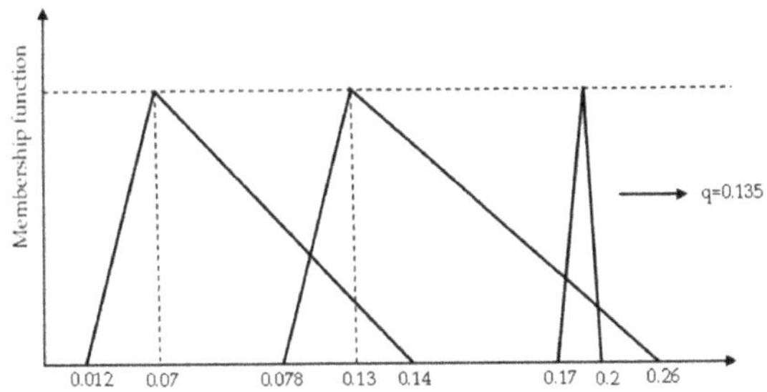

Figure 8. The weighted normalized values of linguistic expressions L1, M3 and L4.

The region of risk for the treated example is obtained by using the procedure (Step 10 of the proposed Algorithm):

max(0.15, 1) = 1, so, it follows, that the level of risk can be described as a moderate risk level.

In a similar way, the level of risk is determined for all combinations of RF values, so that the fuzzy risk matrix is constructed, presented in Table 3:

Table 3. Fuzzy risk matrix.

	Risk Level		Risk Level		Risk Level	
L1-M1-K1	Q1	L2-M1-K1	Q2	L3-M1-K1	Q2	L4-M1-K1
L1-M1-K2	Q1	L2-M1-K2	Q2	L3-M1-K2	Q2	L4-M1-K2
L1-M1-K3	Q2	L2-M1-K3	Q2	L3-M1-K3	Q2	L4-M1-K3
L1-M1-K4	Q2	L2-M1-K4	Q2	L3-M1-K4	Q2	L4-M1-K4
L1-M2-K1	Q2	L2-M2-K1	Q2	L3-M2-K1	Q2	L4-M2-K1
L1-M2-K2	Q2	L2-M2-K2	Q2	L3-M2-K2	Q2	L4-M2-K2
L1-M2-K3	Q2	L2-M2-K3	Q2	L3-M2-K3	Q2	L4-M2-K3
L1-M2-K4	Q2	L2-M2-K4	Q2	L3-M2-K4	Q2	L4-M2-K4
L1-M3-K1	Q1	L2-M3-K1	Q2	L3-M3-K1	Q2	L4-M3-K1
L1-M3-K2	Q2	L2-M3-K2	Q2	L3-M3-K2	Q2	L4-M3-K2
L1-M3-K3	Q2	L2-M3-K3	Q2	L3-M3-K3	Q2	L4-M3-K3
L1-M3-K4	Q2	L2-M3-K4	Q2	L3-M3-K4	Q3	L4-M3-K4
L1-M4-K1	Q2	L2-M4-K1	Q3	L3-M4-K1	Q3	L4-M4-K1
L1-M4-K2	Q2	L2-M4-K2	Q3	L3-M4-K2	Q3	L4-M4-K2
L1-M4-K3	Q3	L2-M4-K3	Q3	L3-M4-K3	Q3	L4-M4-K3
L1-M4-K4	Q3	L2-M4-K4	Q3	L3-M4-K4	Q3	L4-M4-K4

The proposed model is verified by a sample consisting of 100 clients and presented in Table 3 (Step 11 of the proposed Algorithm).

It should be noted that, in the considered insurance company, there is a good record of the values of RFs, at the level of each client. For the purposes of this research, the considered period was last 10 years. The values of these RFs were calculated by using an arithmetic mean operator. Based on thus obtained values each RF, at the level of each client, appropriate linguistic characterizations were joined. Respecting the constructed fuzzy risk matrix, the risk level for a contract extension was determined and presented in Table 3. Furthermore, the risk level determined by the assessment of DMs of the insurance company was presented in the same table.

5. Discussion and Conclusions

The management practice shows that evaluation and enhancement of business effectiveness in the insurance domain represent some of the most relevant issues of competitiveness and sustainability over a long period. The definition of an enhancement strategy should be based on the assessment of the level of risk for a contract extension for each of the insured clients. Insurance companies have different policies and strategies, which depend on the extent to which experts are willing to accept a certain level of risk; it is necessary to determine the current "behavior" of the insured, which includes records of all claims by the same policy (number and the number of claims in an accident year), fulfillment of financial obligations from previous contracts (premium), how often and by which terms the client violated previous contracts, the ratio of premiums and claims.

In practice, risk assessment is mainly based on the application of the risk matrix. The elements of the risk matrix are average values that depend on expert judgment and opinion; it should be noted that DMs can be characterized by a lack of experience, competence, autonomy, as well as a tendency to take risks. Weaknesses and omissions of managers may lead to wrong business decisions that can be immediate but also have long-term consequences for the business and position of the insurer in the market.

Determining the solvency of clients by using the exact method would significantly contribute to reducing the business risk of an insurance company that operates in a changing and competitive environment.

The main contributions of the presented research are:

1. Determines the lists of RFs in compliance with the best practice
2. With respect to the human way of thinking, modeling of existing uncertainties is based on TFNs and TrFNs
3. The aggregated fuzzy pair-wise comparison matrix of the relative importance of RFs is constructed by using the proposed method
4. The weights vector of RFs is determined by FAHP [25]
5. The fuzzy overall risk index at the level of each insured client is calculated by applying fuzzy algebra rules
6. Risk matrix is constructed with respect to all RFs, their weights, and values by using the fuzzy IF-THAN rules.

The proposed model is tested and verified on real-life obtained based on data from 100 clients. The values of RFs were obtained from the data basis within the period from 2009 to 2019.

The practical implications of the proposed methodology are oriented to DMs who need to make a decision about which should enable the liquidity of the company. Based on the results obtained, it can be concluded that 1% of clients have an acceptable risk level. 46% of clients have a Moderate risk level. Based on the results obtained, for about 50% of the insured clients, the company's management may conclude that it could extend the contract. An unacceptable level of risk occurs for 12% of policyholders, which further means that the extension contracts with them may cause a decrease in the liquidity of the company. To extend the contract with these insured clients, it is necessary to do additional research.

The main advantage of the proposed model can be emphasized through the fact that DMs easily extend to the analysis of other management decision-making problems in different areas. The main limitation of the proposed model is that it can make decisions quickly, and at the same time, the obtained decision is less encumbered by the subjectivities of DMs so, this could make it more accurate.

Finally, further research could focus on the development of a software based on the proposed model.

Author Contributions: Conceptualization, J.L., M.M., D.D.M. and A.J.; Data curation, J.L. and A.B.-T.; Formal analysis, J.L., D.D.M., A.B.-T. and A.J.; Investigation, J.L. and A.B.-T.; Methodology, J.L., M.M., D.D.M. and A.J.; Project administration, D.D.M. and A.B.-T.; Resources, M.M. and A.J.; Supervision, M.M. and D.D.M.; Validation, J.L., M.M., D.D.M., A.B.-T. and A.J.; Visualization, J.L., A.B.-T. and A.J.; Writing—original draft, J.L.; Writing—review and editing, M.M. All authors have read and agreed to the published version of the manuscript.

Funding: This research received no external funding.

Data Availability Statement: Not applicable.

Conflicts of Interest: The authors declare no conflict of interest.

References

1. Zhang, C.L. Risk assessment of supply chain finance with intuitionistic fuzzy information. *J. Intell. Fuzzy Syst.* **2016**, *31*, 1967–1975. [CrossRef]
2. Zadeh, L.A. The concept of a linguistic variable and its application to approximate reasoning—I. *Inf. Sci.* **1975**, *8*, 199–249. [CrossRef]
3. Nestic, S.; Lampón, J.F.; Aleksic, A.; Cabanelas, P.; Tadic, D. Ranking manufacturing processes from the quality management perspective in the automotive industry. *Expert Syst.* **2019**, *36*, e12451. [CrossRef]
4. Dubois, D.J. *Fuzzy Sets and Systems: Theory and Applications*; Academic Press: Boston, MA, USA, 1980; Volume 144.
5. Zimmermann, H.J. Fuzzy set theory. *Wiley Interdiscip. Rev. Comput. Stat.* **2010**, *2*, 317–332. [CrossRef]
6. Paksoy, T.; Pehlivan, N.Y.; Kahraman, C. Organizational strategy development in distribution channel management using fuzzy AHP and hierarchical fuzzy TOPSIS. *Expert Syst. Appl.* **2012**, *39*, 2822–2841. [CrossRef]
7. Xu, Y.; Li, M.; Cabrerizo, F.J.; Chiclana, F.; Herrera-Viedma, E. Multiplicative Consistency Ascertaining, Inconsistency Re-pairing, and Weights Derivation of Hesitant Multiplicative Preference Relations. *IEEE Trans. Syst. Man Cybern. Syst.* **2021**, 1–16. [CrossRef]
8. Shapiro, A.F. Fuzzy logic in insurance. *Insur. Math. Econ.* **2004**, *35*, 399–424. [CrossRef]
9. Shapiro, A.F. An overview of insurance uses of fuzzy logic. In *Computational Intelligence in Economics and Finance*; Springer: Berlin/Heidelberg, Germany, 2007; pp. 25–61.
10. De Andres Sanchez, J.; Terceño Gómez, A. Applications of fuzzy regression in actuarial analysis. *J. Risk Insur.* **2003**, *70*, 665–699. [CrossRef]
11. Koissi, M.C.; Shapiro, A.F. Fuzzy Regression and the Term Structure of Interest Rates-A Least Squares Approach. *Actuar. Res. Clear. House Newsl.* **2009**. Available online: https://citeseerx.ist.psu.edu/viewdoc/download?doi=10.1.1.98.6550&rep=rep1&type=pdf (accessed on 8 July 2022).
12. Berry-Stölzle, T.R.; Koissi, M.C.; Shapiro, A.F. Detecting fuzzy relationships in regression models: The case of insurer solvency surveillance in Germany. *Insur. Math. Econ.* **2010**, *46*, 554–567. [CrossRef]
13. Shapiro, A.F. Modeling future lifetime as a fuzzy random variable. *Insur. Math. Econ.* **2013**, *53*, 864–870. [CrossRef]
14. Abul-Haggag, O.Y.; Barakat, W. Application of fuzzy logic for risk assessment using risk matrix. *Int. J. Emerg. Technol. Adv. Eng.* **2013**, *3*, 49–54.
15. Markowski, A.S.; Mannan, M.S. Fuzzy risk matrix. *J. Hazard. Mater.* **2008**, *159*, 152–157. [CrossRef]
16. Zapa, E.; Cogollo, J. Fuzzy-BSC methodology for decision making in indemnity area of insurance companies. *IEEE Lat. Am. Trans.* **2018**, *16*, 2539–2546. [CrossRef]
17. Zhu, S.; Huang, J.; Xu, Y. A consensus model for group decision making with self-confident linguistic preference relations. *Int. J. Intell. Syst.* **2021**, *36*, 6360–6386. [CrossRef]
18. Wang, C. *A Study of Membership Functions on Mamdani-Type Fuzzy Inference System for Industrial Decision-Making*; Lehigh University: Bethlehem, PA, USA, 2015.
19. Ho, W.; Ma, X. The state-of-the-art integrations and applications of the analytic hierarchy process. *Eur. J. Oper. Res.* **2018**, *267*, 399–414. [CrossRef]
20. Jakšić, M.; Moljević, S.; Aleksić, A.; Misita, M.; Arsovski, S.; Tadić, D.; Mimović, P. Fuzzy approach in ranking of banks ac-cording to financial performances. *Math. Probl. Eng.* **2016**, *2016*, 6169586. [CrossRef]
21. Bakır, M.; Atalık, Ö. Application of fuzzy AHP and fuzzy MARCOS approach for the evaluation of e-service quality in the airline industry. *Decis. Mak. Appl. Manag. Eng.* **2021**, *4*, 127–152. [CrossRef]
22. Saaty, T.L. The modern science of multicriteria decision making and its practical applications: The AHP/ANP approach. *Oper. Res.* **2013**, *61*, 1101–1118. [CrossRef]
23. Kahraman, C. A brief literature review for fuzzy AHP. *Int. J. Anal. Hierarchy Process* **2018**, *10*. [CrossRef]
24. Chen, J.F.; Hsieh, H.N.; Do, Q. HEvaluating teaching performance based on fuzzy AHP and comprehensive evaluation approach. *Appl. Soft Comput.* **2015**, *28*, 100–108. [CrossRef]
25. Chang, D.Y. Applications of the extent analysis method on fuzzy AHP. *Eur. J. Oper. Res.* **1996**, *95*, 649–655. [CrossRef]
26. Sultana, I.; Ahmed, I.; Azeem, A. An integrated approach for multiple criteria supplier selection combining Fuzzy Delphi, Fuzzy AHP & Fuzzy TOPSIS. *J. Intell. Fuzzy Syst.* **2015**, *29*, 1273–1287.
27. Sirisawat, P.; Kiatcharoenpol, T. Fuzzy AHP-TOPSIS approaches to prioritizing solutions for reverse logistics barriers. *Comput. Ind. Eng.* **2018**, *117*, 303–318. [CrossRef]
28. Banduka, N.; Tadić, D.; Macuzic, I.; Crnjac Zizic, M. Extended process failure mode and effect analysis (PFMEA) for the automotive industry: The FSQC-PFMEA. *Adv. Prod. Eng. Manag.* **2018**, *2*, 206–215. [CrossRef]
29. Lyu, H.M.; Sun, W.J.; Shen, S.L.; Zhou, A.N. Risk assessment using a new consulting process in fuzzy AHP. *J. Constr. Eng. Manag.* **2020**, *146*, 04019112. [CrossRef]
30. Buckley, J.J. Fuzzy hierarchical analysis. *Fuzzy Sets Syst.* **1985**, *17*, 233–247. [CrossRef]
31. Pomerol, J.C.; Barba-Romero, S. *Multicriterion Decision in Management: Principles and Practice*; Springer Science & Business Media: Berlin/Heidelberg, Germany, 2000; Volume 25.
32. Calabrese, A.; Costa, R.; Levialdi, N.; Menichini, T. Integrating sustainability into strategic decision-making: A fuzzy AHP method for the selection of relevant sustainability issues. *Technol. Forecast. Soc. Change* **2019**, *139*, 155–168. [CrossRef]

33. Wang, Y.M.; Elhag, T.M. A fuzzy group decision making approach for bridge risk assessment. *Comput. Ind. Eng.* **2007**, *53*, 137–148. [CrossRef]
34. Sii, H.S.; Ruxton, T.; Wang, J. A fuzzy-logic-based approach to qualitative safety modelling for marine systems. *Reliab. Eng. Syst. Saf.* **2001**, *73*, 19–34. [CrossRef]
35. Gentile, M.R.W.M.M.; Rogers, W.J.; Mannan, M.S. Development of a fuzzy logic-based inherent safety index. *Process Saf. Environ. Prot.* **2003**, *81*, 444–456. [CrossRef]
36. Tadic, D.; Aleksic, A.; Mimovic, P.; Puskaric, H.; Misita, M. A model for evaluation of customer satisfaction with banking service quality in an uncertain environment. *Total Qual. Manag. Bus. Excell.* **2018**, *29*, 1342–1361. [CrossRef]
37. Aleksić, A.; Stefanović, M.; Tadić, D.; Arsovski, S. A fuzzy model for assessment of organization vulnerability. *Measurement* **2014**, *51*, 214–223. [CrossRef]
38. Tadic, D.; Savovic, I.; Misita, M.; Arsovski, S.; Milanovic, D.D. Development of a fuzzy logic-based inherent safety index for food industries. *Proc. Inst. Mech. Eng. Part E J. Process Mech. Eng.* **2014**, *228*, 3–13. [CrossRef]
39. Zhou, Q.; Thai, V.V. Fuzzy and grey theories in failure mode and effect analysis for tanker equipment failure prediction. *Saf. Sci.* **2016**, *83*, 74–79. [CrossRef]
40. Shariff, A.M.; Leong, C.T.; Zaini, D. Using process stream index (PSI) to assess inherent safety level during preliminary design stage. *Saf. Sci.* **2012**, *50*, 1098–1103. [CrossRef]
41. Xu, Y.; Li, M.; Cabrerizo, F.J.; Chiclana, F.; Herrera-Viedma, E. Algorithms to detect and rectify multiplicative and ordinal inconsistencies of fuzzy preference relations. *IEEE Trans. Syst. Man Cybern. Syst.* **2021**, *51*, 3498–3511. [CrossRef]

mathematics

MDPI

Article

Fuzzy Portfolio Selection in the Risk Attitudes of Dimension Analysis under the Adjustable Security Proportions

Kuen-Suan Chen [1,2,3], Yin-Yin Huang [4], Ruey-Chyn Tsaur [5,*] and Nei-Yu Lin [5]

1 Department of Industrial Engineering and Management, National Chin-Yi University of Technology, Taichung 411030, Taiwan
2 Department of Business Administration, Chaoyang University of Technology, Taichung 413310, Taiwan
3 Department of Business Administration, Asia University, Taichung 413305, Taiwan
4 School of Economics and Management, Nanchang Vocational University, 308 Provincial Road, Anyi County, Nanchang 330500, China
5 Department of Management Sciences, Tamkang University, New Taipei City 25137, Taiwan
* Correspondence: rctsaur@yahoo.com.tw

Abstract: Fuzzy portfolio models have received many researchers' focus on the issue of risk preferences. The portfolio based on guaranteed return rates has been developing and considering the dimension of excess investment for the investors in different risk preferences. However, not only excess investment but also shortage investment to the selected portfolio should be considered for risk preferences, including risk-seeking, risk-neutral, and risk-averse, by different degrees of dimensions in excess investment and shortage investment. A comparison to the degree of dimensions for the excess investment and shortage investment indicates that a risk-seeker would like to have excess investment for securities whose return rates are bigger than the guaranteed return rates and shortage investment for securities whose return rates are smaller than the guaranteed return rates. Finally, we present three experiments to illustrate the proposed model. The results show that the different risk preferences derive different fuzzy portfolio selections under s and t dimensions, where a lower value of s is suggested for a risk-seeker as $t > s$, and we suggest the values of s and t to be smaller than or equal to 3. By contrast, for the risk-neutral investor, we suggest $s = t$; $t < s$ is suggested to the investor who is risk-averse.

Keywords: fuzzy portfolio selection; dimension of shortage investment; dimension of excess investment; guaranteed return rate; adjustable security proportion

MSC: 90B50; 90B60

Citation: Chen, K.-S.; Huang, Y.-Y.; Tsaur, R.-C.; Lin, N.-Y. Fuzzy Portfolio Selection in the Risk Attitudes of Dimension Analysis under the Adjustable Security Proportions. *Mathematics* **2023**, *11*, 1143. https://doi.org/10.3390/math11051143

Academic Editor: Aleksandar Aleksic

Received: 18 January 2023
Revised: 17 February 2023
Accepted: 22 February 2023
Published: 24 February 2023

1. Introduction

Portfolio selection is used to find the combinations of assets, which are used to optimize the objectives of an investor with respect to maximizing the expected return under the constrained risk. The foundation of portfolio selection was laid by Markowitz [1] who proposed the mean-variance model and considered asset returns as random variables in the multi-variate normal distribution. Most of the researchers have devoted themselves to solve some criticisms of the original portfolio models, and then some of the rigid assumptions of Markowitz's model are relaxed to deal with different investment environments or challenges, including the models in mean-absolute deviation, value at risk, conditional value at risk, or semi-variance, which are with respect to portfolio selections [2–8].

In most of the asset markets, we cannot just assume the factors affecting the market are random variables. In order to solve the portfolio selection, the factors which are other than randomness are usually applied in the possibility theory. Then, fuzzy portfolio selection is proposed to consider the knowledge of experts, investors' subjective opinions, or a quantitative and qualitative analysis in portfolio selection problems. For example, the perceived risk of an investor can be shown in different degrees of linguistic descriptions, and then the constrained risk for the portfolio selection cannot be performed by probability distribution. In addition, the investment behaviors to new economy events cannot be precisely evaluated by the previous return rates for the selected securities, because a lot of factors cannot be considered in the portfolio selection, and thus fuzzy portfolio models are another kind of possible method for solving non-probabilistic portfolio selection. Numerous researchers have the objective of maximization of the fuzzy return rates and constrained the upper investment risk using possibility theory, which were modelled and studied for portfolio selection [9–12]. Thereafter, most researchers have focused on the multi-period fuzzy optimization problems for solving the multi-objective problems by genetic algorithm and neural networks [13–15]. Without the self-dual property in the possibility measures, some researchers extended the credibility measures for the uncertain portfolio selection [16–18].

The major studies in fuzzy portfolio models are summarized as possibility or credibility theories to optimal decisions in a single-period or multi-period fuzzy portfolio selection. With respect to portfolio selection, the habitual behaviors of an investor in the field of risk analysis are also important in the vagueness environment. For example, Mehlawat et al. [19] proposed multi-objective risk measures and evaluate the fuzzy portfolio selection. Yue and Wang [20] used the entropy method to formulate a weighted possibility fuzzy multi-objective and higher order moment portfolio model with the efficiency and effectiveness portfolio selections. Guo et al. [21] considered the capital gain tax to fuzzy portfolio selection and formulated a bi-objective mean-variance model solving by an algorithm in time-varying numerical integral-based particle swarm optimization. Li et al. [22] used a skewness fuzzy variable to formulate a mean-variance-skewness fuzzy portfolio selection, by designing the genetic algorithm and fuzzy simulation technique to show the effective algorithm. Zhou and Xu [23] proposed fuzzy portfolio selection for solving qualitative information represented as hesitant fuzzy elements where both the max-score rule and score-deviation trade-off rule were used to distinguish three types of risk behaviors for the investors. It is important to notice that the risk behavior analysis for an investor is an interesting topic in the research field of fuzzy portfolio selections [24].

In fuzzy portfolio selection, we cannot only use a mean-variance model to reflect the measure of risk; by contrast, we also need to consider the will and behavior of an investor with different risk types in the portfolio selection. Lower returns are equivalent to lower risks, in which investors seldom make significant profits from those securities, and thus most investment behaviors intend to make shortage investment for these securities. By contrast, higher risks are equivalent to higher returns where most investors can realize unexpected returns from those securities, and thus most investors would like to make excess investment to those higher risk securities. Based on the concept of risk behavior of an investor, Tsaur et al. [25] proposed the guaranteed return rate to be the threshold of excess investment for each security in portfolio selection, and then Chen et al. [26] revised the model [25] based on the risk behavior of an investor in a different dimension distance between the guaranteed return rate and return rate for each security. However, models [25,26] just consider the risk behavior of an investor in the excess investment. Not only excess investment but also shortage investment should be considered to the securities. Huang et al. [27] proposed the adjustable security proportion for excess investment and shortage investment based on the selected guaranteed return rates for profitable returns, where the mean-variance model was applied for portfolio selection, whereas the risk behavior of an investor in a different dimension distance for shortage investment and excess investment was still not considered. Therefore, we suppose that if an investor prefers to risk, then his degree of intention in excess investment is higher than the degree of shortage investment; if an investor is averse to risk, then his degree of intention in excess investment is lower than the degree of shortage investment; if an investor is neutral to risk, then his degree of intention in excess investment is equivalent to the degree of shortage investment. The research gap of this study is planned to overcome the degree of risk preference investment in the adjustable security proportion, and then we can consider the dimensions of excess investment and shortage investment by the risk attitudes of an investor.

The organization of this article is as follows. In Section 2, we introduce the definition of fuzzy numbers and their operations. Section 3 proposes the dimensional analysis to the adjustable security proportion in the fuzzy portfolio model. In Section 4, an illustration is presented by the proposed model. Finally, conclusions are discussed in Section 5.

2. Preliminaries

In this section, the fuzzy numbers with fundamental algebraic operations and their defuzzification, fuzzy expected values, and fuzzy variances are introduced and defined, and then Section 3 can be easily understood. A fuzzy set $\widetilde{A}$ is characterized by a membership function defined as $u_{\widetilde{A}}(x) : X \to [0,\ 1]$, which maps the elements of the universe of discourse X to the interval [0, 1]. Therefore, we define a fuzzy number as follows.

Definition 1 ([28]). *Let $\widetilde{A}$ be a fuzzy number as any fuzzy subset of the real line R with a membership function $u_{\widetilde{A}}(x) : R \to [0,\ 1]$ satisfying the following conditions:*

(1) *The fuzzy number $\widetilde{A}$ is normal, if there exists an $x \in R$ with $u_{\widetilde{A}}(x) = 1$;*
(2) *$u_{\widetilde{A}}(x)$ is convex, i.e., $u_{\widetilde{A}}(\lambda x + (1 - \lambda)y) \geq min\{u_{\widetilde{A}}(x), u_{\widetilde{A}}(y)\}$, $\forall\, x, y \in R$ and $\lambda \in [0, 1]$;*
(3) *$u_{\widetilde{A}}(x)$ is upper semicontinuous, i.e., $\{\, x \in R\text{: } u_{\widetilde{A}}\,(x) \geq \alpha\} = \widetilde{A}^{\alpha}$ is a closed subset of U for each $\alpha \in (0, 1]$;*
(4) *The closure of the set $\{x \in R\text{: } u_{\widetilde{A}}\,(x) \geq 0\}$ is a compact subset of R.*

Definition 2 ([28]). *A fuzzy number $\widetilde{A}$ is defined as LR-type fuzzy number as $\widetilde{A} = (a,\ c_1,\ c_2)_{LR}$, then the membership function of $\widetilde{A} = (a,\ c_1,\ c_2)_{LR}$ has the following form:*

$$u_{\widetilde{A}}(x) = \begin{cases} L\left(\frac{a-x}{c_1}\right), & if\ x < a \\ 1, & if\ x = a \\ R\left(\frac{x-a}{c_2}\right), & if\ x > a \end{cases}$$

where a is the central value, and c_1 and c_2 are the left and right spread values.

Let $\widetilde{A}$ and $\widetilde{B}$ be fuzzy numbers of the LR-type defined as $\widetilde{A} = (a,\ c_1,\ c_2)_{LR}$ and $\widetilde{B} = (b,\ d_1,\ d_2)_{LR}$, where a and b are the central values, c_1 and d_1 are the left spread values, and c_2 and d_2 are the right spread values of $\widetilde{A}$ and $\widetilde{B}$, respectively. Then,

$$\widetilde{A} + \widetilde{B} = (a,\ c_1,\ c_2)_{LR} + (b,\ d_1,\ d_2)_{LR} = (a+b,\ c_1+d_1,\ c_2+d_2)_{LR}$$

$$\widetilde{A} - \widetilde{B} = (a,\ c_1,\ c_2)_{LR} - (b,\ d_1,\ d_2)_{LR} = (a-b,\ c_1+d_2,\ c_2+d_1)_{LR}$$

Next, the multiplication of both positive fuzzy numbers $\widetilde{A}$ and $\widetilde{B}$ can be derived as

$$\widetilde{A} \bigotimes \widetilde{B} = (a,\ c_1,\ c_2)_{LR} \bigotimes (b,\ d_1,\ d_2)_{LR} = (ab,\ ad_1 + bc_1,\ ad_2 + bc_2)_{LR}$$

Theorem 1 ([29]). *Let $\widetilde{A}$ be a fuzzy number with differentiable membership function with α-level set $\widetilde{A}^{\alpha} = \{\, x\ \ R:\ u_{\widetilde{A}}\,(x) \geq \alpha\} = [a_1(\alpha),\ a_2(\alpha)]$, $0 \leq \alpha \leq 1$ The lower possibilistic mean value of fuzzy number is defined as $M_*\left(\widetilde{A}\right) = 2\int_0^1 \alpha{\cdot}a_1(\alpha)d\alpha$, and the upper possibilistic mean value of fuzzy number is defined as $M^*\left(\widetilde{A}\right) = 2\int_0^1 \alpha{\cdot}a_2(\alpha)d\alpha$. Then, the expected value of a fuzzy number $\widetilde{A}$ is expressed as $M\left(\widetilde{A}\right) = \int_0^1 \alpha{\cdot}[a_1(\alpha) + a_2(\alpha)]d\alpha$.*

By Theorem 1, the lower possibilistic mean and upper possibilistic mean values for $\widetilde{A}+\widetilde{B}$ can be obtained as Equations (1) and (2) as follows:

$$M_*\left(\widetilde{A}+\widetilde{B}\right)=M_*\left(\widetilde{A}\right)+M_*\left(\widetilde{B}\right) \quad (1)$$

$$M^*\left(\widetilde{A}+\widetilde{B}\right)=M^*\left(\widetilde{A}\right)+M^*\left(\widetilde{B}\right) \quad (2)$$

Then, the sum of possibilistic mean value of $\widetilde{A}$ and $\widetilde{B}$ are obtained as follows:

$$M\left(\widetilde{A}+\widetilde{B}\right)=\frac{M_*\left(\widetilde{A}+\widetilde{B}\right)+M^*\left(\widetilde{A}+\widetilde{B}\right)}{2} \quad (3)$$

Next, the lower and upper possibilistic variances of $\widetilde{A}$ are defined as Equations (4) and (5), respectively [29].

$$Var_*\left(\widetilde{A}\right)=2\int_0^1\alpha\left[M_*\left(\widetilde{A}\right)-a_1(\alpha)\right]^2d\alpha \quad (4)$$

$$Var^*\left(\widetilde{A}\right)=2\int_0^1\alpha\left[M^*\left(\widetilde{A}\right)-a_2(\alpha)\right]^2d\alpha \quad (5)$$

In addition, for ranking the return rate of each security to the guarantee return rate, we use a popular ranking method for fuzzy numbers described as follows:

Theorem 2 ([30]). *Let $\widetilde{A}=(a,\ c_1,\ c_2)$ and $\widetilde{B}=(b,\ d_1,\ d_2)$ be triangular fuzzy numbers, the central values be a and b, and the left and right spread values be $c_1,\ c_2$, and $d_1,\ d_2$; then, we define the circumcenter of $\widetilde{A}$ as $S_{\widetilde{A}}=(\overline{x}_0,\ \overline{y}_0)=\left(\frac{6a+(c_2-c_1)}{6},\ \frac{5-c_2c_1}{12}\right)$. The ranking function $R\left(\widetilde{A}\right)$ which maps $\widetilde{A}$ to a real number can be derived as $R\left(\widetilde{A}\right)=\sqrt{(\overline{x}_0)^2+(\overline{y}_0)^2}$. If the ranking value $R\left(\widetilde{A}\right)$ is bigger than $R\left(\widetilde{B}\right)$, then the fuzzy number $\widetilde{A}$ is bigger than fuzzy number $\widetilde{B}$.*

3. The Dimension Risk Analysis in Adjustable Security Proportions

Under the vagueness environment, the fuzzy portfolio model is used to solve the optimal investment proportion for each asset under the maximizing expected return with constrained risk. By considering the s dimension of excess investment and t dimension of shortage investment, we can formulate the fuzzy portfolio model as follows. First, for security j with investment proportion x_j, we define its return rate to be the triangular fuzzy number as $\widetilde{r}_j=\left(r_j,\ c_j,\ d_j\right)$, where r_j is the central value; and c_j, d_j are left and right spreads, $j=1,\ldots,n$, respectively; and then the expected fuzzy return rate is defined as $\widetilde{R}=\sum_{j=1}^{n}x_j\widetilde{r}_j$. In this study, considering the adjustable security proportion in the fuzzy portfolio selection, the degrees of risk preference for the dimension of excess investment and shortage investment between the selected guaranteed return rates and the security return rates are different to different investors. Second, we rank the n securities to their fuzzy return rates and they are assumed as the ordering of the fuzzy return rates as $\widetilde{r}_1<\widetilde{r}_2<\ldots<\widetilde{r}_n$, in which excess investment for the m securities $(m\leq n)$ and the other securities are made a shortage investment based on the selected guaranteed return rate defined as $\widetilde{p}_k=(p_k,\ e_k,\ f_k)$, where p_k is its central value, and $e_k,\ f_k$ are its left and right spread values, respectively. By considering the risk preference of the investor with s dimension of excess investment and t dimension of shortage investment in the expected fuzzy return rate, the following is proposed:

$$\widetilde{R}=\sum_{j=1}^{n}x_j\widetilde{r}_j-\sum_{j=1}^{n1}\sum_{k=1}^{m}x_j\left|\widetilde{r}_j-\widetilde{p}_k\right|^t+\sum_{j=n1+1}^{n}\sum_{k=1}^{m}x_j\left|\widetilde{r}_j-\widetilde{p}_k\right|^s,\ s,t\geq 1 \quad (6)$$

If the fuzzy return rate $\widetilde{r}_j, j = 1, \ldots, n$, is larger than $\widetilde{p}_k$ and $R(\widetilde{r}_j) > R(\widetilde{p}_k)$ [30], then we can make an excess investment on security j; otherwise, we can make a shortage investment. The s dimension of excess investment and the t dimension of shortage investment can be formulated as follows:

$$|\widetilde{r}_j - \widetilde{p}_k|^t = \begin{cases} (\widetilde{p}_k - \widetilde{r}_j)^t & if\ R(\widetilde{p}_k) > R(\widetilde{r}_j) \\ 0 & otherwise \end{cases} \tag{7}$$

$$|\widetilde{r}_j - \widetilde{p}_k|^s = \begin{cases} (\widetilde{r}_j - \widetilde{p}_k)^s & if\ R(\widetilde{r}_j) > R(\widetilde{p}_k) \\ 0 & otherwise \end{cases} \tag{8}$$

Next, the lower and upper possibilistic mean values for the s dimension excess investment $(\widetilde{r}_j - \widetilde{p}_k)^s$ and t dimension shortage investment $(\widetilde{p}_k - \widetilde{r}_j)^t$ are derived as $M_*(\widetilde{r}_j - \widetilde{p}_k)^s$, $M^*(\widetilde{r}_j - \widetilde{p}_k)^s$, and $M_*(\widetilde{p}_k - \widetilde{r}_j)^t$, $M^*(\widetilde{p}_k - \widetilde{r}_j)^t$, $\forall\, k = 1, 2, \ldots, m$, as follows:

$$M_*(\widetilde{p}_k - \widetilde{r}_j)^t = (p_k - r_j)^t - \frac{1}{3}s(p_k - r_j)^{t-1}(c_j + f_k) \tag{9}$$

$$M^*(\widetilde{p}_k - \widetilde{r}_j)^t = (p_k - r_j)^s + \frac{1}{3}s(p_k - r_j)^{s-1}(d_j + e_k) \tag{10}$$

$$M_*(\widetilde{r}_j - \widetilde{p}_k)^s = (r_j - p_k)^s - \frac{1}{3}s(r_j - p_k)^{s-1}(c_j + f_k) \tag{11}$$

$$M^*(\widetilde{r}_j - \widetilde{p}_k)^s = (r_j - p_k)^s + \frac{1}{3}s(r_j - p_k)^{s-1}(d_j + e_k) \tag{12}$$

where $(\widetilde{p}_k - \widetilde{r}_j)^t = \left[(p_k - r_j)^t, t(p_k - r_j)^{t-1}(c_j + f_k), t(p_k - r_j)^{t-1}(d_j + e_k)\right]$ whose α-level set is defined as $\left[(\widetilde{p}_k - \widetilde{r}_j)^t\right]^\alpha = \left[(\widetilde{p}_k - \widetilde{r}_j)^t_{j1}(\alpha), (\widetilde{p}_k - \widetilde{r}_j)^t_{j2}(\alpha)\right]$ for all $\alpha \in [0, 1]$. $(\widetilde{r}_j - \widetilde{p}_k)^s = \left[(r_j - p_k)^s, s(r_j - p_k)^{s-1}(c_j + f_k), s(r_j - p_k)^{s-1}(d_j + e_k)\right]$ whose α-level set is defined as $\left[(\widetilde{r}_j - \widetilde{p}_k)^s\right]^\alpha = \left[(\widetilde{r}_j - \widetilde{p}_k)^s_{j1}(\alpha), (\widetilde{r}_j - \widetilde{p}_k)^s_{j2}(\alpha)\right]$, $\alpha \in [0, 1]$. Then, the expected possibilistic mean values $M(\widetilde{p}_k - \widetilde{r}_j)^t$ and $M(\widetilde{r}_j - \widetilde{p}_k)^s$ are obtained as follows:

$$M(\widetilde{p}_k - \widetilde{r}_j)^t = (p_k - r_j)^t + \frac{1}{6}t(p_k - r_j)^{t-1}[(d_j + e_k) - (c_j + f_k)] \tag{13}$$

$$M(\widetilde{r}_j - \widetilde{p}_k)^s = (r_j - p_k)^s + \frac{1}{6}s(r_j - p_k)^{s-1}[(d_j + e_k) - (c_j + f_k)] \tag{14}$$

On the other hand, the expected possibilistic mean value for the proposed fuzzy return rate in Equation (6) can be obtained as follows:

$$\begin{aligned} M&\left[\sum_{j=1}^{n} x_j\widetilde{r}_j - \sum_{j=1}^{n1}\sum_{k=1}^{m} x_j(\widetilde{p}_k - \widetilde{r}_j)^t + \sum_{j=n1+1}^{n}\sum_{k=1}^{m} x_j(\widetilde{r}_j - \widetilde{p}_k)^s\right] \\ &= \sum_{j=1}^{n} x_jM(\widetilde{r}_j) - \sum_{j=1}^{n}\sum_{k=1}^{m} x_jM(\widetilde{p}_k - \widetilde{r}_j)^t + \sum_{j=1}^{n}\sum_{k=1}^{m} x_jM(\widetilde{r}_j - \widetilde{p}_k)^s \\ &= \sum_{j=1}^{n} x_j\left[r_j + \tfrac{1}{6}(d_j - c_j)\right] \\ &- \sum_{j=1}^{n1}\sum_{k=1}^{m} x_j\left[(p_k - r_j)^t + \tfrac{1}{6}t(p_k - r_j)^{t-1}[(d_j + e_k) - (c_j + f_k)]\right] \\ &+ \sum_{j=n1+1}^{n}\sum_{k=1}^{m} x_j\left[(r_j - p_k)^s + \tfrac{1}{6}s(r_j - p_k)^{s-1}[(d_j + e_k) - (c_j + f_k)]\right] \end{aligned} \tag{15}$$

Then, we can obtain the lower and upper possibilistic variances of the proposed fuzzy return rates shown in Equation (6) as follows:

$$Var_*\left[\sum_{j=1}^{n} x_j\tilde{r}_j - \sum_{j=1}^{n1}\sum_{k=1}^{m} x_j(\tilde{p}_k-\tilde{r}_j)^t + \sum_{j=n1+1}^{n}\sum_{k=1}^{m} x_j(\tilde{r}_j-\tilde{p}_k)^s\right]$$
$$= \frac{1}{18}\left[\sum_{j=1}^{n} c_jx_j - t\sum_{j=1}^{n1}\sum_{k=1}^{m} x_j(p_k-r_j)^{t-1}(c_j+f_k) + s\sum_{j=n1+1}^{n}\sum_{k=1}^{m} x_j(r_j-p_k)^{s-1}(c_j+f_k)\right]^2 \quad (16)$$

$$Var^*\left[\sum_{j=1}^{n} x_j\tilde{r}_j - \sum_{j=1}^{n1}\sum_{k=1}^{m} x_j(\tilde{p}_k-\tilde{r}_j)^t + \sum_{j=n1+1}^{n}\sum_{k=1}^{m} x_j(\tilde{r}_j-\tilde{p}_k)^s\right]$$
$$= \frac{1}{18}\left[\sum_{j=1}^{n} d_jx_j - t\sum_{j=1}^{n1}\sum_{k=1}^{m} x_j(p_k-r_j)^{t-1}(d_j+e_k) + s\sum_{j=n1+1}^{n}\sum_{k=1}^{m} x_j(r_j-p_k)^{s-1}(d_j+e_k)\right]^2 \quad (17)$$

The standard deviation of the proposed fuzzy return rates shown in Equation (6) can be obtained as follows:

$$SD\left[\sum_{j=1}^{n} x_j\tilde{r}_j - \sum_{j=1}^{n1}\sum_{k=1}^{m} x_j(\tilde{p}_k-\tilde{r}_j)^t + \sum_{j=n1+1}^{n}\sum_{k=1}^{m} x_j(\tilde{r}_j-\tilde{p}_k)^s\right]$$
$$= \frac{1}{2}\left\{\left\{Var_*\left[\sum_{j=1}^{n} x_j\tilde{r}_j - \sum_{j=1}^{n1}\sum_{k=1}^{m} x_j(\tilde{p}_k-\tilde{r}_j)^t + \sum_{j=n1+1}^{n}\sum_{k=1}^{m} x_j(\tilde{r}_j-\tilde{p}_k)^s\right]\right\}^{1/2}\right.$$
$$\left.+\left\{Var^*\left[\sum_{j=1}^{n} x_j\tilde{r}_j - \sum_{j=1}^{n1}\sum_{k=1}^{m} x_j(\tilde{p}_k-\tilde{r}_j)^t + \sum_{j=n1+1}^{n}\sum_{k=1}^{m} x_j(\tilde{r}_j-\tilde{p}_k)^s\right]\right\}^{1/2}\right\} \quad (18)$$
$$= \frac{1}{6\sqrt{2}}\left[\sum_{j=1}^{n}(c_j+d_j)x_j - t\sum_{j=1}^{n1}\sum_{k=1}^{m} x_j(p_k-r_j)^{t-1}(c_j+f_k+d_j+e_k)\right.$$
$$\left.+s\sum_{j=n1+1}^{n}\sum_{k=1}^{m} x_j(r_j-p_k)^{s-1}(c_j+f_k+d_j+e_k)\right]$$

The fuzzy portfolio model with s dimension in the excess investment and t dimension in the shortage investment can be formulated as a linear programming model whose objective function is shown in Equation (15), and the constrained risk by the upper bound of an investor's desired value is shown as in Equation (18). Therefore, the proposed possibilistic mean-standard deviation model of portfolio selection in considering the concept of s dimension excess investment and t dimension in shortage investment is obtained as follows:

$$Max \sum_{j=1}^{n} x_j\left[r_j+\frac{1}{6}(d_j-c_j)\right] - \sum_{j=1}^{n1}\sum_{k=1}^{m} x_j\left[(p_k-r_j)^t+\frac{1}{6}t(p_k-r_j)^{t-1}[(d_j+e_k)-(c_j+f_k)]\right]$$
$$+\sum_{j=n1+1}^{n}\sum_{k=1}^{m} x_j\left[(r_j-p_k)^s+\frac{1}{6}s(r_j-p_k)^{s-1}[(d_j+e_k)-(c_j+f_k)]\right]$$
$$s.t.\ \frac{1}{6\sqrt{2}}\left[\sum_{j=1}^{n}(c_j+d_j)x_j - t\sum_{j=1}^{n1}\sum_{k=1}^{m} x_j(p_k-r_j)^{t-1}(c_j+f_k+d_j+e_k)\right.$$
$$\left.+s\sum_{j=n1+1}^{n}\sum_{k=1}^{m} x_j(r_j-p_k)^{s-1}(c_j+f_k+d_j+e_k)\right] \le \sigma \quad (19)$$
$$\sum_{j=1}^{n} x_j = 1$$
$$l_j \le x_j \le u_j,\ j = 1, 2, \ldots, n$$

where the lower and upper bounds on the proportion of security j are defined as l_j and u_j, respectively. In addition, $r_j < p_k$ when jth security is the shortage investment; therefore, a bigger dimension of t implies a smaller value of $(p_k - r_j)^t$. Furthermore, $r_j > p_k$ when jth security is the excess investment; therefore, a bigger dimension of s implies a smaller value of $\left(r_j - p_k\right)^s$. Therefore, the bigger value of s or smaller value of t will derive a smaller objective value of the model (19).

4. Illustrations

4.1. Data Description and Model Explanation

In this study, we use the collected data from April 2002 to January 2004 in Shanghai Stock Exchange, which are the closed prices for each week [31]. By the companies' information offered in the financial reports, there are five securities chosen to formulate the proposed model. The fuzzy return rates for the securities are estimated as $\widetilde{r}_1 = (0.073, 0.054, 0.087)$, $\widetilde{r}_2 = (0.\ 105, 0.075, 0.102)$, $\widetilde{r}_3 = (0.138, 0.096, 0.123)$, $\widetilde{r}_4 = (0.168, 0.126, 0.162)$, and $\widetilde{r}_5 = (0.208, 0.168, 0.213)$, where the first values in the fuzzy return rates are central values, and the second and third values are left and right spread values. In order to range the investment proportion for each security, the lower and upper bounds of investment proportion for security j are derived as $(l_1, l_2, l_3, l_4, l_5) = (0.1, 0.1, 0.1, 0.1, 0.1)$, and $(u_1, u_2, u_3, u_4, u_5) = (0.4, 0.4, 0.4, 0.5, 0.6)$, respectively.

4.2. Results and Discussions

To clearly describe the proposed model, we select the guaranteed return rates to group the fuzzy return rate of the securities. In the first group, we select fuzzy number $\widetilde{p}_1 = (0.1,\ 0.05,\ 0.05)$ which is bigger than $\widetilde{r}_1$; $\widetilde{p}_2 = (0.15,\ 0.1,\ 0.1)$ is bigger than $\widetilde{r}_1$, $\widetilde{r}_2$, and $\widetilde{r}_3$, and $\widetilde{p}_3 = (0.2,\ 0.1,\ 0.15)$ is just smaller than $\widetilde{r}_5$, which are all derived by Theorem 2. Therefore, we can define the first scenario when we select the guaranteed return rate as $\widetilde{p}_1$, where security 1 is set for the shortage investment because its fuzzy return rates are lower than the guaranteed return rate $\widetilde{p}_1$, whereas the other securities 2, 3, 4, and 5 are for excess investment. In the second scenario, securities 1, 2, 3 are the shortage investment because their fuzzy return rates are lower than the guaranteed rate of return $\widetilde{p}_2$; by contrast, the fuzzy return rates of securities 4 and 5 are more $\widetilde{p}_2$ for excess investment. The third scenario shows the securities 1, 2, 3, and 4 to be the shortage investments because their fuzzy return rates are less than the guaranteed return rate $\widetilde{p}_3$, and then we judge security 5 to be the excess investment. In order to clearly state the proposed model, three experiments are conducted for illustration.

4.2.1. Experiment 1

In this experiment, we suppose the risk behavior of an investor is risk-seeking, and he prefers excess investment to shortage investment. The fuzzy portfolio model shown in model (19) assumed the dimensions of shortage to be bigger than excess investments as $t > s$. The fuzzy portfolio selection is proceeded by the following steps.

Step 1: Formulate a linear programming model for the proposed fuzzy portfolio model

First, the guaranteed return rate $\widetilde{p}_1 = (0.1,\ 0.05,\ 0.05)$ is used to group the securities to be shortage or excess investments. Second, the dimension for the shortage investment is set as $t = 2$, and the dimension for excess investment is set as $s = 1$. Third, we formulate model (19) by the collected data, and security 1 is adopted as the shortage investment, and thus the lowermost investment proportion is relaxed from 0.1 to 0. Therefore, the fuzzy portfolio model with $t = 2$ and $s = 1$ can be obtained as follows:

$$\begin{aligned} &Max\ 0.077474x_1 + 0.119x_2 + 0.185x_3 + 0.248x_4 + 0.331x_5 \\ &s.t.\ 0.127986x_1 + 0.454x_2 + 0.538x_3 + 0.676x_4 + 0.862x_5 \leq 6\sqrt{2}\sigma \\ &x_1 + x_2 + x_3 + x_4 + x_5 = 1 \\ &0 \leq x_1 \leq 0.4\ ,\ 0.1 \leq x_2, x_3 \leq 0.4; 0.1 \leq x_4 \leq 0.5; 0.1 \leq x_5 \leq 0.6 \end{aligned} \tag{20}$$

Step 2: Discussion and analysis

After solving model (20), with the constrained risks from 5% to 9%, we can solve the portfolio selections as shown in Table 1. If the constrained risk is smaller than 5%, then the portfolio is infeasible. On the other hand, if the constrained risk is bigger than 9%, then its optimal portfolio remains the same with the optimal portfolio as $x_1 = 0$, $x_2 = 0.1$, $x_3 = 0.1$, $x_4 = 0.2$, and $x_5 = 0.6$, and the expected return rate is 27.86%. With the constrained risk from 5% to 9%, we can find that the investment proportion of security 1 is from its upper bound 0.4 to the shortage investment proportion 0 because the return rate of security 1 is lower than the guaranteed return rate $\widetilde{p}_1$; the investment proportions for securities 2 and 3 are almost the same between the constrained risk from 5% to 9%, and the investment proportion for security 4 finally reaches 0.2 in the increasing process when the proportion of security 5 reaches 0.6. Next, we change the selected guaranteed return rates to $\widetilde{p}_2$ and $\widetilde{p}_3$, respectively. In Table 2, with $t = 2$, $s = 1$, and the guaranteed return rate $\widetilde{p}_2$, the risk of the investment is constrained from 2% to 5%. The optimal portfolio is obtained as $x_1 = 0$, $x_2 = 0$, $x_3 = 0$, $x_4 = 0.4$, and $x_5 = 0.6$ and the expected return rate is 24.78% under the constrained risk of 5%. With the constrained risk from 2% to 5%, we can find that the investment proportion of securities 1, 2, and 3 reach at their lower bounds as 0 in shortage investment because their return rates are less than the guaranteed return rate $\widetilde{p}_2$. In addition, we can find that investment proportions for securities 4 and 5 are increasing between 2% and 5%, because the return rates are higher than the guaranteed return rate $\widetilde{p}_2$. Furthermore, in Table 3, by the constrained risk from 1.5% to 4.5%, we can obtain the optimal portfolio. The optimal portfolio in the maximal expected returns is obtained as $x_1 = 0$, $x_2 = 0$, $x_3 = 0$, $x_4 = 0.4$, and $x_5 = 0.6$, in which the expected return rate is 20.285% under the constrained risk of 4.5%. By the constrained risk from 1.5% to 4.5%, the investment proportions of securities 1, 2, and 3 reach to their lower bounds as 0 in the shortage investment because their return rates are less than the guaranteed return rate $\widetilde{p}_2$. In addition, the investment proportion for security 4 is also relaxed to the shortage investment but reaches the investment proportion of 0.4, since the expected return rate of security 4 is bigger than securities 1, 2, and 3. By contrast, security 5 is in the increasing process between 1.5% and 4.5%, because its return rate is bigger than the guaranteed return rate $\widetilde{p}_3$.

Next, we solve the proposed model with $t = 3$, and $s = 2$, and the results are shown in Tables 4–6 under the guaranteed return rate $\widetilde{p}_1$, $\widetilde{p}_2$, and $\widetilde{p}_3$. Since we add one dimension to the shortage investment and excess investment, by comparing Tables 1–3 to Tables 4–6, we can find that the pattern to obtain the portfolio selection under the constrained risk is similar. However, we can observe two differences in the changed dimension. First, with the increase in the dimension, the maximal expected return rate in the largest constrained risk is lower than the lower dimension results in different guaranteed return rates $\widetilde{p}_1$, $\widetilde{p}_2$, and $\widetilde{p}_3$. Second, the proportion for each security in the shortage investment can be found to be 0 quickly; by contrast, the proportion for each security in the excess investment can be found quickly under the maximum constrained risk.

Table 1. The dimension with $t = 2$, and $s = 1$ with a guaranteed return rate $\widetilde{p}_1$ in the proposed model.

Constrained Risk / Security Proportions	4.5%	5%	5.5%	6%	6.5%	7%	7.5%	8%	8.5%	9%
x_1	Infeasible Solution	0.4	0.4	0.3557	0.2979	0.2401	0.1823	0.1245	0.1245	0
x_2		0.1	0.1	0.1	0.1	0.1	0.1	0.1	0.1	0.1
x_3		0.2096	0.1	0.1	0.1	0.1	0.1	0.1	0.1	0.1
x_4		0.1904	0.1532	0.1	0.1	0.1	0.1	0.1	0.1	0.2
x_5		0.1	0.2468	0.3443	0.4021	0.4599	0.5177	0.5755	0.5755	0.6
Expected Return Rates		0.16198	0.18107	0.19672	0.21138	0.22603	0.4069	0.25534	0.25534	0.27860

Table 2. The dimension with $t = 2$, and $s = 1$ with a guaranteed return rate $\widetilde{p}_2$ in the proposed model.

Constrained Risk / Security Proportions	1.5%	2%	2.5%	3%	3.5%	4%	4.5%	5%
x_1	Infeasible Solution	0.3963	0.2718	0.1474	0.0229	0	0	0
x_2		0.4	0.4	0.4	0.4	0.2093	0	0
x_3		0	0	0	0	0.0907	0.0608	0
x_4		0.1	0.1	0.1	0. 1	0.1	0.3392	0.4
x_5		0.1037	0.2282	0.3526	0.4771	0.6	0.6	0.6
Expected Return Rates		0.12020	0.14624	0.17229	0.19833	0.22371	0.24441	0.24780

Table 3. The dimension with $t = 2$, and $s = 1$ with a guaranteed return rate $\widetilde{p}_3$ in the proposed model.

Constrained Risk / Security Proportions	1%	1.5%	2%	2.5%	3%	3.5%	4%	4.5%
x_1	Infeasible Solution	0.2925	0.008	0	0	0	0	0
x_2		0.4	0.4	0.1417	0	0	0	0
x_3		0.2075	0.4	0.4	0.3264	0.1708	0.0153	0
x_4		0	0.092	0.3583	0.5	0.5	0.5	0.4
x_5		0.1	0.1	0.1	0.1736	0.3292	0.4847	0.6
Expected Return Rates		0.11008	0.13483	0.15429	0.17063	0.18363	0.19662	0.20285

Table 4. The dimension with $t = 3$ and $s = 2$ with a guaranteed return rate $\widetilde{p}_1$ in the proposed model.

Constrained Risk / Security Proportions	2.5%	3%	3.5%	4%	4.5%	5%
x_1	Infeasible Solution	0	0	0	0	0
x_2		0.3827	0.1192	0.1	0.1	0.1
x_3		0.4	0.4	0.1776	0.1	0.1
x_4		0.1173	0.3808	0.5	0.3357	0.2
x_5		0.1	0.1	0.2224	0.4643	0.6
Expected Return Rates		0.14357	0.16199	0.17719	0.19185	0.19854

Table 5. The dimension with $t = 3$ and $s = 2$ with a guaranteed return rate $\widetilde{p}_2$ in the proposed model.

Constrained Risk / Security Proportions	2%	2.5%	3%	3.5%	4%	4.5%	5%
x_1	Infeasible Solution	0.2465	0	0	0	0	0
x_2		0.4	0.2338	0	0	0	0
x_3		0.1535	0.4	0.3485	0.1637	0	0
x_4		0.1	0.2662	0.5	0.5	0.4660	0.4
x_5		0.1	0.1	0.1515	0.3363	0.5340	0.6
Expected Return Rates		0.12427	0.15101	0.17022	0.18450	0.19867	0.20166

4.2.2. Experiment 2

In this experiment, the testing focuses on $t = s$ for risk-neutral. We first solve the proposed model with $t = s = 2$, and the results are shown in Tables 7–9 under the guaranteed return rate $\widetilde{p}_1$, $\widetilde{p}_2$, and $\widetilde{p}_3$. By comparing Tables 1–3 to Tables 7–9, we can find that the expected return rate under the constrained risk is lower when we add one dimension to the excess investment. By increasing the dimension in excess investment, the investment proportions in higher return rate securities 4 and 5 offer more stable results in each constrained risk. Next, by comparing Tables 4–6 to Tables 7–9, we can find that the major difference between (Tables 4 and 5) and (Tables 7 and 8) are Tables 7 and 8 can solve the portfolio under lower constrained risks 2.5% and 2%, respectively. That is, we can solve the portfolio under a lower constrained risk when the dimension of shortage investment is lower. By contrast, compared to Tables 6 and 9, we can find that when we have more securities in shortage investment, a higher dimension in shortage investment enlarges the feasible region of model (19); therefore, we derive portfolio selections under a higher constrained risk than the lower dimension in the shortage investment. On the other hand, we solve the proposed model with $t = s = 3$, and the results are shown in Tables 10–12 under the guaranteed return rate $\widetilde{p}_1$, $\widetilde{p}_2$, and $\widetilde{p}_3$. By comparing Tables 7–9 to Tables 10–12, we can find that higher dimensions to t and s contribute to the narrower feasible region for the portfolio selection, and we can solve the portfolio under smaller constrained risks.

4.2.3. Experiment 3

In this experiment, the testing focuses on $t < s$ for the risk-averse. First, we solve the proposed model with $t = 2$ and $s = 3$, and the results are shown in Tables 13–15 under the guaranteed return rate $\widetilde{p}_1$, $\widetilde{p}_2$, and $\widetilde{p}_3$. It shows that our proposed model can be used to solve model (19) and obtain the efficient portfolio under different constrained risks. Furthermore, by comparing Tables 7–9 to Tables 13–15, we can find three major differences when we add one dimension to the excess investment. First, by increasing the dimension in the excess investment, the maximal expected return rate obtained in the largest constrained risk is lower than the lower dimension results in different guaranteed return rates $\widetilde{p}_1$, $\widetilde{p}_2$, and $\widetilde{p}_3$. From experiments 1 to 3, we can find that the higher dimensions in excess investments derive a lower expected return rate than lower dimensions in excess investments. Second, the lower guaranteed return rate $\widetilde{p}_i$ can obtain a higher expected return rate than the higher guaranteed return rate $\widetilde{p}_j$, $i, j = 1, 2, 3, 4, 5$. Third, through the higher dimension in excess investment, we can find that the higher return rate securities can be quick to reach their maximal investment proportion. On the other hand, we solve the proposed model with $t = 3$ and $s = 5$, and the results are shown in Tables 16–18 under the guaranteed return rate $\widetilde{p}_1$, $\widetilde{p}_2$, and $\widetilde{p}_3$. By comparing Tables 16–18 to Tables 4–6 ($t = 3, s = 2$) and Tables 10–12, ($t = 3$, $s = 3$), we can find that too big of an s dimension for excess investment does not make a significant difference to the portfolio selection. That is, too big of a dimension of s in excess investment makes almost no change to the objective function and the constrained risk, and thus, we suggest the values of t and s are, at most, 3.

Table 6. The dimension with $t = 3$ and $s = 2$ with a guaranteed return rate $\widetilde{p}_3$ in the proposed model.

Constrained Risk / Security Proportions	2%	2.5%	3%	3.5%	4%	4.5%
x_1	Infeasible Solution	0.044	0	0	0	0
x_2		0.4	0.1089	0	0	0
x_3		0.4	0.4	0.2351	0	0
x_4		0.056	0.3911	0.5	0.4934	0.4
x_5		0.1	0.1	0.2649	0.5066	0.6
Expected Return Rates		0.13509	0.15836	0.17754	0.19504	0.19892

Table 7. The dimension with $t = 2$ and $s = 2$ with a guaranteed return rate $\tilde{p}_1$ in the proposed model.

Constrained Risk / Security Proportions	2%	2.5%	3%	3.5%	4%	4.5%	5%
x_1	Infeasible Solution	0.3978	0	0	0	0	0
x_2		0.3022	0.3827	0.1192	0.1	0.1	0.1
x_3		0.1	0.4	0.4	0.1776	0.1	0.1
x_4		0.1	0.1173	0.3808	0.5	0.3357	0.2
x_5		0.1	0.1	0.1	0.2224	0.4643	0.6
Expected Return Rates		0.11918	0.14357	0.16199	0.17719	0.19185	0.19854

Table 8. The dimension with $t = 2$ and $s = 2$ with a guaranteed return rate $\tilde{p}_2$ in the proposed model.

Constrained Risk / Security Proportions	1.5%	2%	2.5%	3%	3.5%	4%	4.5%	5%
x_1	Infeasible Solution	0.3228	0.1273	0	0	0	0	0
x_2		0.4	0.4	0.3569	0.2332	0.1095	0	0
x_3		0	0	0.1	0	0	0	0
x_4		0.1772	0.3727	0.5	0.5	0.5	0.466	0.4
x_5		0.1	0.1	0.1431	0.2668	0.3905	0.5340	0.6
Expected Return Rates		0.11889	0.13898	0.15693	0.17086	0.18480	0.19867	0.20166

Table 9. The dimension with $t = 2$ and $s = 2$ with a guaranteed return rate $\tilde{p}_3$ in the proposed model.

Constrained Risk / Security Proportions	2%	2.5%	3%	3.5%	4%	4.5%	5%
x_1	Infeasible Solution	0.4	0.3924	0.2931	0.1938	0.0944	0
x_2		0.315	0	0	0	0	0
x_3		0	0	0	0	0	0
x_4		0	0.0076	0.1069	0.2062	0.3056	0.4
x_5		0.285	0.6	0.6	0.6	0.6	0.6
Expected Return Rates		0.11854	0.15540	0.16633	0.17726	0.18819	0.19858

Table 10. The dimension with $t = 3$ and $s = 3$ with a guaranteed return rate $\tilde{p}_1$ in the proposed model.

Constrained Risk / Security Proportions	2%	2.5%	3%	3.5%	4%
x_1	Infeasible Solution	0.1996	0	0	0
x_2		0.4	0.1725	0.1	0.1
x_3		0.2004	0.4	0.1496	0.1
x_4		0.1	0.3275	0.5	0.2
x_5		0.1	0.1	0.2504	0.6
Expected Return Rates		0.127134	0.154601	0.173614	0.190217

Table 11. The dimension with $t = 3$ and $s = 3$ with a guaranteed return rate $\widetilde{p}_2$ in the proposed model.

Security Proportions \ Constrained Risk	2%	2.5%	3%	3.5%	4%	4.5%
x_1	Infeasible Solution	0.0045	0	0	0	0
x_2		0.3955	0.0525	0	0	0
x_3		0.4	0.4	0.4	0.0867	0
x_4		0.1	0.4475	0.1337	0.3133	0.4
x_5		0.1	0.1	0.4663	0.6	0.6
Expected Return Rates		0.139588	0.162187	0.180877	0.196336	0.199067

Table 12. The dimension with $t = 3$ and $s = 3$ with a guaranteed return rate $\widetilde{p}_3$ in the proposed model.

Security Proportions \ Constrained Risk	2%	2.5%	3%	3.5%	4%
x_1	Infeasible Solution	0.2001	0	0	0
x_2		0	0	0	0
x_3		0.4	0.3029	0.0783	0
x_4		0.2990	0.5	0.5	0.4
x_5		0.1	0.1971	0.4217	0.6
Expected Return Rates		0.145973	0.172565	0.189003	0.19889

Table 13. The dimension with $t = 2$ and $s = 3$ with a guaranteed return rate $\widetilde{p}_1$ in the proposed model.

Security Proportions \ Constrained Risk	2%	2.5%	3%	3.5%	4%
x_1	Infeasible Solution	0.3155	0	0	0
x_2		0.1	0.1741	0.1	0.1
x_3		0.3845	0.4	0.1523	0.1
x_4		0.1	0.3259	0.5	0.2
x_5		0.1	0.1	0.2477	0.6
Expected Return Rates		0.12935	0.15463	0.17362	0.19030

Table 14. The dimension with $t = 2$ and $s = 3$ with a guaranteed return rate $\widetilde{p}_2$ in the proposed model.

Security Proportions \ Constrained Risk	1.5%	2%	2.5%	3%	3.5%	4%	4.5%
x_1	Infeasible Solution	0.2768	0.0647	0	0	0	0
x_2		0.4	0.4	0.4	0.2759	0.0442	0
x_3		0	0	0	0	0	0
x_4		0.2232	0.4353	0.2003	0.1241	0.3558	0.4
x_5		0.1	0.1	0.3997	0.6	0.6	0.6
Expected Return Rates		0.12309	0.14480	0.16393	0.18060	0.19611	0.19907

Table 15. The dimension with $t = 2$ and $s = 3$ with a guaranteed return rate $\tilde{p}_3$ in the proposed model.

Security Proportions \ Constrained Risk	2%	2.5%	3%	3.5%	4%	4.5%	5%
x_1	Infeasible Solution	0.4	0.3784	0.2791	0.1798	0.0804	0
x_2		0.2902	0	0	0	0	0
x_3		0	0	0	0	0	0
x_4		0	0.0216	0.12	0.2202	0.3196	0.4
x_5		0.3098	0.6	0.6	0.6	0.6	0.6
Expected Return Rates		0.12137	0.15691	0.16784	0.17877	0.18970	0.19855

Table 16. The dimension with $t = 3$ and $s = 5$ with a guaranteed return rate $\tilde{p}_1$ in the proposed model.

Security Proportions \ Constrained Risk	2%	2.5%	3%	3.5%	4%
x_1	Infeasible Solution	0.1682	0	0	0
x_2		0.4	0.1369	0.1	0.1
x_3		0.2318	0.4	0.1064	0.1
x_4		0.1	0.3631	0.5	0.2
x_5		0.1	0.1	0.2936	0.6
Expected Return Rates		0.128984	0.156726	0.176392	0.189312

Table 17. The dimension with $t = 3$ and $s = 5$ with a guaranteed return rate $\tilde{p}_2$ in the proposed model.

Security Proportions \ Constrained Risk	2%	2.5%	3%	3.5%	4%	4.5%
x_1	Infeasible Solution	0	0	0	0	0
x_2		0.3960	0.0460	0	0	0
x_3		0.4	0.4	0.4	0.0474	0
x_4		0.104	0.4540	0.1038	0.3526	0.4
x_5		0.1	0.1	0.4962	0.6	0.6
Expected Return Rates		0.139961	0.162579	0.18199	0.197406	0.198901

Table 18. The dimension with $t = 3$ and $s = 5$ with a guaranteed return rate $\tilde{p}_3$ in the proposed model.

Security Proportions \ Constrained Risk	2%	2.5%	3%	3.5%	4%
x_1	Infeasible Solution	0.2	0	0	0
x_2		0	0	0	0
x_3		0.4	0.3028	0.0781	0
x_4		0.3	0.5	0.5	0.4
x_5		0.1	0.1972	0.4219	0.6
Expected Return Rates		0.145982	0.172575	0.189023	0.19889

Finally, we list two figures under the guaranteed return rate $\tilde{p}_2$ in different dimensions of s and t. In Figure 1, we find that when $t = 2$ is fixed, the increasing dimension of excess investment forms 1 to 3, where $s = 1$ implies under the same risk, and $t = 2$, $s = 1$ have the biggest expected return rate. Therefore, the risk-seeker should select lower dimension s.

Next, in Figure 2, with a guaranteed return rate $\widetilde{p}_2$, we would like to show that when we increase t and fix it to 3, the dimension of excess investment is increasing from 2 to 5, and too big of a value of s cannot offer any useful information for the expected return rate under the same constrained risk, that is, when we adopt $s = 5$ whose results are almost the same as $s = 3$. Therefore, we suggest the values of s and t be smaller than or equal to 3.

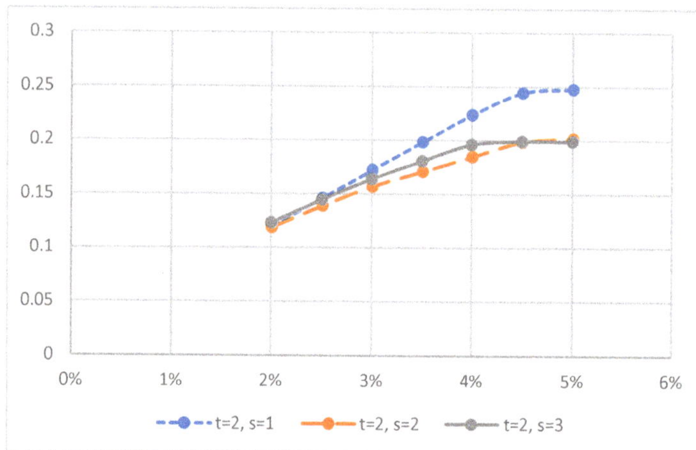

Figure 1. The dimension $t = 2$ with different s under guaranteed return rate $\widetilde{p}_2$.

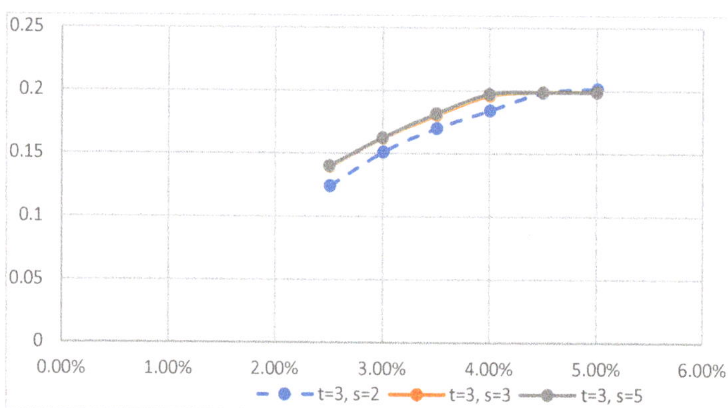

Figure 2. The dimension $t = 3$ with different s under guaranteed return rate $\widetilde{p}_2$.

5. Conclusions

Fuzzy portfolio models have led to a continual increase in the field of single-period or multi-period topics, indirectly resulting in many researchers focusing on the issue of the risk preferences of investors. Some investors might have the challenge of evaluating a better portfolio selection based on the profitable selecting security. Therefore, a method for selecting the most appropriate portfolio based on the guaranteed return rate would be extremely beneficial to these investors, in which any security whose expected return is bigger than the guaranteed return rate will be assumed for excess investment to this security. On the other hand, any security whose expected return is smaller than the guaranteed return rate will be assumed for shortage investment to this security. The present study included different degrees of dimensions to the securities in excess investment or shortage investment that investors expect of maximization of expected return rate and developed a novel decision-making procedure for portfolio selection under the constrained risk. Based on risk preferences, including risk-seeking, risk-neutral, and risk-averse, three kinds of fuzzy portfolio selections comprising different degrees of dimensions in excess investment and shortage investment were established for most of investors. Analysis results indicated that, when using the proposed model, a defuzzy method is required for the ranking between the expected return of each security and the guaranteed return rates. Subsequently, we can decide some securities are for excess investments, and the other securities are for shortage investments. A comparison of the degree of dimensions for the excess investment and shortage investment indicates that a risk-seeker would like to have excess investment for securities whose return rates are bigger than the guaranteed return rates; therefore, a lower value of s is suggested. Then he reduces the security investments whose return rates are lower than the guaranteed return rates; therefore, a bigger value of t is suggested. Next, a risk-seeker will adopt $t > s$, and we suggest the values of s and t to be smaller than or equal to 3. By contrast, for the risk-neutral investor, we suggest $s = t$; and $t < s$ is suggested to the investor who is risk-averse. Lower dimensions in s and t indicate a bigger objective value and feasible region in the linear programming model from the proposed fuzzy portfolio model, and thus we can derive bigger expected return rates from the invested securities. The results suggest that the proposed fuzzy portfolio model can clearly distinguish the relative importance from the ranking results compared to the guaranteed return rate. Finally, using the proposed model, investors could individually select the portfolio for the subjective risk preference, and easily evaluate and analyze the optimized portfolio with ease and convenience, without having to query the experts.

In this study, we expect more investors can be recruited to participate in evaluating and comparing the effects for the dimensions of excess investments and shortage in-vestments. Because risk preferences and standards may differ depending on the perceived risk in the investment, a collaborative discussion involving numerous experts is required to include in the evaluation and selection process for establishing the guaranteed return rate. Therefore, future research should focus on (1) expanding the number of investors, and (2) establishing comprehensive guaranteed return rates according to various experts' opinions on economy trends and business cycling. Furthermore, (3) an investor has a different risk attitude to select a portfolio in a different time period; therefore, multi-period fuzzy portfolio selection in different time periods should be considered with our proposed model.

Author Contributions: Conceptualization, K.-S.C. and R.-C.T.; methodology, R.-C.T. and N.-Y.L.; software, N.-Y.L. and Y.-Y.H.; supervision, K.-S.C. and R.-C.T.; validation, R.-C.T. and Y.-Y.H.; writing—original draft, N.-Y.L. All authors have read and agreed to the published version of the manuscript.

Funding: This research was funded with the financial support from National Science and Technology council with project No. MOST 110-2221-E-032-033.

Institutional Review Board Statement: Not applicable.

Informed Consent Statement: Not applicable.

Data Availability Statement: Not applicable.

Acknowledgments: The authors would like to thank the referees for their helpful comments.

Conflicts of Interest: The authors declare no conflict of interest.

References

1. Markowitz, H. Portfolio selection. *J. Financ.* **1952**, *7*, 77–91.
2. Sharpe, W.F. *Portfolio Theory and Capital Markets*; McGraw-Hill: New York, NY, USA, 1970.
3. Merton, R.C. An analytic derivation of the efficient frontier. *J. Financ. Quant. Anal.* **1972**, *10*, 1851–1872. [CrossRef]
4. Pang, J.S. A new efficient algorithm for a class of portfolio selection problems. *Oper. Res.* **1980**, *28*, 754–767. [CrossRef]
5. Perold, A.F. Large-scale portfolio optimization. *Manag. Sci.* **1984**, *30*, 1143–1160. [CrossRef]
6. Vörös, J. Portfolio analysis—An analytic derivation of the efficient portfolio frontier. *Eur. J. Oper. Res.* **1990**, *203*, 294–300. [CrossRef]

7. Best, M.J.; Grauer, R.R. The efficient set mathematics when mean–variance problems are subject to general linear constrains. *J. Econ. Bus.* **1990**, *42*, 105–120. [CrossRef]
8. Best, M.J.; Hlouskova, J. The efficient frontier for bounded assets. *Math. Method Oper. Res.* **2000**, *52*, 195–212. [CrossRef]
9. Gupta, P.; Mehlawat, M.K.; Yadav, S.; Kumar, A. A polynomial goal programming approach for intuitionistic fuzzy portfolio optimization using entropy and higher moments. *Appl. Soft Comput.* **2019**, *85*, 105781. [CrossRef]
10. Wu, Y.; Xu, C.; Ke, Y.; Tao, Y.; Li, X. Portfolio optimization of renewable energy projects under type-2 fuzzy environment with sustainability perspective. *Comput. Ind. Eng.* **2019**, *133*, 69–82. [CrossRef]
11. Mansour, N.; Cherif, M.S.; Abdelfattah, W. Multi-objective imprecise programming for financial portfolio selection with fuzzy returns. *Expert Syst. Appl.* **2019**, *138*, 112810. [CrossRef]
12. Yue, W.; Wang, Y.; Xuan, H. Fuzzy multi-objective portfolio model based on semi-variance–semi-absolute deviation risk measures. *Soft Comput.* **2019**, *23*, 8159–8179. [CrossRef]
13. Liagkouras, K.; Metaxiotis, K. Multi-period mean–variance fuzzy portfolio optimization model with transaction costs. *Eng. Appl. Artif. Intell.* **2018**, *67*, 260–269. [CrossRef]
14. Gupta, P.; Mehlawat, M.K.; Khan, A.Z. Multi-period portfolio optimization using coherent fuzzy numbers in a credibilistic environment. *Expert Syst. Appl.* **2021**, *167*, 114135. [CrossRef]
15. Gupta, P.; Mehlawat, M.K.; Yadav, S.; Kumar, A. Intuitionistic fuzzy optimistic and pessimistic multi-period portfolio optimization models. *Soft Comput.* **2020**, *24*, 11931–11956. [CrossRef]
16. Garcia, F.; González-Bueno, J.; Guijarro, F.; Oliver, J.; Tamošiūnienė, R. Multiobjective approach to portfolio optimization in the light of the credibility theory. *Technol. Econ. Dev. Econ.* **2020**, *26*, 1165–1186. [CrossRef]
17. Gupta, P.; Mehlawat, M.K.; Kumar, A.; Yadav, S.; Aggarwal, A. A credibilistic fuzzy DEA approach for portfolio efficiency evaluation and rebalancing toward benchmark portfolios using positive and negative returns. *Int. J. Fuzzy Syst.* **2020**, *22*, 824–843. [CrossRef]
18. Mehlawat, M.K.; Gupta, P.; Kumar, A.; Yadav, S.; Aggarwal, A. Multi-objective fuzzy portfolio performance evaluation using data envelopment analysis under credibilistic framework. *IEEE Trans. Fuzzy Syst.* **2020**, *11*, 2726–2737. [CrossRef]
19. Mehralizade, R.; Amini, M.; Gildeh, B.S.; Ahmadzade, H. Uncertain random portfolio selection based on risk curve. *Soft Comput.* **2020**, *24*, 13331–13345. [CrossRef]
20. Yue, W.; Wang, Y. A new fuzzy multi-objective higher order moment portfolio selection model for diversified portfolios. *Phys. A Stat. Mech. Its Appl.* **2017**, *465*, 124–140. [CrossRef]
21. Guo, S.; Ching, W.-K.; Li, W.-K.; Siu, T.-K.; Zhang, Z. Fuzzy hidden Markov-switching portfolio selection with capital gain tax. *Expert Syst. Appl.* **2020**, *149*, 113304. [CrossRef]
22. Li, X.; Qin, Z.; Kar, S. Mean-variance-skewness model for portfolio selection with fuzzy returns. *Eur. J. Oper. Res.* **2010**, *202*, 239–247. [CrossRef]
23. Zhou, W.; Xu, Z. Portfolio selection and risk investment under the hesitant fuzzy environment. *Knowl.-Based Syst.* **2018**, *144*, 21–31. [CrossRef]
24. Michalski, G. Portfolio Management Approach in Trade Credit Decision Making. *Romanian J. Econ. Forecast.* **2007**, *3*, 42–53. [CrossRef]
25. Tsaur, R.C.; Chiu, C.-L.; Huang, Y.-Y. Guaranteed rate of return for excess investment in a fuzzy portfolio analysis. *Int. J. Fuzzy Syst.* **2021**, *23*, 94–106. [CrossRef]
26. Chen, K.-S.; Tsaur, R.C.; Lin, N.-C. Dimensions analysis to excess investment in fuzzy portfolio model from the threshold of guaranteed return rates. *Mathematics* **2023**, *11*, 44. [CrossRef]
27. Huang, Y.-Y.; Chen, I.-F.; Chiu, C.-L.; Tsaur, R.-C. Adjustable security proportions in the fuzzy portfolio selection under guaranteed return rates. *Mathematics* **2021**, *9*, 3026. [CrossRef]
28. Zimmermann, H.-J. *Fuzzy Set Theory—And Its Applications*; Springer: Berlin/Heidelberg, Germany, 2011.
29. Carlsson, C.; Fullér, R. On possibilistic mean value and variance of fuzzy numbers. *Fuzzy Sets Syst.* **2001**, *122*, 315–326. [CrossRef]
30. Rao, P.P.B.; Shankar, N.R. Ranking fuzzy numbers with a distance method using circumcenter of centroids and an index of modality. *Adv. Fuzzy Syst.* **2011**, *2011*, 1–7. [CrossRef]
31. Zhang, W.G. Possibilistic mean–standard deviation models to portfolio selection for bounded assets. *Comput. Appl. Math.* **2007**, *189*, 1614–1623. [CrossRef]

mathematics

MDPI

Article

A Fuzzy AHP-Fuzzy TOPSIS Urged Baseline Aid for Execution Amendment of an Online Food Delivery Affability

Harshitha Urs Ajjipura Shankar [1], Udaya Kumara Kodipalya Nanjappa [1], M. D. Alsulami [2] and Ballajja C. Prasannakumara [3,*]

[1] Department of Mathematics, Reva University, Bangalore 560064, India
[2] Mathematics Department, College of Sciences and Arts at Alkamil, University of Jeddah, Jeddah 21589, Saudi Arabia
[3] Department of Studies in Mathematics, Davangere University, Davangere 577002, India
* Correspondence: prasannakumarabc@davangereuniversity.ac.in

Abstract: The increased demand for this form of food delivery has been expected to drastically alter restaurant patrons' dining habits. As people have been forced to stay indoors to prevent the virus from spreading, food delivery services over the internet are in high demand. As established in this study, the planned ideal is a good executive implementation for online meal delivery services. Food delivery services are rapidly growing in India, opening up several opportunities for a wide range of online food delivery (OFD) platforms while also generating a competitive commercial sector. Following that, the fuzzy technique for order performance by similarity to ideal solution method (FTOPSIS) is used to rank online food delivery (OFD) enterprises based on the characteristics chosen. In this paper, we study the present multi-criteria decision-analysis (MCDA) paradigm based on the fuzzy analytic hierarchy process (FAHP) and the fuzzy technique for order performance by similarity to ideal solution (FTOPSIS) method to achieve the goal. After that, a hierarchy multiple criteria decision-analysis (MCDA) model based on fuzzy sets theory is introduced to deal with the online food delivery Service in the chain system. The fuzzy analytic hierarchy process (FAHP) is a fuzzy set theory technique for generating criteria weights, which are then used to interpret expert phonological evaluation statements.

Keywords: online food delivery; multi-criteria decision-analysis; fuzzy analytic hierarchy process; fuzzy technique for order performance by similarity to ideal solution method

MSC: 90C29; 90C31; 91A35; 91B06

Citation: Ajjipura Shankar, H.U.; Kodipalya Nanjappa, U.K.; Alsulami, M.D.; Prasannakumara, B.C. A Fuzzy AHP-Fuzzy TOPSIS Urged Baseline Aid for Execution Amendment of an Online Food Delivery Affability. *Mathematics* **2022**, *10*, 2930. https://doi.org/10.3390/math10162930

Academic Editor: Aleksandar Aleksic

Received: 19 July 2022
Accepted: 10 August 2022
Published: 14 August 2022

1. Introduction

With the advancement of internet technology, the general trend toward e-commerce, rising urbanisation, and shifting social norms since the middle of the 2000s, the online food delivery (OFD) market has been booming and is predicted to reach USD 200 billion in worldwide output value by 2025. Prior to the alarming COVID-19 becoming widely publicised, online food delivery was already benefiting from increased digitalisation and a plethora of delivery apps. While millions of companies, particularly those in the aviation, tourism, and hospitality sectors, were severely impacted by the COVID-19 crisis and faced the real possibility of significant revenue declines, the global turnover of the online food delivery (OFD) industry increased by about 140 percent as a result of the pandemic. Since the start of the pandemic, contactless delivery has been widespread. Food delivery services are now more swift and quick to acquire momentum among customers thanks to technological advancements.

In India, online food delivery (OFD) has been a popular choice due to a rise in demand for a significant period of time. The COVID-19 pandemic was one of several causes that contributed to the market's expansion, but it also served as an important catalyst for the explosive development of online food delivery (OFD) use during the previous year throughout the nation. These businesses are making significant investments in order to attract retailers and customers. However, as more customers choose to order food online, the expanding business is quickly becoming extremely competitive and difficult for the landscape's current competitors. The four pillars of sustainability-financial, facility value, expertise, societal impact, and environmental friendliness have significant consequences for this.

Analytic hierarchy process is a strategy that may be easily understood and simplifies even complex circumstances, weighting utilising pairwise comparisons that are simple to understand. Analytic hierarchy process also gives the decision-maker the ability to evaluate the consistency of their choices. Analytic hierarchy process is more applicable than other methods in a range of circumstances because of all these advantages. The Analytic hierarchy process also uses fuzzy set theory to address the numerous uncertainties and ambiguities in expert judgement, which are described by linguistic variables.

Fuzzy technique for order performance by similarity to ideal solution method (FTOPSIS) is a novel method that was recently introduced and has enhanced consistency and accuracy for prioritising the options. Additionally, the fuzzy technique for order performance by similarity to ideal solution method (FTOPSIS) turns into a suitable multi-criteria decision-making method for assessing options.

In this paper, we study a methodology based on multi-criteria decision-making (MCDM) to evaluate the long-term growth of the online food delivery (OFD) market in India. First, through a review of the literature and the opinions of experts, evaluation of financial norms (supply rate, operating skill, and risk managing), expertise criteria (network strategy, instantaneous, and e-commerce), societal and environmental criteria (health and living conditions, communication safety, and ecological influence), and facility value (order satisfaction, supply speed, handiness of expense, virtual/offline facility level, and customer response) have been identified. Four significant firms in the OFD market in India-Zomato, Swiggy, Domino's, and Uber Eats are taken into consideration in the assessment to show the applicability of the suggested approach. Here we used the multi-criteria decision-making technique that includes the fuzzy technique for order performance by similarity to ideal solution method (FTOPSIS) and analytic hierarchy process (AHP) employing triangular fuzzy sets. Also, we discussed the analytic hierarchy process method, which is frequently used to calculate the weights of the criterion. In order to produce more accurate findings and weights, the expert opinions were transformed into triangular fuzzy numbers. These triangular fuzzy numbers were then normalised, weighted, and finished in the weighted normalised fuzzy decision matrix. We used a technique that combines the fuzzy analytic hierarchy process (FAHP) with the fuzzy technique for order performance by similarity to ideal solution method (FTOPSIS), as suggested and employed in this paper. The results of this study could serve as a guide for stakeholders and decision-makers in the online food delivery (OFD) and other sectors.

To the best of the authors' knowledge, the online food delivery (OFD) market in India has not been thoroughly evaluated utilising the suggested methodology as previously described. The following goals are set for the case study that is presented in order to close the research gaps. First, the evaluation standards for online food delivery (OFD) are examined, focusing in particular on the Indian market. The weights assigned to the online food delivery (OFD) evaluation criteria are then determined. Third, the online food delivery (OFD) enterprises performing the best in terms of sustainable development are indicated using the deduced weights of the criterion. Finally, a discussion of the suggested work's managerial ramifications follows. The originality of this study may lie in the aims it addressed. The thorough construction of the online food delivery (OFD) market evaluation criteria from the literature and consulting with industry professionals is a significant benefit of this research. Additionally, this is the first study to examine the online food delivery (OFD) market using the advantages of the fuzzy analytic hierarchy process (FAHP) and fuzzy technique for order performance by similarity to ideal solution method (FTOPSIS) approaches. The validity of the suggested integrated framework is demonstrated through a case study from India. Last but not least, the management implications of the employed approach and its analysis would enlighten those in charge of making decisions in the online food delivery (OFD) market, not just in India but also on the international stage.

2. Literature Survey

The COVID-19 epidemic induced by SAR-CoV-2 had a massive worldwide impact. Civic isolates or lockdowns were implemented to limit citizens' drive and also avoid the disease after scattering due to a lack of standardised methodologies and licenced therapies to treat the illness. Whereas the isolation stood vital in halting the increase of COVID-19, it had a substantial effect on the universal wealth and source chains, with a significance that was predicted to go beyond nationwide limits [1]. The restaurant business is one of the pandemic's hardest devastated industries. Lockdowns have seen a significant decline in restaurant patronage, possibly contributing to the closure of a number of eateries [2]. Food delivery has become a standard feature of city life. Since the mid-2000s, the food delivery over the internet, online food delivery (OFD), business has been booming, thanks to the advancement in internet skills and total tendency towards e-marketing, more metropolitan existence, in addition to fluctuating societal behaviours [3]. Since the financial crisis, the endemic has had further effects on the restaurant industry, including changes in food preferences [4]; eating behaviour, and a preference for using digital platforms. Customers can buy meals from a number of restaurants and have them delivered to their door with just a single tap of their phone on online meal delivery services that provide a variety of options and convenience, as well as cashback benefits, rewards, great deals, and savings [5]. Demand for Online food delivery services rose with each fixed fresh case of COVID-19 in Taiwan, for example, with trades and buyers growing by 5.7% and 4.9%, respectively, during the outbreak [6]. Restaurants now have an additional delivery method thanks to the internet food delivery business and a new revenue source has emerged in the form of online food delivery (OFD). In this business model, restaurants sign up for a digital platform that allows clients to order food through an app. The meal will be taken from the restaurant and delivered to the customer by delivery riders. The analogue policy takes into account the number of diner's deals used for each positive transaction. Diners gain from this industry replica, which then allows them to carry on with work in spite of lockdowns, bringing down accumulation and eliminating the necessity of spending on extra labour or bikes/saloons for carters. Global revenues for online food delivery (OFD) were predicted to reach 91 million dollars in 2018 and 107 million dollars in 2019 according to Statista's Analogue Emporium View for virtual carter [7]. The lockdowns might have accelerated buyer and diner approval for this new means of payment as seen by an estimated 11% increase in virtual food delivery revenue in 2020. When restaurant operations transition to a larger focus on meal delivery, more takeout containers and packing are necessary. Unfortunately, this also means more of an ecological problem [8]. As a result of the quick rise of info-communication tools (ICT) and mobiles, wise technologies and software have become widespread in a significant part of routine life [9]. Apps for smartphones and other mobile platforms are developed and designed with the intention of being downloaded and used (e.g., for iPads and tablets). In the first quarter of 2017, there were roughly 178.1 billion apps available for download on mobile devices, with that number is expected to rise to 258.2 billion by 2022 [10,11]. Patron demand for online food delivery (OFD) services has risen dramatically in recent years and is expected to continue to climb steadily in the future. The whole income of the global online food delivery (OFD) service industry is estimated to reach $107.4 billion in 2019 and $182.3 billion by 2024 [12]. Furthermore, due to its contactless ordering and delivery mechanism, the online food delivery (OFD) market has gained even more global interest since the COVID-19 outbreak, and it is expected to continue to attract new customers [13]. Researchers developed a model based on the contingency framework and extended model of IT continuance to find the primary reasons for customers' continuous desire to use online food ordering systems [14]. Customers see such online food ordering systems as making their lives easier as long as they also perceive them to be interesting and engaging, according to researchers, and are thus further expected to obtain additional progressive approaches and inclined to continue. The epidemic has a significant influence on the use of plastic in restaurants. Because of concerns regarding COVID-19 transmission, restaurant guests choose single-use plastic silverware and food containers [15]. The study's objective is to discuss an integrated model based on the fuzzy analytic hierarchy process (FAHP) for evaluating and prioritising selection criteria and the fuzzy technique for order performance by similarity to ideal solution (FTOPSIS) for choosing and developing a reverse logistics partner. With an integrated approach to show how the proposed framework is applied, this study aims to showcase a real problem in the Indian electronics sector. While achieving efficiency and effectiveness in reverse logistics practises, this study seeks to significantly assist electronics businesses in the evaluation and selection of third-party reverse logistics partners [16]. In order to understand the relationships and significance of risks in the development of new products, we developed a research framework for this study based on pertinent literature and expert interviews. We then used the decision-making trial and evaluation laboratory (DEMATEL) and the analytic network process (ANP). The outcomes of a case study demonstrate that the six main risks of product development projects are the following: project completion time, mastery of key technical capabilities, controlling the project's progress, uniqueness and complexity, ability to control the market, and functional integrity of the product [17]. To cope with this assessment process in the fuzzy environment, a novel hybrid multi-criteria decision-making (MCDM) approach is put forth in this paper. For establishing the subjective and objective weights of criteria, we offer fuzzy versions of the SWARA (step-wise weight assessment ratio analysis) and CRITIC (criteria importance through inter criteria correlation) approaches. A new hybrid strategy is suggested based on these extended methodologies and the fuzzy EDAS (evaluation based on distance from average solution) method. In this method, the weights of the subjective and objective criteria are blended to produce more logical weights for the criterion. An evaluation of construction equipment with a focus on sustainability is used to test the proposed methodology [18]. App design has an impact on consumer preference in online food delivery (OFD) prior to customer involvement, individual outlooks, and third-party way, according to previously published studies [19]. The suggested model uses a robust goal programming (RGP) method based on Shannon entropy to address the uncertain multi-objectiveness. With an analysis carried out on various levels of uncertainty, the proposed technique has been applied to a genuine case study from an Iranian green service food production company in order to confirm its applicability [20]. As a result, it was clear that the increasing amount of online food delivery (OFD) users will result in augmented plastic practice. The concern of easily spreading COVID-19 is another issue surrounding the use of online food delivery (OFD), as a result, disposable utensils and food containers are becoming increasingly popular. According to experiments, SARS-CoV-2 may persist on a variety of surfaces for days, including plastic [21]. The odds of getting COVID-19 throughout this pathway are really quite remote [22]. The influence of online food delivery (OFD) product and service developments aimed at improving the consumer's propensity to order food online during the COVID-19 endemic. COVID-19 users' concern is measured and used as a mediator variable [23]. With the use of eight criteria, a model of five cleaner production techniques for the Libyan manufacturing sector has been developed in a study. A novel interval rough the SWARA (step-wise weight assessment ratio analysis) method that applies interval rough numbers to the criteria has been created to assess the significance of the criteria while taking decision-makers' preferences into account [24]. The goal of this study is to present a new, integrated model for creating intellectual capital performance indicators in order to enhance the current IC process model. The suggested model will be used by a company that provides financial shared services. The goal of the study is to create an IC measurement system using IC management and the multi-criteria decision-making approach and to utilise the best-worst method to determine the values of IC performance measures to prioritise Key Performance Indicators [25]. The goal of the paper is to suggest a multi-criteria decision-making approach for analysing the sustainable third-party reverse logistics providers assessment problem using data from hesitant fuzzy numbers. In order to do this, a novel hesitant fuzzy-combined compromise solution strategy is first introduced by fusing the conventional combined compromise solution method with the hesitant fuzzy set operators and discrimination measures in hesitant fuzzy set circumstances. Integrating a proposed discrimination measure-based objective weighting method with a subjective method suggested by experts, the weights of the criteria have been evaluated [26]. This study aims to prioritise the knowledge that students enrolled in cooperative education (co-op) programmes must have, and the findings are used to enhance a study strategy that is directly responsive to the needs of businesses and to improve the human capital of those businesses while working within the constraints of academic institutions. In this study, a rating of the taken-in knowledge is produced using the analytical hierarchy process (AHP). An easy-to-understand map that takes knowledge importance and study effort into account shows the determined priorities and improved opportunities [27].

3. Materials and Methods

As shown in Figure 1, the research approach is divided into two phases. First, based on relevant research and expert interviews, maintainable online food delivery (OFD) estimation norms and descriptions were created (Table 1). Financial, facility value, expertise, societal and eco-friendly factors were all taken into account. fuzzy analytic hierarchy process (FAHP) is a tool that assigns preference weights to criteria using the pairwise comparison concept. Preference weights and alternative ratings for each criterion were expressed as phonological phrases in the shape of triangular fuzzy numbers. To rank all of the alternatives, the fuzzy technique for order performance by similarity to ideal solution (FTOPSIS) was employed. The toughness and comprehensiveness image for computation, use of fuzzified judgement approach, and collection of sustainable online food delivery (OFD) were evaluated through a sensitivity study.

Table 1. Criteria for evaluating and describing sustainable online food delivery.

Main Criteria	Sub-Criteria	Goal	Descriptions
Financial Norms (*f*1)	*f*11: Supply Rate	Minimal	Transportation, labour, and administration costs all add up to a significant amount of money
	*f*12: Operating Skill	Maximal	Value propositions offered by the company, as well as the extension of its operational capabilities
	*f*13: Hazard Managing	Minimal	Investor risk management, cash flow statement, and shareholders' equity
Facility Value (*f*2)	*f*21: Order Satisfaction	Maximal	Order processing time is reduced, order pick-up time is reduced, and packaged food is kept clean.
	*f*22: Supply Speed	Minimal	Arrival of orders in a timely manner
	*f*23: Handiness of Expense	Maximal	Payment options are varied.
	*f*24: Virtual Facility Level	Maximal	SMS response time and customer service employee response time
	*f*25: Offline Facility Level	Maximal	Delivery personnel's attitudes, as well as dealers' responses to consumer concerns
	*f*26: Patron Response	Maximal	Customer behaviour intents, online reviews, and online rating
Expertise (*f*3)	*f*31: Network Strategy	Maximal	Platform that is up to date, has visual impacts on the pages, and is user-friendly
	*f*32: Instantaneous tracking systems	Maximal	Tracking and tracing over the internet, using cutting-edge technologies
	*f*33: Marketing Techniques	Maximal	Digital marketing, as well as digital technologies, are being used to promote products.

Table 1. *Cont.*

Main Criteria	Sub-Criteria	Goal	Descriptions
Societal and Eco-friendly ($f4$)	$f41$: Health and Living quarters	Maximal	Health and safety regulations, food cleanliness, and contactless delivery
	$f42$: Communication Safekeeping	Maximal	Data security for customers, as well as online payment security
	$f43$: Ecological Influence	Minimal	CO_2 emissions from automobiles, solid waste, and traffic noise are all examples of environmental issues

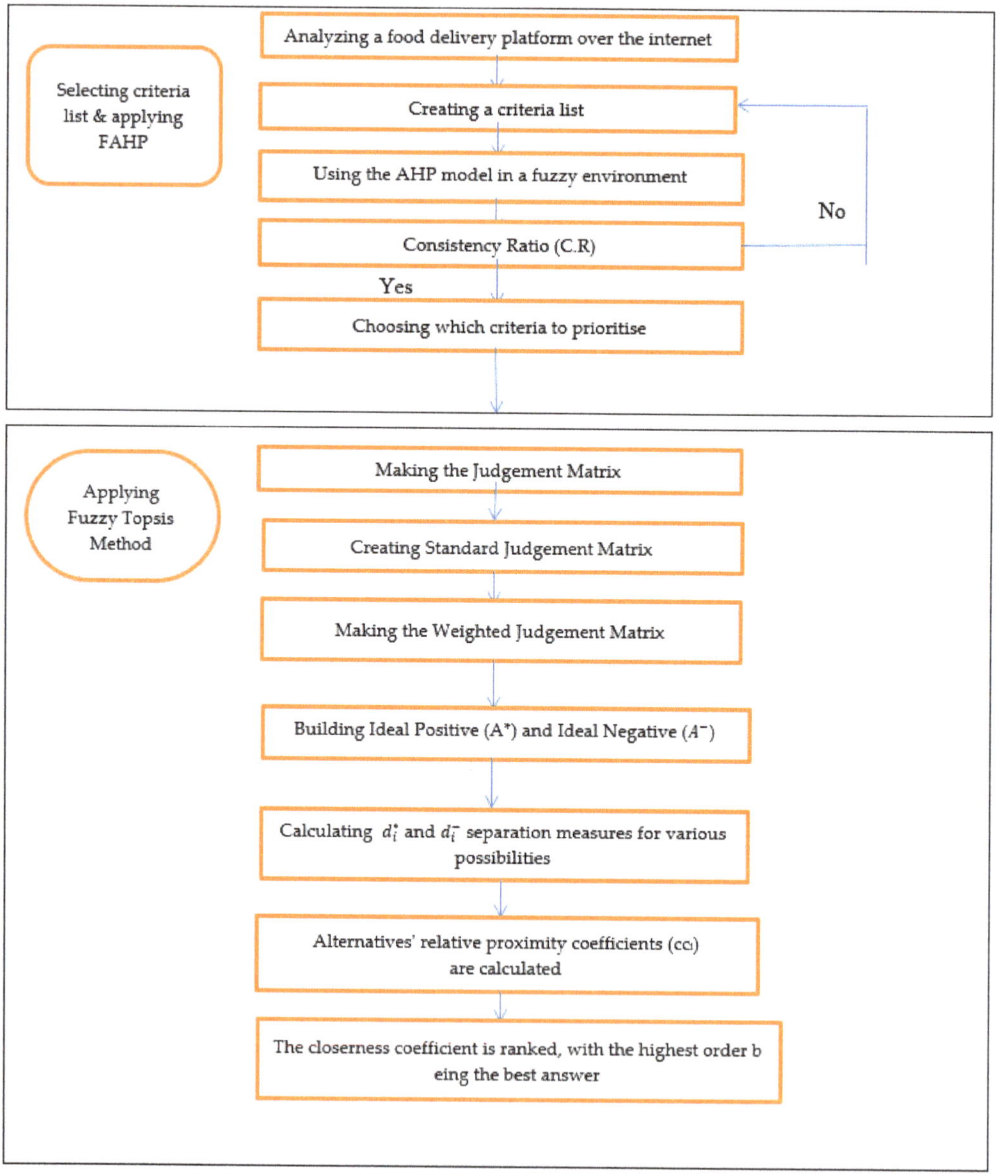

Figure 1. Evaluation process and Selection of online food delivery.

3.1. The Analytic Hierarchy Process Method

The analytic hierarchy process (AHP) is a pairwise comparison measuring technique that creates priority scales based on the opinions of experts. In a multi-criteria decision issue, the analytic hierarchy process discusses how to assess the comparative prominence of a group of actions [28]. The method combines qualitative assessments with quantitative criteria that may be measured [29]. Three principles govern the analytic hierarchy process technique: model construction, option and criteria comparison, and priority synthesis. Analytic hierarchy process has been utilised to tackle a wide range of challenging decision-making situations in the literature [30]. A hierarchy is used to organise a multidimensional outcome problem. Initially, analytic hierarchy process appears to be a complex multi-criteria decision-making problem that may be broken down into a hierarchy of interconnected choice measures and judgement substitutes. The Analytic hierarchy process organises the ideas, principles, and options into a family-tree-like ordered arrangement. There is an order in place at least three stages: top-level goal line of the delinquent, various tiers of criteria in the middle that identify options, and choice substitutes at the end [31]. The comparison of the substitutes and gages is the next phase. After the issue has already been dissected and the hierarchies have indeed been built, the prioritising approach continues by evaluating the significance of the criterion at each level. The paired judgement begins at the second level and continues until the lowest level, which is an alternative. The parameters were evaluated separately at each level based on the upper-level criteria and their influence levels [31]. In the analytic hierarchy process, a nine-level standardised comparison scale is used to make many pairwise comparisons (Table 2).

Table 2. Important scale of nine points of intensity.

Scale Rating	Meaning
1	Equally vital
3	Moderately Crucial
5	Crucial
7	Imperative
9	Very Important
2, 4, 6, 8	Between binary neighbouring decisions, there are values in the middle

The set of criteria is $C = C_j$, $j = 1, 2, \ldots, n$. A (n,n) assessment matrix A can be used to express the results of pairwise comparisons on n criteria, where each member a_{ij} $(i,j = 1, 2, \ldots, n)$ is the quotient of the criteria's weights, as shown:

$$A = [a_{11}\ a_{12}\ \ldots\ a_{1n}\ a_{21}\ a_{22}\ \ldots\ a_{2n}\ \ldots\ \ldots\ \ldots\ \ldots\ a_{n1}\ a_{n2}\ \ldots\ a_{nn}\ ]; a_{ii} = 1,\ a_{ji} = \frac{1}{a_{ij}}; a_{ij} \neq 0 \quad (1)$$

In the final phase, the mathematical procedure begins to normalise each matrix and identify the relative weights. The eigenvector (w) corresponding to the greatest eigen value (λ_{max}) determines the relative weights.

$$(\lambda_{max})w = A_w \quad (2)$$

The matrices A have ranking 1 whereas if assessments are entirely consistent and $\lambda_{max} = n$ if they are not. Weights are obtained by normalising any of A's columns or rows [32]. It is important to note that the reliability of the pair-wise comparison assessments is strongly tied to the accuracy of the analytic hierarchy process output. An association of the entries of A: $a_{ij} \times a_{jk} = a_{ik}$ defines the consistency. The CI (consistency index) is designed as follows:

$$CI = \frac{\lambda_{max} - n}{n - 1} \quad (3)$$

As shown, the total reliability of a measure (CR) is measured as the proportion of the rate constant (CI) to the randomised indices (RI).

$$CR = \frac{CI}{RI} \tag{4}$$

The highest value of CR is 0.1. If the final consistency ratio is higher than the target, the review procedure must be used again to increase consistency. Consistency measurement can be used to assess decision-makers' consistency as well as the overall hierarchy [32].

3.2. The Fuzzy Analytic Hierarchy Process Method

The fuzzy analytic hierarchy process is frequently recommended in conjunction with other multi-criteria decision-making approaches. When addressing ambiguity and vagueness in the given weights to evaluate options, the fuzzy analytic hierarchy process is utilised in conjunction with the multi-criteria decision-making approach. The modelling of decision-making processes based on imprecise and hazy information, such as decision-makers' judgement, has been carried out using fuzzy set theory [33]. Qualitative qualities are represented by linguistic variables, which are qualitatively conveyed by linguistic phrases and quantitatively expressed by a fuzzy set and corresponding membership function in the universe of conversation.

- *Establishing fuzzy number*

The concepts that follow are involved in operations between linguistic variables. Sets with degrees of membership are called fuzzy sets. As an extension of the traditional concept of set, fuzzy sets have been proposed. In traditional set theory, a recombinant criterion determines the inclusion of elements in a collection [33]. A member will either be a member of the set or not a member of the set.

If the membership functions of a fuzzy number A on R is a triangular fuzzy number, then $\mu_A^{\sim(x)} : R \rightarrow [0, 1]$ is equal to the following Equation (5). (l, m, u) is a triangular fuzzy number (TFN), where (l, m, u) are the lowest, mean, and higher values, respectively, as shown in Figure 2.

$$\mu_A^{\sim(x)} = \begin{cases} \frac{x-l}{m-l}, & l \leq x \leq m \\ \frac{u-x}{u-m}, & m \leq x \leq u \\ 0, & \text{Otherwise} \end{cases} \tag{5}$$

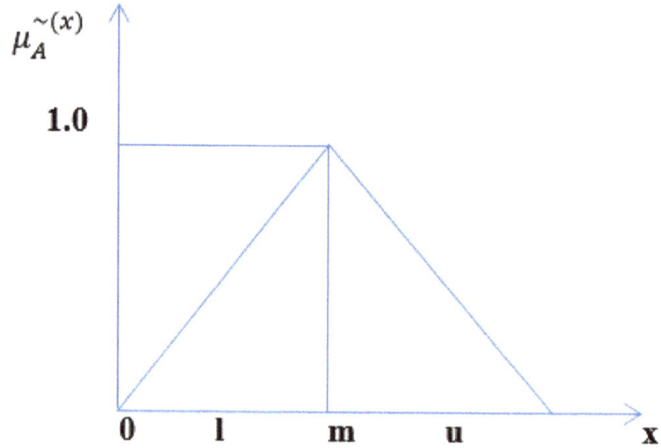

Figure 2. The triangular fuzzy number's Function of Membership.

In Equation (5), The lower and upper boundaries of the fuzzy number A are denoted by the letters l and u, respectively, and m denotes $\widetilde{A}$'s modal value (as Figure 2). $\widetilde{A} = (l, m, u)$ is used to represent the triangular fuzzy number. Triangular Fuzzy Number $\widetilde{A_1} = (l_1, m_1, u_1)$ and $\widetilde{A_2} = (l_2, m_2, u_2)$ possess the following operating laws: Equations (6)–(9).

Addition of the fuzzy number $\oplus$

$$\widetilde{A_1} \oplus \widetilde{A_2} = (l_1, m_1, u_1) \oplus (l_2, m_2, u_2) = l_1 + l_2,\ m_1 + m_2, u_1 + u_2 \quad (6)$$

Multiplication of the fuzzy number $\otimes$

$$\widetilde{A_1} \otimes \widetilde{A_2} = (l_1, m_1, u_1) \otimes (l_2, m_2, u_2) = l_1 l_2,\ m_1 m_2, u_1 u_2$$

for $l_1, l_2 > 0;\ m_1, m_2 > 0; u_1, u_2 > 0$ Equations (2)–(6)

Subtraction of the fuzzy number

$$\widetilde{A_1} \ominus \widetilde{A_2} = (l_1, m_1, u_1) \ominus (l_2, m_2, u_2) = l_1 - u_2,\ m_1 - m_2, u_1 - l_2 \quad (7)$$

Division of a fuzzy number $\varnothing$

$$\widetilde{A_1} \varnothing \widetilde{A_2} = (l_1, m_1, u_1) \varnothing (l_2, m_2, u_2) = l_1/u_2,\ m_1/m_2, u_1/l_2 \quad (8)$$

for $l_1, l_2 > 0;\ m_1, m_2 > 0;\ u_1, u_2 > 0.$

Reciprocal of the fuzzy number

$$\widetilde{A}^{-1} = (l_1, m_1, u_1)^{-1} = \left(\frac{1}{u_1}, \frac{1}{m_1}, \frac{1}{l_1}\right) \text{ for } l_1, l_2 > 0;\ m_1, m_2 > 0;\ u_1, u_2 > 0 \quad (9)$$

- *Identifying phonological variables*

Phonological variables yield the values well-defined by their term set, which is a collection of phonological relations. Language words are phonological variables' personal categorisation. A phonological variable is one whose values are words or phrases in natural or artificial languages. This type of statement is used to compare nine basic phonological concepts, such as "Flawless", "Complete", "Brilliant", "Decent Enough", "Decent", "Better", "Average", "Less benefit", and "Equivalent" with respect to nine equal gauges. The fuzzy numbers mentioned in Table 3 were used to compute the results in this paper by [34]. Each membership function has three symmetric triangular fuzzy number parameters that specify the left, middle, and right points of the range within which the function is defined (scale of fuzzy number).

Table 3. Membership role of phonological gauge.

Fuzzy Numeral	Phonological Variables	Gage of Fuzzy Numeral
9	Flawless	(9, 9, 9)
8	Complete	(7, 8, 9)
7	Brilliant	(6, 7, 8)
6	Decent Enough	(5, 6, 7)
5	Decent	(4, 5, 6)
4	Better	(3, 4, 5)
3	Average	(2, 3, 4)
2	Less Benefit	(1, 2, 3)
1	Equivalent	(1, 1, 1)

- *Fuzzy analytic hierarchy process*

Then, in the following parts, we go over how to perform the fuzzy analytic hierarchy process.

Step 1: Make pairwise comparison matrices for all elements/criteria in the hierarchy system's dimensions. Assign linguistic labels to pairwise comparisons by deciding which of the two dimensions is more important, as shown in matrix $\widetilde{A}$ below.

$$\widetilde{A} = \begin{bmatrix} 1 & \widetilde{a}_{12} & \cdots & \widetilde{a}_{1n} \\ \widetilde{a}_{21} & 1 & \cdots & \widetilde{a}_{2n} \\ \cdots & \cdots & \cdots & \cdots \\ \widetilde{a}_{n1} & \widetilde{a}_{n2} & \cdots & 1 \end{bmatrix} = \begin{bmatrix} 1 & \widetilde{a}_{12} & \cdots & \widetilde{a}_{1n} \\ \frac{1}{\widetilde{a}_{21}} & 1 & \cdots & \widetilde{a}_{2n} \\ \cdots & \cdots & \cdots & \cdots \\ \frac{1}{\widetilde{a}_{n1}} & \frac{1}{\widetilde{a}_{n2}} & \cdots & 1 \end{bmatrix} \tag{10}$$

where $\widetilde{a}_{ij} = \begin{cases} \widetilde{9}^{-1}, \widetilde{8}^{-1}, \widetilde{7}^{-1}, \widetilde{6}^{-1}, \widetilde{5}^{-1}, \widetilde{4}^{-1}, \widetilde{3}^{-1}, \widetilde{2}^{-1}, \widetilde{1}^{-1}, \widetilde{9}, \widetilde{8}, \widetilde{7}, \widetilde{6}, \widetilde{5}, \widetilde{4}, \widetilde{3}, \widetilde{2}, \widetilde{1}, & i \neq j \\ 1, & i = j \end{cases}$.

Step 2: The fuzzified geometrical means and fuzzified weight with each parameter were calculated using the geometric mean approach.

$$\widetilde{r}_i = \left(\widetilde{a}_{i1} \otimes \ldots \otimes \widetilde{a}_{ij} \otimes \ldots \otimes \widetilde{a}_{in}\right)^{\frac{1}{n}}$$

$$\widetilde{w}_i = \widetilde{r}_i \otimes \left[\widetilde{r}_1 \oplus \ldots \oplus \widetilde{r}_i \oplus \ldots \oplus \widetilde{r}_n\right]^{-1}$$

where $\widetilde{a}_{ij}$: the dimension i to criteria j has a fuzzy comparison value, r_i: the geometric mean of each criterion's fuzzy comparison value and w_i: the ambiguous importance of the ith criterion, which is represented by the triangular fuzzy number, $\widetilde{w}_i = (lw_i, mw_i, nw_i)$. The inferior, medium, and superior ideals of the indistinct encumbrance of the ith element are denoted by the letters lw_i, mw_i and nw_i.

Several research studies have used the fuzzy analytic hierarchy process method to handle a variety of managerial issues. To analyse subjective expert judgments made through perception, use an analytical structure with the fuzziness procedure and then a crispy judgement matrix [35]. According to the analytical hierarchical procedure with fuzziness, present an inventory arrangement structure [36]. To create a pairwise comparison matrix that has an additive reciprocal attribute and is consistent, use fuzzy language preference relations [37]. Then calculate the micro and essential factors for an ISO 14001-based environmental management system's successful implementation, as well as the advantages [38]. In this paper, the analytic hierarchy process method is used to optimise delivery network design selection and then make appropriate decisions for home plus distribution centre decision-making [39]. Here, the authors debated over a consequence in the primacy variable derived from the analytic hierarchy process major eigenvalue technique [40]. To provide a judgement-making aid with an intentional collection of choices that combined the analytic hierarchy process and to tackle the issues, use a zero-one optimisation model assortment problematic derived with perspective from a single investor [41]. Here, it is explained how the analytic hierarchy process technique determines priority vectors [42]. In a fuzzy environment, suggested cluster management is founded on a technique for order performance by similarity to ideal solution approach models for the locality sector [43]. Fuzzy analytic hierarchy process and the technique for order performance by similarity to ideal solution are used to evaluate hazardous waste transportation companies [34]. Researchers have created an analogue apparatus for prototyping and small-batch manufacturing processes of industrial goods [44]. The decision-making framework based on an analytic hierarchy process has also been provided with a fuzzy analytic hierarchy process method to estimate the level of risk of mistaken behaviour in work systems [45]. In a multi-criteria judgement setting with fuzziness employed, the fuzzy analytic hierarchy approach was used to establish the weight of the particular/perceptual assessments for each criterion, as well as the generation of fuzzy synthetic utility values for alternatives [46]. To aid designers in identifying customer needs/requirements and design characteristics, as well as achieving an effective evaluation of the final design solution for achieving the desired levels, an outline that combines the analytical hierarchy process and the method for directive inclination by correspondence to perfect elucidation was proposed [47].

3.3. The Fuzzy Technique for Order Performance by Similarity to Ideal Solution Method

The technique for order performance by similarity to ideal solution is commonly utilised in real-world scenarios to solve ranking challenges. This strategy is frequently chastised for failing to account for the inherent ambiguity and imprecision that comes with mapping a decision-perspective maker to precise numbers. Crisp values are used to reflect personal judgments in the conventional technique for order performance by similarity to ideal solution formulation. However, the human preference model is unreliable in many situations, and decision-makers may be hesitant or unwilling to assign precise values to comparison judgements [30]. One of the most difficult aspects of the crisp evaluation process is to employ crisp values. One reason is that decision-makers are more comfortable giving interval assessments than single number values. Because some factors are difficult to quantify with precise numbers, they are frequently overlooked throughout the review. Another factor is the usage of mathematical models that are based on precise values. These approaches are incapable of dealing with the ambiguities, uncertainties, and vagueness that decision-makers face that also cannot be accompanied by a number of explicit values. Judgment can incorporate undefinable information, incompleteness, semi-information, and partly uninformed data in the judgement systems using fuzzy numbers [33,48]. As an outcome, the fuzzified technique for order performance by similarity to ideal solution and its expansions have been created to handle ranking and justification difficulties [49–54]. For the fuzzy technique for order performance by similarity to ideal solution, this paper uses triangular fuzzy numbers. The justification for choosing a Triangular Fuzzy Number is that it is intuitively simple to use and calculate for decision-makers. Furthermore, modelling with triangular fuzzy numbers has proven to be an excellent technique for expressing decision issues where the given knowledge is subjective and inaccurate [55–58]. The triangular form of the membership function is most commonly used in practice to represent fuzzy numbers [59].

Some key fuzzy set definitions are given below.

Definition 1. *A membership function $\mu_{\tilde{A}}^{(x)}$ assigns a real number in the interval [0, 1] to each element x in X in a fuzzy set $\tilde{A}$ in a universe of discourse X. The grade of membership of x in $\tilde{A}$ is denoted by the function value $\mu_{\tilde{A}}^{(x)}$.*

Definition 2. *A linguistic variable is a variable with linguistic terms as values. The concept of a linguistic variable comes in handy when dealing with situations that are too complicated or ill-defined to be adequately expressed using traditional quantitative phrases. "Weight" is a phonological variable with values of little, small, average, huge, elevated, and so on. Fuzzy numbers can be expressed using these phonological values.*

Definition 3. *The vertices approach is used to calculate the route between two triangular fuzzy numbers, $\tilde{a} = (a_1, a_2, a_3)$ and $\tilde{b} = (b_1, b_2, b_3)$.*

$$d\left(\tilde{a}, \tilde{b}\right) = \sqrt{\frac{1}{3}\left[(a_1 - b_1)^2 + (a_2 - b_2)^2 + (a_3 - b_3)^2\right]} \quad (11)$$

Definition 4. *The weighted normalised fuzzy-decision matrix is generated as follows, taking into account the varied relevance levels of each criterion.*

$$\tilde{V} = [\tilde{V}_{ij}]_{n \times j}, \; i = 1, 2, \ldots, n, \; j = 1, 2, \ldots, J \quad (12)$$

where, $\tilde{V}_{ij} = \tilde{x}_{ij} \times W_i$.

A set of $A_j = (j = 1,2,\ldots,j)$ performance ratings in terms of criterion $C_i = (i = 1, 2, \ldots, n)$ called $\tilde{x} = \tilde{x}_{ij}, i = 1,2,\ldots,n, j = 1,2,\ldots,J$.

A list of each criterion's importance weights $W_i = i = 1, 2,\ldots, n$.

A fuzzy technique for order performance by similarity to ideal solution phase is outlined as follows based on the preceding quickly explained fuzzy theory [53]:

Step 1: For criteria alternatives, choose the phonological values $\tilde{x}_{ij}$, $i = 1,2,\ldots,n$, $j = 1, 2, \ldots, J$. The property of normalised triangular fuzzy integers belonging to [0, 1] is preserved by the fuzzy linguistic rating $\tilde{x}_{ij}$; consequently, no normalisation is required.

Step 2: Determine the fuzzy-decision matrix's weighted normalised weights. Equation (12). Calculates the weighted normalised value $\tilde{V}_{ij}$.

Step 3: Determine if the solution is beneficial-perfect (A^*) or deleterious-perfect (A^-). The fuzzy beneficial-perfect solution (FBPS, A^*) and the fuzzy deleterious-perfect solution (FDPS, A^-) is depicted.

The formulas are as follows:

$$A^* = \{\tilde{v}^*{}_1, \tilde{v}^*{}_2, \ldots, \tilde{v}^*{}_i\} = \left\{\left(\max_j v_{ij} \middle| i \in I'\right), \left(\min_j v_{ij} \middle| i \in I''\right)\right\}, \quad i = 1,2,\ldots,n,\ j = 1,2,\ldots,J \tag{13}$$

$$A^- = \{\tilde{v}^-{}_1, \tilde{v}^-{}_2, \ldots, \tilde{v}^-{}_i\} = \left\{\left(\min_j v_{ij} \middle| i \in I'\right), \left(\max_j v_{ij} \middle| i \in I''\right)\right\}, \quad i = 1,2,\ldots,n,\ j = 1,2,\ldots,J \tag{14}$$

where I' denotes benefit criteria and I'' denotes cost criteria.

Step 4: Using the equations below, calculate the distance between A^* and A for each alternative:

$$D^*{}_j = \sum_{j=1}^{n} d\left(\tilde{v}_{ij}, \tilde{v}^*{}_i\right), \ j = 1,2,\ldots,J \tag{15}$$

$$D^-{}_j = \sum_{j=1}^{n} d\left(\tilde{v}_{ij}, \tilde{v}^-{}_i\right), \ j = 1,2,\ldots,J \tag{16}$$

Step 5: Compare your results to the optimum solution.

$$CC_j = \frac{D^-{}_j}{D^-{}_j + D^*{}_j} = 1 - \frac{D^*{}_j}{D^-{}_j + D^*{}_j}, j = 1,2,\ldots,J \tag{17}$$

where $\frac{D^-{}_j}{D^-{}_j + D^*{}_j} = CC_j^-$ is a hazy level of satisfaction and $\frac{D^*{}_j}{D^-{}_j + D^*{}_j} = CC_j^*$ is an indistinct break step that demonstrates how indistinct openings are corrected to meet decision-makers' target levels.

Figure 3 depicts the membership functions of these linguistic values, as well as the triangular fuzzy numbers associated with these variables (Table 4).

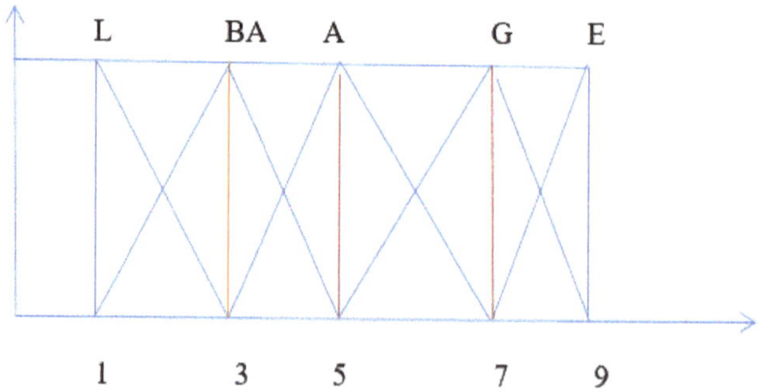

Figure 3. Linguistic values for criteria rating.

Table 4. Phonology grade level substitutes.

Linguistics Rating Level	Allocated Triangular Fuzzy Number
Low	(1, 1, 3)
Below Average	(1, 3, 5)
Average	(3, 5, 7)
Good	(5, 7, 9)
Excellent	(7, 9, 9)

4. Case Study

The efficacy of the suggested approach is examined in this paper using a case study of online food delivery (OFD) platform companies in India. Three specialists worked together to select the top four online food delivery services (OFDs) after conducting preliminary analysis. These firms are Uber Eats, Domino's, Zomato, and Swiggy. The Fuzzy analytic hierarchy process was employed to ascertain the relative preference weight of each criterion. The decision hierarchy for the evaluation and selection of online food delivery (OFD) is shown in Figure 4. It consists of 15 criteria in total, with four main ones being financial norms (supply rate, operating skill, and hazard managing), expertise criteria (network strategy, instantaneous, and e-commerce), societal and eco-friendly criteria (health and living quarters, communication safekeeping, and ecological influence), and facility value (order satisfaction, supply speed, handiness of expense, virtual/offline facility level, and patron response). After ranking all possibilities, a score line of the fuzzy technique for order performance by similarity to ideal solution model is prepared to demonstrate the model's resilience and comprehensiveness.

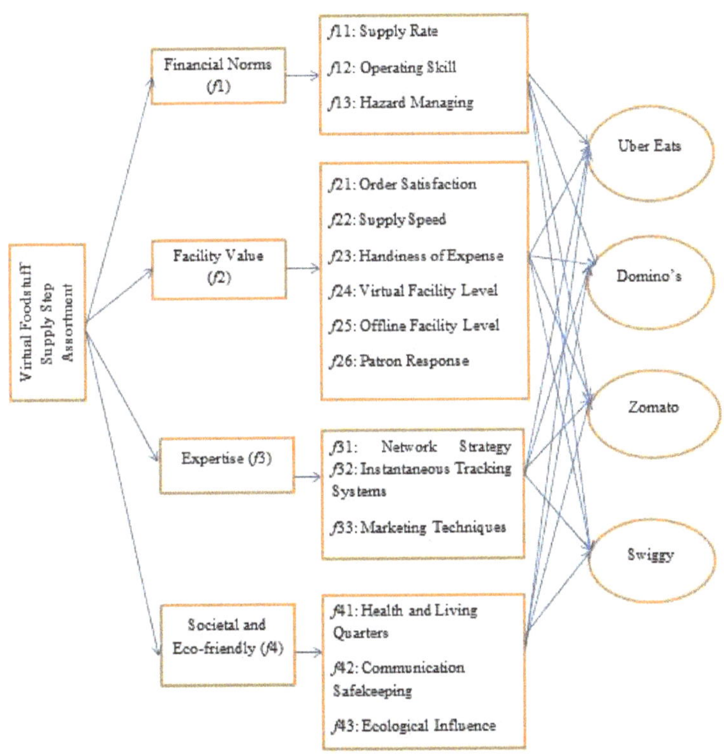

Figure 4. Evaluation and Selection of online food delivery using a decision hierarchy.

5. Results Analysis

With the analytic hierarchy process, a pairwise comparison matrix is created to compare the properties of several food delivery services. Analytic hierarchy process is used to estimate the weights of the main criteria and sub-criteria that take into account the decision-makers' subjective judgements. Four specialists in the field of online food services were recruited to construct the decision matrix.

Normalisation was accomplished using Equation (1) after constructing the evaluation matrix of the primary standards. The primacy vector was worked out. The eigenvalue was calculated with Equations (2) and (3). Equation (4) was used to calculate the consistency indicator and ratio.

The following primary categories are ordered based on estimated weighted values: Financial Norms, Facility Value, Expertise, and Societal and Eco-friendly factors (Table 5). When the same procedure was used to calculate the sub-criterion—Supply Rate, Operating Skill, Hazard Managing, Order Satisfaction, Supply Speed, Handiness of Expense, Virtual/Offline Facility Level, Patron Response, Network Strategy, Instantaneous Tracking Systems, Marketing Techniques, Health and Living Quarters, Communication Safekeeping, and Ecological Influence from the highest to lowest values—the results were as follows. The sub-criteria performance of online food services, on the other hand, was prioritised in the following order. The ranking of online food services ranged from excellent to poor. Table 6 displays the total findings.

Table 5. AHP ranked the main criteria.

Criteria	Weight
Financial Norms ($f1$)	0.4649
Facility Value ($f2$)	0.2086
Expertise ($f3$)	0.2341
Societal and Eco-friendly ($f4$)	0.0924

Table 6. AHP ranked the sub-criteria.

Sub Criteria	Weight	Sub Criteria	Weight
$f11$	0.0728	$f26$	0.0684
$f12$	0.0659	$f31$	0.0842
$f13$	0.0559	$f32$	0.0776
$f21$	0.0789	$f33$	0.0590
$f22$	0.0775	$f41$	0.0581
$f23$	0.0678	$f42$	0.0499
$f24$	0.0726	$f43$	0.0469
$f25$	0.0645		

$f11$: Supply Rate; $f12$: Operating Skill; $f13$: Hazard Managing; $f21$: Order Satisfaction; $f22$: Supply Speed; $f23$: Handiness of Expense; $f24$: Virtual Facility Level; $f25$: Offline Facility; $f26$: Patron Response; $f31$: Network Strategy; $f32$: Instantaneous Tracking Systems; $f33$: Marketing Techniques; $f41$: Health and Living Quarters; $f42$: Communication Safekeeping; $f43$: Ecological Influence.

In the fuzzy analytic hierarchy process, the outcome is significantly influenced by checking the consistency ratio; the following fuzzy analytic hierarchy process technique shows an example of how to calculate the four main criteria. Initial assessments were conducted by a group of specialists to rate the execution of these norms, which include Financial Norms ($f1$), Facility Value ($f2$), Expertise ($f3$), and Societal and Eco-friendly ($f4$). Tables 7 and 8 show the fuzzy analytic hierarchy process model's initial comparison matrix as well as the aggregated fuzzy comparison matrix.

Table 7. The initial comparison matrix.

Main Criteria	(7,8,9)	(6,7,8)	(5,6,7)	(4,5,6)	(3,4,5)	(2,3,4)	(1,2,3)	(1,1,1)	(1,2,3)	(2,3,4)	(3,4,5)	(4,5,6)	(5,6,7)	(6,7,8)	(7,8,9)	Main Criteria
*f*1							*									*f*2
*f*1						*										*f*3
*f*1					*											*f*4
*f*2							*									*f*3
*f*2								*								*f*3
*f*2							*									*f*4
*f*3					*											*f*4

Where * represents the values of combined fuzzy judgment matrix.

Table 8. Combined fuzzy judgment matrix.

Criteria	Financial Norms (*f*1)	Facility Value (*f*2)	Expertise (*f*3)	Societal and Eco-Friendly (*f*4)
Financial Norms (*f*1)	(1,1,1)	(1,2,3)	(2,3,4)	(3,4,5)
Facility Value (*f*2)	(1/3,1/2,1/1)	(1,1,1)	(1,1,1)	(1,2,3)
Expertise (*f*3)	(1/4,1/3,1/2)	(1,1,1)	(1,1,1)	(3,4,5)
Societal and Eco-friendly (*f*4)	(1/5,1/4,1/3)	(1/3,1/2,1/1)	(1/5,1/4,1/3)	(1,1,1)

To transform the phonological terms, the indistinct judgement matrix's sceptical (inferior bound) and expectant (superior bound) assessments were applied. To determine the quality evaluation score's reliability coefficient (CR), convert (i.e., triangle fuzzy number) to crisp values. Table 9 displays the primary criteria's non-fuzzy comparison matrix.

Table 9. Non-fuzzy comparison matrix.

Criteria	Financial Norms (*f*1)	Facility Value (*f*2)	Expertise (*f*3)	Societal and Eco-Friendly (*f*4)
Financial Norms (*f*1)	1	1.7321	2.8284	3.8730
Facility Value (*f*2)	0.5774	1	1	1.7321
Expertise (*f*3)	0.3536	1	1	3.8730
Societal and Eco-friendly (*f*4)	0.2582	0.5774	0.2582	1
Sum	2.1892	4.3095	5.0866	10.4781

Divide each unique value in a cell of the matrices by the column's average to create the normalised pairwise comparisons. This yields the prominence vectors of the fuzzy analytic hierarchy process model's four fundamental standards. As shown in Table 10, the prominence vectors were calculated from the average of the standardised matrix's row members.

Table 10. Normalised judgment matrix.

Criteria	Financial Norms (*f*1)	Facility Value (*f*2)	Expertise (*f*3)	Societal and Eco-Friendly (*f*4)	Priority Vector
Financial Norms (*f*1)	0.4568	0.4019	0.5561	0.3701	0.4462
Facility Value (*f*2)	0.2638	0.2321	0.1966	0.1652	0.2144
Expertise (*f*3)	0.1615	0.2321	0.1966	0.3701	0.2400
Societal and Eco-friendly (*f*4)	0.1179	0.1339	0.0507	0.0954	0.0994
Sum	1.0000	1.0000	1.0000	1.0000	1.0000

The gauge of integrity (CI), the indices at selection (RI), and the measure of dependability (CR) were calculated in this phase by calculating the greatest eigenvector (λ_{max}).

$$[1.8813\ 0.8842\ 0.9971\ 0.4003\,]/[0.4462\ 0.2144\ 0.2400\ 0.0994\,] = [4.2163\ 4.1240\ 4.1548\ 4.0278\,]$$

Four important factors were considered in this research. As a result, we get n = 4. The following formulas were used to calculate λ_{max} and CI.

$$\lambda_{max} = 4.1307$$
$$CI = 0.04357$$

RI = 0.9 was obtained when n = 4 and the consistency ratio (CR) is determined as follows:

$$CR = \frac{CI}{RI} = 0.0484$$

CR = 0.0484 < 0.1, according to the findings. As a result, the fuzzy analytic hierarchy process model's results are satisfactory, and the pair-wise comparison matrices are totally consistent. The same approach was then used to calculate the remaining criteria. Table A1 displays the combined fuzzy comparison matrix with all criteria (Appendix A).

Table 11 illustrates the results of the fuzzy weights computed using the fuzzy geometric mean approach for all criteria in the fuzzy analytic hierarchy process model. Suspicious (lowermost weight), most likely (central weight), and expectant (uppermost weight) are three values in each fuzzy weight (the highest weight). In the fuzzy weight of norms superiority, the suspicious value is 0.0556, the most likely value is 0.0720 and the most expectant value is 0.0928. The remaining conditions are demonstrated in the same way. The subsequent step is the fuzzy technique for order performance by similarity to ideal solution model; these fuzzy preference weights were then applied.

Figure 5 depicts the levels of criterion influence. With 8.3433 percent, 7.9733 percent, 7.81 percent, 7.5466 percent, and 7.4066 percent, respectively, the categories "Network Strategy", "Instantaneous tracking systems", "Order Satisfaction", "Supply Speed", and "Virtual Facility Level" have the biggest influence percentages. The findings reveal that "Network Strategy" is more essential to experts than other cost and quality concerns when it comes to influencing the selection of online food services in the e-commerce industry. Network Strategy is a significant predictor of when managers should replenish stocks in e-commerce businesses, thus it is crucial to consider it while developing an inventory management strategy. It is also important to think about when you are adding new product lines to your online store. In order to survive and grow in India's burgeoning e-commerce sector, e-commerce businesses are focusing more on economic aspects. On the other side, green and resilient development methods have gained popularity. In order to strengthen Indian enterprises' competitiveness, the management inspires them to join successfully in the universal assessment sequence by adopting and assimilating defensible corporate schemes. Thus, financial criteria were highly valued, and other criteria from the three pillars of defensible growth (Societal and Eco-friendly challenges) are also relevant. Among 15 eco-friendly variables, "Supply Rate" and "Patron Response" are placed sixth (7.3466 percent) and seventh (7.2366 percent), respectively. These graphs show how social and environmental elements, in addition to economic considerations, play a considerable effect.

We define CC_j^- as the degree of satisfaction in the jth alternative and CC_j^* as the degree of gap in the jth alternative. We can figure out which gaps should be closed and how they should be closed in order to meet aspirational goals and choose the greatest win-win approach from a hazy collection of viable options. In the fuzzy technique for order performance by similarity to ideal solution model, the intuitionistic fuzzy ratings of criterion are generated using the fuzzy analytic hierarchy process model. According to the fuzzy technique for order performance by similarity to ideal solution process, the hazy normalisation decision problem and hazy weight normalisation decision table are reported in Tables 12 and 13. The satisfaction degrees of each organisation can be determined using the data in Table 14. Uber eats, Domino's, Zomato, and Swiggy's satisfaction degree values are 0.2844, 0.5202, 0.8474, and 0.6196, respectively. Figure 6 shows the online food services, which are, Zomato, Swiggy, Domino's, and Uber Eats, ranked first, second, third, and fourth with scores of 0.8474, 0.6196, 0.5202, and 0.2844, respectively.

Table 11. Fuzzy weights for each criterion.

Major Indicators	Parameters	Goal	Uncertain Parametric Means	Fuzzy Weights
Financial Norms (*f*1)	*f*11: Supply Rate	Minimal	(0.9548, 1.0968, 1.2545)	(0.0556, 0.0720, 0.0928)
	*f*12: Operating Skill	Maximal	(0.9117, 1.0193, 1.1437)	(0.0531, 0.0669, 0.0846)
	*f*13:Hazard Managing	Minimal	(0.8473, 0.9293, 1.0273)	(0.0493, 0.0610, 0.0760)
Facility Value (*f*2)	*f*21:Order Satisfaction	Maximal	(0.9733, 1.1659, 1.3663)	(0.0567, 0.0765, 0.1011)
	*f*22: Supply Speed	Minimal	(0.9293, 1.1268, 1.3299)	(0.0541, 0.0739, 0.0984)
	*f*23: Handiness of Expense	Maximal	(0.8874, 1, 1.1268)	(0.0517, 0.0656, 0.0834)
	*f*24: Virtual Facility Level	Maximal	(1.0472, 1.1268, 1.1801)	(0.0610, 0.0739, 0.0873)
	*f*25:Offline Facility Level	Maximal	(0.9548, 1, 1.0472)	(0.0556, 0.0656, 0.0775)
	*f*26:Patron Response	Maximal	(1.0759, 1.0968, 1.1132)	(0.0627, 0.0720, 0.0824)
Expertise (*f*3)	*f*31:Network Strategy	Maximal	(1.0968, 1.2698, 1.3928)	(0.0639, 0.0833, 0.1031)
	*f*32: Instantaneous tracking systems	Maximal	(1.0675, 1.1978, 1.3299)	(0.0622, 0.0786, 0.0984)
	*f*33:Marketing Techniques	Maximal	(0.7664, 0.9117, 1.1268)	(0.0446, 0.0598, 0.0834)
Societal and Eco-friendly (*f*4)	*f*41: Health and Living quarters	Maximal	(0.7519, 0.8874, 1.0759)	(0.0438, 0.0582, 0.0796)
	*f*42:Communication Safekeeping	Maximal	(0.6277, 0.7267, 0.8705)	(0.0365, 0.0477, 0.0644)
	*f*43:Ecological Influence	Minimal	(0.6158, 0.6754, 0.7725)	(0.0358, 0.0443, 0.0571)

*f*11: 7.3466%; *f*12: 6.82%; *f*13: 6.21%; *f*21: 7.81%; *f*22: 7.5466%; *f*23: 6.69%; *f*24: 7.4066%; *f*25: 6.6233%; *f*26: 7.2366%; *f*31: 8.3433%; *f*32: 7.9733%; *f*33: 6.26%; *f*41: 6.0533%; *f*42: 4.9533%; *f*43: 4.5733%.

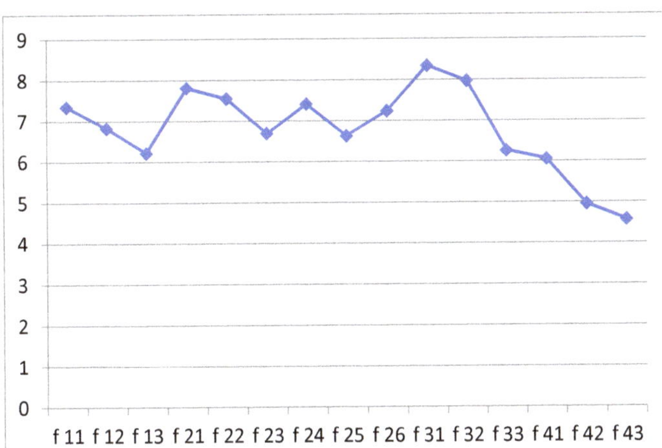

Figure 5. Influence level of FAHP model criterion.

Table 12. Fuzzy normalised decision matrix.

	Financial Norms ($f1$)	Facility Value ($f2$)	Expertise ($f3$)	Societal and Eco-Friendly ($f4$)
Uber Eats	0.1111, 0.3333, 0.7777	0.1111, 0.6296, 1	0.3333, 0.6296, 1	0.3333, 0.5294, 1
Domino's	0.3333, 0.7777, 1	0.5555, 0.8518, 1	0.3333, 0.7037, 1	0.3333, 0.3600, 0.6
Zomato	0.5555, 0.8518, 1	0.3333, 0.7777, 1	0.5555, 0.8518, 1	0.3333, 0.4736, 1
Swiggy	0.3333, 0.6296, 1	0.3333, 0.7777, 1	0.3333, 0.6296, 1	0.3333, 0.4736, 1

Table 13. Fuzzy weighted normalised decision matrix.

	Financial Norms ($f1$)	Facility Value ($f2$)	Expertise ($f3$)	Societal and Eco-Friendly ($f4$)
Uber Eats	0.3333, 1.6665, 5.4439	0.7777, 5.6664, 9	2.3331, 5.6664, 9	1.6665, 3.7058, 9
Domino's	0.9999, 3.8885, 7	3.8885, 7.6662, 9	2.3331, 6.3333, 9	1.6665, 2.52, 5.4
Zomato	1.6665, 4.259, 7	2.3331, 6.9993, 9	3.8885, 7.6662, 9	1.6665, 3.3152, 9
Swiggy	0.9999, 3.148, 7	2.3331, 6.9993, 9	2.3331, 5.6664, 9	1.6665, 3.3152, 9

Table 14. Closeness coefficient of each alternative.

Alternatives	D^{-}_{j}	D^{*}_{j}	Level of Satisfaction	Rank
Uber Eats	2.1883	5.5055	0.2844	4
Domino's	4.1327	3.8111	0.5202	3
Zomato	6.6815	1.2023	0.8474	1
Swiggy	5.5606	3.4128	0.6196	2

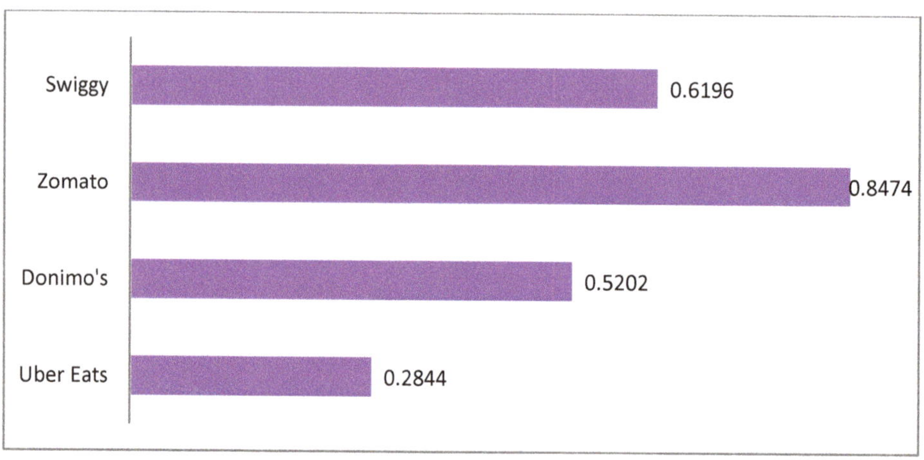

Figure 6. Score line of the FTOPSIS model.

6. Discussion

In the research work that is presented, a hybrid multi-criteria decision-making framework for the evaluation of the online food delivery market in India is established in consideration of a wide range of criteria, including financial norms (supply rate, operating skill, and hazard managing), expertise criteria (network strategy, instantaneous and e-commerce), societal and eco-friendly criteria (health and living quarters, communication safekeeping, and ecological influence), and facility value (order satisfaction, supply speed, handiness of expense, virtual/offline facility level, and patron response). The combination of the fuzzy analytic hierarchy process and fuzzy technique for order performance by similarity to ideal solution method has been initially proposed in the current research to tackle the problem in light of the discussion by thoroughly reviewing the literature. Triangular fuzzy sets in the analytic hierarchy process can translate expert opinions into language terms to get more precise and scientific attribute weights for the criterion. The reliability of the suggested integrated framework is shown by the fact that the offered case study is successfully addressed. The outcomes show that the model in use is able to rank common green online food delivery companies. According to the fuzzy analytic hierarchy process results, the top five online food delivery evaluation factors are Order Satisfaction, Supply Speed, Network Strategy, and Virtual Facility Level.

To meet the customer's needs in the first place, it is crucial in India that online shopping be convenient for payment. The vast majority of Indians still favour cash-on-delivery payment over online transactions, despite the advantages of cashless payment methods like credit or internet banking being established, which include cost savings and numerous conveniences for customers and businesses. Other than that, developing cutting-edge technology solutions that guarantee more efficient order fulfilment while keeping costs down is a competitive edge for online food delivery companies. One way for online food delivery firms to stay alive and attract clients is by merging orders, offering many delivery choices using robotics and drones, and utilising cloud kitchens. Figure 6 shows that Zomato (0.8474), Swiggy (0.6196), Domino's (0.5202), and Uber Eats (0.2844) are in order of top performing online food delivery companies in the current online food delivery market in India according to the chosen evaluation criteria. The outcomes can be used as a benchmark for online food delivery executives and decision-makers as they evaluate their companies' performance while taking into account a wider range of factors and identifying key industry determinants. The current study's chosen evaluation criteria will all aid online food delivery enterprises in overcoming a variety of obstacles and motivate them to consider sustainable development initiatives. The assessment of the online food delivery market in India and other markets has concentrated on a number of variables, including service quality, economic factors, and technology, but it is still difficult to keep in mind social and environmental issues.

7. Conclusions

There are limitations and possible extensions to this study. The author offered a brand-new methodology to offer a quick strategy for evaluating several online food delivery companies and assisting the decision-maker in choosing the finest one. The pairwise comparison procedure is made intuitively by combining enhanced analytic hierarchy process (AHP) with fuzzy set theory, which also helps to lessen or completely eliminate evaluation bias. In order to assist online food delivery services, this paper also introduces a strategy that combines improved analytic hierarchy process (AHP) with technique for order performance by similarity to ideal solution (TOPSIS). To ensure sustained development in this cutthroat industry, it is crucial for online food delivery enterprises to adopt a number of actions and take pertinent factors into account. The managerial implications of the used approach and its analysis would enlighten decision-makers in the online food delivery industry not only in India but also in the international market. The approach put out in this research can be connected to further cutting-edge market influences in subsequent investigations. Multi-criteria decision-making approaches like VIseKriterijumska Optimizacija I Kompromisno Resenje, Preference Ranking for Organisation Method for Enrichment Evaluation, Data Envelopment Analysis, and combinations of these, among others, can be used methodologically. To test the overall validity of the conclusions, additional studies might apply the suggested strategy or relevant approaches to particular situations of industries, particularly those connected to e-commerce.

In this paper, we discussed the fuzzy analytic hierarchy process and fuzzy technique for order performance by similarity to ideal solution methods together and summarised the results as follows. In the first part, we discussed the construction of the online food delivery market evaluation criteria using replies from industry experts that are explained in Figure 4 and Table 1. Next, we discussed a case study to evaluate the online food delivery enterprises in India by using the fuzzy analytic hierarchy process, which is explained with the help of Tables 5–10 and finally executed in Table 11. Also, we discussed the influence level of fuzzy analytic hierarchy process model criteria, which is shown in Figure 5 and the fuzzy technique for order performance by similarity to ideal solution method, which is explained with the help of Tables 12 and 13 and finally executed in Table 14. The final ranking order of online food delivery services from the fuzzy technique for order performance by similarity to ideal solution method is visualised in Figure 6.

Author Contributions: Conceptualisation, H.U.A.S. and U.K.K.N.; methodology, H.U.A.S., U.K.K.N. and B.C.P.; software, H.U.A.S. and U.K.K.N.; validation, U.K.K.N., M.D.A. and B.C.P.; formal analysis, H.U.A.S., U.K.K.N. and M.D.A.; investigation, H.U.A.S., M.D.A. and U.K.K.N.; writing—original draft preparation, H.U.A.S. and U.K.K.N.; writing—review and editing, U.K.K.N. and M.D.A.; supervision, U.K.K.N.; project administration, U.K.K.N.; funding acquisition, M.D.A. All authors have read and agreed to the published version of the manuscript.

Funding: This research received no external funding.

Institutional Review Board Statement: Not applicable.

Informed Consent Statement: Not applicable.

Data Availability Statement: Not applicable.

Conflicts of Interest: The authors declare no conflict of interest.

Appendix A

Table A1. Fuzzy logic combined judgement matrix of 15 criteria.

Parameters		$f11$			$f12$			$f13$	
$f11$	1.00	1.00	1.00	1.00	1.00	1.00	1.00	1.00	1.00
$f12$	1.00	1.00	1.00	1.00	1.00	1.00	1.00	1.00	1.00
$f13$	1.00	1.00	1.00	1.00	1.00	1.00	1.00	1.00	1.00
$f21$	1.00	1.00	1.00	1.00	1.05	1.08	1.00	1.05	1.08
$f22$	1.00	1.00	1.00	1.05	1.08	1.10	1.05	1.08	1.10
$f23$	1.00	1.00	1.00	1.00	1.00	1.00	1.00	1.00	1.00
$f24$	1.00	1.00	1.00	1.00	1.00	1.00	1.00	1.00	1.00
$f25$	1.00	1.00	1.00	1.00	1.00	1.00	1.00	1.00	1.00
$f26$	1.08	1.10	1.11	1.00	1.00	1.00	1.00	1.00	1.00
$f31$	1.00	1.05	1.08	1.00	1.00	1.00	1.00	1.00	1.00
$f32$	1.00	1.00	1.00	1.00	1.00	1.00	1.00	1.00	1.00
$f33$	0.93	0.95	1.00	1.00	1.00	1.00	0.93	0.95	1.00
$f41$	0.91	0.93	0.95	0.93	0.95	1.00	1.00	1.00	1.00
$f42$	0.90	0.91	0.93	1.00	1.00	1.00	1.00	1.00	1.00
$f43$	1.00	1.00	1.00	0.90	0.91	0.93	1.00	1.00	1.00

Table A1. *Cont.*

Parameters		f21			f22			f23	
f11	1.00	1.00	1.00	1.00	1.00	1.00	1.00	1.00	1.00
f12	0.93	0.95	1.00	0.91	0.93	0.95	1.00	1.00	1.00
f13	0.93	0.95	1.00	0.91	0.93	0.95	1.00	1.00	1.00
f21	1.00	1.00	1.00	0.93	0.95	1.00	0.91	0.93	0.95
f22	1.00	1.05	1.08	1.00	1.00	1.00	0.93	0.95	1.00
f23	1.05	1.08	1.10	1.00	1.05	1.08	1.00	1.00	1.00
f24	1.00	1.00	1.00	1.05	1.08	1.10	1.00	1.05	1.08
f25	1.00	1.00	1.00	1.00	1.00	1.00	1.05	1.08	1.10
f26	1.00	1.00	1.00	1.00	1.00	1.00	1.00	1.00	1.00
f31	1.00	1.00	1.00	1.00	1.00	1.00	1.00	1.00	1.00
f32	1.00	1.00	1.00	1.00	1.00	1.00	1.00	1.00	1.00
f33	1.00	1.00	1.00	1.00	1.00	1.00	1.00	1.00	1.00
f41	1.00	1.00	1.00	1.00	1.00	1.00	1.00	1.00	1.00
f42	0.89	0.90	0.91	0.93	0.95	1.00	1.00	1.00	1.00
f43	0.91	0.93	0.95	1.00	1.00	1.00	1.00	1.00	1.00
Parameters		f24			f25			f26	
f11	1.00	1.00	1.00	1.00	1.00	1.00	0.91	0.93	0.95
f12	1.00	1.00	1.00	1.00	1.00	1.00	1.00	1.00	1.00
f13	1.00	1.00	1.00	1.00	1.00	1.00	1.00	1.00	1.00
f21	1.00	1.00	1.00	1.00	1.00	1.00	1.00	1.00	1.00
f22	0.91	0.93	0.95	1.00	1.00	1.00	1.00	1.00	1.00
f23	0.93	0.95	1.00	0.91	0.93	0.95	1.00	1.00	1.00
f24	1.00	1.00	1.00	1.00	1.00	1.00	1.00	1.00	1.00
f25	1.00	1.00	1.00	1.00	1.00	1.00	1.00	1.00	1.00
f26	1.00	1.00	1.00	1.00	1.00	1.00	1.00	1.00	1.00
f31	1.00	1.00	1.00	1.05	1.08	1.10	1.00	1.00	1.00
f32	1.00	1.00	1.00	1.00	1.00	1.00	1.00	1.00	1.00
f33	1.00	1.00	1.00	1.00	1.00	1.00	1.00	1.00	1.00
f41	1.00	1.00	1.00	1.00	1.00	1.00	1.00	1.00	1.00
f42	1.00	1.00	1.00	1.00	1.00	1.00	1.00	1.00	1.00
f43	1.00	1.00	1.00	1.00	1.00	1.00	1.00	1.00	1.00
Parameters		f31			f32			f33	
f11	0.93	0.95	1.00	1.00	1.00	1.00	1.00	1.05	1.10
f12	1.00	1.00	1.00	1.00	1.00	1.00	1.00	1.00	1.00
f13	1.00	1.00	1.00	1.00	1.00	1.00	1.00	1.05	1.10
f21	1.00	1.00	1.00	1.00	1.00	1.00	1.00	1.00	1.00
f22	1.00	1.00	1.00	1.00	1.00	1.00	1.00	1.00	1.00
f23	1.00	1.00	1.00	1.00	1.00	1.00	1.00	1.00	1.00
f24	1.00	1.00	1.00	1.00	1.00	1.00	1.00	1.00	1.00

Table A1. *Cont.*

*f*25	0.91	0.93	0.95	1.00	1.00	1.00	1.00	1.00	1.00
*f*26	1.00	1.00	1.00	1.00	1.00	1.00	1.00	1.00	1.00
*f*31	1.00	1.00	1.00	1.00	1.05	1.10	1.05	1.08	1.10
*f*32	0.93	0.95	1.00	1.00	1.00	1.00	1.00	1.05	1.10
*f*33	0.91	0.93	0.95	0.93	0.95	1.00	1.00	1.00	1.00
*f*41	1.00	1.00	1.00	0.91	0.93	0.95	0.93	0.95	1.00
*f*42	1.00	1.00	1.00	1.00	1.00	1.00	0.91	0.93	0.95
*f*43	1.00	1.00	1.00	0.89	0.90	0.91	1.00	1.00	1.00
Parameters		*f*41			*f*42			*f*43	
*f*11	1.05	1.08	1.10	1.08	1.10	1.11	1.00	1.00	1.00
*f*12	1.00	1.05	1.10	1.00	1.00	1.00	1.08	1.10	1.11
*f*13	1.00	1.00	1.00	1.00	1.00	1.00	1.00	1.00	1.00
*f*21	1.00	1.00	1.00	1.10	1.11	1.13	1.05	1.08	1.10
*f*22	1.00	1.00	1.00	1.00	1.05	1.10	1.00	1.00	1.00
*f*23	1.00	1.00	1.00	1.00	1.00	1.00	1.00	1.00	1.00
*f*24	1.00	1.00	1.00	1.00	1.00	1.00	1.00	1.00	1.00
*f*25	1.00	1.00	1.00	1.00	1.00	1.00	1.00	1.00	1.00
*f*26	1.00	1.00	1.00	1.00	1.00	1.00	1.00	1.00	1.00
*f*31	1.00	1.00	1.00	1.00	1.00	1.00	1.00	1.00	1.00
*f*32	1.05	1.08	1.10	1.00	1.00	1.00	1.10	1.11	1.13
*f*33	1.00	1.05	1.10	1.05	1.08	1.10	1.00	1.00	1.00
*f*41	1.00	1.00	1.00	1.00	1.05	1.10	1.05	1.08	1.10
*f*42	0.93	0.95	1.00	1.00	1.00	1.00	1.00	1.05	1.10
*f*43	0.91	0.93	0.95	0.93	0.95	1.00	1.00	1.00	1.00

References

1. Yu, K.D.S.; Aviso, K.B. Modelling the economic impact and ripple effects of disease outbreaks. *Process Integr. Optim. Sustain.* **2020**, *4*, 183–186. [CrossRef]
2. Song, H.J.; Yeon, J.; Lee, S. Impact of the COVID-19 pandemic: Evidence from the U.S. restaurant industry. *Int. J. Hosp. Manag.* **2021**, *92*, 102702. [CrossRef] [PubMed]
3. Pigatto, G.; Machado, J.G.D.C.F.; dos Santos Negreti, A.; Machado, L.M. Have you chosen your request? Analysis of online food delivery companies in Brazil. *Br. Food J.* **2017**, *119*, 639–657. [CrossRef]
4. Baker, S.R.; Farrokhnia, R.A.; Meyer, S.; Pagel, M.; Yannelis, C. How does household spending respond to an epidemic? Consumption during the 2020 COVID-19 pandemic. *Rev. Asset Pricing Stud.* **2020**, *10*, 834–862. [CrossRef]
5. Kim, J.; Lee, J.C. Effects of COVID-19 on preferences for private dining facilities in restaurants. *J. Hosp. Tour. Manag.* **2020**, *45*, 67–70. [CrossRef]
6. Chang, H.-H.; Meyerhoefer, C. COVID-19 and the demand for online food shopping services: Empirical evidence from Taiwan. *Am. J. Agric. Econ.* **2020**, *103*, 448–465. [CrossRef]
7. Blumtritt, C. Online Food Delivery Report. 2020. Available online: https://www.statista.com/outlook/dmo/eservices/online-food-delivery/worldwide (accessed on 18 July 2022).
8. Li, C.; Mirosa, M.; Bremer, P. Review of online food delivery platforms and their impacts on sustainability. *Sustainability* **2020**, *12*, 5528. [CrossRef]
9. Baabdullah, A.M.; Alalwan, A.A.; Rana, N.P.; Patil, P.; Dwivedi, Y.K. An integrated model for m-banking adoption in Saudi Arabia. *Int. J. Bank Mark.* **2019**, *37*, 452–478. [CrossRef]
10. Statista. Mobile App Usage—Statistics & Facts. 2018. Available online: https://www.statista.com/topics/1002/mobile-app-usage/ (accessed on 15 December 2018).
11. Statista. Number of Mobile App Downloads Worldwide in (2017, 2018 and 2022) in Billions. 2018. Available online: https://www.statista.com/statistics/271644/worldwidefree-and-paid-mobile-app-store-downloads/ (accessed on 15 December 2018).

12. Statista. eServices Report. 2020. Available online: https://www.statista.com/study/42306/eservices-report/ (accessed on 2 December 2021).
13. Maida, J. *Analysis on Impact of COVID-19-Online on-Demand Food Delivery Services Market 2019–2023*; Businesswire: San Francisco, CA, USA, 2020.
14. Yeo, V.C.S.; Goh, S.-K.; Rezaei, S. Consumer experiences, attitude and behavioral intention toward online food delivery (OFD) services. *J. Retail. Consum. Serv.* **2017**, *35*, 150–162. [CrossRef]
15. Hale, R.C.; Song, B. Single-use plastics and COVID-19: Scientific evidence and environmental regulations. *Environ. Sci. Technol.* **2020**, *54*, 7034–7036. [CrossRef]
16. Prakash, C.; Barua, M.K. An analysis of integrated robust hybrid model for third-party reverse logistics partner selection under fuzzy environment. Resources. *Conserv. Recycl.* **2016**, *108*, 63–81. [CrossRef]
17. Chiu, Y.-J.; Hu, Y.-C.; Yao, C.-Y.; Yeh, C.-H. Identifying Key Risk Factors in Product Development Projects. *Mathematics* **2022**, *10*, 1295. [CrossRef]
18. Mehdi, K.G.; Maghsoud, A.; Kazimieras, Z.E.; Jurgita, A. A new hybrid fuzzy MCDM approach for evaluation of construction equipment with sustainability considerations. *Arch. Civ. Mech. Eng.* **2018**, *18*, 32–49. [CrossRef]
19. Gunawan, F.E.; Sondakh, B.L.; Alamsjah, F. Factors affecting the user of online food delivery through mobile apps. *ICIC Express Lett.* **2020**, *14*, 1069–1081. [CrossRef]
20. Tirkolaee, E.B.; Dashtian, Z.; Weber, G.; Tomaskova, H.; Soltani, M.; Mousavi, N.S. An Integrated Decision-Making Approach for Green Supplier Selection in an Agri-Food Supply Chain: Threshold of Robustness Worthiness. *Mathematics* **2021**, *9*, 1304. [CrossRef]
21. Van Doremalen, N.; Bushmaker, T.; Morris, D.H.; Holbrook, M.G.; Gamble, A.; Williamson, B.N.; Tamin, A.; Harcourt, J.L.; Thornburg, N.J.; Gerber, S.I.; et al. Aerosol and surface stability of SARSCoV-2 as compared with SARS-CoV-1. *N. Engl. J. Med.* **2020**, *382*, 1564–1567. [CrossRef]
22. Goldman, E. Exaggerated risk of transmission of COVID-19 by fomites. *Lancet Infect. Dis.* **2020**, *20*, 892–893. [CrossRef]
23. Gavilan, D.; Balderas-Cejudo, A.; Fernández-Lores, S.; Martinez-Navarro, G. Innovation in online food delivery: Learnings from COVID-19. *Int. J. Gastron. Food Sci.* **2021**, *24*, 100330. [CrossRef]
24. Tanackov, I.; Badi, I.; Stević, Ž.; Pamučar, D.; Zavadskas, E.K.; Bausys, R. A Novel Hybrid Interval Rough SWARA–Interval Rough ARAS Model for Evaluation Strategies of Cleaner Production. *Sustainability* **2022**, *14*, 4343. [CrossRef]
25. Lu, M.; Wudhikarn, R. Using the best-worst method to develop intellectual capital indicators in financial service company. In Proceedings of the 2022 Joint International Conference on Digital Arts, Media and Technology with ECTI Northern Section Conference on Electrical, Electronics, Computer and Telecommunications Engineering (ECTI DAMT & NCON), Chiang Rai, Thailand, 26–28 January 2022. [CrossRef]
26. Mishra, A.R.; Rani, P.; Krishankumar, R.; Zavadskas, E.K.; Cavallaro, F.; Ravichandran, K.S. A Hesitant Fuzzy Combined Compromise Solution Framework-Based on Discrimination Measure for Ranking Sustainable Third-Party Reverse Logistic Providers. *Sustainability* **2021**, *13*, 2064. [CrossRef]
27. Wudhikarn, R. An approach to enhancing the human capital of enterprises associated with cooperative education. *Int. J. Learn. Intellect. Cap.* **2015**, *12*, 61–81. [CrossRef]
28. Saaty, T.L. *The Analytic hierarchy process*; McGraw-Hill: New York, NY, USA, 1980.
29. Badri, M.A. A combined AHP–GP model for quality control systems. *Int. J. Prod. Econ.* **2001**, *72*, 27–40. [CrossRef]
30. Chan, F.T.S.; Kumar, N. Global supplier development considering risk factors using fuzzy extended AHP-based approach. *Omega* **2007**, *35*, 417–431. [CrossRef]
31. Albayrak, E.; Erensal, Y.C. Using analytic hierarchy process (AHP) to improve human performance. An application of multiple criteria decision making problem. *J. Intell. Manuf.* **2004**, *15*, 491–503. [CrossRef]
32. Wang, J.J.; Yang, D.L. Using a hybrid multi-criteria decision aid method for information systems outsourcing. *Comput. Oper. Res.* **2007**, *34*, 3691–3700. [CrossRef]
33. Zadeh, L.A. Information and Control. *Fuzzy Sets* **1965**, *8*, 338–353.
34. Gumus, A.-T. Evaluation of hazardous waste transportation firms by using a twostep fuzzy-AHP and TOPSIS methodology. *Expert Syst. Appl.* **2009**, *36*, 4067–4074. [CrossRef]
35. Huang, C.-C.; Chu, P.-Y.; Chiang, Y.-H. A fuzzy AHP application in government-sponsored R&D project selection. *Omega* **2008**, *36*, 1038–1052.
36. Cakir, O.; Canbolat, M.-S. A web-based decision support system for multicriteria inventory classification using fuzzy AHP methodology. *Expert Syst. Appl.* **2008**, *35*, 1367–1378. [CrossRef]
37. Wang, T.-C.; Chen, Y.-H. Applying fuzzy linguistic preference relations to the improvement of consistency of fuzzy AHP. *Inf. Sci.* **2008**, *178*, 3755–3765. [CrossRef]
38. Sambasivan, M.; Fei, N.-Y. Evaluation of critical success factors of implementation of ISO 14001 using analytic hierarchy process (AHP): A case study from Malaysia. *J. Clean. Prod.* **2008**, *16*, 1424–1433. [CrossRef]
39. Sharma, M.-J.; Moon, I.; Bae, H. Analytic hierarchy process to assess and optimize distribution network. *Appl. Math. Comput.* **2008**, *202*, 256–265. [CrossRef]
40. Costa, C.-A.-B.; Vansnick, J.-C. A critical analysis of the eigenvalue method used to derive priorities in AHP. *Eur. J. Oper. Res.* **2008**, *187*, 1422–1428. [CrossRef]

41. Firouzabadi, S.-M.-A.-K.; Henson, B.; Barnes, C. A multiple stakeholders' approach to strategic selection decisions. *Comput. Ind. Eng.* **2008**, *54*, 851–865. [CrossRef]
42. Wang, Y.-M.; Luo, Y.; Hua, Z. On the extent analysis method for fuzzy AHP and its applications. *Eur. J. Oper. Res.* **2008**, *186*, 735–747. [CrossRef]
43. Kuo, M.-S.; Tzeng, G.-H.; Huang, W.-C. Group decision making based on concepts of ideal and anti-ideal points in fuzzy environment. *Math. Comput. Modeling* **2007**, *45*, 324–339. [CrossRef]
44. Armillotta, A. Selection of layered manufacturing techniques by an adaptive AHP decision model. *Robot. Comput.-Integr. Manuf.* **2008**, *24*, 450–461. [CrossRef]
45. Dagdeviren, M.; Yuksel, I. Developing a fuzzy analytic hierarchy process (AHP) model for behavior-based safety management. *Inf. Sci.* **2008**, *178*, 1717–1733. [CrossRef]
46. Chen, M.-F.; Tzeng, G.-H.; Ding, C.-G. Combining fuzzy AHP with MDS in identifying the preference similarity of alternatives. *Appl. Soft Comput.* **2008**, *8*, 110–117. [CrossRef]
47. Lin, M.-C.; Wang, C.-C.; Chen, M.-S.; Chang, C.-A. Using AHP and TOPSIS approaches in customer-driven product design process. *Comput. Ind.* **2008**, *59*, 17–31. [CrossRef]
48. Kulak, O.; Durmusoglu, B.; Kahraman, C. Fuzzy multi-attribute equipment selection based on information axiom. *J. Mater. Processing Technol.* **2005**, *169*, 337–345. [CrossRef]
49. Büyükzkan, G.; Feyzioglu, O.; Nebol, E. Selection of the strategic alliance partner in logistics value chain. *Int. J. Prod. Econ.* **2008**, *113*, 148–158. [CrossRef]
50. Chen, T.Y.; Tsao, C.Y. The interval-valued fuzzy TOPSIS methods and experimental analysis. *Fuzzy Sets Syst.* **2008**, *159*, 1410–1428. [CrossRef]
51. Kahraman, C.; Büyükzkan, G.; Ates, N.Y. A two phase multi-attribute decision making approach for new product introduction. *Inf. Sci.* **2007**, *177*, 1567–1582. [CrossRef]
52. Onüt, S.; Soner, S. Transshipment site selection using the AHP and TOPSIS approaches under fuzzy environment. *Waste Manag.* **2007**, *28*, 1552–1559. [CrossRef] [PubMed]
53. Wang, Y.M.; Elhag, T.M.S. Fuzzy TOPSIS method based on alpha level sets with an application to bridge risk assessment. *Expert Syst. Appl.* **2006**, *31*, 309–319. [CrossRef]
54. Yong, D. Plant location selection based on fuzzy TOPSIS. *Int. J. Adv. Manuf. Technol.* **2006**, *28*, 839–844. [CrossRef]
55. Chang, Y.H.; Chung, H.Y.; Wang, S.Y. A survey and optimization-based evaluation of development strategies for the air cargo industry. *Int. J. Prod. Econ.* **2007**, *106*, 550–562. [CrossRef]
56. Chang, Y.H.; Yeh, C.H. A survey analysis of service quality for domestic airlines. *Eur. J. Oper. Res.* **2002**, *139*, 166–177. [CrossRef]
57. Kahraman, C.; Beskese, A.; Ruan, D. Measuring flexibility of computer integrated manufacturing systems using fuzzy cash flow analysis. *Inf. Sci.* **2004**, *168*, 77–94. [CrossRef]
58. Zimmerman, H.J. *Fuzzy Sets Theory and Its Applications*; Kluwer Academic Publisher: Boston, MA, USA, 1996.
59. Xu, Z.S.; Chen, J. An interactive method for fuzzy multiple attributes group decision making. *Inf. Sci.* **2007**, *177*, 248–263. [CrossRef]

MDPI
St. Alban-Anlage 66
4052 Basel
Switzerland
www.mdpi.com

Mathematics Editorial Office
E-mail: mathematics@mdpi.com
www.mdpi.com/journal/mathematics

www.ingramcontent.com/pod-product-compliance
Lightning Source LLC
Chambersburg PA
CBHW061622220926
51850CB00005B/573
9783725807420